JN299764

英和・和英 エコロジー用語辞典

瀬川至朗
[執筆・監修]

研究社辞書編集部
[編]

研究社

Kenkyusha English and Japanese
Dictionary of Ecology

© 2010 Segawa Shiro, KENKYUSHA Co., Ltd.

研究社編集部
佐藤淳　三谷裕　津田正　根本保行　松原悟

前書き

「エコロジー」は,現代的な意義をもつ,深い言葉である.

本辞典英和篇によれば,エコロジー (ecology) とは「生物とそれを取り巻く環境の相互関係を調べる学問」のことである.「生物」には人間が含まれ,「環境」には生物と無生物のすべてが含まれる.人間をはじめとする生物が,地球上の諸要素,あるいは地球を取り巻く諸要素とどのような相互関係を切り結んでいるのか,そこにある構造やネットワークを明らかにすることが,この学問の中心的な課題である.

グローバルにも,ローカルにも,私たち人間の活動が,地球上のさまざまな環境変化の最大の要因となっている.私たちを取り巻く「環境」について語るとき,自らの存在を抜きにしては議論できないのである.その意味で,エコロジーは,さまざまな環境問題を自らの存在に引きつけて身近に捉え直すことのできる学問的フレームであり,今日の地球異変を語るにふさわしい思想が含意されているのである.

エコロジーは一般に,生物学と同じ自然科学系の学問という印象があるが,その意味するところは,自然科学の領域にとどまるものではない.近年盛んに開かれている環境関連の国際会議や,捕鯨問題に関する国際対立に象徴される政治的・文化的な問題も,エコロジーの枠組みのなかにあり,エコ商品や二酸化炭素 (CO_2) の排出量取引などのエコビジネスに絡む経済的な側面,ごみの分別や行楽といった日常生活・消費活動につながる社会的な側面なども,また同様である.

エコロジー関連の用語は様々な形で日常生活の中に入り込んでいる.

従来からの「環境保護」「自然破壊」「公害防止」などの言葉に加え,近年は「地球温暖化」「生物多様性」「持続可能性」「低炭素社会」などの時事用語を耳にする.さらには「ロハス」「エコ」などの親しみやすい言葉が CM に使われるようになった.

しかしながら,ecology という英語が本来は「生態学」の意味であることがあまり知られていないように,英語で使われるエコロジー関連の用語は,普段カタカナ語として目にするにもかかわらず,日本人にとっては必ずしも明確ではない.環境問題が地球規模の問題であるからには,広く国際的に用

いられる言語の一つである英語においてこれらの用語を正しく理解・使用することがますます重要になってきている．

そのための一助として，本辞典ではまず前半の英和篇において英語圏で用いられる環境保護，生態学，気象学，生物学関連の用語を中心に約2200語句を収集し，特にマスメディアなどで用いられる環境関連の用語に解説を施した．エコロジーという言葉がもつ環境保護のニュアンスを意識しつつ，それに限定することなく，幅広く，環境問題，地球環境問題に関連して，身近に使われる用語を選んだ．

後半の和英篇では上記のような専門用語はもちろん，「マイバッグ」「エコカー」「もったいない」といった，いわばエコライフ用語ともいうべき語句や表現も含めて1700の見出しと7200の下位見出しにまとめ，英訳をつけた．和英辞典としてのみならず，簡便な日本語のエコ語彙集の性格も持たせた．

両篇あわせ，エコロジーに関心やかかわりを持つ人々の活動に役立てば幸いである．

本辞典執筆の基礎データとしては研究社のオンライン辞書データベースであるKOD (Kenkyusha Online Dictionary) から研究社編集部の三谷裕氏が選択・抽出した情報を用い，また一部は研究社刊『辞書のすきますきまの言葉』から著者である清水由美先生，監修者であるトム・ガリー先生のご快諾を得て引用させていただいた．ここに記して感謝申し上げたい．

2010年7月

執筆・監修者　早稲田大学 政治学研究科ジャーナリズムコース 教授

瀬 川 至 朗

凡　例

1. 本辞典は前半の英和篇と後半の和英篇に分かれている．他所への参照は「★⇨」で示してあるが，「★⇨英和」「★⇨和英」とある場合は英和篇，和英篇でその見出しを引く．

2. 和英篇の見出し語順は次の原則による．

2.1 「カー」「オー」「ウィーン」のように延ばす音は「かあ」「おお」「ういいん」のように前の字の母音に置き換えて順番を決める．
例：オイル⇨欧州（おうしゅう）⇨オーガニック（おおがにっく）
農業（のうぎょう）⇨ノーカー（のおかあ）
流出（りゅうしゅつ）⇨リユース（りゆうす）⇨流水（りゅうすい）

2.2 濁音，半濁音，拗音のみが違っている場合は普通の読み方の語句より後に来る．
例：海中（かいちゅう）⇨害虫（がいちゅう）
利用（りよう）⇨漁（りょう）

2.3 CO_2，RPS 法，NPO などアルファベットで始まる見出しは五十音順ではなく，ABC 順にまとめて和英篇末尾「わ」の後に置いた．参照で「★⇨和英篇末尾」とある場合はここに見出しがある．またアルファベット見出しで例えば「**ADI** ⇨英和」とあれば，英和篇にその見出し（ADI）があるのでそちらを引く．

2.4 和英では見出しを使った複合語は原則としてその下位項目としている．例えば「汚染」の項目では，「▶」のマークで区切られた例文の後にある「¶」のマークの後に「汚染管理区域」「汚染源」「汚染原因」など「汚染」で始まる下位項目を五十音順に配列し，「汚染レベル」の後には「油汚染」「化学汚染」「環境汚染」のように「汚染」で終わる下位項目がこれも五十音順に並んでいる．

他の見出しからこれらの下位項目を参照させる場合は原則として「⇨汚染 (汚染原因)」「⇨汚染 (環境汚染)」のように表示してある．このような場合はまず「汚染」という項目を見つけ，その中で「¶」の記号の後に来る下位項目の中から「汚染原因」「環境汚染」を探すと見つかる．

3. 「▶」のマークは見出し語をつかった例文の始まりを表わす．例文では見出し語と同じ部分は原則として「〜」で置き換えて表記するが，見出し語が一文字の場合や，動詞の活用形のように見出しと形が違う場合は「〜」を使わない．下位項目に例文がついている場合も同様．

4. 和英の訳語や例文ではよく使われる前置詞や目的語・動詞・形容詞などの例を《　》で示した場合がある．

例えば「開発」の項目で「exploitation《of resources》」や「〜する develop《backward regions, an industry》」とあるのは，「exploitation of resources　(資源の開発)」「develop backward regions (後進地域を開発する)，develop an industry (産業を開発する)」のように《　》内の位置によく来る典型的な補足語句や目的語を例示したもので，これらはもちろん具体的な文脈によって変わりうる．「原子力安全白書 the《2004》White Paper on Nuclear Safety.」も，具体的な発行年によって the 2010 White Paper on Nuclear Safety (2010年度の原子力安全白書) のように変わることを示す．

5. 分野別・英米別などの表記は以下の通り．

【医】医学　【化】化学　【気】気象学　【建】建築　【鉱】鉱物学，鉱業
【商】商業　【商標】商標名　【植】植物学　【生】生物学　【生態】生態学
【地】地(質)学　【動】動物学　【法】法律　【理】物理学　【労】労働問題
《英》【英】主に英国で使う語句，英国の制度など
《米》【米】主に米国で使う語句，米国の制度など
《文》文(章)語　《俗》《口》俗語，口語　《米俗》《英俗》米国，英国の俗語

6. その他の記号類は以下の通り．

6.1　名 形 動 副 は英和篇でそれぞれ名詞，形容詞，動詞，副詞を示す．

ただし見出しが名詞のみの場合は省略した．

　一つの語が名詞と動詞など複数の品詞として使われる場合は原則として品詞ごとに分けたが，短いものは次のようにまとめた場合もある．
　　Nimby　图形《俗》ニンビー(の)
また和英篇では次のように訳語から派生した形容詞や動詞などを示すのにも用いた．
　　アトピー　atopy.　形 atopic.（「アトピー(体質)の」を意味する形容詞を示す）

6.2　「*」は英和篇に見出しのある語句を示す．
　例：**CT**　Civic Trust*（「Civic Trust という見出しが英和篇にあるのでそちらを参照」の意）

6.3　⇔　反対語を示す．

6.4　置き換え可能な部分は［　］，省略可能な部分は（　）で示した．
　例：CO_2［炭酸ガス］分離［隔離］　（「CO_2 分離」または「CO_2 隔離」または「炭酸ガス分離」または「炭酸ガス隔離」の意）
　　　an association [a society, a group] for ...（an association for または a society for または a group for の意）
　　　a ray of sun(light)　（a ray of sun または a ray of sunlight の意）

6.5　*sb, sth* は somebody, something の略で，それぞれ人・物を指す．
また *one, one's, oneself* は主語によって he, she, his, her, himself, herself など代名詞が変化しうることを示す．

6.6　動詞の主語・目的語，形容詞が修飾する語句を〈　〉内に入れて示した場合がある．
　例：〈政党・政府〉に環境政策を取らせる
　　　〈土地が〉乾燥した

6.7　和英篇の英訳の語句の区切り，英和篇の語義の大きな区切りは ; で示す．また和英篇の文章の訳例が複数ある場合は | で区切る．
　例：**暖房**　この部屋は〜がきき過ぎている．This room is overheated. | The heat is too high in here.（| の前後どちらの文も同義の英訳例）

英和篇
Kenkyusha English and Japanese
Dictionary of Ecology

A

AAUs Assigned Amount Units 初期割当量《排出枠(emission credit*)の一つ；単位は「CO_2 トン」；京都議定書(Kyoto Protocol*)の温室効果ガス削減義務に基づき，あらかじめ同議定書の附属書Bの先進各国(Annex B Countries*)に排出可能量として割り当てられたもの；使用しない分を他国に売ることができる》.

Abalone Alliance [the 〜] アバローニ同盟《サンフランシスコを本拠とした米国の反原子力運動団体；1970〜1980年代にかけてカリフォルニア州南東部 Diablo Canyon の原発建設に反対し，すわり込みなどの直接行動を展開．米反原発運動で最多の逮捕者を出した》.

abatement 《被害などの》軽減，緩和，抑止，《有害物質などの》削減.

abiotic environment 【生態】非生物的環境《水・土壌・空気・太陽光など，生物に影響を与える物理・化学的因子がつくる環境．無機的環境；⇔ biotic environment*》.

ablation 《侵食・風化による》削摩；《溶解・蒸発などによる氷河の雪や氷の》消耗.

Abolition 2000 アボリション 2000《核廃絶を訴える国際的なネットワーク；90か国以上に及ぶ 2000以上の団体が参加している；1995年発足，事務局カナダ Ottawa》.

ABS access and benefit-sharing 遺伝資源へのアクセスと利益配分.

absolute zero 絶対零度《マイナス 273.15°C(摂氏)あるいはマイナス 459.67°F(華氏)》.

absorbing well 《地表の水を地下へ導くための》吸込み排水［下水］孔，吸込み井戸.

abundance 1【生態】数度《一定の調査面積内の種類別の個体数，あるいは個体数を出現枠数で割った値》.
2【理】存在量《宇宙における各元素の存在比や同一元素中の同位体の存在比》.

abyssal circulation 深層循環，深海の海水の流れ.

abyssal zone [the 〜]【生態】深海底帯，深海層《底生生物(benthos*)の生態区分を示す底生区の一つ；漸深海底(帯)(bathyal* zone)の下にある水深 2000〜

6000 m あたりまでの海底；海溝部を除く最深部》．

abyssopelagic 形【生態】深海水層の《遊泳・浮遊生物の生態区分を示す漂泳区分のうち，水深3000[4000] m-6000 m の海洋最深部の層；主な漂泳区分は通例上から epipelagic* (200 m まで) ⇨ mesopelagic*(1000 m まで) ⇨ bathypelagic*(3000[4000] m まで) ⇨ abyssopelagic の順になる》．

ACC adaptive cruise control* 車間距離自動制御．

ACC Antarctic Circumpolar Current* 環南極海流．

acclimatize, acclimate 動〈人・動植物などを[が]〉新しい風土[環境, 気候]に慣らす[慣れる]，順化[順応]させる[する]，馴致(じゅん)する．

 acclimatable 形 順応[順化]可能な．

acclimatization, acclimatation 順化，順応；【生態】《生物個体の環境に対する》順化，《特に》気候順化，気候変動に対する適応．

accommodate 動 適応[順応]させる[する]，合わせる；和解させる[する]，調停する；〈眼が〉遠近の調節をする： *accommodate* one thing *to* another あるものを他に適応[順応]させる / *accommodate* oneself *to* ... 〈境遇などに〉順応する．

accretion 《付着・堆積による》生長，増大，添加，累積；添加物，付加物，付着物，着氷，着雪；生長[添加]を重ねてでき上がったもの；【生】癒着(生長)；【地】付加，アクリーション《新しいリソスフェア(上部岩石圏)がそれまでに存在したリソスフェアの縁辺部に付加すること》： *accretion* cutting【林業】受光伐(ぼ)．

acid cloud 酸性雲《酸性の霧や低い雲；高濃度の硫酸・硝酸を含み，酸性雨より植生に対する影響は大きい》．

acid deposition 酸性沈着《acid rain* の拡張概念；雨以外に微粒子などによる環境汚染も含める》；酸性降下物[沈着物]．

acid dust 酸性塵《大気汚染物質》．

acid fallout 酸性降下物，《特に》酸性雨(acid rain*)．

acid flush 酸性出水《大気汚染による酸性降水(雨・雪など)の河川などへの流出》．

acidify 動 酸っぱくする[なる]，酸敗させる[する]；【化】酸性化する．

 acidifier 名 酸性化するもの；酸性度を高めるもの，土壌酸化剤．

 acidification 名 酸性化《pHが7より小さくなること》；酸敗《腐敗して酸性になること》．

acid mist 《大気汚染による》酸性ミスト《空中に霧状に浮遊する酸性の微粒子》，酸性霧(⇨ sulfuric acid mist*).

acid precipitation 酸性降水《大気汚染による酸性の雨や雪》.

acid rain 《大気汚染による》酸性雨.

Acid Rain Program 酸性雨プログラム《1985年米国環境保護庁(EPA*)が発表した酸性雨対策;硫黄酸化物(SOx), 窒素酸化物(NOx)の排出量削減による酸性雨防止を目指す》.

acid sludge 硫酸スラッジ, 酸性スラッジ《石油精製の際, 硫酸処理によって生じる粘着性の残留物質》.

acid snow 《大気汚染による》酸性雪.

acid soil 酸性土壌.

acoustimeter 騒音測定器《特に交通騒音用》.

acrodynia 《水銀中毒などによる, 特に乳児の》四肢の疼痛; 先端[肢端]疼痛(症).

action 《天然現象・薬などの》作用, 影響, 効果; 【生態】環境作用; (生理学的)作用, 機能: chemical *action* 化学作用.

activate 動 活動的にする, 活動[作動]させる;【理】…に放射能を与える, 放射化する;【化】活性化する;《好気性細菌による汚水分解を助長するために》〈汚水〉の曝気(ばっき)を行なう.

activator 图 正触媒, 活性剤, 賦活体[物質], 賦活剤, 活性体[薬].

activation 图 活性化, 賦活; 活動化.

activated sludge 活性汚泥《排水浄化に使われる》.

activated sludge process 活性汚泥法《下水処理法の一つ》.

activize 動 activate*.

acute effect 急性効果, 急性作用, 急速な影響, 早発性[即発性]効果《放射線に被曝したあと短時間のうちに人体に出る効果や影響; 脱毛や出血など》.

ACW Antarctic Circumpolar Wave* 南極周極波(動).

acyclic 形 周期的でない;【生態】無輪廻の;【化】非環式の.

adaptation 適合, 適応;《感覚器官の》順応, 調節;《環境・文化類型などへの》適応; 適応(して発達した)構造[形態, 習性];《特にclimate change*(気候変動)による温度上昇などへの》適応: *adaptation* syndrome【生理】適応症候群 / *adaptation* policy 適応政策.

adaptive cruise control 車間距離自動制御《ドライバーの負担を軽減する運転支援システムの一つ; 障害物や先行車を検知して速度を制御できるクルーズコントロール》.

adaptive management アダプティブマネジメント, 順応[適応]型管理《野生生物管理や生態系保護など複雑で変化の多いものを対

象とする場合の管理手法；対象を不断にモニタリングしてその結果を実際の管理に反映させる》．

additionality 追加性《京都議定書(Kyoto Protocol*)におけるクリーン開発メカニズム(CDM*)や共同実施(JI*)のプロジェクトが従来とは別の新規計画であり，また環境的に実効性をもつものであること；これが証明できないと認証排出削減量(CER*)，排出削減単位(ERU*)が認められない》．

ADI acceptable daily intake《有害物質の》1日当たりの許容摂取量．

adiabatic 形【理】断熱的な；熱の出入りなしに起こる．
图 断熱曲線．

adjustment 調整，調節，修正，補正；調停；調節装置；(心理的)適応；【生態】適合．

adopt program アダプトプログラム《住民や企業などが，地元の道路や公園，河川敷などを一定区画ごとに分けて取り組む保全・美化のボランティア活動，公共空間の一定区画を adopt(養子)に例えている》．

aeration tank 曝気(ばっき)槽，エアレーションタンク．

aerie, aery 《崖や山頂にある猛禽の》高巣(たかす)；高所にある巣．

aerobic 形【生】好気性の；好気性細菌の；【化】酸素の存在するときにだけ進行する，有気性の．

aerobic digestion 好気的［好気性］消化《好気性細菌と酸素を用いて廃物の量を減少させ，炭酸ガス・水・有機・無機化合物を生成すること》．

aerophyte 【生態】着生［気生］植物(epiphyte*, air plant*)《樹木や岩盤に固着し，栄養分を雨や空気から摂る植物》．

aerosol 【理・化】エーロゾル，エアゾール，煙霧質；煙霧剤；エアゾール容器．

affluent 形 よどみなく流れる；裕福な，［the ～］裕福な人びと；豊富な；流れ込む：*affluent society* 豊かな社会《経済学者 Galbraith が現代社会に関して同名の著書(1958)で用いた》．
图《本流・湖に流れ込む》支流；《下水処理場に入ってくる》下水；裕福な人．

afforest 動〈土地を〉森林にする，…に造林［植林］する．

afforestation 图 造林，新規植林；林野．

afterburst アフターバースト《核爆発後に成層圏からの放射性降下物がもたらす放射能汚染》．

aftermath 《牧草の》二番刈り，二番草，再生草(rowen*)．

AFV alternative-fuel vehicle*代替燃料自動車．

AGBM Ad-Hoc Group for Berlin Mandate ベルリンマンデート・アドホックグループ，京都議定書

準備会合《ベルリンの COP*1（1995 年 3 月）から京都の COP3（1997 年 12 月）までの間に開催された計 8 回の京都議定書交渉；⇨ Berlin Mandate*, Kyoto Protocol*》．

Agenda 21 アジェンダ 21《21 世紀に持続可能な開発を実現するための行動計画（アジェンダ）；1992 年に Rio de Janeiro で開催された「環境と開発に関する国連会議」（United Nations Conference on Environment and Development*）で採択された；前文と「社会的・経済的側面」「開発資源の保護と管理」「主たるグループの役割の強化」「実施手段」の 4 セクション計 40 章から構成される》．

agflation 農産物インフレ，食料インフレ《agriculture と inflation の合成語；特に近年 bioethanol ブームで穀物などが燃料向けに転用され品薄となって起きる農畜産物全般の値上がりを指す》．

agglomeration diseconomies 《過度の人口稠密による》集積の不経済．

aggradation 【地】埋積［進均］作用，アグラデーション《堆積の優勢による平坦化作用》．

aggregate 形【植】〈花が〉集合花の，〈果実が〉集合果の；【地】砕屑物からなる，集成岩の．
名【地】集成岩；《土壌の》団粒；骨材《コンクリート［モルタル］をつくる際に接合剤に混ぜる砂［小石］》．
動 集める；統合する；集まる，集合する．

aggregated distribution 【生態】集中分布《ランダムな分布より局所的に集中した箇所のある分布》．

agonistic, -tical 形【生態】拮抗(きっこう)関係にある．

agrestal, -tial 形〈雑草などが〉耕作地にはびこる，野生の，アグレスタルな．

agribusiness 農業関連産業，アグリビジネス《農業のほか農機具生産，農産物の加工・貯蔵・輸送などを含む》．

agricultural fertilizer 農業用肥料．

agricultural pollution 農業による環境破壊，農業汚染《農業廃棄物や表土流出，農薬汚染など，農業活動によってひき起こされる環境破壊》．

agricultural waste 農業廃棄物．

agrioecology 農業生態学《栽培植物の生態学》．

agroecological 形 農業と環境［農業生態学］に関する．

agroecology 農業生態学《農耕地における自然環境や農作物発育に影響する環境要因など，農業と生態系との関連を研究する学問分野》．

agroecosystem 農業生態系．

agroforestry 農業・林業両用の土

地利用，併農林業．

A-horizon 【地】A層(位)《土壌の最上層》．

aircraft noise 航空(機)騒音．

air freshener 空気清浄スプレー，エアフレッシュナー．

air monitoring 大気汚染監視，空気モニタリング．

air plant 【生態】空気[気生，着生]植物(aerophyte*, epiphyte*)．

air pollution 大気[空気]汚染．

air pollutant 大気汚染物質[汚染源]．

air quality 《汚染度を示すいくつかの指標によって評価される》大気質: *air quality* index 大気質指標 ⇨ 見出し / *air quality* standard 大気質基準, 大気環境基準．

Air Quality Index 大気清浄度指数《米国環境保護局(EPA*)が発表する全国の大気汚染の度合いを示す数値；数字が小さいほど清浄で, 0-50 が良好, 100 までが普通, 150 までが「過敏な人には影響あり」151 以上は明確に不健康ないし有害; 関連機関のホームページではこれに基づき天気予報のように色分けした地図も掲載している》．

air sampling 《大気汚染調査のための》大気試料採取, 空気[エア]サンプリング．

air stripping 曝気(ばっき)処理《汚染された地下水や地表水を噴霧状にして揮発性不純物を除く処理; ⇨ vacuum extraction*》．

airborne particle 浮遊粉塵．

Alaska Pipeline [the ～]アラスカパイプライン(Trans-Alaska Pipeline)《北極海に面する Alaska 州北部の Prudhoe Bay からアラスカ湾の Valdez 港まで 1270 km に及ぶ石油パイプライン；1日 100 万バレルの石油を輸送できる；1977 年に完成したが, 生態系を破壊するとして, 環境保護論者が建設に猛反対した》．

albedo アルベド《太陽系天体で太陽からの入射光の強さに対する反射光の強さの比》．

Alfisol アルフィゾル《鉄含量の大きい表土をもつ湿った土壌》．

algae 藻類，藻．

algal bloom 《水質汚濁による特に有害な》藻の異常発生, 藻類大増殖, 水の華．

algicide 殺藻薬[剤], アルジサイド《硫酸銅など》．

alien 【生態】外来種．

alkali 【化】アルカリ, 塩基性物質；《乾燥地方の土壌に含まれる》アルカリ塩類；《米国西部の》アルカリ性土壌の地域．

alkali soil アルカリ(性)土壌《植物生育に不適》．

alkalitrophic lake 【生態】アルカリ栄養湖．

allelochemical 【生態】他感作用物質, アレロケミカル《他感作用(allelopathy)の原因物質；特に,

植物により産生され他種植物に阻害的な影響を与える化学物質》.

allelopathy 【生態】他感作用, アレロパシー《特に他種の植物体から出る化学物質により植物がうける影響; 特に発育阻害》.

  allelopathic 形 アレロパシーの.

allelotoxin allelochemical*

Allen's rule アレンの法則《同種の恒温動物では, 寒冷地に生息する個体ほど, その体の付属器官(耳, 四肢, 翼など)が, 温暖な地域に生息する個体に比べて小型化するという法則; 体温維持の観点から説明される; ⇨ Bergmann's rule*》

allergen アレルゲン《アレルギーを起こす物質》.

allergenic pollen 《花粉症を引き起こす》花粉アレルゲン.

alleviate 動〈心身の苦痛などを〉軽くする, 緩和する, 楽にする; 〈問題を〉多少とも解決[解消]する.

  alleviator 名 軽減[緩和]する人[もの].

alleviation 軽減, 緩和; 軽減[緩和]するもの.

alliance 《植物の》群団.

Alliance for Zero Extinction 絶滅ゼロ同盟《生物多様性を守ることを目指す, 非政府の自然保護団体の連合体; 生物種が絶滅の危機に瀕している地域を指定し, 保護活動に乗り出すことを目標とする; 略 AZE》.

Alliance of Small Island States 小島嶼国連合《温暖化による海水面上昇を危惧する世界各地の島国が結成し, 温暖化防止などを訴える; 略称 AOSIS》.

allocate 動〈自然保護のため漁獲量などを〉割り当てる.

allocation 〈漁獲量などの〉割り当て.

allocentric 形 他者中心の.

allopatric 形【生態】異所(性)の (⇔ sympatric*): *allopatric* species 異所種 / *allopatric* hybridization 異所性交雑 / *allopatric* speciation 異所的種形成.

allopatry 名 異所性.

allopelagic 形【生態】アロペラジックの《温度以外の影響に応じて海中の種々の深度に生活する》.

alloplastic 形 環境を形成する, 環境を変える: *alloplastic* adaptation 環境変容的適応.

allowable (cut) 許容伐採量(permissible yield, prescribed cut, allowable annual harvest)《ある森林から毎年定期的に伐採できる木材量》.

alluvium 【地】沖積層, 沖積土.

alpestrine 形 アルプス山脈の; 山岳地帯の;【生態】亜高山帯に生える.

Alpine 形 アルプス山脈の; [時に alpine]【生態】高山帯に生える, 高山性の: the *Alpine* flora 高山

植物相.
图 [alpine] 高山植物.

alpine tundra 【生態】高山ツンドラ[凍土帯]《広い平地[斜面]をなす樹木の生育しない高地》.

alteration 〈生息地などの〉変化, 変質.

alternate host 【生態】代替寄主《寄生動植物の本来の寄主以外のもの》.

alternative energy 代替エネルギー《化石エネルギーや原子力エネルギーに代わる, 太陽エネルギー・風力・潮汐・波・地熱, バイオマス(biomass*)など》.

alternative fuel 代替燃料《車両の動力源としてのガソリンやディーゼル油に代わる天然ガス・メタノール・電気など》.

alternative-fuel vehicle, alternative fuel(s) vehicle 代替燃料自動車《二酸化炭素の排出減少を目的に作られた車; ガソリン, ディーゼル油の代りに天然ガス, 電気, 太陽エネルギー, ガソホール(gasohol*)などを使う; 略 AFV》.

alternative technology 代替技術, オルターナティヴテクノロジー.

altimeter 高度計.

altitude 高度, 標高.

aluminium, aluminum アルミニウム.

ambience, -ance 環境, 雰囲気.

ambient 圈 周囲の, ぐるりを取り巻く.
图 環境.

ambient air standard 大気汚染許容限度(値), 大気環境基準.

ambient noise 環境[周辺]騒音《ある地域の騒音の総量》.

ameliorant 土壌改良剤.

amenity 《場所・建物・気候などの》ここちよさ, 快適さ, アメニティー; [複数形で]生活を楽しく[快適に, 円滑に]するもの[設備, 施設, 場所], 生活の便益.

amensalism 【生態】片害作用, 偏害作用《2種の生物が共に生活し, 一方が他方に有害な影響を与えるのに対して, 他方は相手からなんら影響されないという関係; ⇨ commensalism*》.

amino-acid 【生化】アミノ酸.

ammonia 【化】アンモニア《気体》.

Amoco Cadiz [the ~]アモコ・カディズ号《1978年にフランス沖で坐礁し大量の石油流失事故(the Amoco Cadiz disaster)を起こした Amoco International Oil 社所有のタンカー》.

amphibian 圈【動】両生類の; 水陸両性の(amphibious).
图 水陸両生の動物[植物], 《特に》両生類の動物.

amphigaean, -ge- 圈【生】地球の両半球に分布する, 汎存(性)の.

anarcho-green 圈 アナキズム的環境保護主義の.

図 アナキスト的エコロジスト.

angstrom (unit) 【理】オングストローム《長さの単位: = 10^{-10} m; 電磁波の波長などについて用いる; 略, Å, A, AU》.

animal community 【生態】動物共同体, 動物群集.

animal control 動物管理局《動物の行動統制・隔離・処分などに関する法の執行に当たる部局》.

animal control officer 動物管理局員.

animal ecology 動物生態学.

Animal Police アニマルポリス《動物に関わる犯罪を取り締まる民間組織; 米国の多くの地方自治体に置かれており動物虐待を摘発する; 野良犬や迷い犬の保護や飼えなくなった動物の引き取りも行う》.

animal rights 《動物保護に基づく》動物の権利;《虐待されている動物や実験動物などの生存権を守る》動物保護(活動).

Animal Rights militia アニマル・ライツ・ミリシャ《動物実験などに反対する英国の過激派組織; 略 ARM》.

annex 附属書.

Annex A Gases 《Kyoto Protocol* の》附属書 A に記載された温室効果ガス《CO_2(二酸化炭素), CH_4(メタン), N_2O(一酸化二窒素), HFCs(ハイドロフルオロカーボン類(代替フロンの一種)), PFCs(パーフルオロカーボン類(代替フロンの一種)), SF_6(六フッ化硫黄); ⇨ base year*》.

Annex B Countries 《Kyoto Protocol* の附属書 B で温室効果ガス削減数値目標を定められた》附属書 B 締約国.

Annex I Countries / Parties 附属書 I 締約国《気候変動枠組条約(UNFCCC*)の附属書 I に記載される国; 欧米・オーストラリア・日本などの Annex II Countries* および東欧諸国; 2000 年までに温室効果ガスの排出量を 1990 年レベルに減らす義務を負い, その数値目標は Kyoto Protocol* 附属書 B で規定されている》.

Annex II Countries / Parties 附属書 II 締約国《Annex I Countries* から東欧諸国を除いた欧米日豪の先進諸国で, 附属書 II に記載される; 温室効果ガス削減義務に加えて途上国側の温暖化対策への技術移転や資金援助などの義務がある》.

Antarctic Circumpolar Current [the ~] 環南極[周南極]海流《南極大陸の周囲を西から東へ流れる海流; 別名 West Wind Drift》.

Antarctic Circumpolar Wave 南極周極波(動)《環南極海流(Antarctic Circumpolar Current*)の内部にある波動現象; 南極の周囲を 8, 9 年かけて海面下半マイル以上の深さにわたって巡り, 南米,

南アフリカ,オーストラリアなどの気候に影響を及ぼすとされる;略 ACW》.

Antarctic Convergence [the ～]【海洋】南極収束線《亜熱帯水と亜南極水がぶつかってできる収束線;ここでは気象変動が激しい》.

Antarctic Current Antarctic Circumpolar Current*

Antarctic Treaty [the ～]南極条約《南極に対する各国の領有権を認めず,平和的利用に開放することを規定した条約で,1959年に締結された》.

Antarctica 南極大陸(the Antarctic Continent)《南極横断山地(Transantarctic mountains)によって東西に分かれる》.

anthropo- 「人,人類;人類学」の意の連結形.

anthropocentric 形 人間中心主義の(⇨ biocentrism*).

anthropogenic 形【生態】人為起源の;人間活動に由来する.

anthropophyte 人為植物《植物を栽培することによって偶然に紛れ込んだ植物種》.

anthrosphere, anthroposere noosphere*

antibody 《免疫系の》抗体.

anti-environmental group 《大企業などの》環境保護運動に反対する勢力,環境保護規制反対派.

antinuke 形 反原発の,反核の(antinuclear).
名 反原発[反核]派の人.

antipollutant 形 汚染防止[除去]の.

antipollution 形 名 環境汚染[公害]を防止[軽減,除去]するための(物質).

antipollutionist 名 環境汚染[公害]防止論者.

antismog 形 スモッグ防止の.

ANWR Arctic National Wildlife Refuge*

AONB area of outstanding natural beauty*

AOSIS Alliance of Small Island States*

APN Asia-Pacific Network for Global Change Research アジア太平洋地球変動研究ネットワーク.

applied ecology 応用生態学.

aquiclude 難透水層《水を吸収し貯えるが,井戸や泉に供給するほど速くは透過できない地層》.

Arbor Day 植樹の日《米国などで4月から5月にかけて植樹の催しを行う;⇨ Bird Day*》.

archaeology 考古学.

Arche Summit アルシュ サミット《1989年7月フランスのパリ郊外にある Grande Arche センターで開かれた第15回主要先進国首脳会議;経済問題と共に環境問題に関して本格的な議論を行い,経済宣言にもその対策の必要性を盛

り込んだ》.

archibenthos 【生態】中型底生生物(⇨ mesobenthos*).

architectural salvage 《タイルや暖炉など》建築廃材の回収利用.

arcology 完全環境計画都市, アーコロジー《*architectural ecology* を短縮した米国の建築家 Paolo Soleri の造語》.

Arctic 形 北極の, 北極地方の(⇔ antarctic): an *arctic* expedition 北極探検(隊).
名 [the Arctic] 北極地方.

Arctic National Wildlife Refuge 北極圏国立野生動物保護区《米国アラスカ州北東端の国境部にある, 魚類野生動物庁(Fish and Wildlife Service*)管轄下の保護区; 貴重な生態系があり, この地の原油・天然ガス資源をめぐり, 油田開発か自然保護かの論争が続いている》.

Arctic Ocean [the 〜] 北極海, 北氷洋.

area of outstanding natural beauty 自然景勝地域《スコットランドを除く英国内の景勝地で, 国立公園に準ずる指定地域; 49箇所(2009年)あり, 特別保護地域として開発はきびしく規制されている; 略 AONB》.

AR4 Fourth Assessment Report 《IPCC* の》第4次評価報告書《2007年発表; 気候変動が人為的原因で起きていることをほぼ確実とし, 温暖化対策の必要性を訴えている》.

Argo Project [the 〜] アルゴ計画《地球全体の海洋変動をリアルタイムで監視・把握するための計画; 世界気象機関(WMO)・ユネスコ政府間海洋学委員会(IOC)等の国際協力による》. ★⇨ 和英アルゴ・フロート

arid 形〈土地が〉乾燥した, 不毛の; 【生態】乾地性の, 偏乾性の.
aridity, aridness 名 乾燥(状態), 乾燥度.

aridification 【生態】《気候の変化や生態系の破壊などによる土地の》乾燥化.

ARM Animal Rights militia*

aromatic 形 芳香(性)の; 香りの強い[よい]; 【化】芳香族(化合物)の.
名 芳香のもの; 香料; 芳香植物; 芳香薬; 【化】芳香族化合物(aromatic compound).

arrogance of humanity 人類の傲慢さ.

arsenic 【化】砒素《記号 As, 原子番号 33》.

artefact 人工品; 人工遺物《先史時代の単純な器物・宝石・武器など》; 文明の産物; 工芸品; 【生】《細胞・組織内の》人為構造, 人工産物.

article 条項: *Article* 2 of UNFCCC 国連気候変動枠組み条約の第2条(⇨ United Nations

Framework Convention on Climate Change*).

asbestos 石綿, アスベスト.

ASH Action on Smoking and Health アッシュ《英国の禁煙運動推進のボランティア団体; 1971年に設立された; 世界10カ国に同名の禁煙推進団体があるが, 互いに独立し, リンクしていない》.

ASPCA American Society for the Prevention of Cruelty to Animals 米国動物愛護協会《1866年設立; ⇨ RSPCA*》.

aspect 【生態】《植生の》季観; 植物群落の季節的な外観.

assemblage 【生態】《偶然の》群がり.

association 【生態】《ある場所の有機的集合体としての》(生物)群集;《群落単位としての》群集, アソシエーション.

Atlantic Ocean 大西洋.

atmosphere 1 [the 〜]《地球を取り巻く》大気, (大)気圏,《天体を取り巻く》ガス体; 雰囲気《環境ガス・媒体ガス》;《特定の場所の》空気: an inert *atmosphere* 不活性雰囲気 / a moist *atmosphere* 湿っぽい空気.
2【理】気圧《= 1013.25 ヘクトパスカル; 略 atm.》.

atmospheric concentration《有害物質などの》大気中濃度.

atoll 環状サンゴ島, 環礁.

atomic 形 原子力(エネルギー)の(nuclear*).

atomic number【化】原子番号《各元素の原子を軽い順に並べたときの番号; 原子のもつ陽子数に等しい》.

ATSDR Agency for Toxic Substances and Diseases Registry 毒物・疾病登録局《米国保健福祉省の一部局; CERCLA* によって設立された有害廃棄物被害の研究・防止機関; 本部 Atlanta》.

at stake 形 危機に瀕した, 危うくなって, 問題となって.

audio pollution 騒音公害(noise pollution).

Audubon Society オーデュボン協会《米国の著名な自然保護団体; 1905年に全米組織 National Audubon Society として設立され, 鳥類や野生生物の保護, 生物多様性保全を掲げて活動している; 名前は鳥類画家 John James Audubon (1785-1851) に由来する》.

aurora 極光, オーロラ《南極光(aurora australis)と北極光(aurora borealis)がある》.

autecology, autoecology 個[種]生態学(⇨ synecology*).
  aut(o)ecological 形 個[種]生態学の.

autopotamic 形【生態】流水性の《動植物, 特に藻類が淡水の流水中で生活[生育]する; ⇨ eupotamic*, tychopotamic*》.

autosexing 形 誕生[孵化]の際に雌雄の別によるそれぞれの特徴を示す.
名 オートセクシング《鶏などに標識遺伝子を入れて行なう早期雌雄鑑別》.

average life expectancy 平均余命.

avifauna 《ある土地・時期・自然条件における》鳥類相(⇨ fauna*).
avifaunal 形 鳥類相の.

avoidance cost 《環境破壊などの》回避費用, 防止対策コスト.

avoided emissions 《各種対策によって》回避されたと見なされる有害物質などの排出分, 排出削減量.

AZE Alliance for Zero Extinction*

B

Bacillus Thuringiensis BT菌, 卒倒病菌《遺伝子を組み換えた形でマイマイガの幼虫, マメコガネなどの害虫の生物防除に用いられる細菌》.

backcountry 田舎, 僻地; 未開地.

background radiation 【理】バックグラウンド放射(線)《放射線測定で測定対象以外からくるもの; 宇宙線や自然界に存する放射性物質による》.

back-to-the-land movement, 'Back to the Land' movement 大地へ帰れ運動《1960年代から70年代初めにかけて米国で盛り上がった社会運動; 都市から農村へという理念に基づき, 共鳴した人は田舎で農業に従事しながら自給自足の生活をめざした; 大量消費文明や公害問題への批判的な視点を持ち, のちの環境保護運動の先駆けとなったといわれる; ⇨ Nearing*》.

bacteria bed 微生物[酸化]濾床《下水に曝気(ばっき)処理を施すとともに浄化微生物の作用をうけさせるための細砂または砂利の層》.

bag filter 袋濾過器, バッグフィルター《集塵機》; 排出ガス処理装置.

baghouse バッグハウス《bag filter* の設置されている建物》.

Bali roadmap [the ~]バリ・ロードマップ《京都議定書後の温室効果ガスの新たな削減枠組みについて話し合う2009年までの行程表; 2007年12月インドネシアの Bali 島で開催された COP* 13 で採択された》.

ballast water バラスト水《空荷状態の貨物船などに安定のために積まれる海水; 荷積みをすると船外に排出されるが, 排出されたバラスト水に含まれる細菌などの外来の微生物(⇨ marine pest*)がその土地の生態系を破壊する恐れが指摘される; バラスト水を規制する国際条約 the International Convention for the Control and Management of Ships' Ballast Water and Sediments* が2004年に採択された》.

barium 【化】バリウム《アルカリ土類金属元素; 記号 Ba, 原子番号56》.

bark 图 樹皮.
動 …の樹皮をはぐ;《枯らすために》〈樹木を〉環状剥皮する; 樹皮でおおう.

barrier island 砂洲島, 堡礁島《沿

岸洲(†)(barrier beach)の幅が広まったもの；部分的には居住可能で，ハリケーンや津波の時には本土をある程度護る役を果たす》．

barrier reef 堡礁《海岸に並行したサンゴ礁；海岸との間に礁湖が存在する》

basal sediment 《氷河などの》基底堆積物．

BART Best Available Retrofit Technology 利用可能な最良改善技術《古い発電施設などが導入する最良の大気汚染防止装置；⇨ regional haze*, BAT*》．

Basel Convention [the 〜]バーゼル条約《国境を越えて有害廃棄物を他国へ移動[輸出入]することを規制する国際条約；1989 年国連環境計画(UNEP*)により採択され，1992 年発効；先進国で発生した廃棄物が開発途上国に放置され，環境汚染を生む問題を解決するため，各国が廃棄物発生を最小にし，可能な限り自国内で処分することを求める；正式名称は The Basel Convention on the Control of Transboundary movements of Hazardous Wastes and their Disposal》．

baseline emission ベースライン排出量《京都議定書で定められた CDM*, JI* などの対策実効性の基準となる通常稼動時の排ガス・排熱などの量》．

base year 《統計上の》基準年《特に温暖化ガス削減目標数値の相対基準となる年(の排出量)；例えば京都議定書(Kyoto Protocol*)では附属書 A に記載されている Annex A Gases* のうち CO_2, メタン，一酸化二窒素については 1990 年，他の 3 種については 95 年の排出量レベルを基準としてそれに対するパーセンテージで目標を定めている》．

basin 1 ため池，水たまり；陸地に囲まれた港，内湾．
2 盆地；《河川の》流域；海盆；【地】盆状[地]構造《岩層が中心に向かって周囲から下方へ傾斜している地域》；堆積盆地(にある石炭・岩塩などの埋蔵物)：the *basin* of the Colorado River コロラド川流域．

BAT Best Available Technology*

bathyal 圏 漸[半]深海の《大陸棚より下の 200-2000 m 付近の深さ(の海底)についていう》：*bathyal* zone 漸深海底(帯)《底生区の区分の一つ；sublittoral* と abyssal zone* の中間》．

bathypelagic 圏 漸[半]深海水層の《漂泳区の区分で，水深 1000-3000 [4000] m の層；⇨ abyssopelagic*》．

beauty strip 帯状美林[景観林]《伐採した土地を隠すため，幹線道路や川・水路に沿って細長く残した森林》．

becquerel 【理】ベクレル《国際単位系における放射能の単位；放射線核種が1秒につき1個ずつ崩壊するときの放射能の量を1 becquerelとする；記号 Bq》.

bench-terrace 動〈急坂〉に段を築く，段丘にする《侵食を止めるため》.

benthos 水底，《特に》海洋底，深海底；【生態】《水底に居住する》底生生物，ベントス(= benthon).

benzene hexachloride 【化】ベンゼンヘキサクロリド，六塩化ベンゼン《殺虫剤；⇨ BHC*》.

Bergmann's rule 【生態】ベルクマンの法則《同種の恒温動物では，寒冷地に生息する個体の方が，温暖な地域に棲むものより大型化する，という法則；体温維持と体積，体表面の関係で説明される；⇨ Allen's rule*》.

Berkeley Ecology Center バークリーエコロジーセンター《米国カリフォルニア州 Berkeley に1969年に設立された環境保護団体；環境保護関係の書店を運営；リサイクルプログラムに取り組み，地産の農産物市を定期開催》.

Berlin mandate ベルリン・マンデート《1995年の第一回 COP* で採択された決議文書；京都議定書(Kyoto Protocol*)などその後の温暖化[気候変動]対策の道筋を定めたもの》.

Berne Convention ベルン条約《1982年に発効したヨーロッパ諸国の野生生物とその自然生息地の保護条約；正式名 Convention on the Conservation of European Wildlife and Natural Habitats》.

Best Available Technology 利用可能な最善の技術《EU などの環境保護法規・規制で企業側に採用が義務付けられる，可能な限り高度な公害対策技術；略 BAT；⇨ BART*, IPPC*》.

Bethell process ベセル法《真空下でクレオソートを含浸させることによる木材防腐処理；1990年代に EU では禁止》.

BFR brominated flame retardant* 臭素系難燃剤《RoHS Directive* の規制対象であるポリ臭化ビフェニール(PBB*)やポリ臭化ジフェニルエーテル(PBDE*)など》.

BHC benzene hexachloride ベンゼンヘキサクロリド，六塩化ベンゼン《有機塩素系殺虫剤；この物質のうちγ異性体はリンデン(lindane*)として知られ，殺虫作用が非常に強力で，人体に入っても分解されにくい》.

B-horizon B層《土壌層位の一つ；A-horizon* の直下にあり，腐植が少ない》.

bilateral 形 両側の(ある)；【生】左右相称の；相互[互恵]的な，【法律】双務的な；【社会学】双系の，

父母両系の: a *bilateral* contract [agreement] 双務契約[協定]. 图 二者会談[協議];《特に国際貿易に関する》二者協定(bilateral agreement).

bioaccumulation 生物蓄積《生物組織内に農薬などの物質が蓄積される現象》.

bio-aeration 空気接触法《バクテリアによる汚水浄化》.

bioassay 图生物(学的)検定(法), 生物学的定量, バイオアッセイ. 動 …に生物検定をする.

bioavailability 生物学的利用能, バイオアベイラビリティー, 生物吸収性《薬物や有害物質, 栄養素などが生体内に吸収されて利用される度合い》.

bioavailable 形 生物学的に利用可能な.

biocapacity 生物学的生産力, 生物生産力, 生物資源再生能力《土地・海洋が人間に有用な生物資源を生産する総能力; bioproductivity* と物理的な面積との積; ⇨ global hectare*》.

bioc(o)enology 生物群集学《生物群集の生態学》.

bioc(o)enosis, bioc(o)enose 《ある地域の》生物共同体[群集].

bioc(o)enotic 形 生物共同体の.

biocentrism 生物中心主義《人間の権利や必要が他の生物のそれに優先するわけではないという考え方》.

biocentric 形 生物中心主義の.
biocentrist 图 生物中心主義者.

biochar バイオ炭, バイオチャー《間伐材, 農業廃棄物, 汚泥などの biomass* を焼いて作る炭; 農地に埋めれば土壌改良剤となるほか, 炭素隔離(carbon sequestration*)の効果も期待される》.

biocide 生命破壊剤, 殺生物剤《DDT* など生物に有害な化学物質》; 生命の破壊.

biocidal 形 生命破壊性の, 殺生物性の.

biochemicals 生化学製品[薬品].

biochemical oxygen demand 生化学的酸素要求量(biological oxygen demand*)《有機物による水の汚濁度を示す指標の一つ; 略 BOD》.

biochore 生態域《海洋, 地上, 熱帯雨林など》.

bioclean 形 有害(微)生物を含まない: a *bioclean* room 無菌室.

bioclimatic law 【生態】生物と気候に関する法則《北米温帯域での生物事象(開花・落葉など)は, 緯度が1度または標高が400フィート増すごとにほぼ4日ずれる; Andrew Delmar Hopkins (1857-1948)による》.

biocoenology biocenology*

bioconcentration 生物濃縮(bioaccumulation*).

biocontainment 生物学的封じ込め.

biocontrol 生態的防除(⇨ biological control*).

biocycle 1 生物サイクル《biosphere* を陸水・海洋・陸地に分けた下位区分》.
2 生態サイクル, 生体周期《生体で見られる現象の規則的反復》.

biodegradable 形 微生物によって無害な物質に分解しうる, 生物分解性の: *biodegradable* detergents 生物分解性洗剤.
　biodegradability 名 生物分解性.

biodegrade 動《微生物によって》〈洗剤などを〉生物分解する.
　biodegradation 名 生物分解.

biodeterioration 生物劣化《微生物のはたらきによる物質の分解》.

biodiesel バイオディーゼル《ディーゼル油を代用できる, 再生可能で生分解型のバイオ燃料; 特にダイズ油・ピーナッツ油などの植物油から製した代用燃料》.

biodiversification 生物の多様化.

biodiversity 生物(の)多様性; 生物の種類の豊富さ《biological diversity の省略形; 多様性には遺伝子のレベル(同じ種内でも個体によって異なる遺伝子を持つ), 種のレベル(動物から微生物にいたる種), 生態系のレベル(森林, 河川, サンゴ礁など)という三つのレベルがある; 地球環境への適応が生物多様性をうみだしてきたが, 今日では, 生物種の絶滅が進み, かつ多様な生態系が破壊され, 生物多様性に危機が訪れている》.

biodiversity loss 《種の絶滅などによる》生物多様性の損失[低下].

Biodiversity Treaty [the 〜] 生物多様性条約《正式名称は Convention on Biological Diversity*》.

bioecology 生物生態学《動物・植物を合わせた生物群集を扱う》.
　bioecologist 名 生物生態学者.
　bioecological 形 生物生態学の.

bioenergy 《生物由来の物質の燃焼で得られる》生物燃料エネルギー; バイオ燃料(biofuel*).

bioengineering 生体[生物]工学《医学と生物学の分野に工学的知識[機器]を応用すること》;《広く》biotechnology*.

bioenvironmental 形 生物の環境と特にその中の有害な要素に関する, 生物環境の.

bioethanol バイオエタノール《植物由来の多糖から作るアルコールの一種; 特に, サトウキビやトウモロコシを原料とする自動車用燃料; ⇨ E 10*(E の最初の見出し)》.

bioethics バイオエシックス, 生命倫理(学)《遺伝子組み換え・新薬開発・臓器移植・再生医療など生命科学や医療の発展に伴う倫理問題を扱う; 1970 年代初めのアメリカで使われ始めた; 最初に提案した V. R. Potter は, 地球の有

限性を意識した人類生存の学問としての bioethics を考えたが，実際には，生命科学や医療をめぐる倫理(学)を意味する言葉として広く用いられるようになった；「生命」を意味する bio と「倫理(学)」を意味する ethics の合成語》.

bioethical 形 生命倫理上の，生命倫理的な.

bioethicist 名 生命倫理学者.

biofilter バイオフィルター《微生物による分解作用を利用した濾過器具》.

biofiltration 微生物の作用を利用した汚水処理法，バイオフィルター式濾過(法).

biofouling 生物付着《パイプなど水中の機械部位の表面にバクテリア・フジツボなどが付着すること；腐食や機能低下の原因となる》.

biofuel 生物(有機体)燃料，バイオ燃料《食用廃油，有機廃棄物，大豆，サトウキビ，トウモロコシ，木材といったバイオマス(biomass*)に由来する燃料》.

biogas 生物ガス，バイオガス《糞尿や生ごみなどの有機廃棄物から生物分解でつくられるガス；主成分はメタン；燃料として使う》.

biogasification 生物ガス化.

biogeocenology, -coe- 生態系研究.

biogeoc(o)enosis, biogeoc(o)enose 生態系(ecosystem*).

biogeoc(o)enotic 形 生態系の.

biogeochemical cycle 生物地球化学的循環《動植物と無生物の間における窒素・炭素などの物質交換》.

biogeography 生物地理学《生物の地理的分布を研究する》.

biogeographer 名 生物地理学者

biogeographic(al) 形 生物地理学の.

biohazard 生物災害[危害]，バイオハザード《病原体やそれに感染させた動物を扱う研究や医療に伴い，人間や自然界の生物に対して危害を及ぼす生物が外部に漏れ出て起きる災害；危険な病原体，遺伝子組み換えでつくられた微生物などが漏出防止の対象となる》.

biohazardous 形 バイオハザード[生物危機]の.

bio-house バイオハウス《木をはじめとする天然の材料だけで作った家》.

bioinformatics 生物[生命]情報科学，バイオインフォマティクス《コンピューターを用いた分子生物学や遺伝学》.

bioinvader invasive alien species*

biological accumulation bioaccumulation*

biological control 生物的制御[防除] (biocontrol)《天敵の導入などの生物的手段により，害虫や

biological diversity biodiversity*

biological integrity 生物保全(度)《外部からの人為的影響・改変のない自然状態の生物生息状況；どの程度これが保たれているかを生物保全指数(index of biological integrity; IBI)などで示す》.

biological invasion 《外来種・移入種(invasive alien species*)による》侵入，侵略.

biological magnification biomagnification*

biological oxygen demand 生物学的酸素要求量(biochemical oxygen demand*).

biological productivity 生物(学的)生産力.

biological shield 生体遮蔽《人体を放射線から守るために原子炉の周囲に設置される壁》.

biological system 生物システム，生体システム《分子レベルから生態レベルまでのシステム》.

biology 1 生物学；生態学(ecology*)；生物学書.
2 [the ～]《ある地域・環境の》動植物(相)；生態.
  biologist 生物学者.

biolytic tank 生物分解用タンク.

biomagnification 《生態系の食物連鎖における》生物学的(毒物)濃縮.

biomass 生物量《ある時点に，ある特定の地域[領域]に存在する生物体の物質量を，重量あるいはカロリーで表したもの；単位面積[体積]当たりの表示や地域全体での表示がある》；バイオマス《エネルギー資源として利用できる生物体(植物体や動物廃棄物など)；⇨ biofuel*》.

biome 生物群系，バイオーム《熱帯雨林やツンドラなど，主に気候的特性で区分された生態系に生息する生物群集の単位》.

biomedicine 生物[生体]医学《1) 自然科学，特に生物学・生化学の原理を適用した医学 2) 環境(宇宙環境など)の人体に与えるストレスと生存能力とのかかわりを扱う医学》.

bionomics, bionomy 生活誌，生態学(ecology*).
  bionomic(al) 形 生活誌[生態学]の.

biopiracy 海賊的生物探査，バイオパイラシー《発展途上国の動植物を搾取的に探査する生物探査(bioprospecting*)》.

bio-pirate 《貴重な生物資源や希少種などを収奪する》乱獲[盗伐,盗掘]業者，密漁[猟]者，漁[猟]場荒らし，《biopiracy* に加担する》闇(プラント)ハンター.

bioprecipitation 生物沈澱(法)《微生物などの作用を用いた鉱物や汚物の沈澱化》.

bioprocess 図 応用生物学的製

法, バイオプロセス.
⑩ 応用生物学的製法で処理する[作る].

bioproductivity 生物学的生産性, 生物生産性《単位面積の土地・海洋が人間に有用な生物資源を生産する能力; ⇨ biocapacity*, global hectare*》.

bioprospecting 生物探査, バイオプロスペクティング《医薬品などの有用成分を求めて動植物を探査すること》.

bioprospector 生物探査者(⇨ bio-pirate*).

bioreactor バイオリアクター, 生物反応器《酵素や微生物を利用して物質の分解・合成・化学変換などを行なう》.

bioregion バイオリージョン《生態系の特徴から一つのまとまりとして扱うことのできる地域[場所]; biome* より小規模で, より具体的な特徴を持つ》.

bioregional 形 バイオリージョンの.

bioregionalism バイオリージョナリズム, 生命地域主義《人間の活動は政治的国境ではなく, 生態学的・地理学的境界によって束縛されるべきであるとする考え方》.

bioregionalist 生命地域主義者.

bioremediation 生物的(環境)修復, バイオレメディエーション《微生物によって汚染物質を分解し, 環境を修復する技術》.

bioreserve 野生生物保護区域, バイオリザーブ《保護区域に隣接する利用制限のゆるい緩衝地帯を含めた広がり》.

biosafety バイオセーフティー《生物学的研究における安全性; 周辺に悪影響を及ぼさない状態の維持》.

Biosafety Protocol [the 〜] バイオセーフティー議定書, バイオ安全議定書, カルタヘナ議定書 (Cartagena Protocol on Biosafety)《遺伝子組み換え生物の国際取引の規制に関する議定書; CBD* の補完として 2000 年に締結, 03 年に発効》.

biosolids バイオ固形物, バイオソリッド《下水汚物を処理する過程で生成される有機物汚泥; 肥料などに利用できる》.

biosphere 生物圏《地殻を含む地球上および大気中の全生物生存圏; またその全生物》.

biospheric 形 生物圏の.

Biosphere II [2] バイオスフィア 2, 第2の生物圏《地球の生命圏のミニチュア版を密閉巨大ドームに収めた生命圏研究実験施設 ($13{,}000\ \mathrm{m}^2$); 米国南西部アリゾナ州 Tucson の北に造られ, 1990 年代に居住実験が行われた; ⇨ biospherian*》.

biosphere reserve 生物圏保護区域《自然な生態系の営みを守るための区画; しばしば観光客に開放

biospherian 実験バイオスフィアの住人《1993年9月26日, 男女各4名計8名の科学者が約3800種の動植物とともに外界から隔絶された biosphere II* の中での2年の自給自足生活を終えて外界に'帰還'した》.

biospheric cycles 生物圏循環[サイクル]《地球上の生物を維持していくうえで不可欠な, 酸素循環・炭素循環・窒素循環・水循環などを含む自然界の循環プロセス》.

biota 生物相《fauna* と flora* を合わせた, 一地域の動植物》.

biotech 《口》biotechnology*

biotechnical 形 biotechnological

biotechnology バイオテクノロジー《微生物のはたらき・遺伝子操作などの生物学的プロセスを産業・医療・環境対策などに応用する化学技術》.
　biotechnologist 名 バイオテクノロジー学者[技術者].
　biotechnological 形 バイオテクノロジーの[による].

biotecture バイオテクチャー《建築設計の不可欠な要素として生きた植物を取り込んだ建築》.

bioterrorism バイオテロ《生物兵器を利用したテロ行為》.

biotic climax 【生態】生物的極相.

biotic community 生物共同体, 生物群集《一定の地域に生活するすべての個体群の集まり》.

biotic environment 生物的環境《生物に影響を与える, 生物を含む環境; ⇔ abiotic environment*》.

biotic formation biome*

biotic isolation 【生態】生物的隔離.

biotope ビオトープ, 小生活圏《生態系とみなすことができる最小の地理的単位; 特定の生物群集が生存しうる小規模な生息空間; 公園や学校, 河川などの自然環境復元に用いられる; ギリシャ語の bio(生物)と topos(場所)の造語》.

bipolar 形 二極ある; 二つの極圏の, 【生態】両極(性)の; 【生】〈神経細胞などが〉双極の.
　bipolarity 名 二極性; 【生態】両極性.
　bipolarize 動 二[両]極化する.
　bipolarization 名 二[両]極化.

bird band 《野鳥の標識にする》足環(かしわ), リング(bird ring).

Bird Day バードデー《米国の多くの州で植樹の日(Arbor Day*)と同じ日に設けられた野鳥観察・保護の日》.

birdlife 《一地域の》鳥, 野鳥, バードライフ; 野鳥の生態.

bird-watch 動 探鳥[バードウォッチング]する《野鳥の生態を観察する》.

birdwatcher 名 野鳥観察者

(birder), 探鳥者, バードウォッチャー.

birdwatching 图 探鳥, バードウォッチング.

bird ring 《英》bird band*

black box ブラックボックス(システム)《内部の構造を考慮せずユニットとして見た回路網》;《広く》中身の全然わからない装置・体系》.

black carbon すす, 煙煤, 油煙《近年しばしば地球温暖化の要因とみなされる》.

Blacksmith Institute [the ～]ブラックスミス研究所《米国の非営利の環境団体で, 途上国の環境汚染問題に取り組んでいる; 1999年設立; 本部 New York 市》.

black water (工場)廃水, 《トイレからの》汚水, 《台所からの》下水 (waste water); 黒い水[流れ], ブラックウォーター《ピート地帯から流れ出す褐色の水》.

bleach 動 漂白する, しみ抜きする;《サンゴを》白化させる(⇨ coral bleaching*); 白くなる; 白化する.
图 漂白剤; 漂白度; 漂白.

bloom 图 水の華《プランクトンの異常発生; ⇨ algal bloom*》.
動 突然[大量に]現われる, 異常発生する;〈湖沼・海などが〉水の華ができる, 赤潮になる.

blue box ブルーボックス《リサイクル用品収集用の青いプラスチック製の箱》.

BOD biochemical [biological] oxygen demand*

boiling water reactor 沸騰水型軽水炉《原子炉の一種; 略 BWR》.

Bonn Convention ボン条約《1979 年ドイツのボンで採択され 83 年に発効した国際自然保護連合 (IUCN*)の「移動性野生動物種の保全に関する条約」(Convention on the Conservation of Migratory Species of Wild Animals)の通称; 海亀・渡り鳥・鯨類など長距離を移動する野生動物をその生活圏全域に渡って保護する多国間協力体制を定めた; 日本は捕鯨問題などの関係で未加盟》.

boreal 圈 北風の; 北の; (動植物が)北方(帯)の, 亜寒帯の, 北方林[針葉樹林]帯の;【地】氷河期以後の.

boreal forest 北方(樹)林, 針葉樹林.

botanical garden 植物園.

bottle bank 空き瓶回収ポスト[ボックス]《リサイクル用》.

bottle bill ボトル法案, 空き瓶回収デポジット法案《リサイクル推進およびごみ減量のために, ある種の飲料の空き瓶・空き缶についてデポジット制を義務づける法案》.

bottom fauna 底生動物相《海底に生息する動物相》.

bottom gear 海底に仕掛ける[を浚(さら)う]漁具《特に底引き網など移動式のもの(mobile bottom gear)は漁業資源保護のためしばしば規制対象となる》.

bottom trawling 底引き網漁.

Bovine Spongiform Encephalopathy 牛スポンジ様脳症, 牛海綿状脳症(俗称 mad cow disease)《脳組織がスポンジのようになる成牛の神経性疾患; 行動・姿勢に異常をきたし, 死に至る; ヒトに伝播すると, 変異型クロイツフェルト・ヤコブ病を発症する; 餌に含まれる感染性蛋白質粒子プリオンが原因とされている; 略 BSE》.

BP ⇨ Deepwater Horizon*

BPC biological pest control《害虫の天敵などを農薬代わりに利用する》生物的防除.

Bq becquerel*

breeding loan 《希少動物の》ブリーディングローン, 繁殖目的の貸し出し.

brominated dioxin 臭素化ダイオキシン《臭素系難燃剤(brominated flame retardant*)を燃やすと発生する有害物質》.

brominated flame retardant 臭素系難燃剤《電化製品・家具・衣料品などの素材となるプラスチック・ゴム・織物などの可燃性物質に, 燃焼速度を減少・抑制させる目的で混入する臭素化合物; ⇨ brominated dioxin*, BFR*》.

Brower ブラウアー David Ross Brower (1912-2000)《米国の環境保護活動家; Sierra Club* を大組織に育て上げ, Friends of the Earth* を創設するなど環境保護運動の基礎を築き, ノーベル平和賞の候補にもなった》.

brown earth 褐色土, ブラウンアース《湿潤温帯地域の, 主に森林植生下に発達する土壌; brown soil* とは別》.

brown soil 褐色土《温帯乾燥地の成帯性(気候型)土壌》.

BSE Bovine Spongiform Encephalopathy*

bryophyte layer moss layer*

Bt Bacillus Thuringiensis*

bubble concept バブルコンセプト《大気汚染物質について, 工場の個々の排気口からの排出量に対する規制ではなく, 工場全体を一つの排気口をもつドームと仮定して工場全体の排気総量を定めようとする考え方; 米環境保護庁が大気浄化政策で1970年代後半に導入した》.

buffer 图 緩衝動物《ある種の動物を常食とする肉食動物がそれ以外に例外的に, 摂食する他種の動物; これにより常食とされる動物の被害が緩衝される》.

動 …の buffer としてはたらく, …の衝撃を和らげる.

built environment 構築環境, 人

工的環境《環境のうち建造物からなる部分；自然環境に対する》.

built heritage 《一国の》建造物遺産《自然の財産・芸術的財産に対して》.

bund 《アジア諸国の》海岸通り，バンド；《インドの》堤防，築堤；埠頭(とう).

burramys チビヤマポッサム，ブーラミス《豪州産のネズミに似た小型の有袋類；稀少種》.

business-as-usual scenario 趨勢型シナリオ，現状放置予想《各種の予測において経済・社会・技術面の各要素が現状の趨勢のまま推移した場合に考えられる将来図》.

butan ブタン《無色の可燃性のガス状炭化水素》.

butterfly effect バタフライ効果《カオス理論の分野で使われる用語；ある小さな力が長期的にみて，大きな系に大規模な影響を及ぼすこと；たとえば，ある所でのチョウのはばたきが，長期的にみて他の場所で嵐を起こすという考え方》.

BWR boiling water reactor*

by-catch 漁網にかかった目的としない不要な魚(など)，付随漁獲物.

byproduct 副産物，副生成物.

C

CAA Clean Air Act*

cadmium 【化】カドミウム《金属元素；記号 Cd, 原子番号 48》.

CAFE corporate average fuel economy (米国の)自動車メーカーごとの平均燃費規制.

calcicole 好石灰植物.
  calcicolous 形 好石灰の.

calcifuge 嫌石灰植物.
  calcifugous 形 嫌石灰の.

calcify 動 石灰性にする[なる], 石灰化する.

calciphile calcicole*
  calciphilic, calciphilous 形 calcicolous*

calciphobe calcifuge*
  calciphobic, calciphobous 形 calcifugous*

calcium カルシウム《金属元素；記号 Ca, 原子番号 20》.

California Global Warming Act カリフォルニア州地球温暖化対策法《2006年9月に成立；カリフォルニア州の温室効果ガスの排出量を 2020 年までに 1990 年と同等の水準まで減らすことを求めている》.

California League of Conservation Voters [the 〜]カリフォルニア・リーグ・オブ・コンサヴェーション・ヴォーターズ《米国カリフォルニア州サンフランシスコ近郊とロサンゼルスに拠点を置く州規模の環境保護団体》.

Cambrian Explosion 【古生物学】カンブリア爆発《5億〜6億年前のカンブリア紀において現在見られるような主要な動物の門が比較的短期間のうちに急速に発生したとされる現象》.

CAMP continuous air monitoring program 連続大気監視計画.

campshed 動 …に堤防護岸を施す.

CAN Climate Action Network 気候行動ネットワーク《地球温暖化問題に取り組む, 世界の環境 NGO のネットワーク》.

can bank 《リサイクル用の》空き缶収集所, 缶ポスト.

cancer 癌, 癌腫.

candidate species 《endangered species* の》候補種.

canopy 林冠, 樹冠《森林の枝葉が茂っている最上層の広がり； ⇨ crown canopy*》.

CAP Campaign Against Pollution 公害反対運動; Climate Applications Program 気候利用計画; clean air package 汚染防

28

止エンジン.

cap 1 山の頂き,岩石の上層部;表帽《山頂部を覆う小形の氷河》.
2 キャップロック《含油層の上部にある不透水性の岩石;鉱床の上部にあって鉱化作用を妨げている岩石》.
3 (価格・賃金などの)上限,最高限度(⇨ cap and trade*).

cap and trade キャップ・アンド・トレード,上限付き取引《京都議定書(Kyoto Protocol*)が求める温室効果ガス排出削減目標の達成を促進するために設けられた排出量取引の方式;京都議定書が導入した京都メカニズム(柔軟性措置)の一つ;個々の主体(国・企業など)の排出量の上限(cap)を定め,排出許容枠を互いに売買(trade)することで,市場原理を活用した,排出抑制の効率的な達成が図れるとされる;⇨ emission(s) trading*, emission(s) credit*》.

capillary water 毛管水《土壌孔隙の中で重力に抗して保持される水》.

carbon 1 炭素《原子番号6,原子量12.0;自然界で4番目に多い元素;大気,海洋中には二酸化炭素,岩石中には炭酸塩,生物の体には有機物として存在する.炭素だけで構成される物質として,グラファイト,ダイヤモンド,カーボンナノチューブ,フラーレンがある》.
2 《環境用語では特に greenhouse gas* としての》二酸化炭素,炭酸ガス,CO_2 (carbon dioxide*).

carbon-14 炭素14《質量数14の炭素の放射性同位元素;記号 ^{14}C, C_{14};β崩壊半減期5730年;放射性トレーサーにするほか,考古学・地質学における年代決定に用いる》.

carbonate 图 炭酸塩[エステル]:*carbonate* of lime [soda] 炭酸石灰[ソーダ].
動 炭酸塩化する;炭化する;炭酸(ガス)で飽和させる:*carbonated* drinks 炭酸飲料.

carbon credit 炭素クレジット (emission credit*).

carbon cycle 《生物圏における》炭素循環[輪廻(りんね)].

carbon cycle feedback 炭素循環フィードバック《温暖化によって土壌有機物の分解が促進され,CO_2の放出量が増えてさらに温暖化が進む現象》.

carbon dating 放射性炭素年代測定法,炭素14法,カーボンデーティング(= carbon-14 [radiocarbon] dating)《^{14}C(半減期5730年)による年代測定法》.

carbon diet 《二酸化炭素などの》温暖化ガス削減(のための消費生活改善)(⇨ low-carbon*).

carbon dioxide 二酸化炭素,CO_2 《無色無臭;空気中に

0.035%(体積)含まれる；呼吸や有機物分解，化石燃料(fossil fuel*)やバイオマス(biomass*)の燃焼などで生じる；greenhouse gas* の約80%を占める；気体は炭酸ガス(carbon acid gas)とも言う》．

carbon dioxide equivalent 二酸化炭素当量，CO_2当量《さまざまな気体の温室効果を二酸化炭素に換算して，二酸化炭素の量として示す数値；単位はCO_2トン；⇨ global warming potential*》．

carbon dioxide flux 二酸化炭素(拡散)放出量《単位は$g/m^2 \cdot day$など》．

Carbon Disclosure Project [the～]カーボン・ディスクロージャー・プロジェクト《世界の大企業500社に対して投資家・機関が温暖化対策や二酸化炭素排出量についての情報開示を求める，炭素情報開示運動；2002年開始》

carbon footprint カーボンフットプリント《直訳すれば「炭素の足跡」；商品の製造から輸送，販売，廃棄にいたるライフサイクルの各段階において，どの程度，二酸化炭素を排出しているかを数値で表示したもの；企業や消費者に，より排出量の少ない商品の製造や購入を促す狙いがある；いわゆるecological footprint* の一つ；⇨ carbon offset*》．

carbon leakage 炭素リーケージ[漏出]《温暖化ガス削減目標の厳しい先進国から比較的ゆるやかな国へ生産拠点が移動し，結果的に全排出量が増えてしまう現象》．

carbon market 二酸化炭素市場，排出量(権)取引市場《二酸化炭素のemission trading* を行う市場》．

carbon negative 形 カーボンネガティブの《CO_2 の排出量より消費量が多く，carbon neutral* よりもさらに温暖化改善効果があるエネルギー源や技術を言う；このnegativeは「二酸化炭素の量が差し引きマイナスになる」ことを表わすもので，日本で言う「カーボンネガティブ」の用法に散見される「温暖化をさらに悪化させるような」の意味とは逆》．

carbon-neutral 形〈燃料などが〉カーボンニュートラルな《大気中の二酸化炭素の増加につながらない；たとえば木くずなどから作られる木質資源は木の生長過程で大気中の二酸化炭素を消費しているのでカーボンニュートラルであると考える；植林と組み合わせて管理された伐採から得られる木材資源も同様》．

carbon offset カーボン・オフセット，温室効果ガスのオフセット，排出量のオフセット[埋め合わせ，相殺]《市民や企業などが，自らの力では温室効果ガスの削減が難しい部分[量]について，その全

部または一部を，他の場所で取り組まれた排出削減・吸収量を購入するなどして埋め合わせること；emission credit* や carbon-neutral* 燃料による他，専門企業に代価を払って自分の排出量相殺分に見合う植林を代行してもらうサービスもある》．

carbon sequestration CO_2［炭酸ガス］分離［隔離］《温暖化対策として二酸化炭素を排気などから分離吸着し地中・海底に貯蔵したり再利用したりすること；ocean sequestration* や ECBM (⇨ CBM*) などがある》．

carbon sink （二酸化）炭素吸収源《二酸化炭素などの炭素化合物を蓄積し貯留する機能をもつ森林・土壌・海洋など；京都議定書 (Kyoto Protocol*) が carbon offset* の方法の一つとして提示した；⇨ LULUCF*》．

carbon tax 炭素税，二酸化炭素排出税《化石燃料の炭素含有量に応じて課される税金；環境税の一つ》．

carbon value 炭素価値《排出炭素を1トン減らすのに要する費用》．

carbuncle 《英》（景観と調和しない）目ざわりな建造物，見苦しい建物，「おでき」《1984年に英国皇太子 Charles が London の National Gallery の増築計画に対し，'a monstrous carbuncle' とコメントしたことから一般化》．

cardio-respiratory disease 心肺疾患．

carrying capacity 積載量；【生態】(環境)収容力《一地域の動物扶養能力[個体数]》，《牧草地などの》牧養力．

Carson カーソン Rachel (Louise) Carson (1907-64)《米国の海洋生物学者・科学評論家；*Silent Spring* (1962) を出版し，農薬・化学薬品による環境汚染の危険性を知らせた；環境汚染問題を広く社会に告発した最初の書とされる；⇨ silent spring*, DDT*》．

Cartagena Protocol (on Biosafety) [the ~] カルタヘナ議定書 (Biosafety Protocol*)．

CASBEE the Comprehensive Assessment System for Building Environmental Efficiency 建築物[建物]総合環境性能評価システム《建築物を環境性能で評価し，格付けする方法；「建築物の環境品質・性能(Q)」と「建築物の環境負荷(L)」の2つの面で評価する》．

catalytic converter 触媒コンバーター《自動車の排気ガス中に含まれる有害成分を無害化する装置》．

catastrophe 突然の大変動，大災害；《地殻の》激変；《カタストロフィ理論で扱われる》不連続的事象，破局．

catchment 集水, 湛水；集水の場所；(川・貯水池などの)集水地域, 流域(catchment area).

Catchment Board 流域[集水域]管理委員会《ニュージーランドやオーストラリアで一つの河川流域全体の水資源を保護管理する公共機関》.

catch share キャッチ・シェア《漁の解禁前に魚類の生息状況と照らし合わせて一定の全漁獲量(TAC*)を定め, 各漁村・漁協・個人などに割り当て分を分配する漁業資源管理手法；割り当て分は株式のように売買でき, また禁漁日の制限も緩和されるため, 計画的な漁獲・収入が可能になり乱獲防止が期待できる；DAP (dedicated access privilege), LAP (limited access privilege) とも呼ばれる》.

CBD Convention on Biological Diversity*

CBM coal bed methane コールベッドメタン, 炭層メタンガス《地下の石炭層に存在するメタンガス；深炭層に炭酸ガスを注入して吸着させ, 置換されて遊離したメタンガスを資源として回収するCBM増進回収(enhanced CBM recovery; ECBM)は資源開発とcarbon sequestration* を兼ねた技術として研究されている》.

CCAs Community Conserved Areas*

CCD colony collapse disorder 蜂群崩壊症候群《ミツバチの働き蜂の大半が巣箱に戻らず消え失せ, 群れが機能しなくなって養蜂産業や農業授粉に打撃を与える現象；原因については病菌や環境悪化など諸説あるがまだ不明；米国で特に2006年ごろから大発生し問題化》.

CCS carbon (dioxide) capture and storage (二酸化)炭素回収・貯留(carbon sequestration*).

CCSBT Commission for the Conservation of Southern Bluefin Tuna みなみまぐろ保存委員会《漁業資源としての南鮪を保護管理する国際組織で1994年発足；本部はオーストラリアのキャンベラ；⇨ ICCAT*》.

CCX Chicago Climate Exchange《米国内の温室効果ガス排出量取引(emission trading)を行うマーケット》.

CDM Clean Development Mechanism クリーン開発メカニズム《温室効果ガスの削減を定めた京都議定書(Kyoto Protocol*)において, 先進国が(削減の数値目標のない)途上国において実施した排出量削減の一部を自国に移転できる仕組み；京都メカニズムの一つ；⇨ JI*》.

CEB can-eating bird《空き缶をのみ込み, つぶして, たくわえる機械》.

cell from hell [the 〜]地獄の細胞《米国東海岸の渦鞭毛藻類の一種 *Pfiesteria piscicida* の異称；水質汚染が引き金になって活性化し，毒素の分泌によって魚を殺し，人に脳損傷を起こさせて，集中力喪失・記憶短期化を引き起こす》.

Celsius 形 摂氏[セ氏]の.

census 图 人口調査，国勢[市勢]調査，センサス；全数[個体数]調査: take a *census* (of the population) 人口[国勢]調査をする. 動 …の人口を調査する.

CER Certified Emission Reductions 認証排出削減量《排出枠 (emission credit*)の一つ. 京都議定書(Kyoto Protocol*)が規定した途上国へのCDM*によって得られた排出削減量のうちの一定量を，援助した先進国の排出削減量と認証するもの；⇨ additionality*》.

CERCLA Comprehensive Environmental Response, Compensation and Liability Act (⇨ superfund*).

cereal productivity 穀物生産力[生産高].

cerrado セラード《ブラジル中央部に広がる広大な熱帯草原,「やせた土地」とされ，植生はまばらな低木とイネ科の植物など. 1979年に日本が協力したセラード農業開発が始まり，穀倉地帯へと変化した》.

certification 《環境基準などを満たしているという》(正式な)証明, 認証, 認定(⇨ chain of custody*).

certified emission reductions CER*

CFC chlorofluorocarbon*

CFC-free 形 CFC[フロンガス]を使用していない.

chain of custody 《加工・流通過程で順次引き継がれる》管理責任の流れ, COC《木材・木製品について(環境などの見地から)認証を受けた伐採地から最終製品までの全工程を順次追跡すること》: *chain-of-custody* certification COC認証, 加工・流通過程の管理の認証《木製品に使われている木材が認証を受けた伐採地からのものであることを示す, 森林管理協議会(FSC*)による製品認証；略COC; ⇨ forest management certification*》.

channel 图 1 河床, 川底；水路, 河道《河川・港湾・海峡の深水部》；海峡；[the (English) Channel]イギリス海峡. 2 水管, 導管. 動 …に水路をつくる[開く]；…に溝を掘る；〈水路などを〉開く；水路で運ぶ: *channel* off 水路で〈水を〉流す[排水する]；〈資源などを〉消耗する.

chaparral 《米国南西部の》矮性(わいせい)カシの木のやぶ；《一般に》

踏み込みにくいやぶ；【生態】チャパラル《カリフォルニア州南部に特徴的な低木の硬葉半灌木林からなる植生》.

chemical fingerprint 化学の指紋《質量分析器やX線分析器で判断する特別な物質の存在を示すパターン；汚染物質・薬品その他の化学物質が試験対象として含まれているかどうかがそのパターンの有無でわかる》.

chemical oxygen demand 化学的酸素要求量《水中の還元有機物を酸化剤で酸化するときに消費される酸素の量；湖沼や海域などの水質汚濁の指標とされる；単位はppmまたはmg/l; 略COD; ⇨biochemical oxygen demand*》.

chemical toilet 化学的に汚物を処理するトイレット，化学処理トイレット，ケミカルトイレット.

chemosynthesis 化学合成.

Chernobyl nuclear power plant チェルノブイリ原発《旧ソ連のウクライナにあった原子力発電所；Kievの北方約130 km; 1986年4月26日，同発電所内の4号機の原子炉で，出力が急上昇する暴走事故が発生し，原子炉と建屋が爆発で破壊された；大量の放射能が放出され，北半球を中心に広範囲の放射能汚染がもたらされた；国連とロシア・ウクライナなどで構成される国際事故調査機関「チェルノブイリ・フォーラム」によって公式に確認された死者は約60人；原子力発電開発史上最悪の事故》.

Chernobyl factor チェルノブイリ要因[後遺症]《Chernobyl原発事故による放射能汚染に起因する農畜産物価格の下落その他の経済的影響》.

CHESS Community Health and Environmental Surveillance System《米国の》地域健康・環境監視システム，チェス.

China Syndrome チャイナシンドローム《原子炉の炉心溶融の結果，溶融物が地中深く落下して地球の反対側の中国にまで達してしまうという想像上の大惨事》.

chitosan キトサン《キチンの脱アセチル化した誘導体；水中の重金属を吸着するので，廃水処理に使用される》.

chlorides 塩化物.

chlorine 塩素《記号Cl, 原子番号17》.

chlorine demand 【生態】塩素要求量.

chlorocarbon クロロカーボン《塩素と炭素からなる化合物；四塩化炭素など》.

chlorofluorocarbon クロロフルオロカーボン《塩素化・フッ素化された炭化水素[メタンやエタンなど]の総称；慣用名フロン；化学的に安定な人工物質であり，冷媒・噴霧剤・発泡剤に使われてき

たが，大気に放出されたフロンがオゾン層を破壊することから生産禁止となった；略CFC》．

CHM Clearing-House mechanism*

chocolate mousse 《原油が海に流出した時にできる》ムース(mousse*)．

C-horizon C層《B-horizon* の下のやや風化した地層》．

CHP combined heat and power*

chresard 有効水分《植物が吸収しうる土壌中の水分の量；⇨ echard*》．

chromium クロミウム，クロム《金属元素；記号Cr, 原子番号24》．

choromosome 【生】染色体．

chronology 年代学，編年学；年代記，年表；年代順配列．

CI Conservation International*; Consumers International 国際消費者機構《1995年，IOCU (International Organization of Consumers' Unions)から名称変更》．

CIF Climate Investment Funds 気候投資基金《新興国・開発途上国の気候変動対策を促進するために2008年7月に設立された基金；クリーン技術基金(CTF*)と戦略気候基金(SCF*)からなる；世界銀行グループなどの国際金融機関が運営》．

circumboreal 形【生態】(動植物が)北半球の寒帯に産する，周北の．

cirrostratus 巻層雲．

CITES サイテス，Convention on International Trade in Endangered Species of Wild Fauna and Flora 絶滅のおそれのある野生動植物の種の国際取引に関する条約《通称ワシントン条約(Washington Convention)；絶滅のおそれのある野生動植物の保護を目的に，その国際取引の規制を輸出国と輸入国とが協力して実施し，採取・捕獲の抑制を図る．1975年発効》．

Civic Trust [the ～]シヴィックトラスト《都市環境の保全・改善を目的とする英国の民間団体；1957年設立；伝統ある建造物・美観地域の保存に実績がある；啓蒙活動が中心で，National Trust* とは違って土地や建物の買い取り保存はしない；略CT》．

Clean Air Act [the ～]大気清浄法《1963年初めて米国議会を通過し，その後も更新されている法律(特に1970年成立のものをMuskie Act* と呼ぶことがある)；同法は連邦政府に対して，大気汚染を防止するのに必要な補助金の提供，調査研究の実施，種々の規制法の制定・施行といった権限を与えている；中でも自動車の排気ガス中に含まれる汚染物質や大気

汚染地域にある工場の煙・排気ガスを規制するための基準を設定したことが重要である; Muskie Act は当時, 世界で最も厳しい排気ガス規制法であり, 日本の排気ガス規制の整備に大きな影響を与えた; 略 CAA》.

clean weapon 放射能を残さない核兵器.

Clean Water Act ［the～］水質浄化法《水域の化学的, 物理的, 生物学的な清浄さを回復し, 維持する目的で, 1972年に制定された米国の連邦法; 略 CWA》.

clear-cut 動〈森林地を〉皆伐する. 图 皆伐された森林地, 伐採地 (clear-cutting). 形 皆伐地の.

Clearing-House Mechanism クリアリングハウスメカニズム《各種公共機関が持つ情報を共有するためのデータバンク構築システム; 特に Convention on Biological Diversity* に基づく国際的な生物多様性情報交換システムを指す》.

climate 一地域の気候; 《長期にわたる気象の平均状態》; 風土帯《特定の気候の地方》; 屋内の環境状態《温度と湿度など》: a wet *climate* 湿潤な地方 / move to a warmer *climate* 暖かい地方へ転地する.

climate canary 気候のカナリア《気候変動の影響を最も受けやすい人びと・生物種》.

climate change 気候変動 (climatic change)《気候［降水量, 温度などに代表される地球上の大気の平均状態］が, 様々な時間スケールで変化すること; 20世紀末から注目されている地球温暖化のほか, 氷河期に起きた地球寒冷化も含む; 気候変動の要因には, 大気に内在する変動, 海洋の変動, 火山の噴火, 太陽活動の変化といった自然要因と, 温室効果ガスの排出や森林破壊といった人為的な要因とがある》.

Climate Change Convention United Nations Framework Convention on Climate Change*

Climate Gate クライメート・ゲート《IPCC* が採用した, 人為起源の地球温暖化の根拠となるデータに捏造の疑いがあることが2009年11月20日発覚し, その後, 温暖化の科学的データの信憑性についての議論のきっかけになった; ニクソン米元大統領のウォーター・ゲート事件にちなんで名付けられたといわれる》.

climate hazard 気候変動による異常気象などの危険; 気候変動を悪化させる要因.

climate neutral carbon-neutral*

Climate Stewardship Act クライメート・スチュワードシップ法案《対象範囲の2010年の温室効果ガス排出量を2000年レベルに

抑えるために，年間1万トン以上の二酸化炭素を排出する施設などを対象に排出量取引の制度を定めた米連邦法案；2003年6月，リーバーマン上院議員(民主党)とマケイン上院議員(共和党)が提案したが否決された；その後も2回同様の法案が提出されたがいずれも否決された》．

climatic(al) 形 気候上の；風土的な；《土壌よりも》気候による(⇨ edaphic*)．

climatic change climate change*

climatic climax 【生態】気候的極相(⇨ edaphic climax*)．

climatize 動 新風土に順応させる[順応する](acclimatize*)；〈建物などを厳寒・酷暑など特定の気候に適応できるように〉手を加える．

climax 名 極相《生物群集や生態系が変化し，その結果，最終的にたどり着いた安定的な状態のこと；特に植物の群落の安定期》．
動 climaxに到達させる[達する]．

climax forest 極相林《動植物の安定して持続的な共同体からなる森林》．

climax vegetation 極相植生《極相に達した植物群落》．

closed community 閉生[密生]群落《植物が互いに近接して生えている群落》．

closed ecosystem 閉鎖生態系《外界の要因や外部からの助力なしに生命を維持することができるようにした生態系；宇宙船の中など》．

Club of Rome [the ～]ローマクラブ《食糧・人口・産業・環境などをめぐって深刻化する地球全体の問題を人類の危機ととらえ，その回避の道を探るために定期的に提言・研究発表をしている非営利国際的研究団体[シンクタンク]で，経営者・経済学者・科学者らで構成される；イタリア・オリベッティ社の副社長が主導し，最初の会合が1968年にローマで開催されたことで，ローマクラブという名前が付いた；同クラブの報告書として，1972年に発表された *Limits to Growth*（成長の限界)が有名》．

CO₂ carbon dioxide* 二酸化炭素．

coastal ecosystem 沿岸生態系．

coastal stretch 沿岸地域．

coastal waters 沿岸海域《海岸から約20マイル内の水域》．

COC chain of custody*

COD chemical oxygen demand*

codominant 形《生物群集中で》共(同)優占の；《遺伝のヘテロ表現度で》共優性の．
名 共(同)優占種．

cofire 動 混合燃焼させる，混焼する《燃料を別の補助燃料と一緒に燃焼させる；木質ペレットのよう

なバイオマス燃料を石炭と混合燃焼させる方式が注目されている》.

cogeneration 熱電併給, コジェネレーション《発電時の排熱を給湯や冷暖房に利用するなど, 同一の燃料源から同時に2種類の有効なエネルギーを生み出すこと》; combined heat and power*.

coinhabit 動《ひとつ所に》共に住む.

coinhabitant 名 共生者.

cold desert 寒冷(地)砂漠, 寒(地)荒原, ツンドラ.

coliform bacillus [bacterium] 大腸菌群の各種の細菌《人や動物の大腸内に共生する, 特に *Escherichia* (大腸菌属)と *Aerobacter* 属の腸内細菌; 糞便による水の汚染の指標とされる》,《特に》大腸菌(colon bacillus)《学名を略した *E. coli* という名で呼ばれることが多い; バイオテクノロジーの研究材料として広く用いられる》.

coliform index, coli index 大腸菌指数《水の純度を示す指数の一つ》.

colonial 形【生態】コロニー(colony*)の;【生】群体の.

colonization 移住;【生態】住みつき.

colonize 動〈動植物が〉…にコロニーをつくる.

colony 1【生態】コロニー《1) 固形培地上の細菌個体の集落 2) 集合して生活するアリ・ハチの一群 3) ある土地・場所に固定したある種の個体群 4) 2, 3 種の植物の混合小集団など》.
2【生】群体《分裂・出芽によって生じた多数の個体が組織的に結合している集合》.

combined heat and power 熱・電気複合利用《発電所などから出るエネルギーを熱源および電源の両方に用いるエネルギーの効率的利用方式; 略 CHP》; cogeneration*.

combustor ごみ焼却炉, コンバスター;《ジェットエンジン・ガスタービンなどの》燃焼器, 燃焼室.

commensalism 片利[偏利]共生《2種類の生物が共に生活して, 一方だけは他方から利益を得るが, 他の一方には利害関係のないものをいう; ⇨ mutualism*, parasitism*, symbiosis*》;《広く》共生.

Commission on Sustainable Development (国連)持続可能な開発委員会(United Nations Commission on Sustainable Development*).

common but differentiated responsibility 共通だが差異ある責任《1992年の地球サミット(United Nations Conference on Environment and Development*)で採択された「リオ宣言(Rio Declaration*)」や「アジェ

ンダ21 (Agenda 21*)」に盛り込まれた考え方；地球環境問題について世界各国は共通して責任を負うが，その原因やこれからの対処については，途上国に比べ先進国がより大きな責任を担っているとする》．

Commoner コモナー Barry Commoner (1917-)《米国の生物学者・環境問題研究家；*The Closing Circle* (1971), *The Poverty of Power* (1976), *The Politics of Energy* (1979) などの著書によって環境に対するテクノロジーや核エネルギーの脅威を訴え，警告を発した；また民主的社会主義の立場から，米国における企業権力の集中に反対し，1980年の大統領選挙には市民党 (Citizens Party) から立候補した》．

commune 生活共同体，《ヒッピーなどの》共同部落，コミューン (⇨ intentional community*).

community 《生物》群集，《植物の》群落，共同体．

community based conservation 住民参加型保全《環境保全を行政当局が地元に押し付けるだけでなく，例えば自然環境を観光や特産物の形で収入源につなげるなど地元住民の生活利害にも配慮した形で両立を図る保護活動；グラウンドワーク運動 (⇨ 和英篇) なども一例》．

Community Conserved Areas 地域社会保全地域．

community garden 地域住民庭園《New York で，見捨てられた公有地・私有地などに地域のボランティアが植物を植えて維持・管理を行い庭園のようにしたもの；Green Guerillas* が起源》．

companion 【生態】伴生種．

competition 【生態】《個体間の》競争，競合．

competitive exclusion (principle) 競争排除（原理）《同一の生態的地位において2つの種が同時に存在するとやがては一方が絶滅するか追い出される結果になること》．

competitor 【生態】競争者．

compost 图 配合土，培養土；堆肥，コンポスト．
動〈土地〉に配合土[堆肥]を施す；〈草・生ごみなどを〉堆肥にする．

composter 图《家庭用の》コンポスト製造機．

congregate 動 (大勢)集まる[集める].
形 (大勢)集まった；集団的な；コングリゲートの《特に介護の必要な高齢者の集団を対象とするサービス・施設についていう》．

congregative 形 集まる傾向のある，集合的な；集団にうける．

conifer forest 針葉樹林．

coniferous forest 針葉樹林 (⇨ needleleaf*).

coniology koniology*

conservation 1《自然および資源の》保護, 管理, (環境)保全. 2 保全林[地区], 自然[鳥獣]保護地区, コンサーベーション.

Conservation International コンサベーション・インターナショナル《米国に拠点を置く民間非営利の国際環境保護団体; 1987年設立; 本部 Washington, D.C.; 略 CI》.

consociation 【生態】優先種群叢, コンソシエーション.

consocies 【生態】コンソシーズ《特に遷移の途中にある consociation*》.

constancy 【生態】恒存度.

constant 不変なもの, 恒常的なもの;【生態】恒存種.

consumer 【生態】消費者(⇨ producer*, decomposer*).

consumerism 消費者(中心)主義, 消費者保護(運動); コンシューマリズム《健全な経済の基礎として消費拡大を唱える》;《大量生産社会における》大量消費(主義), 消費文明.

consumption 消費(活動), 消費量.

contagious distribution aggregated distribution*

contaminant 汚染物質, 汚染菌;【薬学】混在物.

continuum 【生態】植生連続体.

contranatant 形〈魚の回遊が〉反浮游性の《流れに逆らって泳ぐ》.

Convention on Biodiversity [Biological Diversity] 生物多様性条約《生態系・生物種・遺伝子のそれぞれのレベルでの多様性の保全とその持続的利用を目的とする国際条約; 1992年6月の地球サミット(国際環境開発会議)で150か国以上の署名で採択され, 1993年12月29日発効; 略 CBD》.

cooling degree-day 冷房度日(どにち), 冷房ディグリーデー《冷房期間中の1日ごとの平均気温と, 冷房を開始または停止する基準温度との差を1年間積算した合計; 度日(degree-day)が単位; 建物などの冷房の必要エネルギー算定に用いる; ⇨ growing degree-day*, heating degree-day*》.

COP Conference of the Parties 締約国会議; コップ《加盟国が集まり, 国際条約に関する物事を決める最高決定機関; 代表的な COP に Conference of the Parties to the United Nations Framework Convention on Climate Change 気候変動枠組み条約(COP-FCCC)がある; 生物多様性条約締約国会議(COP-CBD), 砂漠化対処条約締約国会議(COP-CCD)などもある; ⇨ UNFCCC*, COP/MOP*, SBSTA*》.

coplanar PCB [polychlorinated biphenyl], Co-PCB 【化】コプ

ラナー PCB《分子構造がダイオキシン(dioxin*)に似ており,極めて毒性が強いポリ塩化ビフェニール(polychlorinated biphenyl*);オルト位の塩素原子が全くなく,2つのベンゼン環が同一平面に位置する PCB の総称で,13種の異性体がある》.

COP/MOP Conference of the Parties serving as the meeting of the Parties to the Kyoto Protocol 京都議定書締約国会合《COP*の決定事項に基づく運営体制を決定する実務会議;京都議定書(Kyoto Protocl*)では,条約の COP が議定書の実務会合(MOP)を兼ねることになった;第一回会合(COP/MOP 1)は2005年カナダのモントリオールで開催》.

copper 銅《金属元素;記号 Cu,原子番号29;略 cop.》.

coral 図 珊瑚(さんご);サンゴ虫;珊瑚細工.
形 珊瑚(製)の;珊瑚(紅)色の;サンゴを生ずる.

coral bleaching 白化現象《海水温上昇などのストレスによりサンゴから共生する藻類が抜け出て白くなり,しばしば大量に死滅する現象》.

coral reef サンゴ礁.

corporate social responsibility 企業の社会的責任《環境保護・法令遵守・適切な従業員処遇[人権保護]・社会貢献など,企業が社会に対して果たすべき責任;略 CSR》.

COSHH control of substances hazardous for health 有害物質管理規則《英国の衛生安全委員会(Health and Safety Executive)によって導入された法令で,労働者を有害物質の暴露から保護する目的で,雇用主に有害物質の管理を義務づけている》.

cosmopolitan 形【生態】全世界に分布する:*cosmopolitan* species 汎存種.

cosmopolite 图【生態】汎存[広布]種,普通種.

counter-pest 天敵.

Country Code [the ~]カントリーコード《英国で観光地などの旅行者向けに Countryside Commission* により出されている行動規約;農民や自然美観などの保護のためのもの》.

country park 田園公園《英国の地方の公的レクリエーションのために用いられる10ヘクタール以上の土地;しばしば Countryside Commission* の資金で運営される》.

Countryside Commission [the ~]カントリーサイドコミッション,田園部委員会《英国イングランドの非都市地域の保護・改善を目的として1968年に設立された非政府組織;国立公園・景観指定

地域・自然海岸・遊歩道などを管理し, 'the Country Code*' という出版物を作って非都市地域の財産と環境の重要性を訴えている》.

cove 1 入り江.
2 [*The Cove*] ザ・コーヴ《和歌山県太地町のイルカ漁を批判的に取り上げた米国のドキュメンタリー映画 (2009); 隠し撮りなどの手法やイルカ保護の価値観の押しつけといった観点から上映中止を求める運動が進められたことから, 日本では 2010 年の公開時に上映を中止する映画館が相次ぎ, 言論表現の自由という点から大きな議論になった; ⇨ Sea Shepherd*》.

cranberry bog [marsh] クランベリー湿原[沼沢]《定期的に冠水する, クランベリー(蔓苔桃)の生育する低(湿)地》.

CR-DPF continuously regenerating diesel particulate filter 連続再生式ディーゼル微粒子減少装置; 連続再生式 DPF.

crown canopy [cover] 林冠《木々の樹冠(crown)が形成する森の上部[稜線]》.

cruelty-free 圏〈化粧品・薬品などが〉動物実験をせずに[動物に害を与えずに]開発された; 動物性食品を含まない, 菜食主義の.

crumb rubber 粉砕ゴム, クラムラバー《再利用のため細かく砕かれたゴム; 単に crumb ともいう》.

crust 【地】地殻; 凍結雪面, クラスト, アイスバーン.

crust vegetation 殻植生《ものの表面を厚さ 2-3 cm の殻状におおう藻類・蘚苔類・地衣類など》.

cryophyte (雪や氷上に生長する)氷雪植物《赤雪藻など》.

cryptobiont 陰蔽(生活ができる)生物.

cryptobiosis 《超低温下や乾燥状態での》休眠生活; 陰蔽生活; 仮死状態; クリプトビオシス.

cryptobiote 陰蔽生活者[生物].

cryptobiotic 圏 陰蔽生活の《代謝活動なしに生存できる》.

cryptophyte 地中植物《芽が地中, 時には水中で生じる植物》.

CSD United Nations Commission on Sustainable Development*

CSIN Chemical Substances Information Network《米国の》化学物質情報ネットワーク.

CSP concentrating solar power (集光型)太陽熱発電《太陽光を反射鏡で一点に集め, 発生した熱を蒸気などの形で利用する solar thermal* 発電; 飼い葉桶状の横長の凹面鏡(parabolic trough)を並べたものなどがある》.

CSR corporate social responsibility*

CT Civic Trust*

CTF Clean Technology Fund クリーン技術基金《新興国が省エネ

ルギー化および温室効果ガスの排出量を削減するための最新技術を導入できるよう無償資金供与や低利融資を行なう基金; 2008年設立; 気候投資基金(CIF*)の二本立ての一つ》.

cultivated plant 栽培植物.

CWA Clean Water Act*

CWMB catchment water management board《Catchment Board* の別称》.

cyanobacteria 藍色細菌, シアノバクテリア.

cyclone (dust) collector サイクロン集塵機[装置]《火力発電所などで用いる, 遠心力を利用した集塵装置》.

cytoecology 細胞生態学.

D

DAP ⇨ catch share*

dark green 図 ダークグリーンの《環境保護・反核などをスローガンとすることについていう; light green* に比べてより革新的》.

Data Deficient 情報不足《絶滅危惧種カテゴリーの一つで, 情報不足のため判断できない分類をいう; ⇨ Red List*》.

DDD *d*ichloro-*d*iphenyl-*d*ichloroethane ジクロロジフェニルジクロロエタン((ClC_6H_4)$_2$ $CHCHCl_2$)の略称, ディーディーディー《DDT* に似た殺虫剤》.

DDT *d*ichloro-*d*iphenyl-*t*richloroethane ジクロロジフェニルトリクロロエタンの略称, ディーディーティー《戦前に欧米で実用化された, 代表的な有機合成殺虫剤; ヒトへの健康影響は比較的小さいとされ, 第2次世界大戦中は米軍などが疫病対策に使用し, 戦後は, 各国でシラミやハエの衛生駆除, さらには害虫防除の農薬として広く用いられた; しかし, レイチェル・カーソン(Carson*)が「沈黙の春」(1962)で, 鳥類の死滅などDDTによる自然環境破壊を警告し, さらに長期残留性や発がん性が指摘されたため, その後世界的に使用が禁止された; 2004年に発効したストックホルム条約(Stockholm Convention*)でDDTはPOPs(残留性有機汚染物質)に指定されている; 一方で, DDTによるマラリア対策の有効性が見直され, WHOは2006年, マラリア対策に限定して, マラリア流行地域や恒常的な伝染地域において, DDTの屋内残留噴霧を勧める方針を打ち出しており, 安全性をめぐる議論は決着していない》.

DE Department of Energy*

dead time 《米俗》《廃棄物などの》完全に生物分解(biodegradation*)されるに要する時間.

debt-for-nature swap 債務・環境スワップ《累積債務軽減の一方法で, 自然保護団体が債務国の政府債などを購入し, その返還の代わりに債務国内の環境保護事業を求めること》.

decipol デシポル《臭気測定単位; 1 olf* を毎秒10リットルの未汚染の空気と混合したもの》.

decomposer 分解する人[もの]; 【生態】分解者(reducer)《有機物を分解するバクテリア・菌類な

ど；⇨ producer*, consumer*》．

decontaminate 動 …の汚染[放射能汚染]を除去する，除染する．
　decontamination 名 汚染除去(作業)．

deep geological [burial] disposal 《放射性廃棄物の》深地層処分．

deep green 形 ディープグリーンの(⇨ dark green*)．

Deepwater Horizon ディープウォーター・ホライズン《2010年4月に米国メキシコ湾ミシシッピ河口沖で爆発事故を起こした英国の国際石油会社BP(元ブリティッシュ・ペトロリアム(British Petroleum)の石油掘削基地；この結果海底油田から同種の事故で史上最悪とされる規模の原油大量流出が起き，広範囲な海洋汚染が懸念されている》．

deerculler 《ニュージーランドで政府・環境適応協会に雇われた》鹿猟師．

deforestation 森林伐採[破壊]，山林開拓．

DEFRA, Defra Department for Environment, Food and Rural Affairs*

deglaciation 《氷河の》退氷；氷河・氷床の後退・縮小．
　deglaciated 形 (氷河が)縮小[後退]した．

degradation 1 (性質・品質などの)低下，頽廃，堕落．
2 (発達などの)退歩，衰退．
3 【生物】退化，退行．
4 【地】(風化作用・浸食作用による岩石の)崩壊，デグレイデイション；地表の低下；(河底の)減均作用(⇨ aggradation*)．
5 (複合化合物の)分解，減成，変質．
6 【理】(エネルギーの)劣化，散逸．

degree-day 度日(どじつ)，ディグリーデー《ある日の平均気温の標準値(65°F, 18°C など)からの偏差；燃料消費量などの指標；⇨ cooling degree-day*》．

demography 人口統計学，人口学；【生】個体群統計学．

demystify 動 …の神秘性を取り除く，謎を解く，明らかにする；〈人〉の不合理な考えを取り除く．

dendrograph 【林業・生態】(自記)樹径生長計．

dendrometer 【林業・生態】測樹器，生長計《樹高・樹径を測る》．

Department for Environment, Food and Rural Affairs [the ～]《英国政府の》環境・食糧・農村地域省．

Department of Energy [the ～]《米国政府の》エネルギー省《略 DOE, DoE, DE》．

depletion 水分減少(状態)，涸渇，消耗；【生態】消耗《水資源・森林資源などの回復を上回る消費》．

depollute 動 …の汚染を除去する，浄化する．
　depollution 名 汚染除去，浄化．

deposit gauge 降下煤塵捕集装置, デポジットゲージ.

depurate 動 浄化する.

desertification 砂漠化.

desertify 動 砂漠化する.

desert soil 砂漠土(ど)《砂漠の成帯性土壌》.

desilt 動〈川〉から沈泥を除く, 浚渫(しゅんせつ)する.

deteriorate 動《質・価値・はたらきなどの点で》悪化[劣化]させる[する], 〈価値を〉低下させる; 衰退させる[する], 崩壊させる[する].

deterioration 图 悪化, (質の)低下, 劣化, 老朽化.

detritivore 腐泥食性生物, 腐食性生物《ある種の昆虫など有機廃棄物を食糧源とする生物》.

diarrhea, -rhoea 下痢.

dibenzofuran ジベンゾフラン ($C_{12}H_8O$)《有毒で殺虫剤として用いられる》.

dichlorodiphenyldichloroethane DDD*

dicofol ジコホ(ー)ル ($C_{14}H_9Cl_5O$)《DDT* から得られる穀用防虫剤; 商標 Kelthane*》.

dieback《病虫害・寒気・水分不足などのため》枝先から枯れ込んでくること, 胴[枝]枯れ(病), 立枯れ(病), 寒枯れ.

diel 形【生態】一昼夜の, 日周(期)的な.
图 一昼夜.

diesel particulate filter ディーゼル微粒子除去装置《ディーゼルエンジン車用の黒煙除去装置; 略 DPF》.

dioxin ダイオキシン, ジオキシン《二つのベンゼン環を酸素原子でつなぎ, ベンゼンの水素が塩素に置き換わったような構造を持つ有機塩素化合物の総称; 発がん性, 催奇形性, 内分泌攪乱性などが指摘される; 塩素の場所と数で生物毒性は異なる; とりわけ 2, 3, 7, 8-TCDD(四塩化ジベンゾダイオキシン)の毒性が強い》.

discern 動 識別する, 見分ける; (目で)認める; 認識する.

discernible, discernable 形 認められる; 認識[識別]できる.

disclimax 妨害極相《人間や家畜に妨害されて, 生物社会の遷移の進行がある段階で抑えられている相; 水田や畑などの耕作地が好例》.

disease vector 病気媒介生物, ベクター

disintegration 分解, 崩壊, 分裂, 分散;【理】《放射性元素の》崩壊, 壊変;【生態】崩壊統合.

disoperation【生態】《生物間の》相害作用.

disposal, disposer ディスポーザー《流し台に取り付け, 生ごみなどを処理して下水に流す機械》.

dissolved oxygen 溶存酸素《水に溶解している酸素; 略 DO》.

disturb 動 …の自然のバランス[生態系]を乱す[破壊する].

disturbance 【地】擾乱(じょうらん)《軽度の地殻変動》;【気】《風の》擾乱,《特に》低気圧.

DNAPL dense non-aqueous phase liquid 重量非水相液体, 重非水液, 高密度非水相液体《難水溶性で, 比重が水より重い tetrachloroethylene* などの揮発性有機化合物(VOC*); しばしば難透水層の上に溜まって土壌・地下水の汚染源となる; ⇨ LNAPL*》.

DO dissolved oxygen*

DOE, DoE Department of Energy*

dominance, -nancy 優位, 優越; 権勢; 支配; 優勢; 機能的不均斉《手の右利き・左利きなど》;【生態】《植物群落における》優占度,《動物個体間での》優位;《遺伝形質の》優性(⇨ law of dominance*).

dominant 形【生態】優占した;《遺伝的に》優性の, 顕性の: a *dominant* character 優性形質 / *dominant* species 優占種.
名 主要[優勢]なもの; 優性対立遺伝子, 優性形質;【生態】《植物の》優占種,《動物個体間の》優位者.

doomwatch 環境破壊防止のための監視;《特に環境の》現在の状況とその未来についての悲観論, 環境滅亡論.

　doomwatcher 環境監視者,《環境問題に関する》悲観論者.

dosing tank 自動配水タンク《下水処理施設で下水を貯蔵し, 次の処理施設へ一定量ずつ配水するタンク》.

Douglas ダグラス Marjory Stoneman Douglas (1890-1998)《米国の著述家・環境保護活動家; フロリダ州南端に広がる大湿原 Everglades* を乱開発から守って動植物を保護する論陣を張り, *The Everglades: River of Grass* (1947) を出版, その結果, 同湿原は国立公園に指定され, 同書は米国人の自然観に変革をもたらした著作といわれるようになった》.

Douglas scale ダグラス波浪度《1929 年の国際気象学会で推された波浪・うねりの複合階級》.

down 1《鳥の》綿毛(わた); 綿羽(わた)(めん);《綿毛に似た》柔毛, 軟毛; うぶ毛;《タンポポ・桃などの》綿毛, 冠毛.
2 [しばしば複数形で]《広い》高原地.

downdraft, -draught 下向き[下降]気流, 下方流, ダウンドラフト.

downflow 低い方へ流れること[もの]; 下降気流.

downland 傾斜牧草地,《特に》オーストラリア・ニュージーランドの起伏のある草原.

downriver 形 河口に向かって, 河

口に向かう，川下へ[の].
downwind 副 風の吹く方へ，風に沿って，風下に[へ].
形 風向きに沿って動く；風下の.
downzone, down-zone 動 開発制限的に地域指定[線引き]を変更する.
DPF diesel particulate filter*
drainage area drainage basin*
drainage basin 流域，集水域《降水による表流水がある河川に集まる範囲》.
drainfield 排水地《腐敗水タンクの排水などを吸収させる土地》.
dredge 名 曳航(えいこう)式採泥器，浚渫(しゅんせつ)機；浚渫船，ドレッジャー；《カキなどを捕る》けた網，底引網.
動 〈港湾・川を〉浚渫する，さらう；dredge* で取り除く[採る]；水底をさらう；けた網で捕る.
drift 名 1《風力による》緩慢な流れ；海流(の速度)；吹送(すいそう)流《風力による広い海域の流れ》；《潮流・気流の》移動率；流程《風潮による船舶の移動距離》: the *drift* of current 流速.
2 追いやられるもの，押し流される物；《雨・雪・雲・砂ぼこり・煙などの》風に追われて移動する塊り；《雪・砂などの》吹き寄せ，吹きだまり；漂流物；【地】漂積物，《氷河による》漂礫(ひょうれき)(土)；花をつける植物の大群.
動 漂流する[させる]，吹き流される；《風または水に》運ばれて積もる，吹き積もる，吹きだまりになる，漂積する；吹き寄せる；〈野・道などを〉吹き寄せでおおう；〈水の作用が〉堆積させる.
drought 旱[干]魃(かんばつ)，(長期の)日照り続き，渇水.
droughty 形 干魃の，渇水状態の；乾燥した.
DrPH Doctor of Public Health 公衆衛生学博士.
drunken forest [trees] 《特に永久凍土が溶けて》木々が酔っ払ったように傾斜・転倒した森林.
duck squeezer 《俗》自然[環境]保護論者[運動家].
duff 落葉枝，粗腐植，ダフ《森林の地上で腐った枯れ葉や枯れ枝など》.
dugong ジュゴン《インド洋・西太平洋産の海生草食哺乳動物；体長3m》.
dump 動 1〈ごみを〉投げ捨てる，捨て場にどっと落とす，投棄する.
2《商》《商品を》大量に廉売する；《過剰商品を》(国内価格以下で)外国市場へ投げ売り[ダンピング]する.
名 1 ごみ捨て場，ごみの山.
2《低品位の鉱石などを捨てるための装置のある》傾斜路；捨て石の山；(ドサッと)捨てたもの，堆積.
dumpage (ごみの)投棄；投棄されたごみ，塵芥の山；ごみを投棄する権利；ごみ投棄権利金.

dumping 1《放射性物質や毒物のはいった》産業廃棄物の不法投棄. 2 投げ売り, 安値輸出, 不当廉売, ダンピング.

dumping business 産業廃棄物処理業.

dumping ground ごみ捨て場 (dump).

dump power 余剰電力《必要以上に発電され, 蓄電もできない電力》.

dump scow 廃棄物投棄船, ホッパ船.

dumpsite ごみ捨て場, ごみ投棄物, 埋立地.

Dumpster 【商標】ダンプスター《米国の金属製のごみ収集箱》;《一般に》大型のごみ収集容器.

dustbin 《英》ごみ入れ[箱] (trash can, garbage can).

dustbinman 《英》ごみ収集人, ごみ屋さん.

dust counter 塵埃計《空気中の粉塵の濃度を測る》.

dustfall jar 降下煤塵容器《測定や分析のために空中の微粒子を集めるもの》.

dustman dustbinman*

dysphotic 形【生態】弱光性の,《深海などの》ごく弱い光の中で生長する.

dystrophic 形【生態】〈湖沼が〉腐植栄養の《多量の腐植質を含み, 低 pH で, 利用可能な栄養が乏しい状態. 泥炭地に多い》.

dystrophication 【生態】腐植栄養化.

dystrophy, dystrophia 【生態】腐植栄養

E

E 10 ethanol* (特に bioethanol*) 10％の混合ガソリン《ガソホール(gasohol*)の1つ；米国で最も普通のアンチノック用燃料；ガソリンの節約，CO_2 排出の低減などを目的に開発された；フレックス燃料車(flexible fuel vehicle*)ではエタノールの比率を上げた E 85 のような燃料も使える》.

EA environmental assessment*

eagle freak 《米俗》自然［環境］保護主義者，環境オタク．

earth biscuit 《米俗》自然［環境］保護論者［運動家］．

Earth Day 地球の日《1970年に米国で始まった環境保護の日；4月22日；ウィスコンシン州選出のゲイロード・ネルソン上院議員がこの日に全米の環境行動を呼びかけ，その後，地球のことに関心を持ってもらう日として世界に広がった；⇨ Earth Week*》．

earth-friendly 形 地球に優しい，eco-friendly*．

Earth Resources Experiment Package 地球資源実験計画《米宇宙ステーション Skylab (1973-79)を用いた NASA のプロジェクト；地球上の農林業・土地利用・地質・水の循環を研究した；略 EREP》．

Earth Summit [the ～] 地球サミット《1992年ブラジルのリオ・デ・ジャネイロで開催された「環境と開発に関する国連会議」(United Nations Conference on Environment and Development*)の通称》．

Earthwatch 地球監視《United Nations System-wide Earthwatch の略称；国連環境計画(UNEP*)の活動の一つとして 1973 年に設置された》．

Earth Week 地球週間《環境保護週間；Earth Day* を含む4月の1週》．

eat-out 《マスクラット・水鳥などに》植物を食いつくされた沼沢地．

ecad エケード，適応型《生物が外部環境に応じて示す変化形態》．

ECBM ⇨ CBM*

ECCJ Energy Conservation Center, Japan《日本の》財団法人・省エネルギーセンター《1978年に設立された省エネ推進の中核実施機関》．

ece 居住環境，生息地(habitat*)．

ECHA European Chemicals Agency 欧州化学物質庁《REACH* により 2007 年 6 月に

設立された，欧州連合における工業化学物質規制の監督機関；本部ヘルシンキ》．

echard 無効水分《植物が吸収できない土壌中の水分；⇨ chresard*》．

ECLSS environmental control and life support subsystem 環境制御生命維持システム．

eco- 「所帯」「経済」「生息地」「環境」「生態(学)」「自然保護，エコ(ロジー)」の意の接頭辞《ギリシア語の「家」「家政」を意味するオイコス(oikos)が語源；ecology* と economy* はどちらも oikos に由来する》．

ecoactivist 環境保護活動家．

ecoactivity 環境保護活動[運動]．

eco-audit, eco audit 图動 environmental audit* (を行う)．

ecobabble 《俗》環境保護論者のジャーゴン[ごたく]．

eco-bank エコバンク《環境保護意識の強い企業にしか融資しない銀行》．

eco-bomb 《俗》重要な環境問題．

ecocar エコカー，低公害車《天然ガス自動車，電気自動車，ハイブリッド自動車，メタノール自動車，低燃費かつ低排出ガス認定車，LPG 自動車，燃料電池自動車 (fuel cell vehicle*) など；⇨ low-emission vehicle*》．

ecocatastrophe 《環境汚染などによる》大規模な[世界的な]生態系異変．

ecocide エコサイド《環境汚染による生態系破壊》．

ecocidal 形 エコサイドの[につながる]．

ecoclimate 生態気候《生息地の気候要因の総体》．

eco-conscious 形 環境(保護)意識の強い，環境(問題)への関心の高い．

ecocritic ecocriticism* をする人．

ecocritical 形 ecocriticism* の[に関わる]．

ecocriticism エコクリティシズム，環境批評《1980年代から盛んになった，文学と環境・文化と自然の関係，作家の環境意識，エコロジカルな想像力などの問題を中心とする文学・文化批評》．

ecodefender 環境防衛者，自然保護活動家．

ecodevelopment 環境維持開発《環境・経済両面の均衡を保った開発》．

ecodoom 生態系の大規模な破壊．

ecodoomster ecodoom* を予言する人．

eco-fair 環境保護見本市，エコフェア《地球環境保護を訴えるフェスティバル》．

ecofeminism エコフェミニズム《エコロジー・自然環境保護との関連でとらえられたフェミニズム；たとえば，生む性としての母

性を積極的に評価する立場など》.

ecofeminist エコフェミニスト, エコフェミニズム思想家.

ecofreak 《俗》熱狂的な環境保護主義者(ecological freak*).

eco-friendly 形 環境に優しい[無害な].

ecogenetics 環境遺伝学《遺伝的性質と環境との関連を研究する学問》.

ecogenetic 形 環境遺伝(学)の.

ecogeographic, -ical 形 生態地理的な《環境の生態的面と地理的面の両方にかかわる》.

ecography 生態誌.

eco guerrilla 環境保護ゲリラ[過激派], ecoterrorist*.

ecohazard 環境を破壊するもの[活動].

ecol. ecological*; ecology*.

eco-label(ing) エコ表示《環境にあまり負荷をかけずに製造・生産されたことを商品に表示すること》.

ecolaw 環境に関する法律.

ecoliteracy エコリテラシー, 環境保全への理解(ecological literacy).

ecological, -logic 形 生態学の, 生態学的な; 環境(保護)意識をもった.

ecologically 副 生態学的に, 環境保護[エコロジー]の観点から.

ecological allergy 環境アレルギー《プラスチック・石油製品・殺虫剤などに含まれる化学薬品によってひき起こされる》.

ecological balance 《諸部分の均衡に基づく》生態系の安定, 生態系バランス.

ecological displacement 生態的転位.

ecological footprint エコロジカルフットプリント, 生態系を踏みつけた足跡《人間の活動が環境にどの程度負荷をかけているかを知る指標; ある地域の経済活動や消費活動を継続的に支えるために必要な資源量を, その農産物や水産物を生産できる陸水の土地面積として表す; 通例グローバル・ヘクタール(global hectare*)の形で表わされる; 世界自然保護基金(WWF*)の『生きている地球レポート 2008 年版』によると, 世界人口の 4 分の 3 以上が自分たちの土地で生産できる以上の量を消費しており, ecological footprint の観点からは赤字の国に暮らしている; 例えば米国民 1 人あたりの ecological footprint は 9.4 グローバル・ヘクタールで, 世界中の人が同じ消費を続けると, 約 4.5 個の地球が必要と計算される; ⇨ carbon footprint*, food mile*, virtual water*》.

ecological interaction 生態的相互作用《一共同体内に共に生息している種の間の関係, 特にある種に属する個体が他種の個体に及ぼ

す影響》.

ecological niche 生態学的地位, ニッチ(niche)《共同体において, ある有機体が占める生態的役割; 特に食物に関していう》.

ecological pyramid 生態(系)ピラミッド《生態系の生物的構成要素を, 食物連鎖別に, 栄養段階の低いものから高いものへ下から順に積み上げたもの; 栄養段階が高いものほど量が少なく, ピラミッドのように見える; ⇨ food pyramid*》.

ecological succession 生態遷移《ある場所に存在する生物群集が時間とともに変化していき最終的に安定した生物相となること》.

ecological terrorism 環境テロ《(1) 環境破壊行為 (2) 政府・企業などに要求を呑ませる強硬手段として行なう, 環境破壊行為(ecoterrorism*)》.

ecological terrorist 图形 環境テロリスト(の), 環境テロの.

ecology 1 生態学, エコロジー《生物(人間を含む)とそれを取り巻く環境(生物と無生物のすべてを含む)の相互関係を調べる学問; ドイツの生物学者 Ernst Haeckel が 1866 年に初めて使用した; ギリシア語で「家政」を意味する oikos と「学問」を意味する logos を合成した言葉; 近年では, 自然の構造と機能を研究する学問の意味もある》.

2 人間生態学, 社会生態学《人間と環境との相互作用を研究する学問》.
3 生態.
4 自然[生態]環境.

ecologist 生態学者.

ecology freak 《俗》環境問題に異常なほど熱心な輩, 環境マニア (ecofreak*).

ecomanagement 生態[自然環境]管理.

eco-material エコ・マテリアル《環境負荷が少ない素材》.

ecomone 《自然界の均衡に影響を及ぼす》生態ホルモン(environmental hormone*).

ecomonster 環境に深刻な影響を及ぼす大規模プロジェクト[施設].

econiche ecological niche*

economy 1 節約, 倹約, 経済《ギリシア語の「家政」を意味するオイコス(oikos)が語源》: *ecnomy* of scale 規模の利益, 規模の経済.
2 生体内の物質交代, 同化異化作用: the *economy* of the cell 細胞の代謝作用.
3 《自然界などの》理法, 秩序; 有機的組織, 系: the *economy* of nature 自然界の秩序, 自然のつながり.
4 生物体《全体的組織としての動物体または植物体》.

economic(al) 形 経済の, 経済的な.

econut ecofreak*

eco-office エコオフィス《汚染物質など有害な要素のない，環境面で安全なオフィス》．

ecophene 【生態】エケード，ecad*．

ecophenotype 【生態】生態表現形，ecad*．

ecophysiology 生態[環境]生理学．

ecopolicy 生態[自然環境]政策．

ecopolitics 経済政治学；環境政治学．

eco-pornography エコポルノ《環境問題に対する大衆の関心を利用した広告・宣伝》．

ecoradical 過激な環境保護論者．

eco-raider 《環境保護のために暴力による攻撃を行なう》エコ襲撃者．

ecoregion エコリージョン，生態域《ecosystem* より広範囲の生態地理学的単位》．

ecorighteous 形 環境保護的に正しい，環境保護に打ち込んだ．

ecosocialism 生態社会主義《生態学上の問題に関心の高い社会主義》．

ecospecies 【生態】生態種．
  ecospecific 形 生態種の．

ecosphere 《宇宙の》生物生存圏，《特に地球上の》生物圏(biosphere*)，生態圏．
  ecospheric 形 生物(生存)圏の．

ecostore エコストア《環境破壊を招かない商品のみを売る店》．

ecosys 《俗》生態系(ecosystem*)．

ecosystem 生態系，エコシステム《ある地域に生息するすべての生物とそれらに関係する環境を一体としてとらえたもの；とくに物質循環やエネルギーの流れ，情報の維持/伝達といった機能面に着目するが，近年は，生物と無生物環境の全体を指す言葉として使われることが多い；1935年，英国の植物生態学者 A. G. Tansley が提唱した》．

ecotage エコタージュ，環境破壊行為阻止活動《環境保護計画の必要を訴えるため環境破壊者に対しておこなう破壊・妨害活動》．

ecotecture 実用を環境要因に従属させた建築デザイン．

ecoterrorism 環境テロ《1) 環境保護を推進するために行なわれる破壊行為(ecological terrorism) 2) 政治的な関心を高めるため，自然環境を意図的に壊すテロ行為》．
  ecoterrorist 名 形 環境テロリスト(の)，環境テロの．

ecoteur エコタージュ(ecotage*)の活動家[実行者]；環境破壊行為阻止活動家．

ecotone 移行帯，転移帯，推移帯，混交帯，エコトーン《隣接する生物群集間の移行部》．

Ecotopia エコトピア《生態(学)的に理想的な地域[社会形態]》．

ecotopian 形 生態(学)的に理想

的な〈生き方など〉．

图 生態学的理想家, エコトピアン《現代文明の利便を拒否し環境を破壊しない生活をする人》．

ecotopic 圏 特定の局所的生息地条件に適応する, 生息環境適応上の．

ecotour エコツアー《ecotourism* の立場から自然とのふれあいに主眼を置いた体験型観光ツアー》．

ecotourism エコツーリズム;《環境保護の重要性を参加者に認識してもらうためのツアー; 旅行者による環境へのダメージを最小限に抑え, ツアー自体がサステナブルであることを求められる; ツアーによる収入が地域の環境や動物保護に役立てられるケースもある》．

ecotourist エコツアー参加者, エコツーリスト．

ecotoxicology 環境毒物学《汚染物質の環境に与える毒物的影響についての科学的研究》．

ecotron エコトロン《環境因子の環境に対する影響を調べるための密閉容器; 内部に模擬的な環境をつくり研究を行なう》．

ecotype 生態型《同一種が異なる環境に適応して生じ, 遺伝的に固定した形; ⇨ ecospecies*》．

ecotypic 圏 生態型の．

eco-village エコヴィレッジ《intentional community* の考えに基づく田舎や郊外の小集落; しばしば共用建物の周囲に一群の一世帯住宅を建てて住まう共同ハウジングの形を取る》．

ecowarrior 環境保護戦闘員《エコタージュ(ecotage*)活動家》．

ECS environmental control system 環境制御システム．

ECU environmental control unit 環境制御装置．

ecumene 《世界の》居住(可能)地域;《一国の》中核居住地域．

ECV Energy Conservation Vehicle 省エネルギー車．

ED environmental disruption 環境破壊．

edaphic 圏 土壌の;【生態】《気候よりも》土壌による(⇨ climatic*)．

edaphic climax 【生態】土壌的極相(⇨ physiographic climax*)．

edaphon 土壌群集, 土壌生物, エダフォン《土壌中で生活する生物の群集》．

edge effect 周辺[際縁, 辺縁]効果《森林と草地の接点など生物群が推移する境界で生物種の数が多くなったり, 形態的変量を示すものが多く見られる現象》．

edge species 【生態】《生物群集の推移帯の》周辺種．

EE environmental engineering*

EEA European Environmental Agency 欧州環境機関《環境問題を研究する欧州連合の専門機関; スイス, アイスランド, トルコなど

EERE

も加盟している; 1994年から活動開始; 本部コペンハーゲン》.

EERE The Office of *E*nergy *E*fficiency and *R*enewable *E*nergy エネルギー効率・再生可能エネルギー局《米国エネルギー省(Department of Energy)の一部局で, 各種の renewable energy* やエネルギー効率化技術の研究開発を行う》.

effluent 形 流れ出る.
名 流れ出るもの;《川の本流・湖などからの》流出する川, 分流, 流出水; 排出物, 廃棄物《特に環境を汚染する煤煙・工場廃液・下水・放射性廃棄物など》.

EGR exhaust gas recirculation* 排気ガス再循環.

EGT exhaust gas temperature 排気ガス温度.

EHO environmental health officer* 環境衛生監視官.

EHS Environmental Health Service* 環境衛生監視業務.

EIA 1 environmental impact assessment [analysis] 環境影響評価[分析], 環境アセスメント(⇨ environmental assessment*).
2 Electronic Industries Association 米国電子工業会.
3 Engineering Industries Association 英国技術産業協会.

EIR Environmental Impact Report 環境影響報告.

EIS environmental impact statement* 環境影響評価報告.

EIT Economies in Transition 移行経済諸国《中東欧・旧ソ連などの旧社会主義国; 特に京都議定書において共同実施(JI*)の実施対象となる国》.

electrofilter 電子フィルター《煙突に取り付ける公害防止装置; 静電作用により汚染粒子を除去する》.

electromagnetic hypersensitivity 電磁波過敏症.

electronic waste 廃電子・電気機器, 電子・電気機器廃棄物, 電子ごみ(WEEE*).

electroprecipitator electrostatic precipitator*

electrostatic precipitator 静電集塵器, 電気集塵器[機], 電気集塵装置《煙突の排気筒の中に設置される静電気式の粒子除去装置》.

electrostatic precipitation 静電気集塵.

elfin woodland, elfinwood 矮(わい)性低木林, 高山屈曲林《森林限界近くの高山で, 矮性低木や着生植物, コケなどの生えている林地》.

El Niño エルニーニョ《1) 毎年クリスマス前後, 南米エクアドルからペルー北部沿岸にかけてみられる暖水塊の南下現象 2) 数年に一度, ペルー沖の東太平洋赤道海域に発生する大規模な異常暖水塊; 沿岸の農漁業に被害を与え, 発生

年を中心に地球全体にわたって異常気象(El Niño Effect)をもたらす; スペイン語で「男の子, 幼児キリスト」の意; クリスマスのころ訪れることから; ⇨ La Niña*》.

Elton エルトン Charles (Sutherland) Elton (1900-91)《英国の動物生態学者; Julian Huxley の Spitsbergen 調査(1921)に参加, その後 1923, 24, 30 年にも北極圏で調査を行ない, 動物の群集を扱った著書 *Animal Ecology* (1927)は食物連鎖・生態的地位などを論じて現代生態学を誕生させた; 主著は上記のほかに *Animal Ecology and Evolution* (1930), *Ecology of Animals* (1933), *The Pattern of Animal Communities* (1960) など; Oxford に Bureau of Animal Population を創設(1932)》.

ELV 1 end-of-life vehicle 使用済み自動車, 廃車.
2 Expendable Launch Vehicle 使い捨て型打ち上げロケット.

ELV Directive [the ～] ELV 指令, 廃自動車指令《EU による自動車のリサイクル指令; 鉛・水銀・六価クロム・カドミウムの使用を制限する; 2000 年 10 月施行》.

EM effective microorganisms 有用微生物群《酵母菌・乳酸菌など; また, これらを利用した土壌・環境改良剤や病虫害予防剤などの商標としても用いる》.

embark 動 船[飛行機]に乗せる[積み込む]; 〈人を〉(事業に)引き入れる, 〈金を〉(事業に)投資する; 乗り出す, 着手する: *embark upon* an enterprise ある事業に乗り出す.

emergent 巨大木, エマージェント《周囲の森からぬきんでて高くそびえる木》; 抽水[挺水]植物《ハスなど》.

emigrant 《他国・他地域への》移出植物[動物], 移出者.

emission control エミッションコントロール《自動車排ガス内の汚染物質の放出を検知するシステム》.

emission(s) credit (排出)クレジット《京都議定書(Kyoto Protocol*)に基づく温室効果ガス排出枠の単位; 排出量取引(emission trading*)において売買単位となる; AAUs*, CER*, ERU*, RMU* などがある》.

emission standard 《汚染物質の》(最大許容)排出基準.

emission(s) trading 排出量[権]取引《国や企業に温室効果ガスの排出枠を割り当て, 枠が足りなくなった国や企業と, 枠が余った国や企業が, 排出枠を金銭で売買すること; 京都議定書(Kyoto Protocol*)における京都メカニズムの一つ; 略 ET》.

empower 動 …に公的[法的]な権能[権限]を与える; …に能力[資格]を与える.

emulsion fuel エマルジョン燃料《燃料油(廃油や重油・軽油など)に水(と界面活性剤)を混合し乳化させて作る, 農業機械や発電機用のディーゼルエンジン燃料; 燃焼効率が良いため, 排気ガス中の炭酸ガス・窒素酸化物・硫黄酸化物および黒煙が少ない》.

emu parade 《オーストラリアで》グループで行なうごみ拾い.

enclave 大群落の中に孤立する(残存性の)小さな植物群落.

encroach 動 蚕食(さんしょく)する, 侵略する; 侵害する; 〈海が〉侵食する.

　encroachment 図 蚕食, 侵略(地); 不法拡張, 侵害; 侵食地.

endangered species 絶滅危惧[危機]種(⇨ Red List*).

Endangered Species Act [the 〜]絶滅危機種[動物]保護法.

endocrine disrupter 内分泌撹乱物質(environmental hormone*).

endolithic 形 〈ある種の藻類などが〉岩内生の《岩石・サンゴなどの中で生活する》.

endophilic 形 人間環境と関係のある, 内部好性の.

endosymbiont 【生態】内部共生者, 内部共生生物.

enercological 形 エネルギー使用とエコロジーに対する配慮の調和をはかる.

energy agriculture エネルギー農業《アルコール蒸留のための穀物生産など, エネルギー源を作り出すための農業》.

energy budget 《生態系の》エネルギー収支.

energy-conscious 形 省エネ意識の高い, 省エネに配慮した.

energy-efficient 形 エネルギー効率のよい, 省エネの.

energy flow 《生態系における食物連鎖による》エネルギーの流れ, エネルギー流.

energy harvesting 環境発電《自然発生する微弱電流を MEMS (microelectromechanical system 微小電気機械システム; ナノスケールの電子素子などを組み込んだ電子機械システム)などの電源として利用すること; 振動・身体運動・体温などで発生するエネルギーを利用した発電など》.

energy hog エネルギー効率の悪いもの, エネルギーを浪費するもの.

energy recovery エネルギー回収《ゴミを焼却してエネルギーを得たり, 余剰の熱エネルギーを再利用したりすること》.

energy service provider ESP事業《企業のエネルギー関連業務を一括して請け負う事業者; 利用エ

ネルギーの監視と分析を行い，エネルギーコストの削減や管理に役立つサービスや方策を提供する；ESCO* の進化形》.

English Heritage イングリッシュヘリテッジ《正式名称 Historic Buildings and Monuments Commission for England; 英国イングランドの遺跡・歴史的建造物の保護管理のため 1984 年に設立された特殊法人》.

English Nature イングリッシュネイチャー《Natural England* の前身；イングランドの環境保護に従事する政府機関；1991 年 Nature Conservancy Council* が地域別に分割されてできたもの；ウェールズとスコットランドにも同様の機関 Countryside Council for Wales と Scottish Natural Heritage があり，3 者は Joint Nature Conservation Committee を通じて協力体制にあった》.

enshrine 動 神聖なものとして大事にする［保存する，温存する］；〈人権・理想などを公式文書などに〉正式に記す.

entail 動 1 必然的に伴う；必要とする；論理的必然として意味する，内含する.
2〈人を〉《ある状態・地位に》永久的に固定する.
名《性質・信念などの》宿命的遺伝；必然的な結果；論理的な帰結.

enteric fermentation 腸内発酵.

entitlement 資格，権利；《法律・契約で規定された給付の》受給権；【米】エンタイトルメント《特定集団の成員に給付を与える政府の施策；社会保障，恩給，Medicare, Medicaid など》，エンタイトルメントの資金［による給付金］.

entomofauna 《一地域の》昆虫相.

enviro 《俗》environmentalist*.

enviro-extremist 環境保護の過激論者［急進派］.

environ. environment; environmental; environmentalism; environmentalist.

environics 環境管理学.

environment 《生態学的・社会的・文化的な》環境；生物や人間をとりまき，それらと相互作用を及ぼしあうすべての要素群.

environmental 形 (自然)環境の，環境保護の［を目指す］(⇨ environmentally*)：*environmental* pollution 環境汚染 / *environmental* services《地方自治体などの》環境保護［整備］事業(⇨ Environmental Health Service*).

environmental added value 環境付加価値.

environmental assessment 環境アセスメント，環境影響評価《大規模な開発事業，新しい法律や政策，プロジェクトを実施する前

に、それが環境に及ぼす可能性のある影響について公式に調査・予測・評価を実施し、その結果に基づいて、環境への配慮がより適切になされるようにする一連の手続き; 1969年に米国で最初に導入された; 略EA; ⇨ strategic environmental assessment*》.

environmental audit 環境監査《ある企業の環境保全への取り組み方を第三者が調査・査定すること》.

environmental awareness 環境(保護)意識.

environmental biology 環境生物学, 生態学(ecology*).

environmental communication 環境コミュニケーション.

environmental control system 環境管理システム.

Environmental Defense Fund 環境防衛基金《米国サンフランシスコ近郊のBerkeleyにある環境保護訴訟活動グループ; 発癌物質を含む染髪剤やDDT*の使用禁止を実現した》.

environmental design 環境デザイン, 環境を対象とするデザイン《建築学・工学・都市工学・造園学・地域計画など広範な学問領域の共同作業による広域デザイン》.

environmental education 環境教育《環境保護に関する科学的・社会的知識を身につけるための教育》.

environmental engineer 環境工学者《環境保全の専門技術者》.

environmental engineering 環境工学《略EE》.

environmental ethics 環境倫理, 環境倫理学《人と自然の関係を考察し, 自然に対する人の行動規範を明らかにする学問; 応用倫理学の一つ;「自然(生物種, 生態系, 景観など)の生存権」,「世代間倫理」,「有限主義」という3つの視点で語られる》.

environmental governance 環境ガバナンス, 環境統治《さまざまな環境問題を解決し, 持続可能な発展を実現すべく取り組まれている制度・政策的対応》.

Environmental Health Officer 《英国の》環境衛生監視官, 公害防止管理官.

Environmental Health Service [the ～]《英国地方自治体の》環境衛生監視[公害防止]事業[業務]《Environmental Health Officerが実施する大気汚染・騒音の防止や食品衛生の監督など》.

environmental hormone 環境ホルモン《環境中に放出・蓄積され, 生体内のホルモンのバランスをくずすといわれる化学物質の総称; 内分泌攪乱(かくらん)物質(endocrine disruptor)とも言う》.

environmental impact assessment 環境影響評価, 環境アセスメント《略EIA; ⇨ environ-

environmental impact statement 環境影響評価[アセスメント]報告.

environmentalism 環境保護(主義).

environmentalist 環境決定論者; 環境保護論者, 環境問題専門家.

environmental journalism 環境ジャーナリズム.

environmental labelling eco-labelling*

environmental load [burden] 環境負荷(ecological footprint*).

environmentally 副 環境保護の観点で, 環境保護について: *environmentally aware* 環境(保護)を意識した / *environmentally sound* 環境を傷つけない(environment-friendly*) / ⇨ Environmentally Sensitive Area*

Environmentally Sensitive Area 環境上脆弱な地域《英国で環境保全のため, 化学肥料や生産性を向上させる農業技術の使用が制限される地域; 略 ESA》.

environmental protection 環境保護[保全].

Environmental Protection Agency [the 〜]《米国政府の》環境保護局《略 EPA》.

environmental racism 環境人種差別政策《少数民族や低所得者の居住地域に(他所では作りにくい)環境汚染の危険がある施設などを作ろうとする》.

environmental resistance 環境抵抗《個体群に対してはたらく環境制限, 特に人口増加に及ぼす環境条件の制限力》.

environmental science 環境科学.

environmental standard 環境規格, 環境基準: the international *environmental standard* 国際環境規格《ISO14001 など》.

environmental stewardship 環境への責務.

Environmental Study Conference [the 〜] 環境研究会議《米議会の環境保護に関心をもつ議員グループによる会議》.

environmental terrorism [terrorist] ecological terrorism [terrorist]*

environment-conscious 形 環境(問題)への意識が強い, 環境意識をもった.

environment-friendly 形 環境保全に配慮した, 環境親和的な, 環境汚染[破壊]につながらない[を起こさない], 環境[地球]にやさしい.

environment-hostile 形 環境にやさしくない[敵対的な] (environment-unfriendly).

environment-minded 形 環境(問題)に関心がある, 環境に気を配る.

environmentology 環境学.

environpolitics 環境(保全)政策.

EPA Environmental Protection Agency*

EPB Environmental Protection Board《自治体などの》環境保護審議会.

epilithic 形〈植物が〉岩表(がんぴょう)生の.

epipelagic 形 表海水層の《漂泳区の区分で, 光合成をおこなうに十分な量の光が浸透する水深200 mまでの層; ⇨ abyssopelagic*》.

epiphyte 着生植物《他の植物などに付着する; ラン科植物・シダ類・地衣類などに多い》.
  epiphytic, epiphytal 形 着生の.
  epiphytism 名 着生性.

EPNL effective perceived noise level 実効感覚騒音レベル《PNL*の数値に, 騒音の持続時間と特異な高周波音の2つの要素を付加した, 航空機騒音の表現方法》.

EPR ecological planning region 生態計画地域.

equilibrium 釣合い, 平衡, 均衡;《感情の》安定,《心の》平静, 知的不偏;《動物体の》姿勢の安定, 体位を正常に保つ能力;【理・化】平衡.

equitable 形 公正な, 公平な; 衡平法上の, 衡平法上有効な.

eradicate 動 根こそぎにする; 撲滅する, 根絶する, 皆無にする.
  eradication 名 撲滅, 根絶.

eremic 形《動物が》砂漠の[にすむ].

eremophilous 形【生態】砂漠に生息する.

eremophyte 砂漠植物.

EREP Earth Resources Experiment Package*

ericeticolous 形【生態】ヒースの生い茂った環境に生息する, ヒース生の.

erosion 腐食; 侵食;【医】腐食, 糜爛(びらん); 衰退, 低下: wind *erosion* 風食作用.

ERU Emission Reduction Unit 排出削減単位《京都議定書(Kyoto Protocol*)で定められた共同実施(JI*)によって得られる温室効果ガスの排出クレジット(emission credit*)のこと; ⇨ additionality*》.

ESA environmentally sensitive area*

ESCO エスコ《企業や施設に対して省エネルギー診断やエネルギー効率の改善計画を行なう事業者; energy service company の略称; ⇨ energy service provider*》: an *ESCO* project エスコ事業.

ESI 1 environmental sensitivity index 環境脆弱性指標《油流出事故の防除対策ガイドラインとして, 地質・地形・生物生息状況などによって海岸線の汚染危険度をランク付けして地図に示したもの; 米国では商務省海洋大気局

(NOAA*)が作成し，日本でも環境省の脆弱沿岸海域図や海上保安庁の ESI 地図がある》．

2 Environmental Sustainability Index 環境持続可能性指数《環境保護や持続可能な産業構築の達成度を国別にランキング化したもの》．

ESSA 1 Environmental Science Services Administration 環境科学業務局，エッサ《NOAA* の前身》．

2 Environmental Survey Satellite《米国の》環境調査衛星，エッサ．

established 形〈動植物が〉(新しい土地に)定着した．

estuary 《潮の差す》広い河口；河口域，入江．

ET emission(s) trading* 排出量取引，排出権取引．

ETBE ethyl tertiary-butyl ether《ガソリンのオクタン価改良用混入剤；植物由来のため温室ガス削減効果が期待される》．

ethanol エタノール(⇨ bioethanol*)．

EU-ETS 欧州排出量取引制度《2005 年に導入された》．

euphotic 形 真光層の《水面から，光合成の行なわれる限度の深さまで》．

eupotamic 形 流止水性の《動植物が淡水の流水・止水の両方で生育する；⇨ autopotamic*, tychopotamic*》．

Eurobin 《英》wheelie bin*

euroky, -ryo- 広環境性《多くの環境要因に対して広い耐忍範囲をもつこと；⇔ stenoky*》．

eurokous, -ryo- 形 広環境性の．

Europa Nostra ヨーロッパノストラ《1963 年に設立されたヨーロッパの環境保護団体の連合体》．

European Wind Energy Association [the 〜]欧州風力エネルギー協会《1982 年設立；本部はベルギーのブリュッセル；略 EWEA》．

eurybath 広深性生物《さまざまな深度の水低に生息できる；⇔ stenobath*》．

eurybathic 形 広深性の．

euryhaline 形 広塩性の《さまざまな塩度の水に生息できる；⇔ stenohaline*》．

euryhygric 形 広湿性の《さまざまな湿度に耐えられる；⇔ stenohygric*》．

euryoky euroky*

euryphagous 形〈動物が〉広食性の《食物の選択範囲の広い；⇔ stenophagous*》．

eurytherm 広温性生物《さまざまな温度に耐えられる；⇔ stenotherm*》．

eurythermal, -mic, -thermous 形 広温性の．

eurytope 【生態】広(ﾋﾛ)場所性生物．

eurytopic, eurytropic 形 広(う)場所性の《環境因子の広範な変化に適応しうる; ⇔ stenotopic*》.
  eurytopicity 名 広場所性.

eutrophic 形〈湖沼・河川が〉富栄養の(⇨ mesotrophic*, oligotrophic*).

eutrophicate 動〈湖などを〉富栄養化する, 《処理廃水などで》栄養汚染する.
  eutrophication 名 富栄養化, 栄養汚染; 富栄養水.

eutrophied 形〈湖・川が〉富栄養化した, 栄養汚染された.

eutrophy 【医】栄養良好, 正常栄養; 【生態】《湖沼・河川の》富栄養(型).

everglade 1 [通例複数形で] 低湿地, エヴァグレーズ《通例丈の高いスゲの類の草が点在し, 季節によっては水面下に没する》.
2 [the Everglades] エヴァグレーズ《米国フロリダ州南部の大湿地帯; 南西部は国立公園をなす》.

e-waste 電気電子機器廃棄物《バーゼル条約(Basel Convention*)や WEEE* Directive の規制対象となっている》.

EWEA European Wind Energy Association*.

exacerbate 動〈苦痛・病気・恨みなどを〉悪化[増悪, 激化]させる, つのらせる;〈人を〉憤激させる, いらだたせる.
  exacerbation 名 悪化, 激化; 【医】増悪, (病状)再燃; 憤激.

exhaust gas recirculation 排気ガス再循環《吸気に排気ガスの一部を混入して NOx の発生を抑制する; 略 EGR》.

exophilic 形 人間環境から独立した, 外親性の.
  exophily 名 外親性.

exploitation, exploitage 1《天然資源の》開発.
2【生態】《異種生物間での》搾取作用, とりこみ.

ex situ 形 副 (⇔ in situ*) 本来の場所でないところで;《土壌汚染物処理などが》原位置から移動して: *ex-situ* conservation*.

ex-situ conservation 生息域外保存[保全]《生物種を本来の生息域の外で保存すること》.

externality 外部性, 外部効果《一つの経済活動が, 市場の外側で, 商品などの売買とは無関係に, その活動の当事者以外の個人・企業・部門などに影響を及ぼすこと; 有益な場合を外部経済(external economy), 有害なものを外部不経済(external diseconomy)と呼ぶ; 後者の典型が公害; ⇨ Pigovian tax*》.

extinct 1〈生命が〉終息した; 死に絶えた, 絶滅した: *extinct* species 絶滅種 (⇨ Red List*).
2〈火山が〉活動を停止した.

extinction 1 死滅, 絶滅;《家系の》廃絶;《権利・義務・負債など

の》消滅;《条件反射における反射の》消去.
2 消火, 消灯;【理】消光《1) 光のエネルギーの減衰 2) 結晶板の干渉による光の減衰》;【化】《溶液の》吸光度;《地球大気による天体からの光の》減光.

extratropical cyclone 温帯性低気圧.

Exxon Valdez [the 〜] エクソン・ヴァルディーズ《1989年 米国アラスカ州南部の Prince William 湾で大規模な原油流出事故を起こした Exxon 社のタンカー; ⇨ Valdez Principles*》.

F

facies 《動植物の個体(群)の》外観,外見;ファシース《種の量的相違による,植物群落の下位単位》.

facultative 形【生】条件的な,任意の,通性の(⇔ obligate*): a *facultative* parasite 条件的寄生菌.

Fairtrade フェアトレード《途上国の産品の輸入において,適正価格で取引するのみならず,途上国の社会発展に資するよう継続性などの面でも配慮する貿易形態》.

F & WS Fish and Wildlife Service*

FAO Food and Agriculture Organization*

farmland 農地,農耕地.

farm-to-fork 形 農場から食卓[家庭]への,食料生産者から消費者に至る(流通過程の): *farm-to-fork* traceability ⇨ traceability*.

fauna 《一地域または一時期の》動物相,ファウナ(⇨ flora*, avifauna*);動物誌.
   faunal 形 動物相の.

faunule 【生態】小動物相.

FCCC (United Nations) Framework Convention on Climate Change ⇨ United Nations Framework Convention on Climate Change*

FC(E)V fuel cell (electric) vehicle 燃料電池(自動)車.

FDA Food and Drug Administration*

feebate 環境負荷への課金と環境改善への助成《環境に悪い製品に課した金を環境によい製品に支給するなど;課金(fee)と払い戻し金(rebate)の合成語》.

feed-in tariff 固定価格買い取り制度,フィードインタリフ制度《法定の固定買い取り価格により,再生可能エネルギー(太陽光・風力発電による電力など)を,発電コストを上回る価格で買い取る制度; 2000年に再生可能エネルギー資源法でこの制度を導入したドイツが有名;法定買い取り価格は年々,設備導入時期の技術的ハードルの低さに応じて徐々に下げられていくが,ある(個人)事業者に対しては導入時に決定した価格体系が保証される;一般的に再生可能エネルギーの買い取り価格は化石燃料による発電料金に比べて割高になる;買い取りの財源は電気料金の一部としてすべての電力消費者から負担金(tariff)の形で薄く広く

floodplain

徴収される；ドイツが導入して太陽光・風力発電の普及が促進され，他の方式に比して有効性が示されたとして同様の制度を採用する国が増えている；略 FiT》.

feral 《オーストラリアで》森に住んで[キャンプして]伐採阻止などの実力行使をする環境保護主義者（と称する無法者・失業者）.

fertilizability 受精；受精能力，《卵の》受精可能期間；肥沃化可能性.

FFV flexible fuel vehicle*

fidelity 《群落などへの》適合度.

field burning 野焼き.

field capacity 圃場(ほじょう)［現場］容水量《重力水が移動したあと土壌水分量が平衡に達したとき，単位土壌に対する水の容積の百分比》.

field layer 【生態】《植物群落の》草本層(⇨ layer*).

field moisture capacity field capacity*

Fish and Wildlife Service 魚類野生動物庁《米国内務省内の連邦機関；1956年創設され，魚類や野生動物などの自然資源・荒野・原生林・河川流域などの保全と育成に携わっている；水鳥禁猟区や魚類孵卵場を管理し，漁業に関する調査研究を行ない，生存をおびやかされている野生動物を保護し，国際協約を施行するなどが業務；略 F & WS, FWS》.

FiT feed-in tariff*

flag state control フラッグステートコントロール，旗国の検査《港や航行の安全や海洋環境保護を目的に，船籍所属国(flag state)が船の構造・装備などを監督・立ち入り検査すること；略 FSC》.

flex-fuel 名形 flexible fuel vehicle* (方式の).

flexibility mechanism 柔軟性措置《温室効果ガスの排出削減を国際協調で柔軟に実施していくため，京都議定書のもとに定められた仕組みのこと；京都メカニズム (Kyoto mechanism*) とも呼ぶ》

flexible fuel vehicle ガソリン・メタノール両用自動車，FFV《石油消費を節約するためのもの》.

flood 名 1 洪水，大水：in *flood* 満ちあふれて，洪水となって / *floods* of rain 車軸を流すような豪雨.
2 上げ潮，満潮：ebb and *flood* 潮の干満 / at the *flood* 潮が満ちて；よい潮時で[に].
動 氾濫する[させる]；〈潮が〉差す，上げる；〈家屋などが〉浸水する，水浸しになる；みなぎらせる；湛水する，灌漑する：*flooded* districts 洪水被災地 / *flood* out 洪水が〈人を〉家屋から追いやる.

flood level 洪水[浸水]時の最高水位；洪水水位《それを超えると洪水とみなされる川の水位》.

floodplain 氾濫原(げん)《洪水時に流

flora 《一地域または一時期の》植物相(⇨ fauna*, silva*);植物誌,フロラ;《細菌などの》叢(そう):*flora* and fauna 植物相と動物相,動植物.

floral 形 植物(相)の:*floral* zone 植物区系界.

FOE, FoE Friends of the Earth*

FONSI Finding of No Significant Impact 影響希少《環境アセスメントで特に環境への悪影響が認められず,影響評価報告書(environmental impact statement)の提出を免除するという認定》.

Food and Agriculture Organization [the ～]食糧農業機関《1945年設立の国連機関;本部ローマ;略 FAO》.

Food and Drug Administration 食品医薬品局《米国保健社会福祉省[厚生省]の一部局;GRAS*, GLP* など食品・医薬品の許認可を行う;略 FDA》.

food chain 食物連鎖《A は B に,B は C にというように一般に小なるものはより大なるものに順次食われるという生物食性の関連;⇨ food cycle*, food pyramid*》.

food cycle 食物環,食物網(food web)《生物群集内の食物連鎖の全体像;⇨ food chain*》.

food miles 食品マイル,フードマイルズ《生産地から食卓への食糧輸送距離に着目した英国の消費者運動家 Tim Lang が 1994 年,SAFE Alliance (Sustainable Agriculture Food and Environment Alliance; 現 Sustain) を主導しながら,できるだけ近くでとれる食品を食べようと提案した「地産地消」型の消費運動;遠隔地から運ぶほど二酸化炭素の排出も増えることから,その後,フードマイレージ(food mileage*) =「食べ物の量(重量)」×「運ばれてきた距離」という指標が作成された》.

food mileage フードマイレージ《食べ物の輸送がどれだけ二酸化炭素の排出に関係しているかを示す数値;厳密には,輸入食料を対象に「輸入相手国別の食料輸入量(重量)」と「輸入国から消費国までの距離」を乗じて足しあわせた数値として計算される;食料自給率の低い日本のフードマイレージは全体として大きな数字になる》.

food pyramid 食物ピラミッド《食物連鎖を個体数によって示したときの階層関係》.

food safety 食の安全.

Food Safety Commission 食品安全委員会.

food security 食料安全保障.

food web 食物網(もう) (food cycle*).

force main 強制主管《動力により排水がなされる下水系》.

forest 图《広大な》森, 森林, 高木林; 森林の樹木, 樹木《集合的》. 動 …に植林する, 造林する; 森にする, 森林[樹林]でおおう.
   forest(i)al 形 森林の, 林野[森林]に関する.
   forested 形 森林[樹林]でおおわれた.
   forestless 形 森林のない.
forest climate 森林気候.
forest ecology 森林生態学.
forester 1 林業者; 森林官, 林務官, 森林管理者, 森林警備[監視]員.
2 森林の住人; 森の鳥獣.
forest floor 林床(りんしょう)《林地地表面の土壌と有機堆積物の層》.
forest management certification FM 認証, 森林管理の認証《木材を供給する森林が適性に管理されていることの認証; 森林管理協議会(Forest Stewardship Council*)による; ⇨ chain of custody*》.
forest ranger 【米】森林警備隊員(⇨ Smokey*).
Forest Service [the 〜]【米】林野部《農務省の国有林管理部門; 1905 年設置; およそ 77 万 km² に及ぶ国有林と草原のほか, 州・地方自治体・個人に属する 190 万 km² の森林と河川流域が同部局の保全・研究・育成・勧告計画の恩恵をうけている; 略 FS》.
Forest Stewardship Council 森林管理協議会《木材を伐採する森林が環境保護などの観点から適切に管理されているかどうかを審査・認証する非営利の国際組織; 1993 年発足; 略 FSC; ⇨ forest management certification*》.
formation 【生態】《植物の》群系; 【地】《岩相層序区分の》累層(るいそう).
formicivorous 形【生態】アリを食う, 蟻(ぎ)食の.
fossil fuel 化石燃料《石油・石炭・天然ガスなど; ⇨ biofuel*》.
   fossil-fueled 形 化石燃料を使う.
freshwater 形 淡水の, 真水の; 淡水性の, 淡水産の: a *freshwater* fish 淡水魚.
   图 [fresh water とも書く]真水, 淡水; 淡水(域)《淡水の湖沼河川》.
Friends of the Earth [the 〜]地球の友《Brower* らが創設した英国の環境保護団体; 略 FOE, FoE》.
fringing forest gallery forest*
frit フリット《資源再生工場でエネルギー回収のためにごみを焼却して最後に残った滓》.
frit pit フリット(frit)の捨て場.
FS Forest Service*
FSC flag state control*.; Forest Stewardship Council*
fuel cell vehicle 燃料電池車, 燃料電池電気自動車(fuel cell electric vehicle).

fuel cycle 【原子力】(核)燃料サイクル《核燃料物質の再処理・再使用を含む一連の循環過程》.

fuel economy 燃費; 燃料経済性: improve [increase] *fuel economy* 燃費を改善する.

fuel-efficient 形 燃料効率のよい, 低燃費の(⇨ gas-guzzler*): a *fuel-efficient* car 低燃費車, エコカー.

fuel-efficiency 名 燃費のよさ, 低燃費.

fuel farming 《biofuel* を得るための》燃料用作物栽培.

fugitive emissions 漏洩排出, 逸散排出.

fundie, fundy 《特に宗教上[環境保護運動]の》原理主義者(fundamentalist), 熱狂的活動家[支持者], 過激派《ドイツの緑の党(Green Party*)に関してはしばしば Fundi を用いる》.

fusion torch 《核融合エネルギーを利用する》超高温ごみ焼却炉《理論上は1億度の高熱でごみをガスと鉱物元素に分解できるとされる》.

FWPCA Federal Water Pollution Control Administration《米国の》連邦水質汚濁防止局.

FWS Fish and Wildlife Service*

G

Gaia 《有機的組織,一個の生命体としての》地球,母なる大地,ガイア《環境保護論者などが特に好む表現》.

Gaia hypothesis [theory] ガイア仮説[理論]《生命圏・大気圏・海洋・土壌を含む地球を,自己制御機構をもつ一個の生命体としてみる仮説;地球システムにおいて生命の部分と非生命の部分の相互依存性が強いという見方に基づいている;英国の科学者のJames Lovelockが提唱;名前はギリシア神話の大地の女神に由来する》.

gallery forest 拠水林(♮), ガレリア林(♮)《サバンナなどの川沿いの帯状林》.

game ranger 猟鳥獣保護隊員.

gannetry カツオドリ(gannet)の繁殖地.

GAP good agricultural practice 適正農業規範《農業者自らによる農業生産工程管理;農産物の安全確保のほか,環境保全・品質向上・労働安全の確保等を目的とする;ヨーロッパでは統一基準 GLOBALGAP* がある》.

garbage 《米》ごみ,残飯: *garbage* can《屋外に置いて収集してもらうための》ごみ入れ,ごみバケツ / *garbage* collection [collector] ごみ収集[収集員] / *garbage* truck [wagon] ごみ収集[運搬]車.

garbage energy 《米俗》ごみエネルギー《ごみを燃やして資源回収工場で生み出されたエネルギー》.

garbage-to-energy 形 ごみを燃やしてエネルギーを生み出す, waste-to-energy*.

garbology 厨芥研究,ごみ学《特にごみとして廃棄されるものの分析による現代文化研究》;廃棄物[汚物]処理学.

gas-guzzler 《米》燃費が悪くガソリンを食う大型車,高燃費車.

gas-guzzling 形 ガソリンを食う,燃費の悪い.

gas-guzzler tax 高燃費車[ガソリンを食う大型車]の購入価格にかかる税金(⇨ carbon tax*).

gasohol ガソホール《ガソリンとエチルアルコールの混合燃料;アルコール10, ガソリン90のものなど;⇨ E10* (E 最初の見出し)》.

Gause's principle 【生態】ガウゼの原理《生活要求の類似した2種は同じ場所で共存を続けることはできないという考え;旧ソ連の生

物学者 G. F. *Gause* (1910-86) が提唱》.

GBM Green Belt Movement*

GEEREF Global Energy Efficiency and Renewable Energy グローバルエネルギー効率・再生可能エネルギー基金《欧州委員会がアジア・アフリカ途上国の環境支援のために 2008 年に設立した国際基金》.

GEF Global Environment Facility 地球環境資金制度.

genecology 種生態学, ゲネコロギー.

genetic(al) 形 発生の, 起源の; 発生学[遺伝学]的な; 遺伝子の[による]: *genetic* modification 遺伝子組み換え, 遺伝子操作.

genetically 副 遺伝学的に, 遺伝子上[レベル]で.

genetically-altered, genetically-engineered 形 遺伝子が組み換えられた, 遺伝子操作がなされた.

genetically-modified 形 遺伝子組換えの《略 GM》: *genetically-modified* foods [crops] 遺伝子組換え食品[作物] / *genetically-modified* organisms 遺伝子組み換え有機[生命]体《略 GMO》.

geocide 地球の死, 地球崩壊.

geocratic 形 《海洋部分に対して》陸上部分が優勢の[拡大する].

geohygiene 地球衛生学.

geomancy 《米》ジオマンシー, 環境重視建築術《環境や自然資源に配慮して家やビルを設計すること》.

geopark ジオパーク, 地質遺産《地形学的・地質学的・環境的・文化的に価値がある地域; 世界ジオパークネットワーク(Global Geoparks Network*)によって認定される》.

geothermal power 地熱発電.

GGN Global Geoparks Network*

gha global hectare*

GHG greenhouse gas*

GHS Globaly Harmonized System of Classification and Labelling of Chemicals 化学品の分類および表示に関する世界調和システム《2003 年に国連が定めた有害物質の危険性表示の国際基準》.

ghost net 漁船が廃棄した合成繊維製の漁網《魚・海獣・海鳥などがからまって死ぬ》.

glacier 氷河.

global consciousness 《国家・民族などの枠を越えて地球上の一員として社会・経済・環境問題を考える》グローバルな意識.

GLOBALGAP グローバルギャップ《ヨーロッパで青果物(果物・野菜・生花など)を扱うスーパーなど主だった小売業者の主導で発足した作業部会(EUREP)が策定した, 農産物の栽培過程に関わる安全基準認証制度(GAP*)の一つ;

1997年にEUREPGAPとして始まり，2007年より現称》．

Global Geoparks Network [the〜] 世界ジオパークネットワーク《2004年UNESCOの支援により設立；事務局Paris; geopark* の認定を行なう；略GGN》．

global hectare グローバルヘクタール《平均的な生物生産性をもつ地表1ヘクタールの土地；地球環境への負荷(ecological footprint*)や生物資源再生能力(biocapacity*)の単位；記号gha; 耕地と草地で同じグローバルヘクタール数であれば耕地のほうが物理的な面積は小さい；2001年に地球全体で生物学的な生産力のある土地・海洋は約113億ha(地表全体の約4分の1)であり，したがって地球1個分の生物資源再生能力は約113億ghaとなる；人口1人あたりにすると生物資源再生能力は1.8 ghaであるのに対し，環境負荷は2.2 ghaとなる》．

global warming 地球温暖化《地球大気の平均温度が上昇すること．過去には，自然現象として長期的な温暖化や寒冷化が繰り返されてきたが，20世紀後半の急激な気温上昇は自然現象だけでは説明できず，CO_2 など人為起源の温室効果ガス(greenhouse gas*)による可能性がかなり高いとされる；IPCC*の第4次報告書(2007年)によると，過去100年間に，世界の平均気温は0.7℃上昇した．地球温暖化により，21世紀末までに地球の平均気温は1.8〜4.0℃上昇すると予測している；地球温暖化による自然や生活環境への悪影響が懸念され，その対策として緩和策(mitigation*)と適応策(adaptation*)が必要になる；⇨ climate change*》．

global-warming 形 地球温暖化につながる[をもたらす]ような．

global warming potential 地球温暖化係数《それぞれの気体が有する地球温暖化作用の相対的な指標；京都議定書(Kyoto Protocol*)のもとでつくられた；赤外線吸収能に大気中での寿命を加味して算定する；数値が大きいほど温暖化作用が大きい；基準となる二酸化炭素は1，メタンは21, CFC-113(フロン)は5000, HCFC-141b(代替フロン)は630; ⇨ carbon dioxide equivalent*》．

global warming skepticism 地球温暖化懐疑論《「産業革命以降の人為的な温室効果ガスの排出が地球を温暖化させる」という人為起源温暖化説，あるいは温暖化対策の重要性について，さまざまな観点(原因，現象の有無など)から疑問視したり，否定したりする言説》．

GLOBE 地球環境議員連盟 Global Legislators Organization for a Balanced Environment《地球環

境問題の立法者間の国際協力を実現するため，Gore* など欧州，米国，日本の国政レベルの議員が中心となって 1989 年に設立された国際的な議員連盟》．

GLP Good Laboratory Practice《FDA* の》医薬品安全性試験実施基準《OECD や日本もこれに続いて基準を発表した》．

glycophyte 非塩生植物《塩類の少ない土壌に生育する》．

glycophytic 形 非塩生植物．

GM food 遺伝子組み換え食品(genetically-modified* food)．

GM-free 形 遺伝子組み換えをしていない: *GM-free corn* 非遺伝子組み換えトウモロコシ / *GM-free zone* 反遺伝子組み換え地域《遺伝子組み換え作物(genetically-modified crops)の栽培を行わない農地；しばしば農業者が宣言や運動にこの語を用いる》．

GMO 遺伝子組み換え作物，genetically-modified* organism．

God squad 1 [the 〜] 神さま部隊《「絶滅のおそれのある動植物の種の保存に関する法律」の対象種から，ある動植物を営利企業の利益のために除外する権限をもつ連邦政府の委員会》．
2 [the 〜]《病院スタッフに助言を与える》倫理委員会．
3 [the 〜] キリスト教布教グループ．

Gore アル・ゴア Al(bert) Gore (1948-)《第 45 代米国副大統領 (1993-2001)；テネシー州で記者の仕事をしたあと，下院議員，上院議員を経てクリントン政権の副大統領を務めた；地球環境議員連盟(GLOBE*)を結成し，地球環境問題に関心をもって取り組んできた；地球温暖化危機を警告する講演を題材とするドキュメンタリー映画「不都合な真実」*An Inconvenient Truth* (2006; ⇨ Inconvenient Truth*) などの活動が評価され，気候変動に関する政府間パネル(IPCC*)とともに，2007 年度のノーベル平和賞を受賞した》．

GOSAT Greenhouse Gas Observing Satellite 温室効果ガス観測技術衛星．

gradation 【生態】漸進大発生．

graminicolous 形〈寄生性菌類などが〉イネ科植物の上に生育する．

GRAS generally recognized as safe《安全とされる食品添加物の表示として FDA* により用いられる；⇨ 和英篇 グラス》．

grasscycling 芝リサイクル，グラサイクリング《刈り取った芝をごみとして出さず，そのまま庭に放置して土に返すこと；芝の肥料となり，ごみ減量にも有効なので米国やカナダなどで推奨されている》．

grasscycle 動 芝リサイクルを

やる.

gray, grey 形《俗》環境[生態]問題に関心のない[冷淡な](ungreen*).

gray water 中水道(用)水《浄化処理によって再利用される台所・ふろ場などからの排水》.

Great Barrier Reaf [the ~]大堡礁(しょう), グレートバリアリーフ《オーストラリア Queensland 州北東岸沖, 珊瑚海に 2000 km にわたって連なる世界最大のサンゴ礁》.

green 形 環境維持[保護](政策)を支持する, 環境問題意識の強い; [Green で]緑の党(Green Party*)の[に関する]: go *green* エコ路線を取る, (自然)環境(保護)に配慮する[を意識する].
名 [the Green]《ドイツなどの》緑の党の党員; [the Greens]《ドイツなどの》緑の党(Green Party); 緑の党支持者, 環境保護政策支持者.
動《建造物を減らし植物をふやして》〈都会を〉緑化する;〈社会・人びと〉に環境[生態系の]問題に関心をもたせる;〈政党・政府〉に環境保護政策をとらせる.

green audit 環境への影響監査, グリーンオーディット《一企業・一商品の環境に与える影響を徹底的に調査すること; たとえばそのエネルギー・原料の利用, 汚染物質の生成・排出, 廃棄物処理, リサイクル性などについて調査する》.

greenback green 形《米俗》〈消費者などが〉環境保護のためには出費を惜しまない, 環境保護志向の《greenback は裏面が緑色であることからドル紙幣の通称》.

green ban 《オーストラリアで》greenbelt* における建設事業など自然・遺跡破壊につながる事業への就労拒否.

greenbelt 緑地帯, グリーンベルト《1) 地域社会を囲む森林・公園・農地など 2) 砂漠化を阻止するため砂漠の周辺に設けられたもの》.

Green Belt Movement [the ~]グリーンベルト運動《1977 年に Wangari Maathai* と National Council of Women of Kenya (NCWK) が環境保護と生活の向上を目的に, アフリカのケニアを拠点に始めた植樹運動; 女性が中心となって育苗や植樹活動をし, 緑の再生と同時に生活の収入を得ることを目的とする; 本部 Nairobi; 略 GBM》.

green certificate [tag] グリーン電力証書《再生可能エネルギー (renewable energy*) による発電であることを第三者機関が認証するもの; green procurement*, green purchasing* の一環として, 企業などがこのような発電事業体から電力を購入した証明書と

なる；別称 renewable energy certificate [credit] (REC), tradable renewable certificate (TRC) など；英国では電力・ガス供給の監督機関である Ofgem が発行し，電力会社に一定額の購入が義務付けられ renewables obligation certificate (ROC) と呼ばれる》．

green collar グリーンカラー《ブルーカラー，ホワイトカラーなどに対し，農林水産業や環境・自然保護関連の職種を言う》．

green dam 緑のダム，グリーンダム《砂漠の外縁に大量に植林された樹木；砂漠が農地に広がるのを防ぐ》．

greenery 1《特に政治姿勢として》環境問題に対する関心．
2 [集合的] 青葉，緑樹．
3 温室 (greenhouse*)．

green glue 表土流出を防ぐために植える植物．

Green Guerillas, greenguerillas 緑のゲリラ《1973 年に米国ニューヨーク市から起こった市民運動で，見捨てられた公有地・私有地などに植物を植えて緑地化するもの；当初は空き地などに密かに花の種などを蒔くものだった；⇨ community garden*》．

greenhouse 温室；《俗》温室効果ガス (greenhouse gas*)，《広く》地球大気：*greenhouse* pollutant 大気汚染 [温暖化] 物質 / *greenhouse* potential 大気汚染潜在物質 / *greenhouse* warming 《greenhouse effect* による》地球温暖化 (global warming*)．

greenhouse effect [the 〜] 温室効果《CO_2 や水蒸気といった大気中の温室効果ガス (greenhouse gas*) の蓄積が地表面の気温上昇をもたらすこと；地表に届く太陽光のエネルギーの一部は赤外線 (熱) のかたちで宇宙に放射されるが，CO_2 や水蒸気がその赤外線を吸収することで，熱が地表面にたまりやすくなる；CO_2 や水蒸気などの透明なガスが，外からの太陽光を通し，中からの熱の放射を防ぐ温室のビニールシートのような役割をすることから，この名称が付いた》．

greenhouse gas 温室効果ガス，温暖化ガス，GHC《地表から大気圏外に放射される赤外線 (熱) を吸収する性質をもつガス．地球を包む温室のような役割をし，地球温暖化の原因となる；二酸化炭素 (carbon dioxide*)，水蒸気，メタン (methane)，亜酸化窒素 (nitrous oxide*)，フロン (chlorofluorocarbon*) など；⇨ Annex A Gasses*》．

greenie 《俗》環境保護運動家，環境保護論者，グリーニー．

greening 《市街地・砂漠などの》緑化；環境問題に関する意識の向上．

greenism 環境保護主義[運動],グリーニズム.

green labelling グリーン表示,エコ表示(eco-labelling*).

green lung 《英俗》《都市内の》緑地,公園.

green marketing グリーンマーケティング《環境保護の姿勢を売り込むための企業活動》.

green material eco-material*

Green Panther 《米俗》戦闘的[声高な]環境保護運動家,グリーンパンサー.

Green Party 緑の党《地球環境保護運動全般を目標とする政党; 特にドイツなど欧州各国のもの; 英国では 1973 年に the Ecology Party として結成され,他国の同様の政党にあわせるため 85 年に the Green Party と改称》.

Greenpeace グリーンピース《反核運動から出発して反汚染・反捕鯨など幅広く活発な運動を展開している,非暴力直接行動による国際的環境保護団体; 1971 年設立; 本部 Amsterdam; 世界 41 か国に事務所を有し,約 280 万人のサポーターがいる》.

green PEP 《英》グリーンペップ《環境破壊につながるおそれのない投資をする個人株式投資(PEP; personal equity plan)》.

green procurement グリーン調達《環境に配慮した納入元から資材などを優先的に調達すること; ⇨ green certificate*》.

green product エコ製品,環境にやさしい製品.

green purchasing グリーン購入《環境に配慮した販売者から優先的に購入すること; ⇨ green certificate*》.

green revolution 緑の革命,グリーンレボリューション《1) 特に開発途上国での,品種改良などによる穀物の大増産 2) 工業国における環境に対する関心の高まり》.

green rights 緑の権利《豊かな自然環境を確保するうえでの諸権利》.

green seal グリーンシール《環境保護を示すシール》.

greenster 環境保護論者.

green strike 環境保全に反する仕事に対する就労拒否(⇨ green ban*).

green tag green certificate*

Green Thumb グリーンサム《米国ニューヨーク市内にある多くの community garden* にガーデニングのための材料提供や技術指導などを行っている団体; 1978 年設立; 1995 年より市の公園課と直結した団体となる》.

green tourism グリーン・ツーリズム《都市部に住む人が農山漁村に滞在して現地の人たちとの交流や農業[林業・漁業]体験を楽しむ旅行; ⇨ ecotourism*》.

greenwash 图 グリーンウォッ

シュ，企業が熱心に環境保全活動に取り組んでいるように表面を装うこと，企業が自社の商品を実際以上に環境配慮という点を強調して宣伝すること，環境保護に対する支持姿勢を示すために企業が行なう上辺だけの寄付・広報活動，自然[環境]保護への配慮(のポーズ)《欧米でよく使用される言葉；英国の広告基準協議会は2008～2009年，トヨタなど各メーカーの車の広告がエコを過剰に打ち出し，greenwashにあたるとして修正を求めた；例えばLexusは 'perfect for today's economic climate (and tommorrow's)' というコピー表現が禁止の対象となった》.

動 [greenwash *oneself* で]環境保全活動に対する支持姿勢を示す(ための寄付[広報活動]をする).

greenwashing 名形 環境保護のポーズ(を取った)，上辺だけのエコ宣伝活動(の).

greenway 《米》緑道《自然環境を残した歩行者・自転車専用遊歩道路》.

green work station グリーンワークステーション《グレードアップやリサイクリングが容易なため，廃棄されて環境問題を起こす可能性の少ないコンピューター；⇨ electronic waste*》.

grid parity 《太陽光発電の発電コストが》既存の送電線電力と同等になること.

ground cover 地被(植物)，地表植被《裸地をおおう矮性植物》；グラウンドカバー《芝生の代わりに植える装飾用の多年草；キヅタ・フッキソウなど》.

ground pollution 《処理場・埋立地廃棄物による》土壌汚染.

growing degree-day (植物)生長度日(ど)《一日の平均気温から生物活動に必要な最低限の温度を引いた値；これを毎日合計していったもの(有効積算温度)が植物の開花・結実や昆虫の繁殖など各種の生物活動の目安となる；⇨ cooling degree-day*, heating degree-day*》.

GWP global warming potential*

H

habitat 1 【生態】《生物を取り巻く》環境, 居住環境;《特に動植物の》生息地, 生育地, すみか,《標本の》採集地, 原生地;《農業的な》立地;《博物館内展示の》生物環境模型 (habitat group*): *habitat* segregation すみわけ. 2 居住地, 住所; 所在地［場所］;《海洋研究用の》水中家屋.

habitat group 生態類《生息環境を同じくする動物［植物］》;《博物館内展示の》生物環境模型.

habitat loss 生息地の消失［損失］.

half-life (period) 全体の半分がある変成をうけるに要する時間;【理】《放射性元素などの》半減期;【生・医】半減期《生体または生態系に入った化学物質などの半量が消失する, もしくはその実効性［活性］が半減するに要する時間》.

halobiont 【生態】塩生生物.
　halobiontic 形 塩生生物の.

halobiotic 形 【生態】塩分の多い場所に生育する, 塩生の.

haloform ハロホルム, トリハロメタン (trihalomethane)《メタンの水素原子3つがハロゲン原子で置換された物質の総称; クロロホルム (chloroform), ブロモホルム (bromoform), フルオロホルム (fluoroform), ヨードホルム (iodoform) の4種がある; 水道水中に含まれるクロロホルムなどは発癌性が問題になっている》.

halolimnic 形 【生態】淡水湖沼生の.

halophyte 【生態】塩生植物.
　halophytic 形 〈植物などが〉塩生の (salsuginous).

haloplankton 【生態】塩生プランクトン.

HAP hazardous air pollutant, 有害大気汚染物質《大気汚染の原因となり, 継続的に摂取される場合に人の健康を損なうおそれがある長期毒性をもつ物質》.

HAPP 【米】high-altitude pollution project 高高度大気汚染調査計画.

hard release ハードリリース《飼育・繁殖・病気回復などのために人間が世話していた動物を順化期間措置なしに野生に戻す放獣・放鳥形式; ⇨ soft release*》.

hauler 廃品回収業者.

HAW high activity waste 高レベル放射性廃棄物.

hazardous waste 《化学工場などから排出される》有害廃棄物.

Hazchem 【英】ハズケム《化学薬

品などの危険物の表示法; *haz-ardous chem*ical(有害化学物質)の略》.

hazemeter 透過率計, 視程計 (transmissometer*).

Haze Rule ⇨ regional haze*

HCS Hazard Communication Standard 危険有害性周知基準《米国労働安全衛生局が定めた有害物質を含む製品などの表示基準; ⇨ material safety data sheet*》.

heath 《英国などに見られる, ヒースの生い茂った酸性土壌の》荒れ地, 荒野, 荒蕪地;【生態】ヒース《特にヨーロッパの海洋性気候下の温帯寒帯性低木林》;【植】ヒース《荒野に群生するツツジ科エリカ属, カルナ属[ギョリュウモドキ属]などの矮性低木; 鈴形の紫・赤・白の花が咲く》;《オーストラリア産の》ヒースに似たエパクリス属の各種低木.

heat index 熱指数《温度と湿度を組み合わせて人間が体感として受けとる温度を表わす指数; 指数が高くなると熱中症が発生しやすくなる; 略 HI》.

heating degree-day 暖房度日(ど にち)《日平均気温を下回る値の1年間の積算温度を, 度日数(degree-day)を単位として表わしたもの; 建物などの暖房の必要エネルギー算定に用いる; ⇨ growing degree-day*, cooling degree-day*》.

heat island ヒートアイランド, 熱の島《都市部の気温が周辺地域よりも高くなること; 温度の等高線を書くと島状になることから, この名前が付いた; 道路面のアスファルト化, 緑地面積の減少, 都市部へのエネルギー使用の集中による廃熱の増加などが原因として考えられる》.

heat pollution thermal pollution*

heat-related mortality 熱中症関連死亡.

heat-trapping 形 熱をため込む, (地球)温暖化を促進する: *heat-trapping* gas 温室効果ガス (greenhouse gas*).

heat wave 1 長期間の酷暑: in the middle of a sizzling *heat wave* 焼けつくような酷暑のさなかに.

2【気象学】熱波.

hekistotherm 好氷雪性植物, ヘキストサーム《極地や高山の地衣類のような極寒の環境で生育できる植物》.

hekistothermic 形 好氷雪性の.

heliodon ヘリオドン《地上の任意の場所にある特定の建築について, 任意の時期・時刻における日照と影を人工的に測定する装置》.

hemisphere 半球体; 大脳[小脳]半球;《地球・天体の》半球; 半球の住民[国家]; 半球の地図[投影

図]: on the Eastern [Northern] *Hemisphere* 東[北]半球で.

hemispheric(al) 形 半球(状)の.

HEP Habitat Evaluation Procedure ハビタット評価手続き《米国で開発された生態系の数量的評価システム；野生生物生息地内の各種の環境要素・要因を数値化・単位化することにより，異なる時間・場所における生態系・生物多様性の比較評価を可能にし，環境保全と開発とのバランスの目安とするもの；mitigation banking* などに利用される》

herbal medicine 薬草療法；薬草から製した薬(薬草剤)；漢方薬；漢方療法.

herbal remedy 薬草療法.

herbivore 草食[植食]動物《特に有蹄類》.

herb layer 【生態】《植物群落の》草本層(⇨ layer*).

heritage 国民が未来世代に伝えることのできる自然および人工的環境，環境遺産《美しい景観・建造物・史跡など；⇨ National Heritage*, World Heritage Site*》.

heritage coast 遺産海岸《風光明媚で未開発の海岸；英国の非政府機関 Countryside Commission* が指定し，開発は禁止される》.

heterochthonous 形【生態】非土着の，外来の.

heterotopia, heterotopy, heterotopism 【生態】異常生息地.

heterotopic 形 異常生息地の，広範囲に生息する.

HI heat index*; humidity index 湿度指数.

high confidence 《仮説などの》高信頼度.

high moor 【生態】高層湿原.

Hill ヒル Octavia Hill (1838-1912)《英国の自然保護運動家；National Trust* の創設者》.

hive nest 《ミツバチの巣箱のように多くの鳥がコロニーをつくってすむ》大きな共同の巣，集団営巣の巣《シュウダンハタオリドリ(sociable weaverbird)の巣など》.

holobenthic 形 終生海底にすむ.

holomictic 形 〈湖が〉水が湖底まで完全に循環する，完循環の(⇨ meromictic*).

holon ホロン，全体子《より大なる全体の中の一つの全体；ギリシア語で「全体」を表す holos に「粒子」を示す on を組み合わせた造語；1970年，ジャーナリストで哲学者の Arthur Koestler が全体主義と還元主義の矛盾を超えるためにこの概念を提示した》；【生態】生物と環境の総合体.

home range 《定住性動物の》行動圏.

hominization ヒト化《他の霊長類

と人類を区別する特質の進化論的発達》；環境の人間化《環境を人間の利用しやすいように変えていくこと》．

hominize 動〈環境を〉人間化する，人間が利用しやすいように変える；〈ヒト〉の進化論的発達を進める，ヒト化する．

hot water pollution thermal pollution*

human capital 人的資本《人的資源(human resources)に対する投資すなわち教育・訓練によって蓄積された知識・技能など；生産性を高め収益の向上につながる》．

human ecology 人間[人類]生態学．

human-induced 形 人為的な《自然現象と区別して》．

HSUS Humane Society of the United States《米国の動物愛護団体の一つ》．

humicolous 形 土中[土の上]に住む[生育する]生物の．

humiture 気温と湿度の相乗効果による不快感《湿度(humidity)と温度(temperature)の合成語》．

hunger 飢え，飢餓；空腹，ひもじさ；飢饉：die *of hunger* 餓死する．

hunt sab [saboteur] 狐狩り[狩猟]妨害活動家(sab*)．

hydrarch 形【生態】〈遷移が〉湿潤な場所で起こる，湿生の．

hydric 形【化】水素の，水素を含む；【生態】湿潤な(環境に適した)，水生の．

hydrology, hydrogeology 水文地質学《地球上の水の生成・循環・性質・分布などを研究する》．

hydrosere 【生態】湿生(遷移)系列．

hyperparasite 【生態】重寄生者，高次[重複]寄生者《寄生者にさらに寄生する》．

  hyperparasitism 名 重寄生，高次寄生，過寄生．

  hyperparasitic 形 重寄生の[する]．

hypobenthos 底生生物《3000フィート以上の深海底にすむ生物》．

I

IAEA International Atomic Energy Agency*

IAI Inter-American Institute for Global Change Research 北米および中南米地球変動研究所.

IATTC Inter-American Tropical Tuna Commission 全米熱帯まぐろ類委員会《1950年発足；東部太平洋の鮪類漁業資源管理・保護を行う；本部は米国カリフォルニア州ラホーヤ；⇨ ICCAT*》.

ICAP International Carbon Action Partnership*

ICCAT International Commission [Convention] for the Conservation of Atlantic Tunas 大西洋まぐろ類保存国際委員会[条約]，アイキャット《大西洋の鮪類漁業資源保護管理のため1966年にブラジルのリオ・デ・ジャネイロで条約が結ばれ，1969年委員会が発足；本部はスペインのマドリード；鮪資源保護の国際条約・機関には IATTC*, IOTC*, CCSBT*, WCPFC* などがあり，日本はその全てに加入している》.

ice cap 氷冠, 氷帽《高山・極地などの万年(氷)雪[氷原]》.

ice sheet 氷床《南極大陸やグリーンランドの内陸氷》.

ichthyofauna 魚相《ある地域の魚(の生態)；⇨ fauna*》.

ICIPE International Centre of Insect Physiology and Ecology 国際昆虫生理生態センター《1970年ケニアに設立》.

IDA 1 International Dark-Sky Association*
2 Industrial Development Authority《アイルランドの》産業開発庁.

IFAW International Fund for Animal Welfare 国際動物愛護基金.

IGBP International Geosphere-Biosphere Programme 地球圏・生物圏国際共同研究計画.

IHDP International Human Dimensions Programme on Global Environmental Change 地球環境変化の人間社会側面に関する国際研究計画.

IIEC 【米】Inter-Industry Emission Control 産業間排気汚染制御計画.

IIED International Institute for Environmental Development 国際環境開発協会.

illegal dumping 《ごみ・産業廃棄物などの》不法投棄(⇨ indus-

implement

trial waste*).

implement 图 1 道具, 用具, 器具; 手段, 方法; ［複数形で］《家具・衣服などの》備品, 装具: agricultural [farm] *implements* 農具 / stone *implements* 石器.
2《スコットランド法で》履行.
動〈約束などを〉履行する, 実行する;〈要求・条件・不足などを〉満たす; …に道具[手段]を与える.

implemental 形 道具の, 道具[助け]になる;（…の）実現に寄与する.

implementation 图 履行, 実行; 完成, 成就.

imhoff tank インホフタンク《下水を上部の沈澱室で沈澱させ, 汚泥をスロットから下部の腐敗室に導いて消化する二階式の下水処理槽》.

imitation 《動物の》模倣（行動）.
imitational 形 模倣（行動）の.

impoverished 形 動植物の種類[数]が少ない〈地域〉.

inaction 不活動; 休止, 不作為.

incinerator ship ごみ焼却船《海上でごみを焼却する, 海洋型はしけあるいは船》.

Inconvenient Truth ［*An* ～］不都合な真実《地球温暖化の危機を警告したドキュメンタリー映画（2006）; ゴア（Gore*）米元副大統領の講演活動を題材としている; この作品は第 79 回アカデミー賞の長編ドキュメンタリー映画賞を受賞; ゴア氏はこれらの功績によりノーベル平和賞を受賞》.

INCPEN Industry Council for Packaging and the Environment《英国の》包装・環境産業評議会.

Indian Ocean haze 《特にインド沿岸の》インド洋上大気汚染.

indicator (organism) 【生態】《環境を示す》指標生物.

indicator species 【生態】指標（生物）,（環境）指標種（indicator organism)）《生息できる環境条件が限られていることが判明しているため, その分布状況等の調査によって, 特定地域の環境状況の類推・評価の裏付けとなる動植物種》.

indigenous 形 土着の, 原産の, 自生の, 土地固有の;《一定の領域[環境]内で》固有の;《鉱物が》現地成の; 原住民の（ための）; 生来の.

indigenous people 先住民（族）.

industrial development certificate 工業開発証明書《英国において工場の新設・拡張にあたって地方自治体への許可申請に添付が義務づけられている環境省（以前は貿易産業省）の証明書; 工場立地の調整が目的》.

industrial melanism 工業暗化[黒化]《特に工業汚染物質で黒くなった地域の昆虫におこる工業性

黒色素過多変異》.

industrial waste　産業廃棄物, 産廃: dump [dispose of] *industrial waste* illegally 産業廃棄物を不法投棄する.

infectious disease　感染症《細菌・原生動物・真菌・ウイルスなどを原因とする感染[伝染]性の疾患》.

infiltration capacity [rate]　浸透能[率]《ある状態で土壌などに水が浸透する最大の率》.

infiltrometer　浸透能測定器, 浸透率計《土壌などの infiltration capacity* を測る装置》.

influence　影響(力), 感化(力), 作用, 力〈on〉: the *influence* of the moon *upon* the tides 月が潮に及ぼす作用.
動 …に影響[感化]を及ぼす, 左右する, 動かす: Food *influences* our health. 食物は健康を左右する.

influent　影響種《群集全体の調和上重要な影響を及ぼす生物》.

inland　図 内陸, 奥地, 僻地.
形 (海に遠い)奥地の, 僻地の, 内陸の; 国内の, 内国の, 内地の: an *inland* sea 内海, 縁海《大陸棚の上に広がる海》; [the Inland Sea] 瀬戸内海 / *inland* aquiculture 陸上養殖《内陸部に閉鎖循環式の水槽を作って行う魚類や貝類などの養殖》.
副 内陸に, 奥地に.

inland waters　[the 〜] 内水《河川湖沼・湾など一国内の水域, および陸から 3.5 マイル内の領海》.

inshore　形 副 海岸に近い[近く], 沿岸の[に], 沿海の[に], 近海の[に]; 岸に向かう〈波など〉: *inshore* fisheries [fishing] 沿岸漁業.

in situ　副 原位置に, イン・シトゥー(⇔ ex situ*).

instrument　道具, 《実験用の》器械, 器具; 精密器機;《飛行機・ロケットなどの》計器: surgical *instruments* 外科用機械 / nautical *instruments* 航海計器 / fly on *instruments* 計器飛行をする.
動 …に機器を備える.

instrumental record　計測記録.

INTECOL　International Association for Ecology 国際生態学会.

integrated pest management　(総合的)害虫管理, 総合防除《害虫と作物の生態的研究に基づき, 農薬使用を最小限に抑え, 天敵や不妊化法などを応用した害虫管理; 略 IPM》.

intentional (living) community　インテンショナルコミュニティ, 意識的共同体《かつてのヒッピー的 commune* 運動を現代化・穏健化したもの; 物質文明や科学技術を否定せず賢明に利用し, 浪費を避け環境に配慮した持続可能な消費生活と住民の交流・互助を重視する小規模共同体; ⇨ eco-vil-

interfluve

lage*》.

interfluve 河間地域《隣接する河の間》.

Intergovernmental Panel on Biodiversity and Ecosystem Services 生物多様性及び生態系サービスに関する政府間パネル《略称 IPBES》.

Intergovernmental Panel on Climate Change ⇨ IPCC*

Interim Standard Atmosphere 暫定標準大気《高度 50-80 km 上空の仮想大気モデル; 国際標準大気 (International Standard Atmosphere) に対する語》.

intermediate-level waste 中レベル (放射性) 廃棄物.

International Atomic Energy Agency [the 〜] 国際原子力機関《原子力の平和利用と, 軍事目的への転用防止に向けた国際的協力のための国連機関; 略 IAEA; 1957 年設立; 本部 Vienna》.

International Carbon Action Partnership [the 〜] 国際炭素行動パートナーシップ《地球温暖化防止施策として, 温室効果ガスの排出量取引 (cap and trade*) の国際的な共通市場を設けるため 2007 年欧州連合や米国, カナダの一部の州 (米カリフォルニア州, ニューヨーク州) などが設立した協力体制; 排出量取引のルールの共通化により国境を越えた市場での売買が可能になる; 略 ICAP》.

International Civil Aviation Organization [the 〜] 国際民間航空機関《略 ICAO; 1947 年発足の国連機関; 本部 Montreal》.

International Convention for the Control and Management of Ships' Ballast Water and Sediments バラスト水規制条約《正式名称は「船舶のバラスト水及び沈殿物の規制及び管理のための国際条約」; ⇨ ballast water*》.

International Dark-Sky Association [the 〜] 国際暗天協会《市民生活との調和をはかりつつ光害 (light pollution*) のない暗い夜空の保護・回復を目指す団体; 略 IDA》.

international environment law 国際環境法《地球規模の環境破壊を, 2 国間または多国間の条約や協力によって防止するための国際法》.

International Maritime Organization [the 〜] 国際海事機構《略 IMO; 1958 年設立の国連機関の一つ; 本部 London; 旧称 Inter-Governmental Maritime Consultative Organization》.

international standard atmosphere 国際標準大気《略 ISA》.

International Tropical Timber Agreement [the 〜] 国際熱帯木材協定《1983 年採択, 85 年発効; ⇨ ITTO*》.

International Union for the Conservation of Nature and Natural Resources ［the ～］国際自然保護連合《⇨ IUCN*》.

International Whaling Commission ［the ～］国際捕鯨委員会《略 IWC; 1946年の国際捕鯨取締条約に基づき，49年Londonで第1回委員会が開かれた；事務局は英国Cambridge》.

International Year of Biodiversity 国際生物多様性年《国連が2010年を国際生物多様性年と定めた；生物多様性条約(Convention on Biodiversity*)の3つの目的と生物多様性の2010年目標を達成するため，生物多様性の重要さについての認識を高めることが狙い；略称IYB》.

inundation 氾濫，大水，洪水，浸水

invasive alien species 侵略的外来種: 100 of the world's worst *invasive alien species* 世界の侵略的外来種ワースト100《国際自然保護連合(IUCN*)が2000年に発表》.

inventory （商品，財産などの）在庫目録；天然資源調査一覧《特に一定地域にすむ野生生物数》.

IOTC Indian Ocean Tuna Commission インド洋まぐろ類委員会《インド洋の鮪類漁業資源保護管理国際組織；国連FAOの1993年の条約に基づき設置；本部はセーシェルの首都ビクトリア: ⇨ ICCAT*》.

IPBES Intergovernmental Panel on Biodiversity and Ecosystem Services*

IPCC Intergovernmental Panel on Climate Change 気候変動に関する政府間パネル《国連環境計画(UNEP*)と世界気象機関(WMO; World Meteorological Organization)が協力して1988年に発足した会合；各国政府関係者と科学者が参加する；地球温暖化(global warming*)について最も信頼できる科学的な知見が報告書として公表され，各国が温暖化防止のための国際的な枠組みを議論する際の土台となっている；3つの作業部会(Working Group; WG)で構成され，第1作業部会(WG1)は科学的メカニズム，第2作業部会(WG2)は影響評価，第3作業部会(WG3)は緩和策について検討している；ノルウェーのノーベル賞委員会は，温暖化防止の取り組みに貢献したとして，2007年度のノーベル平和賞をIPCCとアル・ゴア(Gore*)元米副大統領に授与した》.

IPM integrated pest management*

IPPC Integrated Pollution Prevention and Control directive 統合的汚染防止・管理指令，IPPC指令《EUの包括的な産業向

け環境汚染防止対策基準；これに基づき，欧州IPPC事務局(European IPPC Bureau; EIPPCB)が，産業分野ごとに取り入れるべき防止対策の指針となる「利用可能な最善技術」(Best Available Technology*)を定めた参照用手引書BAT reference documents (BREF)を作成し，各国監督官庁が工場操業許認可などの目安とする》．

IRENA International Renewable Energy Agency 国際再生可能エネルギー機関《欧州諸国・途上国を中心として2009年に設立されたrenewable energy*開発普及を推進する国際協力機関》．

iron hypothesis 鉄分仮説《海洋学における仮説で，ある種の海洋植物プランクトンに鉄分を供給して増殖させ，二酸化炭素吸収を進めれば地球温暖化を防ぐことができるとする》．

irrigation 灌漑，注流，灌水：an *irrigation* canal [ditch] 用水路．

irrupt 圓【生態】急激に〈個体数が〉増す，集団移入する．

irruption 図急増，集団移入．

ISA international standard atmosphere*

ISES International Solar Energy Society 国際太陽エネルギー学会．

isonome 【生】等頻度線，等数度線．

isophene 【生態】《地図上の》等態線．

ITL International Transaction Log 国際取引ログ《国連が認定した温暖化ガスの排出権を国際間で移転するためのシステム》．

ITTO International Tropical Timber Organization 国際熱帯木材機関《熱帯林資源の保全，利用，取引についての協議機関；1985年4月設立；本部　横浜；⇨ International Tropical Timber Agreement*》．

IUCN International Union for the Conservation of Nature and Natural Resources 国際自然保護連合《1948年に設立された世界最大の自然保護機関；自然保護と生物多様性の保全を目的とし，各国・地域の保全プログラムを支援している；1980年には国連環境計画(UNEP*)などと世界保全戦略(World Conservation Strategy)を作成し，その中に'Towards Sustainable Development'のセクションを盛り込んだ；会員は84カ国から111の政府機関，874のNGOなど(2008年)；本部はスイスのGland; ⇨ red data book*, sustainable*》．

IUU illegal, unreported and unregulated 違法・無報告・無規制の《特に便宜置籍漁船(税金逃れなどのために他国に船舶を登録した船)などによる，公海遠洋漁業の国際的取り決めを無視した操業に

ついていう》: *IUU* fishing
IUU 漁業.
IWC International Whaling Commission*
IYB International Year of Biodiversity*

J・K

JCCCA Japan Center for Climate Change Action 全国地球温暖化防止活動推進センター.

jeopardize 危うくする,危険にさらす.

joom, jhoom, jhum ジューム《焼き畑農法で森などを焼いた区域;インドのベンガル地方東部で行なわれる》.

JSPCA Japan Society for the Prevention of Cruelty to Animals 日本動物愛護協会.

JI Joint Implementation 共同実施《温室効果ガスの削減を定めた Kyoto Protocol* において,先進国が(削減の数値目標をもつ)他の先進国における排出量削減事業に投資をし,それによって得られた排出削減枠を,クレジットとして投資をした自国に移転できる仕組み.京都メカニズムの一つ; ⇨ CDM*》.

Johannesburg Summit ヨハネスブルグ・サミット (⇨ World Summit on Sustainable Development*).

K 【生態】carrying capacity* を表わす記号 (⇨ K selection*).

Keep America Beautiful 1 アメリカを美しく《道路などにごみを捨てないよう呼びかける米国のキャンペーンの標語》.
2 キープ・アメリカ・ビューティフル《米国のボランティア組織;コミュニティの環境改善をめざす,全米最大の実践活動・教育組織》.

Keep Britain Tidy 1 英国をきれいに《道路などにごみを捨てないよう呼びかける英国のキャンペーンの標語》.
2 キープ・ブリテン・タイディ《英国のチャリティ団体;公共の場所の廃棄物汚染防止をめざしている》.

kelp 図 ケルプ《コンブ目・ヒバマタ目の漂着性の大型褐藻》;ケルプ灰《ヨードを採る》
動 ケルプ灰を採るために海草を焼く.

Kelthane 【商標】ケルセン《米国 Rohm and Haas 社が開発した,ハダニ類の退治に用いる散布用農薬;不純物として DDT* を多く含むとの理由で 1985 年一時使用を禁止された; ⇨ dicofol*》.

killer seaweed キラー海藻《大量発生して生態系を破壊する海藻;特に近年地中海などで問題化している岩蔦属の一位蔦(イチイヅタ)

(学名 *Caulerpa taxifolia*)の変異種を指す；繁殖力が強い上に有毒で餌にならず，異常繁茂して他の藻類などを駆逐してしまう；日本でも南西諸島などに見られ，観賞魚の水槽用に「フェザー」などの名で販売されることがあるので警戒されている；⇨ marine pest*》.

killer weed 《異常繁茂などで》有害な(雑)草；killer seaweed*.

kingdom 界《生物分類学上の最上位のカテゴリー；動物界(Animal Kingdom)，植物界(Plant Kingdom)など；⇨ 和英篇 学名》；【生態】区系界.

konimeter 塵埃(じんあい)計，コニメーター《空気中の塵(ちり)，埃(ほこり)の量を測定する装置》.

koniology 塵挨学《大気中の塵その他不純物の動植物に対する影響を研究する》.

konometer konimeter*

krummholz 【生態】高山屈曲林，矮性低木林(elfinwood*).

K selection K選択[淘汰]《安定した生息場所にすむ生物個体群が環境収容力(carrying capacity*)に近い高密度を維持している場合に起こる選択；資源に対する競争能力があり，ゆっくり発育する少産の個体に有利となり，長命の個体の安定した個体群となる；⇨ r selection*》.

Kyoto mechanism 京都メカニズム，柔軟性措置《温室効果ガスの排出削減を国際協調で柔軟に実施していくため，京都議定書のもとに定められた3つの仕組み；クリーン開発メカニズム(CDM*)，共同実施(JI*)，排出量取引(emissions trading*)；flexibility mechanism* とも呼ぶ》.

Kyoto Protocol [the 〜]京都議定書《2008年から2012年における先進国全体の温室効果ガス排出量を，1990年比で5%以上削減することを目標に定めた議定書；日本6%，米国7%，EU8%など法的拘束力のある削減目標が国別に定められた；削減目標達成のため，他国における削減の取り組みを柔軟に利用できる「京都メカニズム」を取り入れたことが特徴；1997年12月11日，京都で開かれた国連気候変動枠組条約 United Nations Framework Convention on Climate Change (UNFCCC) の第3回締約国会議(COP*3)で採択された》.

L

landfill 埋立てによるごみ処理；埋立てごみ；ごみ埋立地.

landfill gas 埋立地ガス《ごみ埋立地に発生するメタン・二酸化炭素などを含むガス》.

land degradation 土壌劣化, 土地荒廃.

land-based pollution 陸上起因汚染.

landscape architecture 景観設計, 造園, ランドスケープアーキテクチュア《人が利用・享有する土地を建造物・道路・植栽の効果的配置によって開発する技術》；造園学.

　landscape architect 造園家.

landscape gardening 造園, 造園学[法]《庭園・公園などにおける開発・植栽の技術》.

　landscape gardener 造園家, 庭師, 庭園師.

La Niña ラニーニャ, 反エルニーニョ現象《エルニーニョ現象とは反対に, 南米ペルー・エクアドルの沿岸から東太平洋赤道域にかけて, 海面水温が平年より低くなる現象；気候変動への影響が研究されている；スペイン語で「女の子」の意で, El Niño* の反対現象であることから命名されたもの》.

LAP ⇨ catch share*

large-statured 形〈森林が〉高木と低木からなる.

laurel forest 照葉樹林《もとスペイン領カナリア諸島のゲッケイジュの林に対して使われた》.

law of dominance [the 〜]優劣[優性]の法則《メンデルの遺伝法則の一：雑種第1代で優性の形質のみが現われる》.

layer 《植物群落を垂直的にみた》階層《高木層(tree layer), 低木層(shrub layer), 草本層(field [herb] layer), コケ層(ground [moss] layer)などに区分される》.

leaf litter 《森林の地表をおおう》落葉落枝(層), 腐葉土.

LEED Leadership in Energy and Environmental Design【商標】《米国の非営利団体 USGBC (U.S. Green Building Council)が認定する環境に配慮した建築基準》.

len(i)tic 形【生態】静水の[にすむ], 静水性の(⇔ lotic*).

lestobiosis 盗食共生《社会性昆虫の間の共生で, 小型種のコロニーがより大きな種の巣にすみ, 幼虫を食べたり貯蔵した餌を盗んだり

LEV low-emission vehicle*

life-form 生活形, 生物形態《成熟した生物の種の特徴を示す形態》; 生き物, 生物.

life-support 形 生命維持の(ための);《環境などの》(野生)生命を養う能力に関する.

life-support system 《宇宙船内・水中・坑内などの》生命維持装置, 環境保全装置; 生物圏(biosphere*)《地球上の生物がすんでいる範囲》.

life system 生活系《個体群とその主体的環境を合わせた系》.

light green 形 ライトグリーン(の)《政治的目的から環境保護を標榜することについていう; ⇨ dark green*》.

light pollution 光害《天体観測などに支障をきたす, 都市のネオン, ビル照明, 街路灯などの夜光; ⇨ International Dark-Sky Association*》.

liman (河口の)潟, 入り江; おぼれ谷《陸上の谷が海面下に沈んだ河口付近の浅瀬》.

limewater 石灰水, 炭酸カルシウム[硫酸カルシウム]を多量に含む自然水.

limicoline 形 水辺[海辺, 湖沼畔, 川原]に生息する, 泥地生の.

limiting nutrient 制限的栄養物質《湖水の富栄養化を遅らせる物質》.

limnetic, limnic 形 淡水の[にすむ], 沖帯の[にすむ].

lindane リンデン《主にBHC*のγ-異性体からなる殺虫剤; 難分解性》.

liner ゴミ箱の中敷きにするごみ袋, ごみ用ポリ袋.

lithophilous 形【動・植】石上に生ずる, 石の多い場所で生活する.

lithophyte 岩生植物《地衣類・コケ類》; 樹状生物《サンゴ類など》.
　lithophytic 形 岩生[樹状](生物)の.

lithosere 岩上[石]遷移系列《岩石上で始まる植物群落の生態遷移》.

litoral littoral*

litter [散らかった]くず, ごみ(⇨ leaf litter*).

litterbag 《自動車の中などで使う》ごみ袋.

litterbin, litterbasket 《英》《街路に置く》くず物容器, くずかご.

litterbug 名 動《米》《街路・公園など公共の場所を》紙くず・廃物などを捨ててよごす(者).

litterer,《英》**litterlout** 名 litterbug*

litterpick 《生徒の一団などによる特定地域の》協調的清掃活動.

littoral 形 沿岸[沿海]の;【生態】沿岸にすむ, 沿岸性の.
　名 沿海地方;【生態】沿岸帯,《特に》潮間帯.

livestock 家畜《集合的》: *livestock* farming 牧畜, 畜産.

Living Planet Index 生きている

地球指数《世界自然保護基金(WWF*)が発表する,世界各地の生物多様性の減少状況を示す指数;略 LPI》.

LMMAs locally managed marine areas*

LNAPL light non-aqueous phase liquid 軽量非水相液体,軽非水液《難水溶性で比重が水より軽いベンゼンなどの揮発性有機化合物(VOC*);しばしば帯水層に浮かんで土壌・地下水の汚染源となる;⇨ DNAPL*》.

local authority 《英国などの》地方行政当局,地方自治体.

local community 地域社会,地域コミュニティ.

locally managed marine areas 地域主導型管理海域.

locavore 地産のものを食べる人,ローカルフード派,地産地消主義者《環境・健康に配慮してなるべく地元で生産されたものを食べようとする人;⇨ food miles*》.

LOHAS ロハス《アメリカから始まった健康や持続可能な社会を重視するライフスタイル(Lifestyles Of Health And Sustainability)の頭文字からつくられた造語;1998年,アメリカの社会学者らの調査で,健康で無理のない生活様式を志向する人々の存在が指摘され,彼らが環境に配慮した自然志向の製品を消費する傾向があることから LOHAS ビジネスが生まれた.大量生産,大量消費を反省し,スローライフにつながる考え方をしている》.

long term perspective 長期展望,長期見通し.

lotic 形 動く水[流水]の[に住む],動水性の(⇔ lentic*).

Love Canal 周辺地域に脅威を与えている毒性産業廃棄物の捨て場《米国ニューヨーク州 Niagara Falls 市近くの,毒性産業廃棄物の捨て場である Love Canal という地名にちなむ;1978年に Love Canal から漏れた毒物が周辺の飲料用水源に到達する汚染事件があった》.

low-carbon 形 低炭素の《二酸化炭素(carbon dioxide*)などの温暖化ガス排出量が少ない[抑制された];⇨ low-carbon society*》: *low-carbon* energy 低炭素エネルギー《石油などの化石燃料(fossil fuel*)に対して,carbon-neutral* なエネルギーや renewable energy* など》/ *low-carbon* diet 低炭素生活(carbon diet*)《carbon offset*, slow food*, locavore* といった形で,環境に配慮した消費・食生活を心がけること;しばしばカロリー計算風に日常生活の carbon footprint* を計算する》.

low-carbon society 低炭素社会《二酸化炭素などの温室効果ガスの排出を低く抑える仕組みをそな

えた社会；産業構造やライフスタイルの変革が必要になる》.

low-emission vehicle 低排ガス車, 低公害車《排気ガス中の大気・環境汚染物質を大幅に低減した自動車；在来の車とは燃料が異なるメタノール車・天然ガス車や電気自動車, あるいはガソリン車でもエンジンが改良されたり排気系に有害物質低減装置が取り付けられた車にもいう；略 LEV》.

lower yield 低収量, 低収穫, 低収率.

low-impact 形《体・環境などに》負担をかけない, 影響の少ない, ローインパクトの.

low-lying coastal area 低地の沿岸地域.

LPI Living Planet Index*

LULUCF Land-Use, Land-Use Change and Forestry 土地利用, 土地利用変化および林業《京都議定書(Kyoto Protocol*)において, 温室効果ガスの人為的吸収源(carbon sink*)として規定された活動》.

lysocline 【生態】溶解層《その層より深所では水圧によりある種の化学物質が溶解を起こす深海の層》.

M

Maathai マータイ Wangari Maathai (1940-)《ケニアの生物学者・環境活動家，元ケニア環境副大臣；Green Belt Movement* をはじめとする持続可能な開発・民主主義・平和への貢献が評価され，アフリカの女性として初めて Nobel 平和賞を受賞した (2004)；翌 2005 年に来日し，毎日新聞社での初インタビューで「もったいない」という日本語に共鳴し，Reduce, Reuse, Recycle という 3R 運動 (⇨ three R's*) につながる言葉として，Mottainai の国際的な使用を呼びかける運動に携わった》.

MAB Man and the Biospace Program《ユネスコの》人間と生物圏計画.

macrobenthos 【生態】大型底生生物.

  macrobenthic 形 大型底生生物の.

macrobiota 《ある地域の》大型生物相《肉眼で見えるくらいの大きさの生物の》.

macroconsumer 大型消費者《他の生物または粒状有機物を食う動物》.

macrofauna マクロフォーナ《広い区域の動物相》；《ある場所の》大型動物類《肉眼で見えるくらいの大きさ》.

macroflora マクロフローラ《広い区域の植物相》；《ある場所の》大型植物類《肉眼で見えるくらいの大きさ》.

macrophyte 【生態】大型水生植物.

  macrophytic 形 大型水生植物の.

malnutrition 【医】栄養不良[失調，障害].

mammal 哺乳動物.

mangrove ヤエヤマヒルギ，マングローブ，紅樹《熱帯海浜・河口に密生する常緑高木[低木]；幹の下部から多数の気根をおろす》；ヒルギダマシ《熱帯アジア原産》.

mangrove swamp マングローブ湿地《熱帯・亜熱帯の内湾・河口にみられる，マングローブが生育した汽水域(海水が混じる最下流域)の湿地》.

Manhattanization 高層ビルによる都市の過密化，マンハッタン化.

manure (有機質)肥料，こやし；《特に》厩肥(きゅうひ)，堆肥：artificial *manure* 人造肥料 / barnyard *manure* 厩肥 / farmyard

manure 堆肥.
動〈土地〉に肥料を施す.

marine ecosystem 海洋生態系.

marine pest マリンペスト, (外来の)有害海洋生物《異常繁殖・食害で漁業や生態系に悪影響を及ぼす海洋生物; 特に問題化しているのは外来種のヒトデ, 蟹, クラゲ, 貝などの卵・幼生や赤潮プランクトンといった微生物が船舶安定用のバラスト水(ballast water*)に紛れて持ち込まれるケースで, ワカメが外国で異常繁茂した例もある; ⇨ killer seaweed*》.

Marine Stewardship Council [the 〜] 海洋管理協議会《海洋環境の保全と持続的な海洋資源の利用を推進するための国際的なNPO法人; 環境に配慮した漁業団体の認証事業を行う; 本部London; 略MSC(⇨ 和英篇末尾)》.

marine turtle 海亀(⇨ sea turtle*).

market imperfection 市場の不完全性.

marsh gas 沼気《沼や湿地の土中で有機物が分解する過程で発生する, 主にメタンからなる燃焼性ガス》.

material flow cost accounting マテリアルフローコスト会計《環境負荷の低減と, コスト削減による企業利益追及との両立を目指す会計手法; 原材料から完成品に至る工程(material flow)で出るロス・廃棄物なども含めて投入材料費・加工費・設備償却費などを総合的に算定し, 資源やエネルギーの節約に活用する; 略MFCA》.

material safety data sheet 化学物質等安全データシート, 製品安全データシート《指定された化学物質を含む化学製品の毒性や安全な取り扱いに関する説明書; 米国ではHCS*によって定められ, また日本など他の多くの国でも添付が義務付けられている; 略MSDS》.

MDGs Millennium Development Goals ミレニアム開発目標《2000年New Yorkで開催された国連ミレニアム・サミットで採択された国連ミレニアム宣言と1990年代のサミット等で採択された国際開発目標を統合したもの; 2015年までに達成すべき8つの目標を含む》.

meadow 草地, 採草地, (永年)牧草地, 草刈地《特に川辺の草の生えた未開墾の低地; 樹木限界線近くの緑草高地》: a floating *meadow* 容易に水をかぶる(牧)草地.

medium confidence 《仮説などの》中信頼度《⇨ high confidence*, minimal confidence*》.

mega delta 《アジアの大河などの》大三角州.

megafauna 《一地域の》大型動物相《肉眼で確認できる地上動物

群》.

meiofauna 【生態】中型底生動物.
  meiofaunal 形 中型底生動物の.

melt down 動〈原子炉が〉炉心溶融する.
  meltdown 名《原子炉の》炉心溶融, メルトダウン.

MEF Major Economies Forum on Energy and Climate エネルギーと気候に関する主要経済国フォーラム《2013年からのいわゆるポスト京都議定書(Kyoto Protocol*)期間の温暖化ガス排出規制の枠組みを話し合うため, 2007年9月, 米国のブッシュ大統領の主導により, MEM*という名称で始まった国際会議; Kyoto Protocolを離脱した米国が2013年以降に参加するための重要な枠組みとなった; 先進7カ国(G7)や新興国ブラジル, ロシア, インド, 中国など主要排出国が参加; MEFは2009年7月, オバマ大統領が呼びかけてMEMを改組するかたちで立ち上がった多国間協議の枠組みで, MEMにデンマークが加わっている》.

MEM Major Economies Meeting (on Energy Security and Climate Change)エネルギー安全保障と気候変動に関する主要経済国会合, 主要排出国会合《MEF*の前身》.

meromictic 形〈湖が〉秋季に水が部分的に循環する, 部分循環の(⇨ holomictic*).

mesarch 形 1【植】中原型の《後生木部が四方へ発達する》.
2【生態】〈遷移が〉適潤状態の生育地で始まる.

mesic 形【生態】中湿[適湿]性の, 中生の.

mesobenthos 中型底生生物《200-1000 mの海底にすむ動植物: ソコミジンコ・有孔虫など》.

mesofauna 【生態】中型動物類(⇨ macrofauna*, microfauna*).

mesopelagic 中深海水層の《漂泳区の区分で, 水深 200 m-1000 mの層; ⇨ abyssopelagic*》.

mesophyte 【生態】適潤植物, 中生植物(⇨ xerophyte*).
  mesophytic 形 適潤植物の.

mesosere 【生態】中生(遷移)系列.

mesothermal 形〈鉱脈・鉱床が〉中熱水(生成)の《中程度の深度で熱水溶液から中温・中圧条件のもとで生じた》; 中温植物の.

mesotrophic 形【生態】〈湖沼・河川が〉中栄養の(⇨ eutrophic*, oligotrophic*).

metabolism (新陳)代謝, 物質交代《物質代謝およびエネルギー代謝》;《ある環境における》代謝総量: constructive *metabolism* 同化作用 / destructive *metabolism* 異化作用.

methyl isocyanate イソシアン酸メチル《猛毒無色の可燃性液体;

特に殺虫剤の生産に用いる；略 MIC》.

MFCA material flow cost accounting*

microclimate 小気候《一局地の気候》；微気候《小気候よりさらに小さな地点の気候》；《広く》小範囲の(地域)環境, 局地的条件[状態].
  microclimatic 形 小気候の.

microclimatology 小[微]気候学.
  microclimatologist 小気候学者[研究者].
  microclimatologic(al) 形 小気候学(上)の, 小気候学的な.

microcosm(os) 【生態】微小生態系, ミクロコズム.
  microcosmic 形 微小生態系の.

microdistribution 微小分布《生態系の一部分や微小生息域における生物の精確な分布》.

microecology 【生態】ミクロ[狭域]生態学.
  microecological 形 ミクロ生態学的な.

microenvironment 微環境 (microhabitat*).
  microenvironmental 形 微環境(で)の.

microfauna 微小動物相《非常に狭い区域の動物相》；【動】微小動物類.
  microfaunal 形 微小動物(相)の.

microflora 微小植物相《非常に狭い区域の植物相》；【植】微小植物類.
  microfloral 形 微小植物(相)の.

microhabitat 微小生息域, ミクロハビタット, 微環境(microenvironment)《草の茂み, 岩の間など微小な生物の生活の場として特有な環境条件をそなえた場所》.

microorganism 微生物《細菌・原生動物・酵母・ウイルス・藻類など顕微鏡によって観察される大きさの生物》.

microphyte 《顕微鏡によらなければ見えない》微細植物；《悪条件による》矮小植物.
  microphytic 形 微細[矮小]植物の.

micropopulation 《特定環境内の》微生物集団；狭域生物集団.

microtremors 常時微動《高感度の地震計が感じる, 振幅が非常に小さくて周期 0.1〜数秒程度の, 正体が特定できない地面の揺れ；地上のあらゆる場所において見られ, 原因は様々で, 交通機関・工場などが生み出す人工的振動と, 風・波浪などに励起される非人工的振動の 2 種に大別される》.

mileage, milage 《車のガソリン一定量当たりの》走行マイル数, 燃費: a used car with a small *mileage* 燃費がよくない中古車.

Millennium Development Goals ⇨ MDGs*

Millennium Ecosystem Assessment [the 〜] ミレニアム生態

系評価《国連の提唱により2001-5年に行われた地球規模の生態系調査》.

minimal confidence 《仮説などの》最小信頼度.

miombo 【生態】ミオンボ(林)《東アフリカの乾燥性疎林植生》.

mitigation 緩和(策)《環境への悪影響を回避, あるいは最小限にするために採用される行動; 気候変動問題においては, 温室効果ガスの人為的な排出を減らして, 大気中の温室効果ガスの濃度を安定化させる方策を意味する(⇨ global warming*)》.

mitigation banking 【米】ミティゲーション・バンキング《開発に伴う湿地帯などの環境破壊緩和を図る環境ビジネス制度の一つ; 当局の認可を受けた環境保全事業体であるミティゲーション・バンク(mitigation [conservation] bank)が保全管理する自然環境をHEP* などの手法で数値化してemission credit* と同様にクレジットの形にし, 他所の開発で失われる自然環境の代償として同量のクレジット購入を開発事業者に義務付けるもの; 土地開発においてcarbon offset* に似た役割を果たす》.

mitigation policy 《環境破壊の》緩和策(⇨ mitigation*).

mitigation potential 《温室効果ガスの排出などに対する》緩和ポテンシャル, 軽減ポテンシャル.

m.m. mutatis mutandis*

mobilize 動 …に移動性をもたせる;〈富などを〉流通させる;《軍隊を》動員する,〈産業・資源などを〉戦時体制にする;【生】〈染色体を〉起動させる.

MOC the meridional overturning circulation 子午面循環, 熱塩循環《中深層の世界的海洋循環の一つ》.

monocyclic 形【生態】単輪廻(りんね)性の.

monocycly 名 単輪廻.

Montreal Protocol [the ～]モントリオール議定書《オゾン層保護のための国際的な取り決め(1987); 正式名称は the Montreal Protocol on Substances that Deplete the Ozone Layer (オゾン層を破壊する物質に関するモントリオール議定書); オゾン層保護の基本的な枠組みであるウィーン条約(1985; Vienna Convention for the Protection of the Ozone Layer)のもとで策定された; 議定書の目的は, オゾン層を破壊するおそれのある物質を特定し, 当該物質の生産, 消費, 貿易を規制して人の健康や環境をまもること; 議定書により, 特定フロン, ハロン, 四塩化炭素などが1996年以降全廃となっている》.

montane 形〈動植物などが〉山地

の，低山帯に生育する；山地動植物の．
图《森林限界より下の》低山帯 (montane belt).

morbidity 病的状態[性質]；《一地方の》罹病[罹患]率 (morbidity rate).

Morris Minor 【商標】モリスマイナー《英国 Austin Morris 社が1950-70年代に製造した小型乗用車で，現在もしばしば高い値で取引されている；このエステートカーモデル[ワゴン型]の Morris Minor Traveller は環境保護運動の活動家がよく使っている》.

mortality 《地方別・年齢別・病気別の》死亡数；死亡率，《家畜の》斃死(へいし)率 (mortality rate).

moss forest コケ林，蘚苔(せんたい)林《熱帯地方の山岳部にみられる，湿度が高くてコケの着生の多い森林》.

moss layer 《植物群落の》コケ層，蘚苔(せんたい)層 (⇨ layer*).

MOU memorandum of understanding 覚書，協定書.

mousse ムース《海上石油流出の際に生ずる暗褐色の乳状流出油；chocolate mousse ともいう》.

MSC Marine Stewardship Council*

MSDS material safety data sheet*

multilateral 多国間の．

Mumford マンフォード Lewis Mumford (1895-1990)《米国の文明・社会批評家，都市研究家》.

Muskie Act [the 〜] マスキー法《米国の Edmund S. Muskie 上院議員が提案した Clean Air Act* of 1970 (1970年大気清浄法)の通称；とくに自動車の排気ガス規制で知られる》.

mutatis mutandis 副 必要な変更を加えて，準用して《略 m.m.》；個々の違いを考慮して: apply *mutatis mutandis* regarding changes in the Kyoto Protocol Target Achievement Plan 京都議定書目標達成計画の変更について準用する．

mutualism 【倫理学】相互扶助論；【生態】《2つの種間の》相利共生，相利作用 (⇨ commensalism*).

mutualist 图 相互扶助論者；相利共生生物.

mutualistic 形 相互扶助[相利共生]の.

myrmecophile 【生態】蟻巣(ぎそう)生生物，好蟻(こうぎ)生生物，蟻動物《アリと共生する，特に昆虫》.

myrmecophilous 形 アリを好む，アリと共生する．

myrmecophily 图 好蟻性，蟻共生．

N

namby 環境問題となりうる政策に断固反対している人, ナンビー《軍事施設や原子力発電所などの反対運動のスローガン not in anyone's backyard を頭字語ふうにもじったもの; ⇨ NIABY*, nimby*》.

NAPL non-aqueous phase liquid 非水相液体[流体]《油などの難水溶性の液体; 特に DNAPL* や LNAPL*》.

natant 圏 浮遊性の, 水に漂う, 遊泳する.

National Biodiversity Strategies and Action Plans 生物多様性国家戦略《生物多様性保持のため各国政府が定める長期的政策; 略 NBSAPs》

National Heritage ナショナルヘリテージ《英国の博物館などの機関が歴史的に価値のある物品を購入したり歴史的建造物を保守したりするのを援助するための基金 (the National Heritage Memorial Fund) を運用する組織》.

National Park Service [the 〜] 国立公園局《1916年に設置された米国内務省の一部局; 国立公園・文化財などを管理する; ⇨ National Park System*》.

National Park System [the 〜] 国立公園制度《米国の国立公園局 (National Park Service*) の管轄下にある制度; 風光にすぐれ, あるいは歴史的に意義のある土地を「国民の利益と楽しみのために」保護するのが目的; 国立公園思想は米国に発した; すなわち Wyoming 州 Yellowstone 川の源流地帯に関する 1870-71 年の記述が発端となって, 1872 年に Yellowstone 国立公園 (8990 km^2) を創設する法律が制定されたのが始まりである; 次いで 1890 年には California 州 Sequoia と Yosemite が国立公園に指定された; 今日主なものの内訳は, 単に national parks と命名されているもの 50 か所, 国定記念物 (national monuments) 約 80 か所, 国立歴史公園 (national historical parks) 31 か所, 国有記念物 (national memorials) 26 か所など》.

National Rivers Authority [the 〜] 全国河川局《英国で 1989 年に設置された, イングランドとウェールズにおける河川の管理を受け持つ政府の独立機関; 水質調査・汚染管理・水資源の運用・漁

業などについて管轄権を有する; 略 NRA》.

National Trust [the 〜]ナショナルトラスト《英国の自然保護, 史跡などの保存のための民間組織; 美しい自然や景観がのこる土地の所有者となり, その環境を永久に管理し守ることを目指している; ピーター・ラビット(Peter Rabbit)の生みの親である絵本作家ベアトリクス・ポター(Beatrix Potter)は湖水地方の土地を買い取り, National Trust に管理を委ねた; Hill* が 1895 年設立; 略 NT》.

National Trust for Scotland [the 〜]スコットランドナショナルトラスト《National Trust* の姉妹組織で, 1931 年にスコットランドのために設立された; National Trust のイングランド・ウェールズ・北アイルランドでの活動と呼応している; 会員数 10 万余》.

natural 形 天然の, 自然(食品)の: *natural* food 自然食品 / *natural* yeast 天然酵母 / *natural* enemy 天敵.

Natural England ナチュラル・イングランド《English Nature* が 2006 年に統合改組した英国イングランドの自然保護を司る独立政府機関; 国立公園, AONB*, NNR* などの指定・管理に携わる》.

natural regulation 自然調整《1) 人為を最低限にとどめた景観・生態系管理政策 2) 生態系の正常な発達による人口推移》.

natural regulator 自然という調整者.

Natural Resources Defense Council 天然資源保護協議会, 自然資源防衛協議会《環境や核問題に詳しい米国のシンクタンク》.

natural system 自然分類《生物の形質に基づいて自然群に分けた分類; 特にリンネ種を否定する 18 世紀フランスの植物学者ジュッシュー(Bernard de Jussieu)による植物自然分類》.

Nature Conservancy [The 〜]ネイチャーコンサーバンシー, 自然管理委員会《1951 年米国で設立され, 世界規模で自然保護活動を行なう非営利団体; 略 TNC》.

Nature Conservancy Council [the 〜]自然保護審議会《英国政府の環境保護機関; 1949 年設立; 1991 年 English Nature* などに分割された》.

nature conservation 自然保護.

nature reserve [preserve] 自然保護区.

nature trail 自然(遊)歩道, 自然研究[探勝]路.

NBSAPs National Biodiversity Strategies and Action Plans*

NCC Nature Conservancy Council*

NCRPM National Council on Radiation Protection and Measurements《米国の》全国放射能防護測定委員会.

Nearing ニアリング
(1) Helen Nearing (1904-95)《米国の作家;夫の Scott と共に「大地へ帰れ運動」を提唱》.
(2) Scott Nearing (1883-1983)《米国の社会主義的経済学者・思想家・文明批判家;第一次大戦時に反戦平和を唱え大衆運動を組織;第二次大戦以後は妻の Helen と共に「大地へ帰れ運動」(back-to-the-land movement)を提唱,環境保護運動の先駆者となる》.

near threatened species 準絶滅危惧種《現時点では絶滅危険度は小さいが,生息条件の変化によっては絶滅危惧種(threatened species)に移行する可能性のある生物種; ⇨ Red List*》.

necton ⇨ nekton*

needleleaf 形 針葉樹の繁茂する;針状葉の生えた: *needleleaf* trees 針葉樹.

nekton, nec- 遊泳生物, ネクトン《魚・鯨などのようにプランクトンに比べ遊泳力の強い大型水中動物》.
　nektonic 形 遊泳する,ネクトンの.

nemoral 形 森の[に関する];〈動物が〉森にすむ[よく行く].

neo-Malthusian 形 新マルサス主義の《産児制限などによる人口調節を主張する》.
　图 新マルサス主義者.
　neo-Malthusianism 图 新マルサス主義.

NEPA National Environmental Policy Act 国内環境政策法,ネパ《米国で 1969 年制定》.

neritic 形 浅海の,沿岸性の.

net zero (energy) 形《住宅・工場・施設などが再生可能エネルギーや資源リサイクルによって相殺して》化石燃料消費や温暖化ガス排出を(ほぼ)ゼロにした,エネルギー自給自足の,環境負荷のない.

neuston 【生態】水表生物.
　neustic 形 水表生物の.

NIABY not in anybody's backyard 誰の裏庭でもご免だ《環境破壊の恐れがある施設の建設に地域を越えて反対する場合のスローガン; ⇨ namby*, NIMBY*》.

niche ecological niche*

NIMBY, Nimby, nimby 形 图《俗》ニンビー(の),近所での建設反対者(の)《原子力発電所・軍事施設・刑務所など地域環境にとって好ましくないものが近所に設置されるのは反対という人,またそういう住民[地域]エゴの態度についていう; *not in my backyard* ((よそはともかく)うちの裏庭はだめだ)の略; ⇨ NIABY*》.
　nimbyism 图 地域住民エゴ,ニ

ンビー主義.
nimbyist 名形 地域エゴ的な(人).

Niño El Niño*

nitrate 名 硝酸塩[エステル]; ニトロセルロース(製品); 硝酸肥料《硝酸カリウム[硝酸ナトリウム]を主成分とする化学肥料》.
動 硝酸(塩)で処理する; 硝化する.

nitration 名 硝化, ニトロ化.

nitrous oxide 一酸化二窒素, 亜酸化窒素《無色の気体; 吸入すると快活になり, 無痛状態をもたらすので笑気とも呼ばれ, 歯科の麻酔剤とする; 燃焼によって生じる大気汚染物質》.

NMC, NMM nuclear material control [management] 核物質管理.

NNR National Nature Reserve 国定自然保護区《英国の国立公園やAONB*に準ずるものとして政府の自然保護機関が管理下に置く自然環境・野生生物保全地区; 全国で350箇所ほど指定されている》.

NOAA National Oceanic and Atmospheric Administration 海洋大気局《米国商務省の一局; 天候・海洋環境の観測・調査を行う; ⇨ ESI*》.

noise meter 騒音測定器.

noise pollution 騒音公害.

noise rating number NR数, NRN《工場・寝室などある条件下で人が耐えうる知覚騒音レベル(PNdB*)の上限値》.

non-climatic drivers 《環境に影響する》非気候的要因.

non-fluorocarbon 形 ノンフロンの, フロン(chlorofluorocarbon*)を用いない.

nonhost 非宿主植物《他の生物による攻撃・寄生をうけない》.

non-point 形 発生地を特定できない, 【環境】汚染源を特定できない (⇨ point source pollution*).

nonpolluting 形 汚染しない, 無公害性の.

nonrenewable 形〈資源などが〉再生不可能な(⇨ renewable source*).

no-nuke antinuke*

nonylphenol ノニルフェノール《界面活性剤・有機原料; 内分泌攪乱性が疑われている》.

noosphere 【生態】人智圏, 人間生活圏《人間の活動による変化が著しい生物圏; anthrosphere, anthroposere とも言う》.

no-return 形 使い捨ての〈瓶〉, 返却する必要がない.

NPS 1 National Park Service. 2 nuclear power station [system] 原子力発電所[発電システム].

NPS non-point source (pollution) ⇨ point source pollution*

NRA 1 National Recreation

Area 国定レクリエーション地域《米国で貯水池・渓谷・海岸などに設けられ, National Park Service* などが保護管理する保養地区》.
2 National Rivers Authority*.

NRC Nuclear Regulatory Commission*

NRDC Natural Resources Defense Council*

NT National Trust*

nuclear 形 原子力の, 原子力に関係する, 核(分裂)の, 核兵器の: *nuclear* energy 核エネルギー, 原子力 / *nuclear* war 核戦争 / *nuclear* (power) plant 原子力発電所, 原発 / *nuclear* reactor [pile] 原子炉

nuclear autumn 核の秋《核戦争による大規模な核爆発や火災によって, 大気圏に煤(すす)や塵(ちり)が巻き上げられた結果, 太陽光がさえぎられ, 地球全体の気温が低下して, 農業に大打撃を与える状態; 1983年, 氷点下に気温が下がるとする'核の冬'(nuclear winter*)の仮説が提示されたが, その後の研究の結果, より正確には'核の秋'というべきものの到来が予測されるようになった》.

nuclear-free 形 非核の: *nuclear-free* zone 非核地帯.

nuclearize 動 …に核兵器[原子力]を備える, 核兵器[原子力]使用に転じさせる, 核保有国とする.

nuclearization 名 核化, 核保有.

Nuclear Regulatory Commission [the 〜]《米国の》原子力規制委員会《1975年発足; 略 NRC》.

nuclear reprocessing 《原子力発電所から出る》使用済み核燃料の再処理.

nuclear safeguard 核保障措置《IAEA* により考案された, 原子力(核)の軍事利用への転用を防止する国際的保障措置; 原子力の平和利用に関する定期的報告書の提出と, IAEA による原子力(核)施設への立ち入り査察を柱とする; 核拡散防止条約によりこの措置の枠組みが与えられている》.

Nuclear Test-Ban Treaty [the 〜] 核実験禁止条約《1963年に英国・ソ連・米国の間で調印された大気圏内・宇宙空間・水中での核実験を禁止する条約; 加盟118か国, フランス・中国は未加盟》.

nuclear waste 核廃棄物, 放射性廃棄物(radioactive waste);《原子炉から出た》使用済み核燃料.

nuclear winter 核の冬《核戦争が起きてしまうと, 大規模な核爆発や火災が生じさせる多量の大気中の塵によって太陽光線が地表に届かなくなり, 気温が氷点下に極度に低下し, 生物が死に絶えるという仮説; ⇨ nuclear autumn*》.

nucleus counter 塵埃計(dust

counter*), 計塵器.

nuisance 害, 有害物, 困った事情, 迷惑な行為; 生活［不法］妨害, ニューサンス(⇨ public nuisance*)《煤煙, 騒音, 悪臭などによって他人の財産上の利益や安寧な生活を侵害する行為》: Commit no *nuisance*!《掲示》小便無用, ごみを捨てるべからず.

nutrient loading 《富栄養化の原因となる》栄養塩負荷, 栄養塩流出.

O

OBD on-board diagnosis 車載式故障診断システム《自動車の排ガス対策としてエンジンや排気部の自己診断・モニターを行う装置；異常を感知するとデータが車載コンピューターに蓄積され，警告ランプが点灯する；車内のコネクターにテストツールを接続するか，無線で携帯電話やパソコンへ送信してデータを読み取り，故障を迅速に診断できる；1990年代から第二世代システム OBD II の設置義務化が進み，日本でも2008年を皮切りに順次搭載が義務化されつつある》.

obligate, obligatory 形 〈寄生菌・寄生虫などが〉ある特定の環境にのみ生活しうる，無条件的な，絶対的な，偏性の，真正の(⇔ facultative*)：an *obligate* parasite 絶対寄生体[菌].

occupant 【生態】占有[居住]動物，現住種.

occurrence 【生態】出現.

ocean acidification 海洋酸性化.

ocean incineration [burning] 海洋ごみ焼却《ごみをはしけで海上に運び，はしけの焼却炉で処理すること》.

ocean sequestration 海洋隔離《大気中の二酸化炭素濃度上昇を抑えるために二酸化炭素を海に封じ込める技術；⇨ carbon sequestration*》.

OD oxygen demand (⇨ biochemical oxygen demand*).

ODA Official Development Assistance 政府開発援助.

oecology ⇨ ecology*

offgas 图 発生気体，オフガス《化学反応の際に排出される気体》. 動 (有毒)ガスを発生する.

off-grid, off-the-grid 形 副 電気[水道など]の来て[引いて]いない，自家発電[自家用水など]で(自給自足して).

O horizon O層《森林下地表面に落葉・落枝が集積して20-30％以上の有機物を含む層；O は Organic* の略》.

oiling, oil spill 流出石油による汚染(⇨ OPRC convention*).

oil slick 《海洋・湖水などに浮いた石油の》油膜.

old growth 1 原生林，処女林. 2 老齢林.

olf オルフ《タバコの煙や呼気など人が吐き出す臭気や体臭の測定単位；平均1日0.7回入浴する事務職労働者によって放出される空気

oligotrophic 形【生態】〈湖沼・河川が〉貧栄養の(⇨ eutrophic*, mesotrophic*).

oligotrophy 名 貧栄養.

Only One Earth かけがえのない地球《1972年の国連人間環境会議 (United Nations Conference on the Human Environment*)での合い言葉》.

OPRC convention [the ～] OPRC条約, 油濁事故対策協力条約(the International Convention on Oil Pollution Preparedness, Response and Co-operation)《1989年に米国アラスカ沿岸で起きた大型タンカー, エクソン・バルディーズ号(Exxon Valdez*)の油流出事故を契機につくられた; 油による汚染についての準備・対応・協力が柱; ⇨ Valdez Principles*》.

organic 形 有機体の, 生物の, 有機化学の; 有機農法の[による], 有機飼育の[による], 自然(食品)の《化学肥料・農薬・生長促進剤・抗生物質などを用いない》; 〈商店が〉自然食品を扱う; 〈生き方が〉簡素で自然に即した: *organic* evolution 生物進化 / *organic* compound 有機化合物 / *organic* fertilizer 有機肥料 / *organic* farming 有機農法 / *organic* vegetables 有機野菜 / *organic* foods 自然食品 / *organic* chicken [meat] 有機飼育の鶏[食肉].
名 有機肥料; 有機農薬[殺虫剤].

organically 副 有機的に; 有機栽培[農法]によって.

organicity 名 有機性.

organotin 形 有機スズの: *organotin* compound 有機スズ化合物《農薬・プラスチックの安定剤などとして用いられる》.

orphan site 孤立地《地主などが責任をとらない汚染地域; 自治体当局などが汚染除去を肩代わりせざるを得ない場合が多い》.

Osborn オズボーン(Henry) Fairfield Osborn (1887-1969)《米国の環境保護運動家; 同名の父(古生物学者)の跡を継いで New York Zoological Society の会長をつとめながら *Our Plundered Planet* (1948) や *The Limits of Our Earth* (1953) などの著作によって環境破壊の禁止を世に訴えた》.

OTEC 《海洋の表層と深層の温度差を利用した》海洋熱エネルギー変換 (*O*cean *T*hermal *E*nergy *C*onversion) (方式の発電所), 海洋温度差発電, オテック.

Our Common Future 我ら共通の未来《国連の「環境と開発に関する世界委員会」(WCED*)が1987年にまとめた報告書の名前;「持続可能な開発」(sustainable development*)の概念が提唱さ

overcut 動 切り[カットし]すぎる；《特に》〈森から〉木を(年間の生長や割当量以上に)切り[伐採し]すぎる，過剰伐採する．

overdraft 《地下水の》汲み上げすぎ．

overdrafting 過剰排出《雨や融雪水によって供給されるよりも多くの水を地中・ため池から排出すること》．

overexploitation 《天然資源の》乱開発，乱獲．

overfish 動 〈漁場〉の魚を乱獲する，〈特定種を〉乱獲する．

overfishing 名 乱獲

oxidant smog オキシダント[光化学]スモッグ(photochemical smog*)．

oxygen cycle 酸素循環《大気中の酸素が動物の呼吸で二酸化炭素になり，光合成によって再び酸素となる循環》．

oxygen demand biochemical oxygen demand*

ozone オゾン《酸素原子3つが結合している分子》．

ozone-benign 形 ozone-friendly*

ozone depleter オゾン破壊物質．

ozone depletion オゾン[量]減少，オゾン層破壊《冷媒として使用しているクロロフルオロカーボン類(CFC*s；通称フロン)などが廃棄され大気中に放出されると，オゾンと反応を起こしてオゾンの量が減少する；オゾン層が破壊されると，有害な紫外線から地上の生物を守るオゾン層の機能が弱くなる；人類が産業活動で利用してきた便利な物質にオゾン層破壊採用があることから，フロンなどの生産・使用を禁止する国際的な動きにつながった；⇨ Vienna Convention*, Montreal Protocol*》．

ozone-friendly 形 〈工業製品が〉オゾン層を破壊しない化合物を使用している，オゾン(層)にやさしい，オゾン層を破壊しない《フロン(CFC*)などを使っていない》．

ozone hole オゾンホール《フロン(CFC*)などが原因で生じる，成層圏オゾン層のオゾン濃度が極端に低下した部分；濃度分布の図を描くと穴があいているように見えることから，この名前が付いた；オゾン濃度が低いと，有害な紫外線を吸収することで地上の生物を保護するオゾン層の機能が弱くなる；オゾンホールは春先(南半球)の南極上空で観測される》．

ozone layer オゾン層《地上10～50キロの上空に分布する，オゾン濃度の高い領域》．

ozonometry 《大気中の》オゾン測定．

P

Pachauri ラジェンドラ・クマール・パチャウリ Rajendra Kumar Pachauri (1940-)《インド出身の環境学者；気候変動に関する政府間パネル(IPCC*)の第3代議長(2002-)；在職時の2007年，IPCCがゴア(Gore*)米元副大統領とともにノーベル平和賞を受賞した》.

Pacific Decadal Oscillation [the 〜] 太平洋十年規模変動[振動]《北太平洋の海面温度の周期的な変動；1998年に提唱された；略PDO》.

paleoclimate 古気候《地質時代の気候》.

paleoecology 古生態学.
  paleoecologist 古生態学者.
  paleoecologic(al) 形 古生態学(上)の，古生態学的な.

paleoenvironment 古環境《人類出現前の海洋および大陸の環境》.
  paleoenvironmental 形 古環境の.

paleopedology 古土壌学.

paleosol 古土壌《地質時代にできた土壌》.

paleotemperature 古温度《先史時代の海洋などの温度》.

paludicolous, -dicoline,
paludicole palustrine*
paludose 形 沼地にすむ[生える].
palustral 形 沼地の[にいる]；沼地の(多い)；沼地から発生する.
palustrine 形 沼地に生息する，沼地生の.

parabolic trough ⇨ CSP*

paralimnion 周辺湿地部《水底に根を有する植物の占める湖の植生帯》.

parallelism 1【生態】《単系統的に分かれた2系統間の》平行現象. 2【生物】平行進化.

parasitism 【生態】寄生(生活) (⇨ symbiosis*).

particulate matter 粒子状物質, PM, 固体や液体の粒《工場などからの煤塵やものを砕いて発生する粉塵，ディーゼル車排ガス中の煤(すす)など；特に粒の大きさが10ミクロン以下の微小粒子を浮遊粒子状物質(SPM*)と呼ぶ；しばしば大気汚染源となる》.

party 当事国，当事者.

passive smoking 《他人のタバコの煙を吸ってしまう》受動[間接]喫煙(⇨ secondhand smoke*).

Passmore パスモア John (Arthur) Passmore (1914-2004)《オーストラリアの哲学

者・歴史家；著作には近現代西洋哲学史の名著とされる *A Hundred Years of Philosophy*(1957; 補遺版 *Recent Philosophers* (1985) のみ邦訳『哲学の小さな学校』)や環境倫理学を論じた *Man's Responsibility for Nature* (1974, 邦訳『自然に対する人間の責任』) など).

Patterson パターソン Clair Cameron Patterson (1922-95)《米国の地球化学者；1953年に地球の年代を46億年と最初に測定した；鉛汚染などの環境問題の研究でも高い評価がある》.

pay-as-you-throw 形《ごみ収集が，量に応じて》有料の《略 PAYT》.

PAYD Pay As You Drive【商標】PAYD(ペイド)，走行距離連動型保険《GPSなどを応用して自動車の走行距離や運転状況を把握し，契約者の運転リスクに応じて保険料を定める方式の自動車保険；短時間・短距離・低速度といった低リスク運転であれば保険料が下がり，また環境面からも車の利用抑制につながるといった効果が期待されている》.

PBB polybrominated biphenyl*

PBDE polybrominated diphenylether*

PCB polychlorinated biphenyl*

PCF Prototype Carbon Fund*

PCT 1 polychlorinated terphenyl ポリ塩化テルフェニール《殺虫剤；PCB* とともに環境汚染物質の一つ》.

2 peak clad temperature《原子炉の》燃料被覆最高温度.

PDO Pacific Decadal Oscillation*

peak oil ピークオイル《21世紀になって石油産出量がピークに達したあと徐々に減り，現在のように安価な石油エネルギーに頼った文明は立ち行かなくなる恐れがあるという説；2010～20年の間か，あるいは2030年頃にピークが訪れるという見方がある一方で，ピークは来ないという見方もある；石油の究極可採埋蔵量をどう予測するかで悲観論と楽観論に分かれる》.

perceived noise decibel PNデシベル《知覚騒音レベルの単位；略 PNdB, PNdb》.

perceived noise level 知覚騒音レベル(⇨ PNL*).

periphyton 【生態】植物表面生物，付着生物《水中の基物に付着》.
  periphytic 形 表面[付着]生物の.

permafrost 《寒帯・亜寒帯の》永久凍土層.

permissible yield allowable cut*

pest 害虫，病害虫，有害生物[小動物]：a garden *pest* 庭を荒らす

害獣［害虫］．

pesticide-free 形 無農薬の：*pesticide-free* vegetables ［farming］無農薬野菜［農業］．

PHEV plug-in hybrid* electric vehicle.

phosphate 名 燐酸塩［エステル］；燐酸肥料；一種の炭酸水《燐酸を少量含む》．
動 燐酸(塩)で処理する．

photobathic 形 〈海水が〉太陽光線の届く深さの．

photochemical smog [haze] 光化学スモッグ．

photodegradable 形 〈プラスチック・殺虫剤などが〉光分解性の．

photoenvironment 【生態】光環境，明環境．

phreatophyte 地下水植物《地下水やそのすぐ上の地層から水を得る根の深い植物》．
  phreatophytic 形 地下水に根を張る，地下水植物の．

PHV plug-in hybrid* vehicle.

phylloplane 葉面《微生物などの生息環境としての葉の表面》．

physiognomy 【生態】相観；《物などの》外的特徴，外観．

physiographic climax 【生態】地形的極相(⇨ edaphic climax*)．

physiological atmosphere ecosphere*

phytobenthos 底生植物《水底に生活する植物》．

phytobiology 植物生物学［生態学］．

phytoremediation ファイトレメディエーション《カドミウムなどの重金属に汚染された土壌に，重金属を吸収する植物を栽培して土壌を浄化［改善］する技術；植物が生育した後，収穫して焼却処分する；二次汚染の心配が少ない，比較的コストがかからないという利点を持つ；ファイトは「植物の」を表す接頭語》．

picloram ピクロラム《除草剤，強力で持続性の高い枯れ葉剤；米軍がベトナム戦争で使用》．

Pigo(u)vian tax ピグー税，厚生経済学的な税《external diseconomy (⇨ externality*) などの社会的費用(social cost)を税の形で負担させるもの；carbon tax*, pollution tax* など；英国の経済学者 Arthur Cecil Pigou (1877-1959) の厚生経済学に由来》．

pioneer 《裸地に最初に侵入定着する》先駆動物［植物］．

PLA polylactic acid*

plagioclimax, plagioclima 偏向的極相《人為などなんらかの外的要因によって，動植物の群集の一次遷移系列がゆがめられたまま安定した極相》．

planetism 地球環境保護(主義)．

pleuston 浮表生物，プロイストン《水面に生活するウキクサ・カツオノエボシなど》．

pleustonic 形 水面に浮かんで生きる，浮漂生物の．

plot 小地面，小区画；《生態学などの》調査区．

plug-in hybrid 充電式ハイブリッド車，プラグインハイブリッド車 (plug-in hybrid (electric) vehicle; 略 PHV, PHEV)《家庭用のコンセントで直接充電できるハイブリッド車；普通のハイブリッド車よりバッテリー容量が大きく，電気だけで長距離を走行でき，ガソリンを節約できる；電気自動車により近い； ⇨ V2G* (V の最初の見出し)》．

PM particulate matter* 粒子状物質．

PNdB, PNdb perceived noise decibel(s)*

PNL perceived noise level 感覚騒音レベル《航空機騒音の最も一般的な表現方法；各周波数幅の瞬間の最高値を示す； ⇨ EPNL*, WECPNL*》．

poach 動 1 密猟[密漁]する；《密猟・密漁のため》〈人の土地など〉に侵入する: *poach* hares 野ウサギを密猟する / go out *poaching* 密猟[密漁]に出る / *poach* for game 猟獣[猟鳥]の密猟をする / *poach* on a neighbor's land 隣人の土地に侵入する．
2 〈地面を〉踏み荒らしてぐしゃぐしゃ[穴だらけ]にする；〈粘土など〉に水を加えて均質にする；〈道などが〉踏みつけられてぬかる[穴などができる，へこむ]；〈足などが〉ぬかるみに沈む．

podzolization, podosolization ポドゾル化《湿潤寒冷気候のもとで，土壌の表層において塩基や鉄・アルミニウムが溶脱し，これらが浸透水とともに下層に運ばれて集積する作用》．

podzolize 動 ポドソル化する[させる]．

point source pollution [polluter] 《工場など》特定(できる)汚染源(による環境汚染・公害)《これに対し一般家庭など不特定の汚染(源)を non-point source pollution [polluter] と呼ぶ》．

pollination 授粉[受粉](作用)，花粉媒介，送粉．

pollutant 汚染物，汚染(物)質．

Pollutant Standards Index 《米国の》汚染基準指標．

pollute 動 〈大気・水・土壌などを〉よごす，不潔にする；(特に廃棄物などで)〈環境を〉汚染する．

polluter 图 汚染者，汚染源(⇨ point source polluter*, PPP*)．

pollutive 形 汚染を引き起こす，汚染源となる．

pollution よごすこと，よごれ，汚染，環境汚染，公害，汚濁，不潔；汚染物質(pollutant)《他に contamination, public nuisance* などが日本語の「公害」に当たり，pollution は主に化学物質などに

よる環境の汚染・破壊一般, contamination は特に放射性物質・有毒物質などによる見えにくい汚染, public nuisance は騒音・振動・悪臭・糞尿などによる生活環境破壊を指す》: air *pollution* 大気汚染.

pollution tax 公害税(⇨ Pigovian tax*).

polybrominated biphenyl ポリ臭化ビフェニル《毒性の強い汚染物質; 略 PBB》.

polybrominated diphenylether ポリ臭化ジフェニルエーテル《有害化学物質である難燃剤; 略 PBDE》.

polychlorinated biphenyl ポリ塩化ビフェニル, PCB《ビフェニルの1個以上の水素を塩素で置換して得られる合成有機化合物; 複写用油剤, 変圧器の絶縁油として用いられたが, 有害な環境汚染物質で生体内に蓄積される; 略 PCB》.

polylactic acid, polyactide ポリ乳酸《乳酸がエステル結合で重合した生物分解性プラスチック; 原料のデンプンがトウモロコシなどから得られる植物由来プラスチックでもあり, もともと大気中の二酸化炭素を固定してつくられている(carbon-neutral*)ために, 燃やしても大気中の二酸化炭素増加につながらない; 乳酸を環状二重エステルにし, その開環重合によりつくられる; 略 PLA》.

polysaprobic 形【生態】分解した有機質が多く遊離酸素がない水の中で生活する, 多腐生水中生活の.

polyuronide ポリウロナイド《土地改良に役立つ腐植(土)中の天然樹脂》.

POPs Convention ストックホルム条約, POPs 条約(⇨ Stockholm Convention*).

population 《ある地域内などの生物の》個体群, 生息数, 個体数, 集団.

populational 形 個体群[集団]の.

population control 【生態】個体群制御.

population dispersal 【生態】個体群散布[分散].

population dynamics 【生物・遺伝学】個体群ダイナミックス, 個体群動態学[論], 集団動態.

postclimax 【生態】後極相, 後安定期《局部的好気候によって, そこの群落が周囲の安定群落よりも少し進んだ段階にある状態; ⇨ preclimax*》.

poverty alleviation 貧困の解消.

poverty eradication 貧困の撲滅.

PPD paraphenylenediamine パラフェニレンジアミン《フェニレンジアミンの異性体で, 毛髪染料などに使われる; アレルギー原性が強い》.

PPP polluter pays principle 汚染者負担の原則《公害防止や環境対策の費用は,汚染の原因者が負担すべきとする考え方; OECD(経済協力開発機構)が1972年に「環境政策の国際経済面に関する指導原理」で提唱し,その後,各国の環境政策の基本として採用された》.

precautionary principle 予防原則《化学物質や人間活動などが環境に与える影響について,深刻な,あるいは不可逆的な被害のおそれがある場合には,その因果関係を示す決定的な科学的な証拠がない場合でも,環境への悪影響の防止を確保しようという原則; Rio Declaration* に盛り込まれた; ⇨ prevention principle*》.

precipitation 投下,落下; 降水(量),降雨量; 沈積; 沈澱; 析出; 【免疫学】沈降.

Precipitron 【商標】プレシピトロン《静電気を利用した空気清浄機; ほこり・花粉・煙などの空中浮遊粒子をイオン化して器内の帯電吸着板に凝結する》.

preclimax 【生態】前極相,前安定期《局部的な気候の不適な方への変化によって,そこの群落が周囲の安定群落よりも一歩前の段階にある状態; ⇨ postclimax*》.

precycling プレサイクリング,予備リサイクル《無駄な買い物・消費を控え,またごみの出にくい商品を選ぶなど消費行動を改善してごみや環境への悪影響を減らすこと》.

predation pressure 【生態】弱小動物がこれを食する捕食者のために種の保存をおびやかされること.

predator 捕食者.

predator-prey 形 捕食者と被食者との(生態的平衡に関する): *predator-prey* relationships [interactions] 食うものと食われるものの関係[相互作用], 捕食者-被食者関係[相互作用] (⇨ food chain*, primary [secondary] consumer*).

prescribed 定められた,規定された.

prescribed cut allowable cut*

preservator 自然観光資源保護官.

preserve 動〈危害・腐朽などから〉保護する,保存する; 禁猟(地)とする.
图 禁猟地[流域]; 《米国の》自然資源保護区域; 生簀(いけす).

Pressurized Water Reactor 加圧水(型原子[軽水])炉.

prevention principle 未然防止《化学物質や人間活動などが環境に与える悪影響についてその因果関係が科学的に証明されている場合に,悪影響を未然に防ぐための規制を実施すること; 予防原則 (precautionary principle*) とは意味が異なるとされる》.

primary consumer 第一次消費者《植物を食べる草食動物；食物連鎖のなかで第二次消費者に食べられる》.

primary forest 一次林, 原生林.

primary producer 《食物連鎖における》一次生産者《光合成により有機物を生産する緑色植物；primary consumer* に食われる》.

primary production 一次生産《光合成生物による有機物の生産》.

primitive area 《米国国有林内の》原生林保護地域《防火措置以外の一切の人工処置をしない地域》.

prisere 【生態】一次遷移系列《植物の全くない裸地から極相に至るまでの遷移；⇨ subsere*》.

proactive 形 1 順向[前進]の《前に記憶[学習]したことによって次の記憶[学習]が妨害されること》. 2 先を見越して行動する[行なう].

proclimax 【生態】準極相《極相に近い形の植物群落》.

producer 【生態】生産者(⇨ consumer*, decomposer*).

production 【生態】(生物学的)生産(量).

profundal 形 深底帯の《水温に成層現象がみられる湖水で, 深水層の上部から湖底まで, 特に深い湖では水深 600 m までについていう》.

projected 予定の, 計画的な; 見積もられた.

protocol 議定書(⇨ Kyoto Protocol*, Montreal Protocol*).

protocooperation 【生態】原始協同《双方にとって必須でない相利共生》.

Prototype Carbon Fund [the〜] (プロトタイプ)炭素基金《地球温暖化防止の取り組みのために, 2000 年世界銀行内に設立された炭素基金; 先進国の政府・企業の拠出金を途上国や移行経済国の温室効果ガス削減プロジェクトに投資する; 温室効果ガスの削減分は, 炭素クレジット(carbon credit*)として出資者に返還される; 略 PCF》.

PRTR pollutant release and transfer register 環境汚染物質排出移動登録[届け出](制度)《工場などが環境中に排出・移動[運び出しや流し出し]をした有害化学物質の量のデータを自ら把握して行政機関に届け出, 行政が集計・公表する仕組み》.

psammon 砂地間隙生物群集, 砂地生物群《淡水の水際の砂粒の隙間にすむ水生生物》.

psammosere 【生態】砂地遷移系列《不安定な砂地から始まる遷移系列》.

PSD Prevention of Significant Deterioration 重大な環境悪化の防止《米国環境保護局(EPA*)の環境汚染測定の基準として用いる》.

public awareness 市民の問題意

識.

psychrometry （空気)湿度計測学[法]，サイクロメトリー《空気中の水蒸気の含有量を測定し，それを支配している自然の法則を研究する》.

public nuisance 1《個人でなく地域社会全体に対する》公的不法妨害《公道の通行妨害・騒音[悪臭]公害など； ⇨ pollution*》. 2《俗》世間の厄介もの．

pyrithione zinc ピリチオン亜鉛，ジンクピリチオン (⇨ zinc pyrithione*).

pyrolytic incinerator 熱分解焼却炉《酸素の消費が少なく，汚染を減じ燃焼に適したガスを発生する》.

Q・R

quadrat 方形区，枠，コドラート《植生調査などで設定される方形の区画あるいは面積》.

quantified emission limitation 《温室効果ガスの》数量化された排出量抑制.

quiescent tank （下水などの）沈澱槽《汚物を一時留めて置くタンク》.

r 【生態】intrinsic rate of natural increase 内的自然増加率.

radiative forcing 放射強制力《地球に入射するエネルギーと地球から宇宙に放射されるエネルギーの均衡を破る効果；正なら地球の温暖化，負なら寒冷化につながる効果を表わす；温室効果ガス(greenhouse gas*)は正の放射強制力をもつ；単位 W/m^2》.

radioactive 形 放射性の，放射能をもった: *radioactive* contamination 放射能汚染 / *radioactive* rays 放射線.

radioactivity 1 放射能[性]《不安定な原始核が放射線を放出して安定な原子核になろうとする性質》2《慣用的に》放射能を持つ物質，放射性物質(radioactive material).

radioactivity [radioactive] concentration 放射性物質濃度，放射能濃度《単位体積または単位重量中の放射能；単位 Bq/cm^3, Bq/g など；特定の放射性物質による汚染を考えるときには cesium concentration (セシウム濃度) などともいえる》.

radioactive fallout 放射性降下物《死の灰》.

radioactive waste 放射性廃棄物(nuclear waste*).

radioecology 放射線生態学《生物に対する放射能の影響を研究する》.

Rainbow Warrior [the 〜] 虹の戦士号《Greenpeace* の旗艦；1985 年 7 月 10 日，ニュージーランドの Auckland 港停泊中に，フランスの情報機関員によって不法に爆破された》.

rain forest 多雨林，降雨林，《特に》熱帯雨林(tropical rain forest).

rainwater harvesting 《水資源としての》雨水貯蔵.

raised bog 隆起沼，隆起湿原《ミズゴケなどの湿地性植物が酸性水の中で繁殖し凸状に隆起した湿原》.

Ramsar Convention [the 〜] ラ

ムサール条約,国際湿地条約《1971年にイランのRamsarで採択された「特に水鳥の生息地として国際的に重要な湿地に関する条約(the Convention on Wetlands of International Importance Especially as Waterfowl Habitat)」;水鳥や渡り鳥の生息地として重要な湿地の保全を目的とする;日本では,釧路湿原,谷津干潟,琵琶湖,藤前干潟などが登録湿地となっている》.

Randolph ランドルフ Theron G. Randolph (1906-95)《米国のアレルギー専門医;環境医学の分野の草分け》.

ratification 《条約の》批准.

raw sludge 生スラッジ,生汚泥.

RCRA Resource Conservation and Recovery Act 資源保護[保全]回収法《有害廃棄物の処分などを規制する米国の法律;1976年成立》.

RDF Refuse-Derived Fuel ごみ固形燃料,廃棄物固形燃料《廃棄物等から可燃性物質を取り出し,粉砕・乾燥・成形などの加工をする》.

REACH Registration, Evaluation, Authorisation and Restriction of Chemicals REACH (リーチ)《2007年に発効した欧州連合の工業化学物質の総合規制法;欧州連合域内で規制対象となる化学物質製造・輸入の登録・安全性評価・認可・規制の手続きを定め,実施機関としてECHA* を設立した; ⇨ SVHC*》.

reafforest(ation) 《英》reforest(ation)*

reclamation 1 耕地化,開拓,干拓,土地改良.
2 (廃棄物の)再生利用(のための回収).
3 《動物の》馴致.

recreationist 《国立公園・海岸・湖沼などの》レクリエーションのための自然保護主義者.

REC(S) renewable energy certificate [credit] (system) グリーン電力証書《再生可能エネルギーによって得られた電力であることを第三者機関が認証する証書;企業などがグリーン電力を購入した証明書となる; ⇨ green certificate*》.

recyclable waste 資源ごみ《加工して再利用が可能なごみ》.

recycle 動 再循環させる;循環[再生]処理する,再生利用する;回収[再生]する;再利用[再使用,リサイクル]する,修復[改修]する《特に「再資源化」,すなわち使った製品や容器を,新しい製品の原料になるように処理すること; ⇨ three R's*, recyclable waste*》.
图 再循環,リサイクル.

recycler 图 リサイクル実施者.

recycled paper 再生紙.

recycling 再循環, 再利用, リサイクリング, リサイクル: a *recycling* plant リサイクル工場.

red-bag 形《感染性の病菌で汚染された》医療ごみ[廃棄物](用)の. 動〈医療ごみを〉赤い袋に入れて廃棄する.

REDD Reducing Emissions from Deforestation and Degradation (in developing countries)(途上国における)森林減少と森林劣化からの(温室効果ガス)排出削減《森林の減少や劣化を防止して二酸化炭素などの温室効果ガスの排出を削減しようとする地球温暖化対策の1つ; 排出削減量は炭素削減クレジット(emission credit*)とみることができる》.

red data book 『レッドデータブック』《絶滅のおそれのある野生動植物の生息状況・減少原因などを種別に記した資料集; 国際自然保護連合(IUCN*)が中心となって詳しく記した資料集; 日本版は環境省が作成》.

Red List [the 〜]レッドリスト《red data book* を拡充して国際自然保護連合(IUCN*)が発表する絶滅のおそれがある野生動植物のリスト; 正式名称は the IUCN Red List of Threatened Species; 初回公表は 1966 年で, 2006 年以降は毎年公表されている; 絶滅の危険度によって EX (extinct 絶滅), EW (extinct in the wild 野生絶滅), CR (critically endangered 絶滅危惧 IA 類), EN (endangered 絶滅危惧 IB 類), VU (vulnerable 絶滅危惧 II 類), NT (near threatened 準絶滅危惧), LC (least concern 軽度懸念)のカテゴリーに区分される; 他の組織や国・自治体が発表するものもある》.

Red List Index レッドリストインデックス《Red List* に基づいて作成される生物種別の危機状況指標; 略 RLI》.

red tide, red water 赤潮(⇨ bloom*).

reduce 動《廃棄物を》減らす, 減量する, リデュース[発生量抑制]する《⇨ three R's*》.

reducer 【生態】還元者;(有機物の)分解者(decomposer*).

reduction 1 縮小,《排出物などの》削減, 減少量;《細胞の》還元[減数]分裂《特にその第 1 段階》. 2【化】還元(法).

reduction commitment 《温室効果ガスなどの排出》削減の約束.

reed bed ヨシ湿原《ヨシが優占する沼地[水域]》.

reforest 動《米》〈土地〉に森林を復活させる, 再び森にする.
  reforestation 图 森林復活[再生].

refuge 鳥獣保護地区;《街路の》安全地帯.

refugium (複数形 refugia) レフュジア《氷河期のような大陸全体の気候の変化期に, 比較的気候の変化が少なくほかの場所では絶滅した種が生き残った地域》.

refuse ごみ, くず, 廃棄物.

refuse-to-energy 形 ごみを燃やしてエネルギーを生み出す (waste-to-energy*; ⇨ RDF*).

refuse worker 《英》廃棄物処理作業員, ごみ収集人.

regenerative brake 回生ブレーキ《電車や電気自動車, ハイブリッドカーなどに搭載した電気モーターを発電機として駆動することで得られる回転抵抗を利用したブレーキ; 発生した電力は熱として放散せず, 電線や蓄電池に回収して再利用する》.

regime 《気候・自然現象などの》(一定の)型;《河川における》水量の状況;《ある現象[相]が支配的になるための温度・周波数などの》領域.

regional haze 地域煙霧《大気汚染により広範囲にわたって広がる靄; 米国では国立公園などの視界・景観を損ねるとして環境保護局(EPA*)が Haze Rule を定めて規制し, 各州に発電所などの BART* と称する改良技術導入計画を義務付けている》.

regionality 局所性;【生態】部域性.

regreen 動〈土地を〉再緑化する.

relict, relic 残存生物, 遺存[残存]種《環境の変化で限られた地域に取り残された生物[種]》.

relocate 動 再び配置する, 配置しなおす, 移転[移動]させる[する].
relocation 图 移転, 移動, 再配置.

renewable energy 再生可能エネルギー《核燃料・化石燃料などではなく, バイオマス(biomass*)・太陽・風・波力・水力などから得られる自然エネルギー; 自然に由来し, 理論上は無尽蔵; ⇨ green certificate*》.

renewable portfolio standard 新エネルギー利用の割合基準《電力会社に対し一定割合の太陽・風力など再生可能エネルギーによる発電買い取りを義務付ける基準; 日本でも RPS 法(⇨ 和英篇末尾)で制定されている; ⇨ feed-in tariff*》.

renewable resource 再生可能資源《木材・生物・淡水などの, 理論的に繰り返し使用できる資源; ⇨ renewable energy*》.

reproductive isolation 生殖隔離《生理学的・行動学的相違や地理上の障害のために潜在的に交配可能な集団どうしの交雑が妨げられること》.

repulp 動〈パルプ製品を〉再パルプ化する.

reserve 特別保留地, 保護区(⇨ nature reserve*); reservation*: a forest *reserve* 森林保護区, 保護林.

reservation 《アメリカ先住民・オーストラリア先住民などのための》政府指定保留地, 居留地;《学校・森林などに用いる》公共保留(地);禁猟[休猟]地, 保護繁殖地: a grass *reservation*《英国の中央分離帯または道路わきの》芝生(用)地帯.

reservoir 图 1 貯蔵所;貯水池, 給水所;貯蔵器;(貯)水槽, ガスだめ: a depositing [settling] *reservoir* 沈澱池 / a receiving *reservoir* 集水池 / a storing *reservoir* 貯水池.
2《体内の》貯蔵器;病原体保有者, 保有宿主《病原体を宿しているが, 自分は病に冒されない》. 動 貯水所[貯蔵地]にたくわえる;蓄積する.

resilience 《自然環境の》回復力.

resource recovery 《米》資源回収《廃棄物からエネルギーを回収すること;通例 ごみを燃やして電力を起こすこと;⇨ refuse-to-energy*》.

retaining wall (土砂の崩壊などを防ぐ)擁壁, 支え壁.

reuse 動 再利用する, リユース[再使用]する《回収した製品や容器を洗ったり修理したりして, 繰り返し使うこと;⇨ three R's*》.

rewild 動〈一度開発された土地を〉再び未開地[原野]に戻す, (手付かずの)自然状態に復元する〈野生では見られなくなった動物を〉(他所から連れて来て)放獣[放鳥, 放流]する, 人為的に再野生化[野生復帰]させる.

rheophile, rheophilic, rheophil 圏 流水を好む[にすむ], 好流性の.

rheophily 图 好流性.

rhizoplane 【生態】根面《土壌のついた根の表面》.

rhizosphere 【生態】根圏《土壌中で植物の根の影響が及ぶ範囲》.

rhizospheric 圏 根圏(内)の.

RIA 1 regulatory impact analysis [assessment] 規制影響分析《環境行政などの規制を導入するに当たりその影響と費用対効果などを客観的数値として算出する評価手法》.
2 reactivity initiated accident《原子炉の》反応度事故.

Rio+20 summit リオ+20会議《2012年に開催予定の, 地球環境問題について討議する大規模な国際会議;リオの地球サミット(Earth Summit*)から20年後になることから, この名称で呼ばれている》.

Rio Declaration on Environment and Development (環境と開発に関する)リオ宣言《1992年にブラジルのリオ・デ・ジャネイロで開催された地球サミット(United Nations Conference on

Environment and Deveropment*)で採択された；「持続可能な開発(sustainable development*)」や「共通だが差異ある責任(common but differentiated responsibility*)」,「予防原則(precautionary principle*)」などの理念が27の原則に盛り込まれた》.

rising damp 上昇水分, 上昇湿気《地中から建物の壁に染み込む水分》.

risk assessment 危険性事前評価, 危険度評価, リスクアセスメント, RA《化学物質や放射線などが健康や環境に与える影響を短期・長期的に, あるいは質的・量的に事前に評価する手法；広く社会システムに潜む危険性を対象とすることもある》.

risk factor 危険因子, リスクファクター《疾病の発現を促す要因；たとえば肺癌に対する喫煙など》.

river basin 《河川の》流域.

RLI Red List Index*

RMU Removal Unit 除去単位《先進国内の植林などによって得られる排出枠(emission credit*)の一つ；京都議定書(Kyoto Protocol*)で定められた, 植林, 再植林によるCO^2吸収量を指す》.

RO(C) renewables obligation (certificate) ⇨ green certificate*

RoHS Directive [the ～] RoHS指令, ローズ指令, 特定有害物質使用制限指令《EUが2006年7月に施行した, 電気・電子機器に対する特定有害物質の使用を制限する指令；鉛・水銀・カドミウム・六価クロム・ポリ臭化ビフェニール(PBB*)・ポリ臭化ジフェニルエーテル(PBDE*)が規制対象；正式名は Directive on Restriction of the Use of Certain Hazardous Substances in Electrical and Electronic Equipment》.

roof [rooftop] gardening 屋上緑化.

room temperature 《通常の》室温, 常温《20°Cくらい》.

ROPME Regional Organization for the Protection of Marine Environment 湾岸海洋環境保護機構《ペルシア湾の汚染防止のため, 1978年クウェートを本部に設立された機関》.

rowen 牧草用に夏の終わりまで耕さないでおく刈り株畑；《牧草の》二番刈り.

Royal Society for the Protection of Birds [the ～] 王立野鳥保護協会《英国で1889年に創立された野鳥保護・保存のためのボランティア組織；20万人余の会員をもつ；略 RSPB》.

RPAR Rebuttable Presumption Against Registration《米国環境保護局(EPA*)の危険物質の使用規制に先立って必要とされる手続

き；危険の立証責任は規制側にあるとする》.

RPF refuse paper and plastic fuel《古紙や廃プラスチックを原料とする発電用の固形燃料》.

r selection r 選択[淘汰]《不安定な生息場所にすむ生物個体群が環境収容力(carrying capacity*)をはるかに下回る場合に起こる選択；つかの間の資源を利用して子孫の存続を確保するために, 早く大量に繁殖する個体が選択される；⇨ K selection*》.

RSNC Royal Society for Nature Conservation《英国の》王立自然保護協会.

RSPB Royal Society for the Protection of Birds*

RSPCA Royal Society for the Prevention of Cruelty to Animals 英国動物愛護協会(⇨ SPCA*).

RT room temperature*

ruderal 形 荒れ地[廃物]に生育する.
图 荒れ地[人里(ひとざと)]植物.

runoff 《地中に吸収されないで流れる》流去水, 表面流去；《液体の》あふれ出し

rural science [studies] 《英》田園科学[研究]《農学・生物学・環境学などの総称》.

S

sab 《俗》图 サップ, 狐狩り妨害運動家, 狩猟破壊活動家;《妨害活動を行なう》動物保護運動家.
動〈狩猟〉の妨害活動を行なう, 狩猟破壊活動をする.

sacred grove 聖地.

safety phrase 安全フレーズ《欧州連合の基準による製品安全データシート(material safety data sheet*)で, Sを冠した番号で示される定型の取り扱い注意事項》.

Sahel [the ～] サヘル《サハラ砂漠に南接する半砂漠化した広大な草原地帯; モーリタニア・セネガル・マリからチャドに及ぶ; 最貧国が集中し, 砂漠化が進行している; Sahel はアラビア語で縁の意味》;《一般に》砂漠に隣接するサバンナ[ステップ]地域.

　Sahelian 形 サヘル(地域)の.

saline water 塩水.

salinity 塩分, 塩分濃度, 塩度, 鹹度(かん).

salinize 動 塩で処理する;…に塩を染み込ませる.
　图 salinization 塩(類)化(作用), 塩処理.

salsuginous 形 (植物などが)塩生の.

salt marsh 塩性沼沢(地), 塩湿地, 塩生草原《牧草地や製塩に利用》.

salvage yard 《使用されなくなった機械・自動車などからの》部品回収場.

sand binder 砂止め植物《砂丘などに生育して, 根などで砂地を固定する植物》.

sapropelic 形【生態】腐泥にすむ.

Saro-Wiwa サロ＝ウィワ Kenule ['Ken'] Beeson Saro-Wiwa (1941-95)《ナイジェリアの著述家・環境保護論者; 著述活動を通じて, ナイジェリア軍事政権と Shell 石油を環境破壊の責任者として糾弾; 仲間の活動家8人と共に絞首刑となる》.

saturation level 【生態】carrying capacity*《(牧草地などが養える)動物の数》.

saxicolous, saxicoline, saxicole, saxatile 形【生態】岩石間[表面]に生息する, 岩生の.

SBS sick building syndrome* シックビル症候群.

SBSTA Subsidiary Body (for Scientific and Technological Advice)(科学的・技術的助言に関する)補助機関(会合)《国連の気候変動枠組み条約締結国会議

(COP*)の下部機関(の会合)》.

scavenger 清掃動物《特にハゲタカ・カニ・アリなど》；くず屋，廃品回収業者；《英》市街清掃夫；掃除夫；不純物除去剤，殺菌[消毒]剤，掃鉛剤，スカベンジャー．

scavengery 街路掃除，掃除屋の仕事．

scenic reserve 《ニュージーランドなどの》風致地区，景観保護区．

SCF Strategic Climate Fund 戦略気候基金《開発途上国の森林保全を支援するために2008年に設立された基金；英国や日本などが出資し，世界銀行が管理する；気候投資基金(CIF*)の二本立ての一》．

SEA strategic environmental assessment* 戦略的環境アセスメント．

seabird 海鳥《かもめ，海雀など》．

sea grass 海辺[海中]の植物，《特に》アマモ．

sea level 海水面；平均海面：1000 meters above *sea level* 海抜千 m．

Sea Shepherd シー・シェパード《海洋生物保護を主張する団体Sea Shepherd Conservation Society の短縮・通称；捕鯨・フカヒレ採取・アザラシ乱獲などに反対する過激な運動を展開している；1981年，グリーンピース(Greenpeace*)に所属していた Paul Watson が米国で設立(前身は1977年にカナダで設立)，本部は米ワシントン州，活動拠点はオーストラリア；日本に関しては2003年に和歌山県太地町のイルカ漁を「イルカ虐殺映像」として世界に配信，論争を巻き起こした；また，入り江に潜入し，捕獲されたイルカを網を切って逃そうとしたメンバー2人が逮捕された；こうした活動がドキュメンタリー映画 'The Cove' (2009; ⇨ cove*) の撮影・制作のきっかけとなった》．

sea turtle ウミガメ《暖海に広く分布する大型のウミガメの総称；オサガメ(leatherback turtle)，タイマイ(hawksbill turtle)，アカウミガメ(loggerhead turtle)，アオウミガメ(green turtle)などを含む》．

Secchi disk セッキ円板《海水の透明度を測定するのに用いる白いまたは色のついた平円板》．

secondary consumer 《草食動物を食する》二次消費者《キツネ・タカなど；⇨ primary consumer*》．

second generation biofuel 第二世代バイオ燃料《作物の枝葉や藁，廃油，非食用植物などを原料とし agflation* を起こしにくいバイオ燃料》．

second growth 《原生林破壊後の》二次林，再生林．

secondhand smoke 二次喫煙《非喫煙者が他人のタバコの煙を

吸い込む状態；喫煙者が吐き出す煙と，タバコの点火部から立ち上がる副流煙とがあり，喫煙者が吸い込む主流煙に比べ有害物質が多い；⇨ passive smoking*, third hand smoke*》.

sediment 图 沈降[沈澱]物，おり；【地質学】堆積物.
動 沈殿[沈降]する[させる].

sedimentation 沈降[沈積，堆積] (作用)；沈降分離，沈澱法.

seepage pit 浸出孔，浸透ます《浄化槽からの汚水を徐々に地面に吸収させるための多孔性の石積みで作られた孔；排水地の代用にされる》.

Sellafield セラフィールド《イングランド北西部 Cumbria 州西部の Sellafield にある核燃料再処理施設；1983 年に近くの海への廃液漏れを起こし有名になった；1981 年までは Windscale の名で知られていた》.

semi-arid 半乾燥の，非常に雨の少ない；《動植物が》半乾地性の.

semiparasitism 【生態】半寄生(生活).

semiterrestrial 形【生態】沼地生の，半陸地生の.

sensible temperature 体感温度.

seral 形【生態】遷移系列(sere*)の.

sere 遷移系列《生物群集や生態系がある状態から別の状態に変化していく過程；裸地から草原などの遷移(succession)の過程》.

sewage 下水汚物，下水；(化学処理を受けていない)生(き)汚水.

sewage farm 下水処理場《しばしば肥料を製造する》；下水畑(ばたけ)《下水を灌漑に利用する畑》.

sewage gas 汚水[下水]ガス《メタンが約 2/3，二酸化炭素が約 1/3 含まれる》.

sewage plant 《米》水処理場 (sewage farm*).

sewage sludge 下水汚泥，下水スラッジ.

sewage treatment 下水[汚水]処理.

sewer gas sewage gas*

shellfish 貝；甲殻類の動物《エビ・カニなど》；ハコフグ.

shelterbelt 《農作物保護の》防風林；《土壌保全用の》保安林.

shelter wood 防風林，保安林 (shelterbelt).

shrub layer 《植物群落の》低木層 (⇨ layer*).

sick building syndrome シックビル症候群，閉め切りビル症候群《断熱性能が高く新鮮な空気の導入が少ないオフィスビルで働く人にみられる症状；頭痛・眼の炎症・かぜに似た症状・無気力などを伴う；エアコンの普及による換気不足が影響する；空気の汚れや一酸化炭素・二酸化炭素の増加，ビル建材・内装材に含まれる化学物質などが原因と考えられる 略

SBS; ⇨ tight building syndrome*》.

sick house syndrome シックハウス症候群《新築や改修後の住宅において起きる，頭痛や吐き気，目やのどの痛みなどの様々な体調不良；建材中に含まれるホルムアルデヒドなどの化学物質が原因と考えられている 略 SHS》.

Sierra Club [the 〜] シエラクラブ《1892年に設立された米国の環境保護団体；本部は San Francisco；全国的な問題や地元の問題（同市の下水道問題など）に関してロビー活動を実施；国立公園の保全や野生動物保護などの運動で知られる；全米各州に支部がある；⇨ Brower*》.

silent spring 沈黙の春《有害化学[環境汚染]物質により将来起こりうる，動植物が一切死に絶えた春（が来るような時代）；Carson* の同名の著書から》.

silva 《特定地域の》高木林，樹林，樹木；樹林誌《ある森林中の樹木に関する記述；⇨ flora*》.

silvex シルベックス《木質化する植物の除草剤》.

silvicolous 形 森林に生息[生育]する．

silvics 森林生態学；森林の樹木の特徴，森林の生態.

single stream 形 名 非分別型の（ごみ収集・処理）《ごみ分別自動化技術によって手間を軽減する方式》.

singularity 気象異常日，特異日；【天文学】特異点《密度が無限大となる》；

sink 下水だめ，汚水溝(⇨ carbon sink*).

site 【生態】《植物の》立地.

six pack ring [yoke] シックスパックリング《三つずつ二列に繋げた六連のプラスチックリングで，飲み物の缶・瓶を輪に一本ずつ嵌め込み纏めて持ち運べる；ポイ捨てされたものが釣り糸などと同様にしばしば魚や動物の体に絡まって食い込み環境問題化している》.

skin protection factor sun protection factor*

skip 《英》建築廃棄物搬出用の大型容器.

SLOP smog, litter, overpopulation, pollution スモッグ・ごみ・過剰人口・汚染.

slow food スローフード《画一的で大規模生産のファーストフード(fast food)に対し，手間ひまをかけた伝統的な料理[食べ物]を指す；小生産者が有機農法などで提供する食材を用い，食文化の多様性を尊重する運動；1989年にイタリアの Bra という地域で始まった；ゆっくり食事を楽しむ意味もある》.

Slow Food International Association 世界スローフード協会《イ

sludge

タリアに本部がある，食に関する意識改革を提唱している団体》．

sludge 图 軟泥，泥，（下水）汚泥，スラッジ，ヘドロ；ぬかるみ；《鉱石の》泥滓(でいさい)；《ボイラー・水槽などの》沈澱物，スラッジ；《鉱油の》不純物；活性汚泥；微粉末と水の混合液；半解けの雪；海面氷．📖 軟泥［スラッジ］化する［を生じる］；軟泥でふさぐ；…から軟泥を除く，汚泥浚渫(しゅんせつ)［処理］する．

sludger 图 軟泥を除去する装置，汚泥処理装置《サンドポンプなど》．

smart grid スマートグリッド《情報技術を使って電力の需要と供給を効率的に制御する次世代の送電網；太陽光発電，風力発電と，既存の火力，原子力発電などとの有効な組み合わせ，自家発電によって余った電力の売電，（アメリカなどの）雑多な電力網の間の連携向上，電力需要に合わせて課金を変動させることによるピーク負荷の分散促進など，さまざまな用途が考えられる；オバマ米大統領がグリーンニューディール政策の一つとして掲げた》．

smart meter スマートメーター《家庭内電力メーターに通信・管理機能を持たせて家電全体をネットワーク(home area network, HAN)化し，電力会社が機器稼働状況などを把握しつつ効率的な給電を行って省エネルギーをはかる仕組み；AMI (advanced metering infrastructure)とも呼ぶ；⇨ smart grid*》．

smoke abatement 《都市の》煙突排煙規制．

smokeshade スモークシェード《大気中の粒子状汚染物質；その計量単位》．

Smokey, Smoky 熊のスモーキー(Smokey the Bear)《米国の森林警備隊員(forest ranger)の服装をした漫画の熊；山火事防止・環境保護のシンボルで，モットーは"Only you can prevent forest fires"（山火事を防ぐのは君だ）》．

snow crust 雪殻《積雪表面の氷結した硬い雪の層》．

sociability 【生態】群度．

socially responsible investing 社会的責任投資《投資対象の企業が，環境保護・法令遵守・適切な従業員処遇［人権保護］・社会貢献などの社会的責任にどれだけ配慮しているかを見定めた上での投資；略 SRI；⇨ CSR*》．

society 生物社会，《群落内の亜優占種による》社会，ソサエティー；《群落単位としての》群集；《ミツバチなど》社会的単位をなす一対の昆虫の子孫: Ants have a well-organized *society*. アリには非常に組織立った社会がある．

socioecology 社会生態学．

sociology 社会学；群集生態学(synecology*)．

sociologist 图 社会学者．

soft release 图 ソフトリリース《繁殖・飼育・病気回復などのために人間が世話していた動物を，屋外ケージなどで徐々に野生に慣らしつつ放獣・放鳥すること；⇨ hard release*》．
動《動物を》保護しつつ徐々に野生に戻す．

soil binder 粘土［ローム］質の土壌の流失を防止する植物，土(ど)止め植物．

soil erosion 土壌浸食．

soil formation 土壌生成，土壌形成．

soil group 土壌群《同じような湿度・温度の状態の下にあり同一層位をもつ土壌の集まり》．

soil stack 垂直の汚水管，汚水スタック．

solar energy ソーラーエネルギー，太陽エネルギー《地球に降り注ぐ太陽の輻射エネルギーから得られる各種のエネルギー；元をただせば化石燃料，バイオマス，水力など多くのエネルギーがソーラーエネルギーに分類できる》．

solar power generation 太陽光発電《太陽電池をもちいて太陽光のエネルギーを直接電力に変換するシステム；再生可能な自然エネルギーギーとして，CO_2 排出削減への効果が期待される》．

solid waste 固形廃棄物．

solar thermal 图 太陽熱利用の．

sootfall 煤煙の降下(量)．

sound-level meter 音圧レベル計，音位計，音量計，騒音計．

sound pollution 騒音公害(noise pollution)．

soy ink ソイインク，大豆インキ《大豆(soybeans)から作る印刷用インク；油性インクと違い環境にやさしく発色も良い》．

SPCA Society for the Prevention of Cruelty to Animals 動物愛護協会，動物虐待防止協会《米国では ASPCA*，英国では RSPCA*，日本では JSPCA* など》．

species 图 (複数形も species)《生物学上の》種(しゅ)(⇨ 和英篇学名)：*The Origin of Species*『種の起源』《進化論を説いたダーウィンの著書(1859)》/ birds of many *species* 多くの種の鳥 / extinct [vulnerable, rare] *species* 絶滅［危急，希少］種(⇨ Red List*) / the [our] *species* ヒト，人類(the human race)
形《園芸上でなく生物学上の》種に属する．

species barrier 種の障壁，種の壁，種間障壁《病気がある種から別の種に転移するのを妨げると考えられている自然界の機構》：cross [jump] the *species barrier*《病気が》本来ならうつらない他の種に感染する．

species-group《いくつかの種からなる》種群；《種・亜種を含む》

種グループ.

species population 【生態】種個体群.

speciesism 1《動物に対する》種(による)差別, 種偏見《たとえば犬や猫に対するのと実験動物に対するのとで態度に差のあること》. 2 動物蔑視, 人間優位主義.

SPF sun protection factor* 太陽光線保護指数.

SPM suspended particulate matter 浮遊粒子状物質《particulate matter* の一種で, 工場からの煤煙や排ガスなどで放出され, 大気中に浮遊している微小粒子; 日本の環境基準では直径10ミクロン以下のものを SPM と呼ぶ; 直径が小さいほど健康への影響は大きいと考えられている》.

spoil 発掘[浚渫(しゅんせつ), 採掘など]で廃棄される土[石など], 廃石, 捨土, ぼた: *spoil* bank《不要な》土砂などの山,《鉱山などの》不用堆積物の山, ボタ山. / *spoil* ground 浚渫土砂の指定投棄海域.

sprawl 動〈都市などが〉不規則に[無秩序に][無計画に]広がる, スプロール化する, はみ出す. 名 不規則に延び広がること; ばらばらな集団[集まり];《都市などの》スプロール現象(⇨ vertical sprawl*).

sprawlingly 副 無秩序に, スプロール状に.

sprawly 形 (不規則に)延び広がった.

SRES the IPCC Special Report on Emissions Scenarios《IPCC* が将来の温暖化ガス排出状況をいくつかの異なる場合に分けて予測した》温暖化ガス排出シナリオに関する特別報告書.

SRI socially responsible investing* 社会的責任投資.

stabilization 安定化, 安定処理.

stakeholder 《企業などの》利害関係者《最近では, 企業に限らず, 広く社会のなかでのさまざまな利害関係者を指すことが多い; 行政, 企業, NPO, 専門家, 市民など》.

stand 立ち木, 立木(りゅうぼく);《一定面積に対する》立ち木の数[密度], 株立本数;【生態】植(林)分, 林分; 生えたままの草本, 植生, 草生《雑草・穀物などの》.

starfish ヒトデ.

START 1 Global Change System for Analysis, Research, and Training 地球変化の分析・研究・研修システム《国際科学会議が定めた IGBP*(地球圏−生物圏国際共同研究計画)における連携研究の一つ》.
2 Strategic Arms Reduction Treaty [Talks] 戦略兵器削減条約[交渉]《START I は1991年7月, 米国と旧ソ連により署名された核兵器を減らす軍縮条約》.

steam generator 蒸気発生器[装置].

stenobath 狭深性生物《狭い範囲の水深でのみ生息できる；⇔ eurybath*》.
stenobathic 形 狭深性の
stenohaline 形【生態】狭塩性の，狭鹹(きょうかん)性の(⇔ euryhaline*).
stenohygric 形【生態】狭湿性の(⇔ euryhygric*).
stenoky 【生態】狭環境性(⇔ euroky*).
stenokous 形 狭環境性の.
stenophagous 形【生態】〈動物が〉狭食性の.
stenotherm 【生態】狭温性生物(⇔ eurytherm*).
stenothermy 狭温性.
stenothermal, -thermic, -mous 形 狭温性の.
stenotopic, -tropic 形【生態】狭場所性の(⇔ eurytopic*).
stenotropic 形 環境変化に対する適応の幅が狭い，狭環境性の.
Stern Review スターン・レビュー《英国財務省が 2006 年 10 月に公表した，気候変動(climate change*)の経済的側面についてのレビュー；元世界銀行チーフエコノミストのニコラス・スターン氏に依頼して実施したため，スターン・レビューと呼ばれる；今行動を起こさなければ毎年，GDP の 5～20% に相当する被害を受けるが，行動を起こせば気候変動の最悪の影響は避けることができるという内容；その場合の対策コストは GDP の 1% 程度だと試算している》

Stockholm Convention （残留性有機汚染物質に関する）ストックホルム条約，POPs 条約《2001 年にストックホルムで締結され 2004 年に発効した国際条約；PCB* やダイオキシン(dioxin*)など難分解性で生物に蓄積されやすい残留性有機汚染物質(Persistent Organic Pollutants ＝ POPs)の製造・使用・輸出入，保管などを規制する；正式名は Stockholm Convention on Persistent Organic Pollutants》.
storm surge 風津波，高潮.
straddling 〈漁業資源が〉複数の経済水域にまたがる：*straddling* (fish) stock ストラドリング魚類資源《回遊魚など》.
stranger 【生態】客員種.
strata stratum* の複数形.
strategic environmental assessment 戦略的環境アセスメント《欧州で始まった制度で，事業計画がまだ柔軟な初期立案段階で環境アセスメントを行い，時に抜本的な事業見直しも含む計画変更を容易にする；略 SEA》.
strategy 戦略《ある環境で生活・繁殖するために個々の種がとる行動・代謝・構造面での適応》.
stratum （複数形 strata, stratums）層；《大気・海洋の》層；【生態】《植生の》層.

streamflow 河川流, 河川流量.

streamway 《河川の》流床;《河川流の》主流.

subalpine 形 (アルプス)山麓の;【生態】亜高山帯の.

subarid 形【生態】亜乾燥の.

subclimax 【生態】亜極相, 亜安定期, サブクライマックス.

subdominant 名形【生態】亜[次]優占種(の).

subdrainage 地下排水.

subinfluent 亜影響種《生物群集の中で影響種(influent*)よりも影響を与える度合いが少ない生物》.

sublittoral 形 岸辺近くの水中にある, 低潮線から大陸棚の間の, 亜沿岸[潮間](帯)の. 名 亜沿岸帯, 亜潮間帯.

subordinacy subordination*

subordinate 【生態】劣位者[種].

subordination 【生態】劣位.

subpopulation 【生態】《混系中の》特定生物型群, 副次集団, 分集団.

subsere 【生態】二次遷移系列《伐採などによる人為的裸地における遷移系列; ⇨ prisere*》.

Subsidiary Body for Scientific and Technological Advice ⇨ SBSTA*

subsidy 《政府の》補給金, 助成金, 補助金;交付金, 寄付金

subtersurface 形 地表下の, 水面下の.

subtropical forest 亜熱帯林.

succession 【生物】系列;【生態】遷移, サクセッション(⇨ sere*)

sulfuric acid mist 硫酸ミスト.

sun-powered 形 動力を太陽光線から得ている, 太陽エネルギー利用の.

sun protection factor 太陽光線保護指数 (skin protection factor)《日焼け防止用化粧品が紫外線などの太陽光線の悪影響から皮膚を保護する効果を表わすもので2から12またはそれ以上の数値で表現される; 略 SPF; ⇨ UV index*》.

Superfund 有害産業廃棄物除去基金, スーパーファンド《米国で1980年制定の総合環境対策補償責任法(Comprehensive Environmental Response, Compensation and Liability Act)で創設された, 放置有害廃棄物除去のための基金; 元化学廃棄物投棄地に立てられた住宅地の汚染問題などが対象となった》.

superorganism 超個体《社会性のある動物などの群落》.

superparasite 【生態】重寄生者(hyperparasite*).

superparasitic 形 過寄生の.

superparasitism 過寄生《一寄主に, 通例同一種の複数の寄生者が寄生すること》; hyperparasitism*.

supralittoral 形 潮上帯の《湖・海

などの沿岸部で常に水上にあるが波しぶきや毛管現象によって湿っている生物地理学的地域の》. 図 潮上帯.

suspended particulate matter 浮遊粒子状物質, SPM《⇨ SPM*, particulate matter*》.

sustainable 形〈開発・農業などが〉資源を維持できる方法の, 持続可能な;〈社会などが〉持続可能な方法を採用する生活様式の(⇨ sustainable development*).

sustainability 图 持続可能性, サステイナビリティー《自然の生産物やエネルギーを枯渇させないで人間活動が将来にわたって継続していく可能性; sustainable の概念導入については, 経済学者の J. R. Hicks の論文(1939)が貢献したといわれ, 消費しながらも一定量を残して維持していく sustainable consumption といった概念が確立していった》.

sustainable development 持続可能な開発[発展]《環境と資源の維持と両立できるような開発; 国連の環境と開発に関する世界委員会(WCED*)が 1987 年に発表した 'Our Common Future*' という報告書のなかで sustainable development が最重要のキーワードとして提唱され, その後の普及のきっかけとなった; 定義は「将来の世代のニーズを満たす能力を損なうことなく, 現在の人のニーズを満たすような開発」; ⇨ United Nations Commission on Sustainable Development*》.

sustainable fisheries 持続可能な漁業.

sustainable forest management 持続可能な森林経営.

sustainable future 持続可能な未来.

sustained yield 持続的生産《木材や魚など生物資源の収穫した分が次の収穫以前に再び生育するように管理すること》.

sustained-yield 形 持続的生産の.

SVHC substances of very high concern 高懸念物質《REACH* 規制の対象となる化学物質; 次のような主要分類がある》.

  CMR (carcinogens, mutagens or toxic to the reproductive system 発癌性・変異原性・生殖毒性物質).

  PBT (persistent, bioaccumulative and toxic 残留性・生体蓄積性・有毒性物質).

  vPvB (very persistent and very bioaccumulative 高残留性・高生体蓄積性物質).

swamp gas 沼気(marsh gas*).

swarm 图 群れ,《特に》巣別れ[分封(ぶんぽう)]するハチの群れ, 一つの巣のハチの群れ;【生】遊走群: a *swarm of* ants アリの群れ.
動 群がる, たかる; 群れをなして

移住[移動, 集合]する；〈ハチが〉巣別れ[分封]する；【生態】群飛する, 群泳する.

sylvics silvics*

symbiont, -on 【生態】共生者, 共生生物.
  symbiontic 形 共生する, 共生者の.

symbiosis (複数形 symbioses)【生態】(相利[双利])共生, 共同生活(⇨ parasitism*).

symbiote symbiont*

sympatric 形【生態】同所(性)の(⇔ allopatric*)：*sympatric* species 同所種 / *sympatric* hybridization 同所性交雑 / *sympatric* speciation 同所的種分化.
  sympatry 名 同所性.

symphily, symphilism 【生態】友好共生.
  symphilous 形 友好共生の[する].

synanthropic 形【生態】ヒト(の生活)と関連して存在する, 〈ハエが〉共ヒト生の.
  synanthropy 名 共ヒト性.

synecology 群(集)[群落]生態学(⇨ autecology*).
  synecologic(al) 形 群集生態学上の[的な].

syn(o)ecious, synoicous 形《植物が》雌雄同株[同体, 共株]の；【生態】(運搬[住込み])共生の.
  syneciousness 名 雌雄同株[同体].

T

TAC total allowable catch*

tallgrass prairie トールグラスプレーリー《米国中西部ミシシッピ川流域の肥沃な大草原》.

TAR Third Assessment Report《2001年にIPCC*が発表した》第三次評価報告.

tar ball 《海岸に打ち上げられたりする》タールの塊り.

taxon 《生物学上の》分類群; 類名.

taxonomy 分類学; 分類; 分類法.
  taxonomist 图 分類学者.
  taxonomic(al) 形 分類学上の[的な].

TBS tight building syndrome* 気密ビル症候群.

TBT tri-*n*-butyl tin《船舶用塗料に用いてフジツボなどの生物付着を防止するための殺生物剤》.

TCDD tetrachlorodibenzo-*para* [*p*-] dioxin テトラクロロ・ジベンゾ・パラ・ダイオキシン, 四塩化ジベンゾダイオキシン《このうち 2,3,7,8–四塩化ジベンゾダイオキシンは残留性があり, 発癌性がきわめて強い; アメリカ軍がベトナム戦争で使用した枯れ葉剤にも含有されていた》.

temperate 形〈気候・温度など〉穏やかな, 温和な;〈地域など〉温帯性の;〈生物が〉温帯の

temperate rain forest 【生態】温帯多雨林.

temperate zone [the 〜] 温帯.

tensiometer 張力計, 土壌容水分張力計, 水量計《土壌中の水分量をその表面張力で測定する計器》.
  tensiometry 图 張力測定(学).
  tensiometric 形 張力計の.

tension zone 【生態】移行帯, 推移帯(ecotone*).

tephra テフラ《噴火によって放出され, 空中を飛んで堆積した火山砕屑(さいせつ)物, 特に火山灰》.

terraqueous 形 水陸からなる, 水陸の;〈植物など〉水陸両生の.

terrestrial 形 地球(上)の; 陸(上)の, 陸からなる; 土の, 土質の;《生物が》地上生の, 陸生の; 陸生生物の;〈惑星が〉地球型の: *terrestrial* heat 地熱 / *terrestrial* transportation 陸上輸送.
图 陸[地球]上の生物.

terrestrial ecosystem 陸上[陸域]生態系

territorial 形【生態】なわばり行動をする, なわばり制の.

territorial imperative 【生態】なわばり意識.

territorialism 【生態】なわばり制.

territoriality なわばり意識；【生態】なわばり制.

territory 《動物の》なわばり，テリトリー.

tertiary 形 三次(処理)の《微粒子・硝酸塩・燐酸塩の除去による汚水浄化についていう》.

tertiary consumer 三次消費者《小型肉食動物を食う大型肉食動物；⇨ primary consumer*》.

tetrachloroethylene 四塩化エチレン，テトラクロロエチレン《洗浄剤・ゴムやタールの溶剤などに用いる；⇨ DNAPL*》.

TFAP Tropical Forest Action Plan 熱帯林行動計画《熱帯林を保存しつつ合理的に利用するために，1985年にFAO*(国連食糧農業機関)の熱帯林開発委員会によって作成された計画》.

thermal pollution 熱(廃棄物)汚染，熱公害《火力・原子力発電所などの廃水が昇温したまま河川や海に放流され，環境・生態系に変化をもたらし，また動植物に被害を与えること；heat pollutionともいう》.

thermokarst 熱カルスト，サーモカルスト，アラス《永久凍土が溶けてできた窪地・池；「アラス」はシベリアの現地語 alas に由来》.

third hand smoke 三次(受動)喫煙《煙草を消した後も，後からその場へ来た人(特に子供)が，喫煙者の髪や衣服，部屋のクッションや床などに付着している残留有害物質を吸い込むこと；⇨ second-hand smoke*, passive smoking*》.

threatened 形 〈野生動植物の種が〉《直ちに絶滅するまでには至らない程度ではあるが》絶滅の危機に直面して(⇨ endangered species*).

Three Mile Island スリーマイル島《米国北東部ペンシルベニア州 Susquehanna 川にある島；1979年3月ここにある原子力発電所の2号炉で一次冷却水の減少により部分的な炉心溶融事故が起きた；放射能が漏れ，周辺住民が一時避難した；チェルノブイリ (Chernobyl*)原発事故に次ぐ2大原発事故の一つ》.

three R's 1 3Rの原則《リデュース(reduce* 廃棄物量の削減)，リユース(reuse* 再使用)，リサイクル(recycle* 再生利用・再資源化)という3つの言葉の頭文字から名付けられた；環境への負荷を減らす循環型社会を実現するための行動方針；リサイクルよりもリデュースの方が重要とされる》.2 動物実験の倫理的原則《実験数・個体数を最小限に削減(reduce)，苦痛軽減のための手法改良(refine)，他の方法への代替化(replace)；日本でも2006年施行の動物愛護管理法に盛り込まれ

threshold 《ある効果が生じ始める》限界, 閾(いき), 域, しきい(値).

tidal energy 潮汐エネルギー.

tidal power 《発電に使われる》潮汐力, 潮力.

tide pool 《岩場などの》潮だまり, 潮汐池.

tight building syndrome 気密ビル症候群《ビルの換気不良に起因する一群の身体的不調; 略 TBS; ⇨ sick building syndrome*》.

tile field 互いにつながっていないいくつかの排水管を張りめぐらした地域《湿地排水・排液吸収などのため》.

timberline 《高山・極地の》高木限界線, 樹木限界線(tree line).

TN total nitrogen 全窒素《水中に含まれる窒素化合物の総量》.

TNC The Nature Conservancy (⇨ Nature Conservancy*).

TNR trap neuter return 捕獲・去勢ないし不妊手術・解放《野良猫などを殺処分せずに頭数管理する愛護活動》.

tolerance 《食品中における殺虫剤など有害物質の》残留許容限界量; 《薬物・毒物に対する》耐性, 許容(度); 《生物の環境に対する》耐性.

Torrey Canyon [the ～]トリーキャニオン号《1967年イングランド南西端沖のScilly諸島近くで坐礁して大規模な原油流出事故を起こしたリベリアのタンカー》.

total allowable catch 漁獲可能量《略 TAC: ⇨ catch share*》.

traceability トレーサビリティー《食品の生産履歴の追跡可能性》: a food *traceability* system 食品生産履歴追跡システム.

trade liberalization 貿易自由化.

trade waste 《英》産業廃棄物 (industrial waste*).

traditional medicine 伝統医学.

Trans-Alaska Pipeline [the ～] トランスアラスカパイプライン (Alaska Pipeline*).

transboundary 形 国境を越えた, 越境的な: *transboundary* air pollution 越境大気汚染.

transect トランセクト《植物群落の種類・組成, 植生の変化などを調べるため, 植生を横切って作った帯状横断面》.

transfer station ごみ集積所《家庭ごみや産業廃棄物などを埋立地などに運ぶために, 圧縮して車に積み込む場所》.

transformer 1 性転換遺伝子;《生態系内の》転換者;《遺伝的な》形質転換体.
2 変圧器, トランス.

transmissometer 《大気の》透過率計, 視程計.

trash can ごみ入れ, ごみバケツ (⇨ garbage*).

trash compactor 《米》《台所用の》ごみ圧縮器.

trash-to-energy 形 ごみを燃やしてエネルギーを得る(garbage-to-energy*).

TRC tradable renewable certificate ⇨ green certificate*

tree hugger (特に森林保護の)環境保護運動家.

tree layer 【生態】《植物群落の》高木層(⇨ layer*).

tree line timberline*

trichloroethane トリクロロエタン《無色の不燃性の液体, 2種の異性体がある: 1) = 1, 1, 1-trichloroethane; 金属の脱脂洗浄剤など溶剤として用いる; methyl chloroform ともいう: 2) = 1, 1, 2-trichloroethane; 溶剤・有機合成原料》.

trickling filter 散水濾床《下水・工場廃水などを生物学的に濾過する設備》.

troglobiont, troglobite 真洞窟性動物《洞窟の中にのみ生存する動物; 眼や体の色素を失い, 感覚器が発達する》.

  troglobitic 形 真洞窟性の.

trophic level 栄養段階《生態系を構成する生物をエネルギーや物質の動きの面から生産者・消費者・分解者などに分けたもの》.

tropical cyclone 熱帯低気圧.

tropical rain forest 熱帯(多)雨林(rain forest*).

tropophyte 【生態】天候順応性植物, 季節的落葉植物.

tropophytic 形 天候順応性(植物)の.

true-blue green 形《俗》環境保護運動に献身している.

TRU transuranic 超ウラン元素《ウランよりも元素番号が大きい放射性元素》.

TRU waste 【原子力】TRU[超ウラン元素]廃棄物《プルトニウムなどの超ウラン(transuranium)元素を含む廃棄物》.

TSCA Toxic Substance Control Act 毒性物質規制法《1976年制定の米国連邦法》.

TTS temporary threshold shift 一過性閾値変動《航空機騒音による一時的聴力低下》.

tuna マグロ(⇨ ICCAT*).

tundra 凍土帯, 凍原, ツンドラ.

tychopotamic 形 止水性の《動植物が主に静止した淡水中で生育する; ⇨ autopotamic*, eupotamic*》.

U

UDC Urban Development Corporation*

ULL uncomfortable loudness level 不快音レベル.

UNCD United Nations Conference on Desertification 国連砂漠化防止会議.

UNCED United Nations Conference on Environment and Development*

UNCHE United Nations Conference on the Human Environment*

UNCSD United Nations Commission on Sustainable Development*

undermine 動 1 …の下を掘る, …の下に坑道を掘る;《侵食作用で》…の根元[土台]を削り取る. 2 徐々に衰えさせる, 害する.

underpin 動〈構造物〉につっかいをする; 補強する, 支持する, 応援する; 実証する.

underscrub 《樹木の下の》下生え, 亜低木; 小低木,《特に》木本性の地表植物.

understory 【生態】《植物群落の》下層, 下層植生.

underwood 下生え, 下草, 下木 (undergrowth, underbrush),【生態】下層木.

UNEP United Nations Environment Program* 国連環境計画, ユネップ.

UNEP-WCMC UNEP 世界動植物保全監視センター (the World Conservation Monitoring Center of the United Nations)《UNEP* の一部門; 世界の生物多様性情報収集や, 生物保護政策支援などを行う》.

UNFCC(C) United Nations Framework Convention on Climate Change*

-unfriendly 形「…にとって役立たない[有害な]」の意: environment-*unfriendly* 環境にやさしくない[有害な].

ungreen 形 環境に対する十分な意識[配慮]のない, 環境に有害な.

United Nations Commission on Sustainable Development (国連)持続可能な開発委員会, CSD《1992 年にリオ・デ・ジャネイロで開催された地球サミット (United Nations Conference on Environment and Development*)を受けて, 同年 12 月に国連の経済社会理事会に設置され

た；略(UN)CSD: ⇨ sustainable development*》.

United Nations Conference on Environment and Development [the ~] 環境と開発に関する国連会議《通称「地球サミット」(Earth Summit*), 略称 UNCED; 1992年6月に, 国連が世界の政府関係者や経済界, 市民団体に呼びかけてブラジルのリオ・デ・ジャネイロ(Rio de Janeiro)で開催した, 地球環境問題についての大規模な国際会議; 1972年6月の国連人間環境会議(United Nations Conference on the Human Environment*)から20年目の節目にあたり, 世界の多くの首脳が出席した(日本の首相は欠席); 地球サミットでは「環境と開発に関するリオ宣言」(Rio Declaration on Environment and Development*), 21世紀に向けた行動計画であるアジェンダ21(Agenda 21*), 森林保全に関する原則声明が採択され, また気候変動枠組み条約(United Nations Framework Convention on Climate Change*), 生物多様性保全条約(Convention on Biodiversity*)の2条約について署名・採択がなされた; さらに, 地球サミットでの合意を受け, 国連の経済社会理事会のもとに「持続可能な開発委員会」(United Nations Commission on Sustainable Development*)が設置された》.

United Nations Conference on the Human Environment 国連人間環境会議, ストックホルム会議《環境問題と人類社会をテーマとし国連が初めて催した国際会議;「かけがえのない地球」('Only One Earth')を合い言葉に, 1972年6月スウェーデンのストックホルムで開催された; 世界114カ国が参加し,「人間環境宣言」と「環境国際行動計画」を採択した; その結果を実施に移す機関として, 同年, 国連環境計画(United Nations Environment Program*)が誕生した; 略称 UNCHE》.

United Nations Environment Program [the ~] 国連環境計画《1972年にストックホルムでの国連人間環境会議(United Nations Conference on the Human Environment*)で採択された「人間環境宣言」と「環境国際行動計画」を実施に移す機関として, 同年の国連総会の決議にもとづき設立された; 環境分野にかかわる諸機関の活動を総合調整しながら, 国連活動, 国際協力活動を実施している; 本部はナイロビ; 略称 UNEP》.

United Nations Framework Convention on Climate Change [the ~] (国連)気候変動枠組み条

約，気候変動に関する国際連合枠組条約，温暖化防止条約《大気中の温室効果ガスの濃度を，気候に深刻な影響が及ばないように安定化させることを目標とする国際条約；略称UNFCCC；1992年に採択され，1994年3月に発効した；「共通だが差異のある責任」(common but differentiated responsibility*)という考え方が盛り込まれており，先進国により大きな削減の責任が求められる；この枠組みのもとで具体的に削減に取り組むための京都議定書(Kyoto Protocol*)が1997年12月に採択され，2005年2月に発効した》．

unsite 動〈発電所・工場・各種施設などを〉移転する．

   unsited, un-sited 形〈施設などが〉建設場所が未定の．

UN Special Committee on Environment [the 〜]国連環境特別委員会(⇨ World Commission on Environment and Development*)．

UN's International Year of Biodiversity [the 〜] 国際生物多様性年(⇨ International Year of Biodiversity*)．

unsustainable consumption 持続可能でない消費(⇨ sustainable*)．

untreated 形 治療[対処，処理]しない(ままの)．

UOD ultimate oxygen demand 究極酸素要求量．

UPOV 新品種保護国際同盟《1961年のUPOV条約によって植物新種開発者の育成・販売権を保護する国際機関；フランス語 L'Union international pour la protection des obtentions végétables の略》．

upwelling わき上がること；【生態】《栄養塩に富む深海水などの》湧昇(ゆうしょう)．

Urban Development Corporation 都市開発公社《都市開発[再開発]のために英国政府が設立した組織；1981年にロンドンの港湾再開発地区Docklandsとリバプール近郊の2か所に設立されたのが最初；略 UDC》．

urban heat island 都市熱，都市ヒートアイランド《一日の最低・最高気温ともに周辺地域より高い都市部；⇨ heat island*》．

urban planning 都市計画．

Urban Programme [the 〜]《英国政府の》都市再開発計画，アーバンプログラム．

USCAP U(nited) S(tates) Climate Action Partnership 米国気候行動パートナーシップ《2007年に温室効果ガス排出削減を目指して設立された米国の自動車，電力，化学などの企業と環境保護団体・シンクタンクとの提携団体；大幅な排出削減を求める連邦法の実現を目標にしている》．

UV-A, UVA ultraviolet-A 長波長紫外線《波長 320-400 nm》.

UV-B, UVB ultraviolet-B 中波長紫外線《波長 290-320 nm; 俗に日焼け光線と言われ，皮膚紅斑生成の主因》.

UV-C, UVC ultraviolet-C 短波長紫外線《波長 200-290 nm; 大気によって反射され地表に到達しない》.

UV index 紫外線指数，UV 指数《紫外線の強さを表す国際的な数値で，人体皮膚への損傷の警告と適切な防護行動をとるための指標; 世界保健機関(WHO)・国連環境計画(UNEP*)・世界気象機関(WMO)などが 1997 年に共同で研究開発; ⇨ SPF*》.

V

V2G vehicle-to-grid《電気自動車やソーラーカーなどで使い切れなかった余剰電力を家庭電源から送電線に戻し電力会社が買い取る方式; ⇨ plug-in hybrid*》.

vacuum extraction 真空抽出(法)《真空ポンプにより土壌中の気体・揮発性の汚染物質などを除くこと; ⇨ air stripping*》.

vagile 形《生物が》自由に動く, 移動性の.

   vagility 名【生態】《生物の》散布力.

Valdez Principles [the ~] ヴァルディーズ原則《環境問題に企業がどう対応すべきかの判断基準となるよう1989年米国で作成された一連の指針; 同年3月アラスカ州の沖合いでオイルタンカーExxon Valdez*号が起こした原油流出事故がきっかけとなって生まれたもの; ⇨ OPRC convention*》.

validate 動 (法的に)有効にする, 批准する; 押印して認可する; 実証[確証]する; 確認する, 検証する.

   validation 名 確認, 批准.

valuation 評価, 値踏み; 見積もり[査定]価格, 評価額.

VAWT vertical axis wind turbine《発電用などの》垂直軸風力タービン[風車].

VDF vegetable diesel fuel*

vegetable diesel fuel ベジタブルディーゼルフューエル《使用済みのてんぷら油などの廃食油を原料として作るディーゼル燃料; 黒煙排出量は軽油の3分の1以下, 硫黄酸化物はゼロで環境に優しい; 略VDF》.

vegetated 形 植物の生育している, 草木のある.

vegetation management 【生態】植生管理.

vertebrate 形 脊椎のある; 脊柱のある; 脊椎動物門に属する. 名 脊椎動物.

vertical sprawl 垂直スプロール《高層建築の乱立や家賃高騰など無秩序な都心部再開発(の弊害)》.

vicar 分断分布(vicariance*)をなす姉妹種.

vicariance 分断分布, 異所的不連続化(vicariance biogeography)《山脈・海洋などの障壁によって姉妹種が地理的に隔てられて分布すること》.

Vienna Convention for the Protection of the Ozone Layer

（オゾン層保護のための）ウィーン条約《オゾン層の保護を目的とする国際協力のための基本的枠組み；1985 年に採択された；オゾン層を破壊する物質の特定や規制などの具体策を進めるためのモントリオール議定書(Montreal Protocol*)は 1987 年に採択された》.

view tax 景観税《景勝地や歴史的建築物などの保存や維持のためにその場所を展望できる建物の窓ガラスの数に応じて求められる寄付金》.

virtual water 仮想水, バーチャルウォーター《農産物や食料品などの生産に使われ, その生産品の使用により間接的に消費される水；外国の生産品を日本が輸入することは, それだけの水を日本の消費者が使用していることになる；例えば牛肉 1 キロの輸入には, それを生産するための水約 20 トンの消費が伴う；London 大学の J. Anthony [Tony] Allan が 1990 年代末に提唱した考え方； ⇨ global hectare*, food miles*》.

visual pollution 視覚公害《広告物・建物・ネオンサインなどによる美観の破壊》.

vitality 1 生命力, 活力；生活力；活気, 生気, 元気；持続力, 存続力.
2【生態】活力度.

volatile organic compound [chemical] 揮発性有機化合物《略 VOC》.

voluntary simplicity 自発的な質素《最小限の消費と環境に対する責任とを特徴とする, 物質主義拒否の哲学[生き方]》.

W

Waldsterben 《大気汚染による》森林の枯死《ドイツ語の Wald(森林), sterben(死ぬ)から》.

Washington Convention [the 〜] ワシントン条約《CITES* の通称》.

waste disposal 廃棄物処理, 廃棄物投棄;《英》生ごみ処理機 (disposer*, waste disposal unit).

waste-end tax 《特に有害な》産業廃棄物税.

   waste-end taxation 産業廃棄物課税.

waste heat 廃熱: a *waste heat* boiler 廃熱利用ボイラー.

waste incineration 廃棄物[ごみ]の焼却処理.

Waste Management ウェイストマネジメント社《世界最大の廃棄物処理会社; 1968 年設立; 本社米国イリノイ州 Oak Brook》.

wasteplex 廃棄物再循環処理施設.

waste product 1《生産工程で出た》廃棄物, くず, 廃産物, 産廃. 2 [通例複数形で]老廃物, (糞尿など生物の)排泄物.

waste stream 《最終的なごみ処理場に至るまでの》ごみの流れ.

waste-to-energy 形 ごみ焼却により電力を起こす, ごみのエネルギー転化の.

waste water (工場)廃水, 廃液, 下水: *waste water* treating 廃水処理.

water conservation 水保全.

water emulsion fuel エマルジョン燃料(emulsion fuel*).

water pocket 《岩などの》水のたまるくぼみ;《水の涸れた川の, 流水時には滝の落下点だったところにできた》河床のくぼみ; (大)滝つぼ.

water resourses 水資源

water scarcity 水不足

waterlog 動〈船を〉水浸しにして航行不能にする; 水のしみ込みで〈木材〉の浮力を奪う;〈土を〉水浸しにする.

water pollution 水質汚染, 水質汚濁.

wave-power station 波力発電所.

WBCSD World Business Council for Sustainable Development 持続可能な開発のための世界経済人会議.

WBGT wet bulb globe temperature*

147

WCED World Commission on Environment and Development*

WCPFC Western and Central Pacific Fisheries Commission [Convention] 中西部太平洋まぐろ類委員会[条約]《2000年ホノルルで採択され2004年発効した鮪類漁業資源保護を目的とする条約またはその実施組織；広い海域を回遊するマグロ類の資源管理を，日本に近い中西部大西洋海域で実施するもの；条約の正式名は「西部及び中部太平洋における高度回遊性魚類資源の保存及び管理に関する条約」(Convention on the Conservation and Management of Highly Migratory Fish Stocks in the Western and Central Pacific Ocean)；本部はミクロネシア連邦のポナペ島；⇨ICCAT*》.

web-foot 图形《米俗》環境[自然]保護論者(の)，環境保全に熱心な(やつ).

WECPNL weighted equivalent continuous perceived noise level 加重等価持続感覚騒音レベル，うるささ指数《空港周辺の航空機騒音についての国際的な評価単位；PNL*の数値に，離発着などで繰り返される騒音の回数を，1日の時間帯による重みづけを付加して計算したもの》.

WEEE Waste Electrical and Electronic Equipment 電気電子機器廃棄物，電子ごみ，ウィー《⇨WEEE Directive*》.

WEEE Directive Waste Electrical and Electronic Equipment Directive WEEE 指令《使用済み電子機器・電化製品の回収・リサイクルを義務づける欧州連合(EU)の指令；2005年に施行され，廃棄に際し有害物質を出さないこと，一定量の廃棄物リサイクルを行うことなどが定められた；RoHS*指令とともに欧州の環境規制策の柱となっている》.

weird beard, weirdie-beardie 《英俗》《反核運動・環境保護運動などに》入れ込みの激しい活動家，エコおたく.

West Wind Drift 西風海流(Antarctic Circumpolar Current*).

wet bulb globe temperature 湿球黒球温度，暑さ指数《気温のほかに湿度や放射熱も加味して，人が感じる暑さの目安とした指標；略 WBGT；環境省が熱中症などへの注意を促すために「熱中症予防情報サイト」で「暑さ指数」として掲載している；湿球温度 $T\text{wb}$, 黒く塗った中空銅球内の温度 $T\text{g}$, 通常の気温(乾球温度)$T\text{db}$ を 7:2:1 の割合で加重平均する；湿球温度 $T\text{wb}$ は湿度が低く風通しがよければ気化のため通常の気温よりかなり低くなる；また気温そのものが高くなくても熱線があ

たっていれば Tg を通じてその効果が反映される;》.

wetland 湿地, 湿地帯(⇨ Ramsar Convention*).

whale watching 鯨観察(会), ホエールウォッチング《鯨に近づいて船上などから観察する》.

wheelie bin 《英国でごみ収集作業を容易にするために地方税納税者に自治体が供与する》車輪付き大型ごみ箱.

wilderness area 《米国の》原生[自然]環境保全地域, ウィルダネスエリア.

Wildfowl and Wetlands Trust [the ～] 英国水禽湿地協会《カモなど水鳥の研究・保護のために1946年に創設》.

wildlife conservation park 《米》野生生物保護公園《zoo (動物園) の言い換え》.

wind energy, wind power 風力(エネルギー); 風力発電.

wind generator [plant] 風力発電機[所].

wind park, wind(mill) farm 風力発電地帯《多数の風力タービンが一面に設置されている》.

windthrow 風が木を根こぎにする[吹き倒す]こと; 風倒木.

wise use 「賢い利用」主義《絶対的な環境保全主義に対して, 自然環境をもっと多面的な賢い方法で利用しようという(企業寄りの)主張》.

wise-use 形「賢い利用」主義[派]の.

witchweed モロコシ・トウモロコシなどの根に寄生するゴマノハグサ科の有害植物.

withstand 動 抵抗する, (環境に)耐える.

Wohnbereich Woonerf*

Woodsy Owl ウッジー・アウル, 森のフクロー《梟の姿をした米国林野部 Forest Service* のマスコット; 森林環境保護を呼びかける "Give a hoot. Don't Pollute." (注意ホー, ゴミをホーるな), "Lend a Hand — Care for the Land!" (君も手伝え自然を救え) がスローガン》.

Woonerf 《ドイツやオランダで高速・多量の交通を拒否し住環境を護るように設計された》住宅地域内の道路(Wohnbereich).

World Commission on Environment and Development 環境と開発に関する世界委員会, ブルントラント委員会《日本政府の提案により, 1984年, ノルウェーの政治家(当時首相) G. H. Brundtland を議長とする「賢人会議」の形で国連に設置された委員会; 略称 WCED; 通称「国連環境特別委員会」(UN Special Committee on Environment); 世界の有識者21人が委員となり, 1987年の東京における最終会合で「われら共通の未来」('Our

Common Future')という報告書を公表；21世紀に向けて，持続可能な開発(sustainable development*)を最優先目標とすることを盛り込み，地球環境問題に取り組む際の中心概念を打ち立てた》．

World Environment Day 世界環境デー《6月5日；ストックホルムで1972年に開催された国連人間環境会議(United Nations Conference on the Human Environment*)を記念し，日本が設置を提案した；日本では「環境の日」》．

World Heritage Committee 世界遺産委員会《毎年1回開催；世界遺産の新規登録案件などを審議》．

World Heritage Site 世界遺産登録地《ユネスコ(UNESCO 国連教育科学文化機関)が世界遺産条約(Convention Concerning the Protection of the World Cultural and Natural Heritage)に基づいて，普遍的な価値のある自然遺産と文化遺産を審査・登録；⇨ 和英篇　世界(世界遺産)》．

World Summit on Sustainable Development 持続可能な開発に関する世界首脳会議，ヨハネスブルグ・サミット《南アフリカのヨハネスブルグで2002年8月〜9月に開催された国連主催の会議；地球サミット(United Nations Conference on Environment and Development*)で採択されたアジェンダ21(Agenda 21*)の見直しや，新たに生じた環境と開発の議題についての議論が目的；「持続可能な開発に関するヨハネスブルグ宣言」と「ヨハネスブルグ実施計画」が採択された》．

World Water Council [the 〜] 世界水会議《世界の水資源問題の解決策をさぐるために1996年に設立されたシンクタンク；世界水フォーラム(World Water Forum*)を主催する；本部フランスのマルセイユ；略 WWC》．

World Water Forum [the 〜] 世界水フォーラム《1997年から3年ごとに開催される水問題に関する国際会議；世界水会議(World Water Council*)が主催；略 WWF；会期は世界水の日(3月2日)を含む約1週間》．

World Wide Fund for Nature [the 〜] 世界自然保護基金《1961年創立の国際自然保護団体 World Wildlife Fund* の本部(WWFインターナショナル)が1986年改称(ただしアメリカとカナダの組織は旧称のまま)したもので，略称は現在も WWF；500万人以上のサポートを集め，100カ国以上で活動する世界最大の自然保護団体；本部はスイスのグラン；野生生物の保護から生態系保全，さらには持続可能な社会づくりに向けて活動している》．

World Wildlife Fund [the 〜] 世

界野生生物基金《野生生物保護を目的に1961年に設立された国際組織；英国のロンドン動物園に中国からパンダがやってきた時期で，同基金のわかりやすいシンボルとしてパンダがロゴマークに採用された；国際本部の初代総裁はオランダ女王の夫君ベルンハルト殿下，英国組織の初代代表はエリザベス女王の夫君フィリップ殿下がつとめた；日本でも1971年に正式にWWFの組織(現在のWWFジャパン)が発足している；のちに世界自然保護基金 World Wide Fund for Nature* に改称》.

WQC water quality criteria 水質規準.

WTO World Trade Organization 世界貿易機関《GATTを引き継いで1995年に発足した国連の関連機関》.

WTW well to wheel 油田から車輪まで《石油などの一次エネルギー原料採取から貯蓄・精製を経て実際に自動車を動かすため消費されるまでの製造工程や輸送行程を指す文句；消費エネルギー効率や排ガスなどの環境負荷を検討する際に言う》.

WWC World Water Council*

WWF 1 World Water Forum*
2 World Wide Fund for Nature*

XYZ

xerarch 形【生態】〈遷移が〉乾燥した場所で起こる，乾生の.

xeric 形〈土壌などが〉乾燥した；〈植物などが〉好乾[耐乾]性の，乾生の.

xeriscape 乾景観《水やりが少なくて済む耐乾性植物でつくった景観》；[Xeriscape で]【商標】ゼリスケープ《乾燥地で節水をしながら行なう造園法》.

xeriscaping, xerigscaping 乾景観づくり.

xerophyte 【生態】乾生植物(⇨ halophyte*, mesophyte*).

xerophytic 形 乾生的な.

xerosere 【生態】乾生(遷移)系列《乾燥した生育地に始まる遷移系列》.

yellow algae 黄藻《単細胞プランクトンの一種；大増殖すれば黄潮となり海水を酸欠状態にして魚類・アシカ・アザラシなどの生存をおびやかす》.

yellowcake イエローケーキ，ウラン精鉱《核燃料である金属ウランをつくる原料ウラン鉱の粗精錬産物》.

yellow earth 《湿潤亜熱帯常緑樹林の》黄色土.

yellow sand 黄砂.

ZBA zoning board of approval. (米国などの)土地用途規制委員会.

zinc pyrithione ジンクピリチオン，ピリチオン亜鉛(pyrithione zinc)《抗菌薬・抗真菌薬；水にほとんど不溶で表皮細胞に対する細胞増殖抑制活性がある；ふけや脂漏性皮膚炎防止のためにシャンプーなどの頭髪用製品に多く使われてきた；内分泌攪乱(⇨ environmental hormone*)作用の疑いが報告されている；略 zpt, ZPT》.

Zip team ジップチーム《英国の地方自治体に雇われた，地区美化監視チーム；ごみを投げ捨てる者に反則キップを切ることができる》.

zone of saturation 飽水帯，飽和帯《地下水面より下の地下水に満ちた部分》.

zoning 《都市計画において工場・住宅地帯などの》地帯設定，地域制，ゾーニング.

zoobenthos 【生態】底生動物.

zoogeographic region 【生態】動物地理区.

zoology 動物学；動物学論文；動物相(fauna*)；動物生態.

zoologist 動物学者.

zooparasite 【生態】寄生動物，動物性寄生体．
  zooparasitic 形 寄生動物の．
zoophytology 植虫学．
zpt, ZPT zinc pyrithione*

和英篇

Kenkyusha English and Japanese
Dictionary of Ecology

アースデイ ⇨英和 Earth Day

愛護 〔(野生生物などの)保護・保存〕protection; preservation; conservation;〔(動物などを)いたわること〕kind(ly) treatment; tender care. ▶～する protect; preserve; conserve; treat《an animal》kindly [with tender care]; be kind to《an animal》/ 動物～の精神 concern for the welfare of animals.

愛鳥 *one's* pet bird. ▶～家 a bird fancier [lover] / 5月10日から16日までは～週間である．Bird Week is observed [held] from May 10 to 16.

アイドリング 〔エンジンの空転状態〕idling. ▶ガソリンの節約と排気ガスの抑制のためにも停車中の～は控えるべきである．Idling when a vehicle is stopped ought to be refrained from, both to save gasoline and to cut down on exhaust fumes. ¶**アイドリング・ストップ** stopping [switching off] the engine when a vehicle is not moving; "idling stop." **アイドリング・ストップ運動** a campaign to encourage drivers to stop [switch off] their engines when not moving; an "idling stop" campaign. **アイドリングストップ車** 〔車が停止すると自動的にエンジンも停止する機構を備えた自動車〕an idle-stop vehicle. **アイドリングストップ宣言車** a vehicle that is taking part in an "idling stop" campaign;〔掲示〕This Vehicle Shuts Off Its Engine When Stopped.

相乗り ▶～する ride together;〔オートバイなどの後ろに〕ride pillion. / タクシー[車]の～ taxi [car] pooling. / タクシーに～する share a taxi [cab]《with a stranger》. / ～させる give《a stranger》a lift [ride]《in a taxi》; take《a stranger》to his home《in *one's* taxi》. / たくさんの人がタクシーを待っていたが，幸い親切な人が～させてくれた．A lot of people were waiting for taxis, but luckily a kind man let me share his [gave me a lift in his].

亜鉛中毒 zinc poisoning.

青粉(あお) 〔水を緑に濁らせる微生物の総称〕blue-green algae; water bloom. ▶～が発生して，湖が

緑色に見える．Because of the growth of blue-green algae, the lake looks green.

青潮(あおしお) 〔低酸素水塊の上昇による海面の変色〕(a) blue tide.

青立ち ▶〜の稲田 a field of rice plants not yet in ear / 8月になっても低温続きで，稲は〜の状態である．Although it is already August, due to the continued low temperatures, the rice plants are not yet in ear.

青の革命 〔緑の革命(⇨英和 green revolution)に対し，水危機克服のための取り組み〕a blue revolution.

赤潮 〔プランクトンの大量発生による海面の変色〕(a) red tide; red water. ▶相模湾と駿河湾で大量の〜が発生した．(A) red tide occurred [was generated] in Sagami Bay and Suruga Bay. / 〜の被害 red tide damage; damage done by (a) red tide.

空き缶，空き瓶 an empty [a used, an old] can [tin, bottle]. ▶空き缶を拾う pick up empty cans. / 空き缶[空き瓶]を回収する collect empty cans[bottles]. / 空き缶回収 empty can collection. / 〔標語として〕空き缶はリサイクルへ．Recycle empty cans. | Return empty cans for recycling. / 空き缶は空き缶入れに．〔掲示〕Empty cans into the empty-can receptacle, please! | Please put empty cans in the receptacle provided. / 空き缶ポイ捨て禁止条例 an ordinance making it illegal to drop empty cans.

ア(ッ)クア・アルタ 〔イタリアのヴェネスなどで，冬の満潮時に海面上昇により陸地が冠水する現象〕acqua alta ★元来はイタリア語で「高い(alta)潮(acqua)」の意；high water [tide].

悪臭 a bad [foul, nasty] smell; an offensive odor [smell]; a stench; a stink; a reek;《文》(a) malodor;《文》〔廃棄物などから発する〕(an) effluvium《複数形 effluvia》;《文》a fetor. ★⇨異臭，臭気 ▶猛烈な〜 a powerful stench [stink]. / 〜のある stinking; smelly; foul-[bad-, evil-, ill-]smelling. / 〜がする．Something smells (bad) [stinks]. / その部屋に入ると〜が鼻を突いた．When we entered the room, there was a horrible smell. / 〜を放つ smell (bad); stink; give off [out] a (bad) smell; reek《of cigarettes》;《文》emit an offensive odor. / 〜を除去する[消す] remove [get rid of] a bad smell; deodorize. ¶**悪臭公害** olfactory pollution; (an) olfactory nuisance; a bad smell constituting a public nuisance;

《口》"smell pollution." **悪臭防止法** the Malodor Prevention Law.

アクティブレンジャー 〔レンジャーの補佐役〕an acting [assistant] ranger.

アグリビジネス ⇨英和 agribusiness

アグロフォレストリー 〔農業・林業両用の土地利用〕agroforestry. 図〔その従事者〕agroforester.

亜高山帯 a subalpine zone. ▶～の動植物 subalpine animals and plants.

アゴニスト 〔作用薬[物質]〕agonist.

亜酸化窒素 〔全身麻酔薬〕nitrous oxide.

アジア太平洋統合モデル the Asian-Pacific Integrated Model (略: AIM).

アジェンダ21 ⇨英和 Agenda 21

足尾(銅山)鉱毒事件 〔1890年代の〕the Ashio Copper Mine Pollution Case.

亜硝酸性窒素 a nitrite nitrogen.

足輪 〔生態調査のために鳥の脚につける〕a leg band [ring]; a bird band. ▶私たちは渡り鳥に～をつけて放した．We put leg bands on migratory birds and then released them. / ～をつけているハトが見つかった．A pigeon with a leg band was found.

アスファルト asphalt. ▶～を敷く，～で舗装する surface [pave, lay] 《a road》 with asphalt; asphalt 《a road》. ¶**アスファルト固化** 〔核廃棄物の〕asphalt solidification; bituminization. **アスファルト道路** an asphalt [a blacktop] road.

アスベスト asbestos. ★⇨石綿(せきめん) ¶**アスベスト金網** wire gauze with asbestos. **アスベスト含有建材** asbestos-containing building materials. **アスベスト含有廃棄物** asbestos-containing waste (略: ACW). **アスベスト公害** pollution by asbestos; asbestos pollution. **アスベスト症** ⇨石綿(石綿沈着症). **アスベスト新法** ⇨石綿(石綿健康被害救済法). **アスベスト肺** ⇨肺石綿症. **アスベスト被害[被曝]** 《suffer from》asbestos exposure [exposure to asbestos]. **アスベスト被害者** an asbestos exposure victim. **アスベスト粉塵** asbestos particles. **吹き付けアスベスト** (a) sprayed asbestos coating.

アセス(メント) 〔評価・査定〕assessment. ¶**環境アセスメント** 〔環境影響評価〕environmental assessment; environmental impact study. **アセス法** ⇨環境(環境影響評価法)

アセトアルデヒド acetaldehyde.

暑さ指数 〔環境省が発表する〕the WBGT index. ★⇨英和 wet bulb globe temperature

アトピー atopy. 圏 atopic. ¶**アトピー患者** an atopy sufferer; a patient suffering from atopy; an atopy case. **アトピー性喘息** atopic asthma. **アトピー性皮膚炎** atopic dermatitis. **アトピー体質** a predisposition to atopy. ▶〜体質の人 a person with [who has] a predisposition to atopy; a person constitutionally predisposed to atopy.

穴あきダム a flood control dam.

アニマル・セラピー 〔動物療法〕animal(-assisted) therapy（略：AAT）.

アニマル・ライツ ⇨英和 animal rights

亜熱帯 a subtropical zone [region]; the subtropics. ▶この島は〜に属する．This island is located in a subtropical zone [region]. / 〜の subtropic(al); near-tropical. ¶**亜熱帯気候** a subtropical climate. **亜熱帯高圧帯** a subtropical high-pressure belt. **亜熱帯高気圧** a subtropical high [anticyclone]. **亜熱帯ジェット気流** a subtropical jet stream. **亜熱帯植物** a subtropical plant. **亜熱帯多雨林** a subtropical rain forest. **亜熱帯動物** a subtropical animal.

油，脂 ⇨石油 ▶油にまみれている be covered in [with] oil; be filthy with oil; be all oily / 油にまみれた鳥 birds smothered in oil / 油による汚染 oil pollution. ★⇨汚染，油濁 ¶**油漏れ** an oil leak; oil leakage. **油流出事故** an oil spill accident. **油火災** an oil [a flammable-liquid] fire; a Class B fire. **油除け** an oil guard. **油[脂]汚れ** oil [grease] stains; greasy dirt. ▶油汚れに強い洗剤 a detergent good at removing grease [greasy dirt]. | a detergent that's good for getting out oil stains. / 油汚れを落とす remove grease.

雨水 rainwater. ★⇨雨水(うすい) ▶有害物質が〜に流されて川に流入した．Harmful materials were washed away by the rainwater. / 〜の利用 use of rainwater; rainwater usage. / 森林には〜を貯留する働きがある．The forests function to store [accumulate] rainwater.

アメニティー 〔快適な環境・設備〕(an) amenity; amenities. ▶〜重視の設計 an amenity-based design; a plan [design]《of a building》that emphasizes convenience; a design that focus on (providing) convenient amenities. ¶アメニティー空間 an amenity space; a space for the amenities《in a house [an apartment]》.

アモコ・カジズ号 ⇨英和 Amoco Cadiz

アラス ⇨英和 thermokarst

亜硫酸 sulfurous acid; sulphurous acid. ¶**亜硫酸ガス**〔二酸化硫黄〕sulfurous acid gas; sulfur dioxide. ▶〜ガス中毒 sulfurous acid gas poisoning. / 〜ガス漂白 stoving; sulfurous acid bleaching.

アルカリ alkali. ▶〜性の alkaline / 〜性にする alkalize; alkalinize; alkalify. / 弱〜の alkalescent. ¶**アルカリイオン水** alkaline (ionized) water. ▶〜イオン整水器 an alkaline ionized water filter. **アルカリ性欠乏** alkaline deficiency. **アルカリ性食品** alkaline food. **アルカリ性体質** an alkaline body; alkaline blood; alkalosis. **アルカリ性土壌** alkali [alkaline] soil; basic soil. **アルカリ石けん** alkaline soap. **アルカリ泉** an alkaline spring. **アルカリ中和剤[中和物]** an antalkali; an antalkaline (agent). **アルカリ電池** an alkaline battery [cell]. **アルカリ土類** alkaline earths. **アルカリ土類金属** an alkaline earth metal. **アルカリ標準液** standard alkali solution. **アルカリマンガン電池** an alkaline manganese cell. **廃アルカリ** waste [spent] alkali.

アルキル鉛 alkyllead.

アルゴ計画 ⇨英和 Argo project ¶**アルゴ・フロート**〔アルゴ計画用の水深 2000 メートルから海面までの海水の組成を約 10 日間ごとに計測する装置〕an Argo float.

アルコール〔総称〕an alcohol;〔エチルアルコール〕ethyl alcohol; ethanol; alcohol. ▶〜で消毒する disinfect [sterilize]… with alcohol. / 〜を分解する酵素 an enzyme that breaks down alcohol [breaks alcohol down]. ¶**工業用アルコール** industrial alcohol. **醸造用アルコール** brewing alcohol. **消毒用アルコール** ethanol [alcohol] for disinfection. **変性アルコール** denatured alcohol. **アルコール飲料** alcohol; an alcoholic drink [beverage]; a drink containing alcohol. **アルコール温度計** a spirit thermometer; an alcohol thermometer. **アルコール系合成洗剤** alcoholic synthetic detergent. **アルコール自動車** an alcohol car; a car [vehicle] that uses alcohol as fuel. **アルコール燃料** (an) alcohol fuel.

アルシュ・サミット ⇨英和

荒れる

Arche Summit

荒れる （土地などが）run to waste; be desolate [neglected]；（建物が）go [fall] to ruin; fall into decay; be ruined [ruinous, dilapidated, tumbledown]; be in bad repair. ▶畑が荒れている．The field lies waste./ 庭は荒れて草がぼうぼうとしている．The garden's been neglected and is overgrown [the weeds have run riot]./ 沖縄の珊瑚の海は，急速な開発によってすでにかなり荒れている．Okinawa's coral seas are already fairly devastated as a result of rapid development.

アレルギー an allergy. ▶〜性の allergic. ★⇨アレルギー誘発性/ 〜性の炎症 an allergic inflammation./ 〜性疾患 an allergic disease; an allergy./ 〜性鼻炎[皮膚炎，喘息] allergic rhinitis [dermatitis, asthma]./ うちの子供たちはみんな卵に対して〜がある．Our children are all allergic to eggs./ 〜になる，〜を起こす develop an allergy./ 社会的ストレスの影響で都会の人々は〜を起こしやすくなっている．As a result of social stresses people (who live [work]) in big cities become readily susceptible to allergies./ ストレスが〜を引き起こす場合もある．In some cases stresses can cause an allergy [allergies]. / 〜を治す cure an allergy./ 私は金属〜だ．I have an allergy to certain metals./ 花粉〜 an allergy to pollen. ★⇨花粉症 / 大豆が原因の食物〜 a food allergy caused by soybean.

¶ **アレルギー学** allergology. **アレルギー患者** an allergy sufferer. **アレルギー原因物質** ⇨アレルギー物質. **アレルギー検査** an allergy test. ▶病院で 10 種類の〜検査をした結果，スギ花粉だけに反応した．I underwent tests for ten different types of allergy in the hospital and had a reaction only to cedar pollen. | I was tested at the hospital for ten different types of allergy, and my only positive reaction was to cedar pollen. **アレルギー試験** allergy [allergen] testing;〔1 回の〕⇨アレルギー検査. **アレルギー症(状)** an allergic symptom; an allergy condition. ★⇨アレルギー反応 ▶〜症状に苦しむ suffer from an allergy condition./ 〜症状を呈する develop an allergy condition; exhibit allergic symptoms./ 薬や化粧品により〜症状を起こしたことのある人は本製品を使用しないでください．Do not use this product if you have previously developed allergic symptoms from medicines or cosmetic products.　**アレルギー対応食** food (suitable) for a person with a food allergy [food allergies]; meals for people with food allergies.　**アレルギー体質** a (physical) constitution susceptible to allergy; allergic predisposition.　▶私は〜体質だ．I have a (genetic) susceptibility [predisposition] to allergic reactions.　**アレルギー物質** an allergen; an allergenic substance.　▶〜物質表示〔食品などに表示すること〕labeling of allergens; allergen labeling;〔そのラベル〕an allergen label./ 〜物質を含む食品 foods containing allergenic substances [allergens].　**アレルギー専門医** an allergist.　**アレルギー素質** allergic diathesis [predisposition].　**アレルギー反応** an allergic reaction.　▶〜反応を調べる test for any allergic reaction./ スギ花粉に強い〜反応を示す exhibit [show] a strong allergic reaction to cedar pollen.　**アレルギー誘発性** allergenicity.　圏 allergic.　**I型アレルギー反応** type I allergic reaction.　**寒冷アレルギー** cold allergy [urticaria].　**蕎麦(そば)アレルギー** an allergy to buckwheat; a *soba* allergy.　**猫[犬, 鳥]アレルギー** (an) allergy to cats [dogs, birds]; (a) cat [dog, bird] allergy.　**薬物アレルギー** an allergy to a medication.

暗渠(あんきょ)　a culvert; a (covered) drain; a blind ditch; a closed [covered] conduit; an underdrain; a duct.　▶〜で排水する drain by (a) culvert; underdrain./〔川を埋めて〕〜にする cover a river over; channel [divert] a stream underground./ 〜排水 drainage by (a) culvert; underdraining;〔農地の〕(agricultural) under-drainage.

安全　safety; security; freedom from danger.　▶〜な safe; secure; free from danger.

¶ **安全意識**　《lack》safety awareness.　**安全運転** careful [safe] driving.　▶〜運転をする drive carefully [with care]./ 〜運転を心がける make a point of driving safely; always drive safely./ 〜運転を心がけているドライバー a safety-minded driver.　**安全運転義務違反** (a) violation of safe-driving practices.　**安全運転管理者** a driving safety manager.　**安全運転支援システム** 〔路上の危険情報などを車内のナビゲーションに伝えるシステム〕a [the] Driving Safety Support System (略:

安全

DSSS). **安全運転宣言車**〔輸送トラックなどに貼ったステッカー表示〕a safe-driving declaration vehicle. **安全確認**《do, carry out》a safety check. ▶〜確認を怠る fail [neglect] to do a safety check. **安全確保命令**〔事故などを起こした運送業者などに国土交通省が発する〕an order [instructions] to improve safety standards. **安全監査**〔行政当局による,交通・食品・医療・衛生などについての〕a safety audit. **安全監査員** a safety auditor. **安全基準** safety standards. ▶国の〜基準を上回る水質汚染レベル water contamination levels exceeding government safety standards. **安全規則** a safety regulation [rule]. **安全教育** safety education; education in safety. **安全距離** a safe distance. **安全区域** a safety zone. **安全神話** a myth about [surrounding] the safety of《a nuclear power plant》; the myth that *sth* is completely safe. **安全性** (a degree of) safety [security]. ▶食品の〜性 food safety; the safety of food. / 原発の〜性に絶対の自信を持っていた. They were absolutely convinced of the safety of nuclear power plants. / これらの施設の〜性に問題がある. There are problems with the security of these facilities. / 薬品の〜性を確認する verify the safety of a drug. / 輸入食品の〜性を確保する guarantee [ensure] the safety of imported foodstuffs. / それは高速増殖炉の〜性を保証する根拠の1つだった. That was one of the principles that guarantee the safety of a high-speed breeder reactor. / 残留農薬の〜性評価 an assessment of the (degree of) safety of residual agricultural chemicals. **安全性評価指針** a 《food》safety assessment guideline;〔厚生労働省が定める,組み換え DNA 技術応用食品・食品添加物の安全性評価指針〕the Guideline for Safety Assessment of Foods and Food Additives Produced by Recombinant DNA Techniques. **安全設備** a safety device [installation]; safety facilities. **安全宣言** a declaration of safety [the safety of《beef》]. ▶〜宣言をする issue a declaration of safety; declare《beef》safe. **安全操業** safe operation《of fishing equipment, etc.》. **安全措置** a safety [fail-safe] device; a safeguard. **安全対策** a safety measure [precaution]; a safeguard. **安全地帯** a safety zone;〔街路上の〕a safety island [isle]; a street [road, traffic, pedestrian] island; an island. **安全点検**《carry out》a safety check [inspection]. **安全灯** a safety lamp;〔航空機の〕a blaze orange. **安全ネット** a safety net. **安全配慮義務** responsibility [the duty] to ensure [maintain] safety; responsibility for safety;〔雇用者の労働者に対する〕《the employer's》responsibility for workplace [worker] safety. ▶～配慮義務違反 a breach of responsibility to ensure safety; failure to maintain safety. **安全評価** (a) safety assessment [evaluation]. **安全評価基準** standards for (a) safety assessment [evaluation]. **安全評価委員会** a safety assessment committee. **安全評価報告書**〔原発などの〕a safety assessment report. **安全標識** a safety sign [notice]. **安全ベルト** a safety belt;〔座席の〕《put on, wear》a seat belt. **安全帽** a safety [crash] helmet; a hard [crash] hat. **安全棒**〔原子炉緊急停止用の制御棒〕a safety rod. **安全問題** ▶～問題の研究家 a safety researcher; an expert on safety problems [matters]. **安全率** a safety factor.

アンドロゲン〔男性ステロイドホルモンの総称〕androgen.
アンブレラ種(しゅ)〔地域の生態ピラミッドの頂点に位置し,個体群維持のために広い生息地を必要とする動物〕an umbrella species.
安眠妨害 disturbance of sleep; a nuisance in the middle of the night. ▶～だからラジオを止めてくれ. Turn off the radio. It prevents us from sleeping.
アンモニア ammonia ▶～臭 the smell of ammonia./この溶液は～臭がある. This solution smells of ammonia./～濃度 ammonia concentration; an ammonia level.

い

イエロー・ケーキ 〔ウラン精鉱〕 yellow cake.

硫黄(いおう) sulfur; sulphur; brimstone. ▶〜分の多い[少ない] high-[low-]sulfur《fuels》. / 〜分[質]を抜く desulfurize. / 〜臭い stink [smell] of sulfur. / 〜で処理する sulfuret; sulfurize; sulfurate; treat with sulfur. / 〜の, 〜を含む sulfurous; sulfureous; sulfuric; sulfury; thionic. / 〜を含ませる[と化合させる] sulfurize; sulfur. ¶**硫黄栄養性** 〔生物の〕thiotrophy. **硫黄温泉華** a sulfurous hot spring deposit. **硫黄華** flower of sulfur; sublimed sulfur. **硫黄細菌**[バクテリア] a sulfur bacterium. **硫黄サイクル** 〔地球環境中での〕the sulfur cycle. **硫黄酸化物** a sulfur oxide. **硫黄泉** a sulfur [sulfurous, sulfureous] spring.

イオン交換 (an) ion exchange. ¶**イオン交換樹脂** ion exchange resin. **イオン交換水** ion exchanged water; deionized water. **イオン交換体** an ion exchanger. **イオン交換膜** an ion exchange membrane. **イオン交換膜法** the ion-exchange-membrane method.

維管束(いかんそく) a vascular bundle. ¶**維管束植物** a vascular plant.

閾値(いきち) a threshold. ▶〜が高い[低い] have a high [low] threshold《for pain》. / 〜下の subliminal. ¶**閾値刺激** threshold [liminal] stimuli. **閾値線量** a threshold dose (of radiation).

生き物 〔命あるもの〕a living thing [being]; a living creature; 〈集合的に〉life; the living world; 《文》animate nature. ▶干潟には種々様々な〜がすんでいる. The tidal flats provide a home for a wide variety of living creatures [things]. | Many forms of life live on the mud flats. / 理由もなく〜を殺すことは重大な罪だ. To kill a living creature [To destroy life, The taking of life] for no reason is a grave sin.

移行帯 ⇨英和 ecotone

異臭 ⇨悪臭, 臭気 ¶**異臭騒ぎ** a scare caused by a foul smell.

異常 ▶この夏の暑さは〜だね. The heat this summer is extraordinary. ¶**異常乾燥** abnormal dryness; a spell of unusually dry weather. **異常乾燥注意報** a warning of abnormally dry conditions. **異常寒波** an unusual cold spell. **異常気象** abnormal weather; freakish weather. ★⇨因果関係 **異常高温[低温]** unusually high [low] temperatures. **異常潮位** an anomalous high tide. **異常放射能** abnormal radiation. **異常隆起** 〔地面の〕an abnormal upthrust. **異常繁殖** abnormally rapid reproduction; rampant [unchecked] growth.

石綿(いしわた) ⇨石綿(せきめん)，アスベスト

磯焼け 〔コンブ・テングサなどの有用な海藻が沿岸岩礁地域で一斉に枯死する現象〕rocky-shore denudation.

遺存種 ⇨英和 relict

イタイイタイ病 〔カドミウム慢性中毒症〕itai-itai [ouch-ouch] disease.

一次 ¶**一次宿主** 〔寄生虫などの媒介における〕a primary host; an original host. **一次消費者** 〔生態系の〕a primary consumer. **一次処理** 〔汚水などの〕the primary treatment. **一次遷移** a primary succession. **一次遷移系列** a primary sere; a prisere. **一次BOD** 【土木】first-stage BOD. **一次冷却水** 〔原子炉などの〕primary [first-stage] cooling water. **一次冷却装置** 〔原子炉などの〕a primary cooling system.

一日摂取許容量 〔有毒物質の〕acceptable daily intake（略：ADI）.

一酸化炭素 carbon monoxide. ¶**一酸化炭素検知管** a carbon monoxide detector tube. **一酸化炭素中毒** carbon monoxide [CO] poisoning. ▶〜中毒を起こす〈物が主語〉cause carbon monoxide poisoning;〈人が主語〉become ill from [suffer] carbon monoxide poisoning /〜中毒死する die of [from] carbon monoxide poisoning. **一酸化炭素ヘモグロビン** carboxyhemoglobin. **一酸化二窒素** nitrous oxide; dinitrogen monoxide.

一般 ▶〜家庭から出るごみの量も増え続けている．The amount of garbage coming out of ordinary households continues to grow. ★⇨家庭ごみ ¶**一般環境大気測定局** 〔大気汚染防止法に基づき各地に設置される〕an ambient air pollution monitoring station. **一般廃棄物** general [municipal, nonindustrial, normal household, domestic] waste. **一般廃棄物処理業** the general-waste management [disposal] industry; a general-waste disposal [management] business.

遺伝

一般廃棄物処理施設 a general-waste treatment [disposal, management] facility.

遺伝 heredity; (hereditary) transmission; (genetic) inheritance. ★⇨遺伝子 ▶〜する be inherited; be transmitted 《from the parents》; be handed down. / その病気は〜する．That disease is hereditary. / 〜の，〜的な hereditary;〔遺伝した〕inherited. / 放射能による〜的障害 genetic damage caused by radioactivity. / 〜的多様性 genetic diversity. / 〜的危険率 genetic risk. / 〜的素因 hereditary predisposition. / 〜的多型 genetic [DNA] polymorphism. / 〜のメカニズム a genetic mechanism. / その魚はよく〜の実験に用いられる．That type of fish is often used in experiments regarding heredity.

¶ **遺伝暗号** the genetic code. **遺伝カウンセリング** genetic counseling. **遺伝資源** genetic resources. / 植物〜資源 plant genetic resources（略：PGR）**遺伝疾患** (a) genetic disease; a genetic disorder. **遺伝型** a genotype. **遺伝基質**【生物】an id. **遺伝形質**[特性，性向] an inherited [a hereditary] character [quality, tendency]. **遺伝質滲透** introgression. **遺伝情報** genetic information. / 〜情報の解読 decoding of genetic information. **遺伝素** a determinant. **遺伝毒性** genotoxicity; genetic toxicity. / 〜毒性試験 a genotoxicity test [assay]. / 〜毒性物質 a genotoxin. **遺伝病** a hereditary [family, congenital] disease. **遺伝標識** a genetic marker. **遺伝物質** genetic material. **遺伝法則** the laws of heredity. / 血液型は〜法則によって決まる．Blood type is determined by the rules of heredity. **隔世遺伝** atavism（形 atavistic）; (a) reversion; (a) throwback. ★⇨優性遺伝，劣性遺伝

遺伝子 a gene. ★⇨DNA（和英篇末尾）▶ヒトの〜 a human gene. / 〜の［に似た，に起因する］genic; genetic. / 〜の損傷 damage [injury] to a gene [genes]. / 〜を合成する synthesize a gene. / 〜を組み入れる insert [introduce] genes into 《plants》. / 〜を操作する manipulate genes. / 赤ん坊は両親の〜を受け継いで生まれてくる．A baby inherits the genes of both its parents. | A baby inherits its parents' genes. / ダイオキシンなどの有害物質が人間にどういう影響を与えるかを〜レベルで研究する study the effects of dioxin and other harmful substances on human beings at the genetic level; study the genetic effects on humans of harmful substances such as dioxin.

¶ 遺伝子暗号　the human genetic code.　遺伝子移植　gene transplantation.　遺伝子医療　medical genetics; genetic medicine.　遺伝子汚染　genetic pollution [contamination].　遺伝子解析　(a) genetic analysis; (a) gene analysis.　遺伝子型　a genotype.　遺伝子記号　a genetic symbol.　遺伝子組み換え[改変]　genetic [gene] recombination; genetic modification (略: GM); recombination of genes. ★⇨カルタヘナ法 / ～組み換え技術 gene recombination technology. / ～組み換え植物[動物] a transgenic [recombinant, genetically engineered] plant [animal]. / ～組み換え生物[体] genetically modified organisms (略: GMOs). / ～組み換え作物 (a) transgenic [genetically modified, genetically engineered] crop; (a) GM crop. / ～組み換え大豆 genetically modified [engineered, altered] soybeans; GM soybeans. / ～組み換え品種 a genetically modified variety. / ～組み換えを施した genetically-modified《corn》. / ～組み換え食品 (a) genetically modified [engineered, altered] food; (a) GM food. / ～組み換えをしていない，非～組み換えの GM-free《corn》; non-genetically modified《rice》; non-GM [GMO]《soybeans》農産物の～組み換えに規制の網をかける institute a complex of restrictions on genetic recombination of agricultural products ★⇨英和 GM-free　遺伝子群　a gene cluster.　遺伝子検査　a genetic [gene] test.　遺伝子工学　genetic engineering.　遺伝子工学者　a genetic engineer.　遺伝子構造　genetic structure [makeup].　遺伝子座　a gene [genetic] locus《複数形 loci》.　遺伝子差別　genetic discrimination.　遺伝子資源　gene resources.　遺伝子重複　gene duplication.　遺伝子障害　a genetic defect.　遺伝子情報　genetic information.　遺伝子情報解析　genetic information analysis.　遺伝子診断　(a) gene diagnosis; genetic testing.　遺伝子スプライシング[接合]　gene-splicing.　遺伝子刷り込み　genomic imprinting.　遺伝子操作　gene [genetic] manipulation; genetic engineering. / ～操作された genetically engineered (略: GE)《plants》.

遺伝子増幅　gene amplification. 遺伝子損傷　genetic damage; a gene injury; a genetic hazard. 遺伝子ターゲティング　(gene) targeting; targeted gene replacement. 遺伝子多型〔DNA の配列の個体差〕genetic [DNA] polymorphism. 遺伝子地図[マップ]　a genetic [gene] map. 遺伝子地図作製　gene mapping; genetic mapping. 遺伝子治療[療法]　gene therapy. 遺伝子治療薬　a gene therapy product [drug]. 遺伝子導入　transgenics; gene transfer. 遺伝子導入植物[動物]　a transgenic plant [animal]. 遺伝子ドーピング　genetic doping. 遺伝子特許　a gene patent. 遺伝子(突然)変異　(a) gene [genetic] mutation. 遺伝子内組み換え　intragenic recombination. 遺伝子ノックアウトマウス　a gene knockout mouse. 遺伝子配列　gene sequence [arrangement]. 遺伝子バンク[銀行]　a gene bank. 遺伝子病　a genetic disorder [disease]; a hereditary disease [disorder] 単一～病 a monogenic disease [disorder]; a single-gene disorder [disease]; a Mendelian disease [disorder] 多～病 a polygenic [multigenic] disease [disorder]; a multifactorial disease [disorder]. 遺伝子頻度　(a) gene frequency. 遺伝子プール[給源]　the gene pool. 遺伝子浮動　genetic drift; the Sewall Wright effect. 遺伝子分析　(a) gene analysis. 遺伝子密度　gene density. 遺伝子ライブラリー　a gene library. がん[腫瘍形成]遺伝子　an oncogene. がん抑制遺伝子　a tumor suppressor gene. 偽遺伝子　a pseudogene. 原遺伝子　a protogene. 合成遺伝子　a synthesized gene. 人工遺伝子　an artificial gene. 性決定遺伝子　a sex-determining gene. 対立遺伝子　an allele. 治療用遺伝子　a therapeutic gene. 複対立遺伝子　multiple alleles. 優性遺伝子　a dominant gene. 劣性遺伝子　a recessive gene.

移動　migration. ▶～する〔鳥などが〕migrate. ¶移動禁止(措置)〔家畜伝染病が発生した際の家畜の〕a ban on moving《poultry across prefectural borders》; an《animal》movement ban. 移動式トイレ　⇨仮設トイレ　移動養蜂　migratory beekeeping. 移動性野生動物の種の保全に関する条約　⇨英和 Bonn Convention

移入　〔生物の〕immigration; introduction《of new plant or animal life》. ▶～する〔生物を〕immigrate; introduce《new plant or animal life》. ¶**移入種** [生物, 動物, 植物] an immigrant.

医療廃棄物 medical waste.

因果関係 a relation(ship) of cause and effect; a cause-and-effect relation(ship); cause and effect; a causal relation(ship); a causal link. ▶異常気象とエルニーニョ現象には強い〜が認められる. It is clear that there is a strong cause-and-effect relation(ship) between El Niño and abnormal weather. | A strong (causal) relation(ship) has been established between El Niño and abnormal weather conditions. / この工場の排水と地域住民の病気との間には, 確かに〜が存在する. Waste from this factory is certainly involved [definitely plays a role] in the health problems among the local residents. / 一部の野生生物の生殖能力の衰えは, 環境ホルモンとの〜が疑われている. Endocrine disrupters are suspected of being involved [playing a (causal) role] in the declining fertility of certain kinds of wildlife. / 彼は, 喫煙とがんとの〜をまだ認めていない. He still doesn't accept that smoking has anything to do with (causing) cancer [there is any link between smoking and cancer].

インタープリター 〔自然ガイド・体験学習の指導員〕an interpreter.

インバース・マニュファクチャリング 〔資源循環工程を組み込んだ生産システム; 逆生産〕inverse manufacturing.

インベントリー ⇨英和 inventory

飲用 ▶川の水を〜する場合は必ず沸騰させてから使ってください. Always boil river water before drinking it. / ここの温泉は〜すれば胃腸病にも効く. Drinking the water from this hot spring also helps to cure an upset stomach. / 〜に適している be fit [good] to drink; be drinkable. / 〜に適さない be unfit to drink; be undrinkable. / この井戸の水はそのままでは〜に適さない. The water in this well is not fit to drink (as is [without being treated]). / 〜の〔飲むための〕for drinking; potable;〔内服用の〕for internal use;〔飲める〕drinkable; fit to drink. / 汚染源が特定されるまでこの井戸水の〜を禁止する. Until the source of contamination is identified, drinking from this well is prohibited.

飲料水 drinking water; water to drink; potable water;〔掲示で〕Fit to drink. ▶この井戸水は水

飲料水

質に問題があって～には適さない．The water from this well is unfit for drinking because of water quality problems. / ～の安全基準 safety standards for drinking water. / ～の汚染 (the) contamination [pollution] of drinking water. / ～の確保が目下の急務である．Acquiring [Finding] a secure potable water supply is now a pressing issue. / ふだんから雨水をためておけば，非常時には～としても使える．If effective steps were taken to store [save] rainwater, it could also be used for drinking in emergencies.

う

ウィーン条約 ⇨オゾン(オゾン層保護条約)

ウインド・ファーム 〔風力発電機を多数設置した地域〕a wind park [farm].

ウォームビズ "Warm Biz"; the Japanese government's program to encourage people to wear warm clothes and use less heating during the winter.

ウォーレス線 〔動物分布上の〕Wallace's Line.

魚付き林(うおつきりん) 〔海・湖沼・河川の沿岸に魚を集めるために育成した林〕a fish-breeding forest; a riverside [coastal] forest that encourages fish breeding.

雨期 the rainy [wet, moist] season;〔熱帯の〕the rains; the monsoons. ▶熱帯に〜が来た．The rains [monsoons] came to the tropics. / 半月前からこの地域は〜に入った．Here in this region the wet [rainy] season set in half a month ago.

雨水(うすい) rainwater. ★⇨雨水(あまみず) ¶ 雨水溝 a (rainwater) drainage ditch. 雨水浸透ます[桝], 雨水浸透施設 a rainwater seepage pit. 雨水滞水地 a rainwater reservoir. 雨水排水システム storm water drainage systems. 雨水利用設備[システム] rainwater utilization equipment; a rainwater utilization system. 雨水タンク a rainwater tank. 雨水ます[桝] a rainwater pit. 雨水利用ビル a building (equipped) with a rainwater utilization system.

宇宙船地球号 Spaceship Earth.

海 〔海洋〕a [the] sea; an [the] ocean;〔海域〕waters;〔海辺〕the beach; the seashore; the seaside; the coast. ★⇨海岸, 海洋 ▶この魚は北の海で捕れる．This fish can be caught in northern seas. / 海に囲まれた国 a country surrounded by (the) sea; a seagirt [seabound] country. / 海にすむ魚 a marine fish; a fish that lives in the sea. ★⇨海水(海水魚) / この辺の海にはイカがたくさんいる．There are a lot of squid in the sea around here. | There are many cuttlefish in these waters. / 海の生物 a sea creature; a marine organism;〈集合的に〉marine [sea] life. / 海の幸 seafood; products of the sea; marine products. / 海の幸山の幸 food from the land and sea. / この時間にはいつも陸から海へ向かって風が吹きます。At this time, the wind always blows from the land to [toward] the sea. / 工場の排水が海を汚している。Industrial effluent is polluting the sea [ocean]. ¶**海風**〔海から吹いてくる風〕a sea breeze; a wind blowing (landward) from the sea;〔海上を吹く風〕a wind at sea. **海亀** a sea [marine] turtle; a turtle. / この海岸には,海亀が卵を産みにやってくる。On this beach, the turtles come to lay their eggs. ★⇨産卵 **海鳥** a seabird; a seafowl. / あの島には海鳥の群れが生息している。Flocks of seabirds live on the island. **海のエコラベル** ⇨ MSC ラベル(和英篇末尾)

埋め立てる〔堀,池を〕fill in《a moat, a pond》;〔海の一部を〕reclaim [recover]《land from the sea》. ▶海を埋め立てて空港を建設する計画がある。There are plans to fill in the sea and build an airport. / 埋め立て計画に反対する大規模な運動が展開された。A large-scale campaign developed against the reclamation project. ¶**埋め立て工事** reclamation work. **埋め立て作業** a landfill operation. **埋め立て処分場** a landfill site; a dumping site. / このあたりは産業廃棄物の埋め立て処分場に決定した。It was decided to use this area as a landfill [dumping] site for industrial waste. **埋め立て地** reclaimed [filled-in] land [ground]; land reclaimed from a lake [the sea]; a (land) reclamation site;〔廃棄物を埋めてできた〕a landfill. / 埋め立て地がいっぱいになり,行き場を失ったゴミが町にあふれている。There is no more room in the landfill site, and there is nowhere to dispose of the garbage, so it is piling up in the town. **埋め立て予定地** a proposed landfill site. / 自然保護のため,埋め立て予定地の位置を少しずらすことになった。For the sake of conservation, the location of the proposed landfill site is to be moved a little. **海岸埋め立て** reclamation of the foreshore.

ウラン uranium. ▶〜の〔四価の〕uranous;〔六価の〕uranic. / 〜(の)分裂 uranium fission. ¶**ウラン核燃料加工施設** a uranium-fuel processing plant [facility]. **ウラン化合物** a uranium compound. **ウラン型**〔核の〕a uranium-based [uranium-fueled]《nuclear bomb》. **ウラン鉱石** uranium ore. **ウラン・ガラス** uranium glass. **ウラン系列** the uranium(-radium) series. **ウラン鉱** uranium ore. **ウラン酸** uranic acid. **ウラン酸塩** uranate. **ウラン酸ナトリウム** sodium uranate. **ウラン残土** waste from uranium ore. **ウラン試験** 〔核燃料再処理施設での〕uranium testing; a uranium test. **ウラン精鉱** ⇨ イエロー・ケーキ **ウラン転換** uranium conversion. **ウラン転換施設[工場]** a uranium conversion facility [plant]. **ウラン235** uranium-235; U(-)235 (記号：^{235}U). **ウラン238** uranium-238; U(-)238 (記号：^{238}U). **ウラン濃縮** uranium enrichment. **ウラン濃縮装置** uranium enrichment equipment. **ウラン爆弾** a uranium bomb; a U-bomb. **ウラン・プルトニウム混合酸化物** uranium and plutonium mixed-oxide (略：MOX). **ウラン崩壊** uranium decay. **ウラン溶液** a uranium solution. **ウラン・ラジウム系列** the uranium-radium series. **加工ウラン** processed uranium. **酸化ウラン** uranium oxide. **重ウラン酸ナトリウム** sodium diuranate. **超ウラン元素** a transuranic element (略：TRU)；a transuranic. **天然ウラン** natural uranium. **濃縮ウラン** enriched [concentrated] uranium; uranium concentrate. **兵器級ウラン** weapons-grade uranium. **硫酸[酢酸，硝酸]ウラン** uranium sulfate [acetate, nitrate].

雨量 (amount of) rainfall; (amount of) rain; precipitation. ▶〜が少ないので貯水池が減水した．Because of little rainfall the level of water in the reservoir is low [has fallen]. / 屋久島は年間を通じて〜が多い．Yakushima has a lot of rainfall right through the year. / 〜を計る measure [gauge] the rainfall [(amount of) precipitation]. / 1時間に110ミリという記録的な〜を観測した．A record rainfall of 110 millimeters an hour was recorded. / 昨日の〜は30ミリであった．We had thirty millimeters of rain yesterday. ¶**雨量観測所** a rainfall station. **雨量計** a rain gauge; a pluviometer. **雨量図** a precipitation [rain] map; a pluviograph; a pluviogram; a hyetograph. **雨量測定** pluviometry. 形 pluviometric(al). **雨量分布** rainfall distribution. **雨量レーダー** precipitation radar. **時間雨量**

雨緑(樹)林

the hourly rainfall [precipitation]. **積算[累積, 総]雨量** the total rainfall 《over [during] a 24-hour period》. ▸降り始めからの総〜が900ミリに達した. The total rainfall was [came to] 900 millimeters. **土壌雨量指数** 〔土中に溜まっている雨水の量を示す〕a soil water index (略: SWI). **年間雨量** an annual rainfall《of about 1,000 millimeters》. ▸東京の年間〜 the annual rainfall in Tokyo. **平均雨量** mean precipitation; average rainfall. **予想雨量** an expected amount of rainfall. **流域雨量指数** 〔河川の流域に降った雨水の量を示す〕a basin rainfall [precipitation] index. **レーダー・アメダス解析雨量** amount of precipitation analyzed by radar-AMeDAS. **連続雨量** the total amount of continuous rainfall.

雨緑(樹)林 a rain-green forest; a tropical deciduous forest.

うるささ指数 an aircraft noise index;〔加重等価平均感覚騒音レベル〕⇨英和 WECPNL

上澄み supernatant fluid [liquid]; supernate. ▸廃水処理場では凝固剤を使って汚水を汚泥と〜に分離する. A coagulant is used at waste-treatment plants to separate sewage into sludge and supernatant fluid.

上乗せ条例 〔国の環境基準より厳しい都道府県などの規制〕a stiffer regulation; an addition to the rules.

運転音 〔機器などの〕(an) operating sound; (騒音) an operating noise. ▸〜を42 dBに抑えた掃除機 a vacuum cleaner which keeps (operating) noise levels down to 42 decibels. ▸この掃除機は〜がかなりうるさい. This vacuum cleaner is pretty noisy when it's working [operating].

運動 ⇨活動, 環境(環境保護)

雲霧林 a cloud forest.

エアコン 〔空気調節〕air-conditioning; air con（略：AC, a/c）;〔空調装置〕an air conditioner. ★⇨クーラー，暖房，冷房，温度 ▶～付きの〔乗用〕車 an air-conditioned car; a car with air-conditioning. / 東京の夏は～がないと，暑くて仕事もできない. If you don't have air-conditioning in Tokyo in the summertime, you can't do any work, that's how hot it is. / ～の使い過ぎで電気代が跳ね上がった. Overuse of air-conditioning has led to a big jump in the electricity bill [electricity bills]. / ～の普及とともに消費電力が増大した. With the diffusion of air conditioners there has been an increase in consumer electricity. / 適度に～の利いた部屋だと仕事の能率が上がる. In a suitably air-conditioned room working efficiency goes up. | I work more efficiently in an appropriately air-conditioned room. ★⇨温度 / ～を24時間フル稼働させる use the air conditioner at full strength around the clock [24 hours a day]. ★⇨つけっぱなし，消費 / 時々フィルターを掃除しないと～は機能が低下する. If you don't occasionally clean the filter, the performance of an air conditioner drops [goes down].
¶ **エアコンフィルター** an air conditioner filter.

エアロゾル ⇨英和 aerosol

衛生 hygiene; sanitation; cleanliness. ▶～的な hygienic; clean; germ-free; free of infection. / ～的な包装 hygienic wrapping [packaging]. / ごみを家のまわりに放置するのは～上よくない. Leaving rubbish lying around the outside of one's home is hygienically bad [unsanitary]. / ～によい be good for the health; be beneficial to health; be healthful [health-giving, salubrious]. / ～に悪い，非～的な be unhealthy [insanitary, unhygienic]; be bad for the [*one's*] health. ¶ **衛生化学** hygienic chemistry. **衛生学** hygienics; hygiene [sanitation] (studies). **衛生学者** a hygienist; a doctor specializing in public health medicine. **衛生管理** hygiene maintenance;

〔行政〕health administration [control]; 〔施設〕sanitation. **衛生管理者** 〔労働法上の〕a health supervisor. **衛生技師** a sanitary engineer [technician]; a sanitarian; a hygienist. **衛生教育** health [hygiene] education; instruction in hygiene. **衛生検査技師** a medical technologist; a clinical [medical] lab(oratory) technician. **衛生工学** sanitary [public health] engineering. **衛生昆虫[害虫]** 〔衛生上の害を与える昆虫〕a disease-carrying insect; an insect pest; a public health pest. **衛生材料** sanitary supplies. **衛生試験所[研究所]** an institute of hygienic sciences. **衛生試験場** a hygienic laboratory. **衛生施設** a sanitary [hygienic] facility. **衛生条例** hygiene [health, sanitation] regulations. **衛生設備** health facilities; 〔下水など〕sanitation; sanitary facilities. / 〜設備が完備している be completely hygienic; be fully equipped in terms of hygiene. **衛生統計** sanitation [hygiene] statistics. **衛生費** sanitation expenses. **衛生部** 〔役所の〕Sanitation Department. **衛生証明書** 〔輸入動物による人の感染症を防ぐために厚生労働省が定める, 輸出国政府発行の証明書〕a health certificate (for an imported animal). **学校衛生** school hygiene; children's health care. **環境衛生** ⇨環境 **公衆衛生** public hygiene. **食品衛生** food hygiene; keeping food clean [uncontaminated]. **労働衛生** labor hygiene; hygiene in the workplace.

営巣(えいそう) nest building; nidification. ▶〜する build a nest; nest. ¶**営巣地** a nesting place; nesting grounds.

液状化(現象) 〔地震による〕liquefaction. ▶〜する liquefy.

エクソン・バルディーズ号 ⇨英和 Exxon Valdez

エコ 〔環境に優しい・環境保護志向の〕eco-. ¶**エコカー** an eco-car. ★⇨英和 ecocar **エコカー減税** an ecocar tax break; a reduction in automobile taxes for fuel-efficient and low-emission vehicles **エコグッズ[商品]** eco-friendly [ecofriendly] goods. **エコ効率** eco-efficiency. **エココンシャスな** 〔環境問題への関心の高い〕environmentally aware; ecoconscious. 图 environmental awareness; ecoconsciousness. **エコシティー** 〔環境共生都市〕an eco(-)city [Eco(-)city]. **エコステーション** a service station for cars that

エコ

run on alternative fuels.　**エコセメント**　ecocement; ecoconcrete.　**エコツアー**　an ecotour;〈総称〉ecotourism.　★⇨英和 ecotour ▶エコツアー参加者 ecotourist.　**エコツーリズム**〔自然保護や先住民文化との共存を目指す観光〕ecotourism.　**エコテクノロジー**〔地球環境の保護に役立ち経済効用もある科学技術〕ecotechnology.　**エコデザイン**〔環境問題を考慮した製品デザイン〕ecological design.　**エコテロリズム**　⇨英和 ecoterrorism.　**エコトイレ**〔環境に負荷をかけないトイレ〕an ecological toilet; an environment(ally) friendly toilet.　**エコドライブ**〔環境に配慮した車の運転法〕eco-friendly driving (techniques).　**エコハウス**〔環境共生住宅〕environmentally friendly housing.　**エコバッグ**〔ごみになるスーパーのレジ袋の代わりに持参する買い物袋〕an ecological (shopping) bag; a reusable (cotton or canvas) shopping bag.　★⇨マイバッグ　**エコバンク**〔環境優先の企業への融資を行う銀行〕an ecobank.　**エコビジネス**〔環境を破壊しない製品の開発をめざす産業〕(an) ecobusiness.　**エコファーマー**〔環境保全型農業に取り組む生産者〕an ecofarmer.　**エコファーミング**　ecofarming.

エコフィード【商標】〔食品残渣飼料〕Ecofeed; animal feed made from food scraps.　**エコ・ファンド**〔企業の環境対策を基準に銘柄選択をする投資信託〕an eco fund; a green [an environmental] (mutual) fund.　**エコフェミニズム**　⇨英和 ecofeminism　**エコブリッジ**〔森林を切り開いて道路を作る場合に，動物が道路を越えて渡れるように作る橋〕an "ecobridge"; an animal overpass.　**エコポイント**　an ecopoint; a point-based rebate system for purchasers of energy-efficient products.　**エコマーク**〔環境保全型の製品の表示マーク〕an "ecomark," a symbol that shows that a product is environmentally friendly; an ecolabel.　**エコマテリアル**〔環境負荷が少ない素材〕an environmentally-friendly material; an eco-material; a green 《building》material.　**エコミュージアム**〔生活環境博物館〕an ecomuseum.　**エコライト**〔温室効果ガス排出権〕an [a carbon] emission(s) credit; a carbon credit; the right to emit greenhouse gases.　**エコラベル**〔EU の環境保全運動のシンボルマーク〕an ecolabel. / 海のエコラベル⇨ MSC ラベル(和英篇末尾)，水産エコラベル　**エコラン**

⇨エコノミーラン　**エコリーフ環境ラベル**　〔環境ラベル(⇨環境)の一種〕an EcoLeaf environmental label.　**エコレール・マーク**　〔鉄道貨物輸送を活用している商品・企業であることを表示するマーク〕an Eco Rail Mark.

エコトーン　⇨英和 ecotone

エコノミー・ラン　〔燃費の低さを競う自動車レース〕an economy run.

エコロジー　〔自然環境保護主義〕environmentalism;〔生態学〕ecology.　¶**エコロジー・ビジネス**　ecology [environment] business.　**エコロジー・ファッション**　ecological fashion.　**エコロジー・カー**＝エコカー(⇨エコ).　**エコロジー住宅**　an eco-house; an ecological house; an eco-friendly house.

エコロジカル・アート　〔自然の土・砂・氷などを素材にした芸術〕ecological art.

エコロジカル・フットプリント　⇨英和 ecological footprint

エコロジスト　〔生態学者〕an ecologist;〔自然環境保護主義者〕an environmentalist.

エストロゲン　〔卵胞ホルモン〕(an) estrogen; (an) oestrogen. ▶～(様)の estrogenic.　¶**エストロゲン作用**　the action of estrogen; estrogenic action.　**結合型エストロゲン**　〔女性ホルモン剤〕(a) conjugated estrogen.　**植物(性)エストロゲン**　a phytoestrogen.

エタノール　ethanol; ethyl alcohol.　¶**エタノール・アミン**　〔炭酸ガスなどの吸収剤・フェノール抽出溶剤〕ethanolamine.　**エタノール混合ガソリン**　ethanol-blended gasoline.　★⇨英和 E(E10)　**エタノール混合燃料**　ethanol-blended fuel.

枝打ち　pruning; trimming; lopping. ▶～する trim [lop] the branches of a tree.

エチル・ターシャリー・ブチル・エーテル　〔ガソリン混入剤〕ethyl tertiary-butyl ether. ★⇨英和 ETBE

越境　crossing [infiltration of, violation of] a border. ▶～する cross a frontier [border]; violate a frontier [border]　¶**越境公害[汚染]**　transboundary [cross-border, transborder] pollution.　**越境酸性雨**　〔国境を越えて広域に降る酸性雨〕cross-border [transborder] acid rain.　**越境大気汚染**　cross-border air pollution.

越冬　▶～する stay over the winter; winter 《in the Antarctic》;〔冬眠する〕hibernate. / ランを露地で～させるのは難しい．It's difficult to get an orchid to survive the winter outdoors. / 数羽

の傷ついた渡り鳥を保護し，当地で～させた．They took several injured migratory birds into protective care and made them spend the winter here. / 熊は穴の中で～する．Bears hibernate in holes for the winter. ¶ **越冬芽** a winter bud. **越冬蛹** 〔昆虫の〕an overwintering [a wintering, a hibernating] pupa. **越冬燕** an overwintering swallow. **越冬地** a wintering place [area]. / この湖はいろいろな水鳥の～地になっている．This lake is a wintering place for all sorts of water birds.

エネルギー energy; strength; power. ★核～，化石～ などは ⇨ 核，化石，再生，省エネルギー，代替，太陽，自然の各項目を参照．¶ **エネルギー安全保障** energy security; security in energy. **エネルギー開発** energy development. **エネルギー革命** an energy revolution; a revolutionary change in the form of energy used. **エネルギー管理士** a certified energy manager. **エネルギー危機** an energy crisis [crunch]. **エネルギー企業** an energy corporation. **エネルギー起源 CO_2 [二酸化炭素]** CO_2 [carbon dioxide] emissions from fuel combustion; energy-derived [fossil fuel-derived] CO_2 ▶非～起源 CO_2 [二酸化炭素] non-energy derived CO_2. **エネルギー・ギャップ** an energy gap. **エネルギー協力** energy cooperation; cooperation in the field of energy. / 中国とロシアは～協力に関する協定に調印した．China and Russia have signed an agreement concerned with cooperation in the field of energy. **エネルギー源** an energy source; a source of energy. **エネルギー効率** 《improve》energy efficiency; the efficiency of energy use. **エネルギー効率化** making《homes》more energy-efficient; increasing the energy-efficiency《of homes》. **エネルギー・サービス・プロバイダー** 〔電力・ガスなどエネルギーを一括供給・管理する業者〕an energy service provider (略: ESP). **エネルギー・サイクル** an energy cycle. **エネルギー産業** the energy industry. **エネルギー支援** energy aid [assistance]《to North Korea》. **エネルギー自給国** a country [nation] that is self-sustaining in terms of energy. **エネルギー自給率** 《Japan's》energy self-sufficiency rate. / ～自給率を上げる[高める] increase energy self-sufficiency. **エネルギー資源** energy resources. **エネルギー事情** the energy situation 《in Japan》. **エネルギー自由化** energy deregulation. **エネルギー集約的な** energy-intensive 《industries》. **エネルギー需要** energy demand [needs]; the demand for energy. **エネルギー消費** 《high》energy consumption. /～消費効率 energy consumption efficiency;〔消費電力当たりの冷却・加熱能力を表す数値; COP〕the [a] coefficient of performance (略: COP). /～消費量 (total) energy consumption; the amount of energy consumed. **エネルギー植物** plants (grown [suitable]) for bioethanol. **エネルギー政策** (an) energy policy. **エネルギー政策基本法** the Basic Law on Energy Policy. **エネルギー政策法**〔米国の〕the Energy Policy Act of 2005. **エネルギー生産** 《the nation's annual》energy production. **エネルギー節約** 《promote》energy conservation. **エネルギー選別器** an energy selector. **エネルギー損失** energy loss. **エネルギー代謝** energy metabolism. /～代謝を促進する promote energy metabolism. /～代謝率 the rate of energy metabolism; the relative metabolic rate (略: RMR). **エネルギー多消費型産業** an energy-intensive industry.

エネルギー白書 a white paper on energy;〔経済産業省資源エネルギー庁の〕Energy in Japan 《2005》. **エネルギー変換工学** energy conversion engineering. **エネルギー変換効率** energy conversion efficiency. **エネルギー問題** an energy problem; energy problems;〔国家レベル・全地球的規模での〕the energy problem. **エネルギー利用効率** 《improve》energy utilization efficiency. **エネルギー林** trees (grown [suitable]) for bioethanol. **エネルギー・ロス** energy loss; (a) loss of energy.

エル・ニーニョ ⇨英和 El Niño ▶いつになく強い～ an unusually strong El Niño. /～が起こりそうな予兆があった．There were warnings of an impending El Niño. / 1997年から98年にかけての～は世界の気候パターンをひどく狂わせた．The 1997-98 El Niño severely disrupted global weather patterns. ★⇨因果関係, 温度 ¶反エル・ニーニョ現象 ⇨ラ・ニーニャ

塩害(えんがい) damage [harm] caused by salt. ▶～を受ける〔畑などが〕be harmed [badly affected] by seawater;〔電気の碍子(がいし)などが〕be damaged by (the salt in) sea air. /畑の作物が～を受けた．The crops in fields were damaged by sea air.

煙害 smoke pollution; damage [harm] caused by (cigarette) smoke. ★⇨煙毒 ▶銅山の〜 smoke pollution from a copper mine.

塩化ビニル vinyl chloride; chloroethylene. ★⇨ポリ塩化ビニル

沿岸 〔水域に沿った陸地〕the coast; the shore; 〔海の〕the seashore; 〔陸に沿った水域〕coastal [inshore] waters; a coastal area [zone]. ▶タンカーから漏れた重油は九州〜に達した. The oil that leaked from the tanker reached the coast [shores] of Kyushu. / 太平洋〜の地域は津波の危険がある. There is a danger of tsunami [tidal waves] in Pacific coastal areas. ¶**沿岸海** 〔海岸線から3海里以内の海域〕the marginal sea. **沿岸回遊** 〔魚の〕littoral migration. **沿岸魚** a littoral [coastal] fish. **沿岸漁業** coastal [inshore, longshore] fishery. **沿岸洲** 〔海岸線に平行する長い洲〕an offshore bar; a barrier beach [bar]. **沿岸水** 〔外洋水に対して〕coastal water. **沿岸線** a coastline; a shoreline. **沿岸帯** the littoral zone. **沿岸動物** a littoral animal. **沿岸流** a littoral [coastal] current.

園芸 gardening; horticulture; 〔花樹・草花栽培〕floriculture; flower-growing. ▶趣味の〜 gardening (as a hobby). / 家庭〜 gardening; growing [cultivating] a garden at home. ¶**園芸家** a gardener; a horticulturist; a horticultural expert. **園芸学校[試験所]** a horticultural school [experiment(al) station]. **園芸暦**(れき) a gardener's calendar. **園芸作物** garden produce; things grown in a garden. **園芸雑誌** a gardening magazine. **園芸植物** a garden plant. **園芸植物培養場** a nursery. **園芸種** a cultivated species《of...》. **園芸農家** a horticulturist; a (commercial) plant-grower. **園芸農業** horticulture; the horticultural industry. **園芸用具** a gardening tool. **園芸療法** horticultural therapy; hortitherapy. **園芸療法士** a horticultural therapist.

エンジン an engine. ★⇨アイドリング ▶その車は〜がかかったままだった. That car was left with the engine running. | The engine of that car had been left running. / あの車の〜音はうるさい. That car engine is noisy [gets on my nerves]. / 〜をかける start an engine; turn [switch] an engine on; get an engine going [moving, start-

ed]. / ～をかけっぱなしにしておく leave the engine on [going, running]. / ～を止める[切る] stop an engine; turn [switch] an engine off. / ～を切ってお待ちください．Please wait here, with the engine off. / ～をふかす race [gun] an engine. / ～を空ぶかしする race an engine.
¶エンジン効率　engine efficiency.　エンジン車　〔バッテリー車に対して〕an engine-driven car [vehicle].

塩水化(えんすいか)　〔地下水などの〕salinization [salination]《of groundwater》; saltwater [sea water] intrusion《into coastal groundwater》.

塩生植物　a halophyte; a halophytic plant.　¶塩生植物群落　a halophyte community.

塩素　chlorine.　▶…から～を除去する dechlorinate…. / ～化する[させる] chlorinate; chloridize.
¶塩素ガス　chlorine gas.　塩素系殺虫剤　a chlorinated insecticide.　塩素系漂白剤　(a) chlorine bleach.　塩素除去剤　〔風呂用などの〕a chlorine remover; an antichlor.　塩素消毒〔殺菌〕chlorine disinfection [sterilization];〔水の〕chlorination. 動 chlorinate.　塩素消費量　〔水道水処理における〕chlorine consumed.　塩素処理　〔水の塩素殺菌〕chlorination. / ～処理剤 chlorine microbiocide. / ～処理施設 a chlorination plant.　塩素注入量　〔水処理における〕a chlorine dose.　塩素量　〔海洋の〕(a) chlorinity.　脱塩素　dechlorination.

煙毒(えんどく)　a toxic substance in smoke.

鉛毒　〔鉛の毒〕a toxic substance in lead; lead toxicity;〔鉛中毒〕lead poisoning; plumbism; saturnism. ★⇨鉛(なまり) ▶～にかかる suffer from [be affected by] lead poisoning. / ～にかかった saturnic; affected by [suffering from] lead poisoning. / ～の徴候[患者] a saturnine [lead-poisoning] symptom [patient].

援農　volunteer work on a farm.

塩ビ　⇨塩化ビニル

塩分　salt; (a) salt content; saline matter; salinity. ▶この温泉は～を多量に含んでいる．This hot spring contains a lot of salt [has a high salinity]. / ～を含んだ水 water containing salt; salt water; saline water. / ～を除く remove the salt《from…》; desalinate [desalt]《seawater》. / ～を控えなさい．Keep your salt intake down. |《口》Go easy on the salt.　¶塩分計　a salinometer.　塩分濃度　(a) salinity《of 3 ppm》.

遠洋 the open sea; the deep sea. ¶ **遠洋漁業** deep-sea [pelagic] fishing [fishery]. **遠洋動物** a pelagian.

塩類 a salt; salts. ¶ **塩類泉** a mineral (salt) spring. **塩類腺** 〔海鳥などの〕a salt gland.

お

オイル 〔一般に油〕oil;〔石油〕petroleum. ¶**オイル・サンド**〔タール状の重油質を含む砂岩〕oil sand; tar sands. **オイル・シェール**【鉱】〔油母頁岩〕(ゆぼけつがん) oil shale. **オイルショック**〔石油危機〕an oil shock; an oil crisis. ▶1973 年には〜ショックが全世界を襲った. In 1973 there was an oil crisis that affected the whole world. / 石油を輸入に頼っていた日本は第一次〜ショックで大混乱に陥った. In Japan, a country dependent on imports for oil, the first oil shock caused chaos. **オイル・スリック** an oil slick. **オイル・タンカー**〔石油輸送船〕an oil tanker. **オイル・タンク**〔燃料タンク・油容器〕an oil tank. **オイル・フェンス**〔流出油の移動・拡散を食い止めるための囲い柵〕an oil fence; a fence boom. / 座礁したタンカーから流出した原油が湾内に入るのを防ぐために〜フェンスを張った. Fence booms were set up to prevent crude oil from the stranded tanker escaping into the bay. ★⇨重油 **オイル・ボール**〔海洋に浮遊する廃油のかたまり〕a congealed ball of (crude) oil. **オイル漏れ** an oil leak; oil leakage; a leakage of oil.

欧州エネルギー取引所〔ドイツにある〕the European Energy Exchange (略: EEX). ★⇨気候(気候取引所)

欧州化学物質庁 ⇨英和 ECHA

欧州環境機関 ⇨英和 EEA

欧州食品安全機関 the European Food Safety Authority (略: EFSA).

欧州風力エネルギー協会 the European Wind Energy Association (略: EWEA).

オーガニック〔有機(農業)の〕organic. ★⇨有機, 英和 organic ▶〜食品 organic foods. / 〜野菜 organic vegetables; organically grown vegetables. / 〜コットン〔無農薬栽培された綿花で作られたコットン〕organic cotton.

屋上緑化 rooftop [roof] gardening [greening]; decorating a rooftop with greenery [plants]. ★⇨緑化

オクチルフェノール octylphenol. ¶**4-オクチルフェノール**〔抗酸化剤; 樹脂中間体;

内分泌攪乱物質とされる〕4-octylphenol.

汚水 dirty [filthy, foul] water; polluted water [liquid(s)];〔下水〕sewage; (liquid) waste; effluent; (liquid) discharge;〔台所からの〕dishwater; liquid kitchen waste;〔工場からの〕(liquid) industrial waste; industrial effluent; waste effluent. ★⇨廃液 ▶あちこちから〜が流れ込み，川を汚染した．Waste from all over the place flowed in and polluted the river. / 〜処理施設 a sewage [liquid waste] treatment plant. / 当工場では浄化した〜を近くの川に流しています．At our factory we put purified [decontaminated, treated] liquid waste [effluent] into the nearby river.
¶ **汚水管** a waste water [soil, waste] pipe; a sewer; a sewage pipe. **汚水槽** a cesspool; a cesspit; a sink. **汚水浄化装置** sewage treatment equipment.

汚染 pollution; contamination. ▶〜する pollute; contaminate; taint. / 空気を〜する pollute the air. / 放射能で〜された空気 air contaminated with radioactivity. / 海水浴場の海水は下水で〜されている．The seawater at the bathing resorts is polluted with sewage. / 化学物質で〜された魚 fish contaminated with chemicals; contaminated fish. / 工場廃液による河川の〜 industrial pollution of a river; contamination of a river with industrial effluent. / 〜の進行が懸念される地域 areas where it is feared that pollution is increasing [pollution is feared to be growing worse]; areas threatened by [with] pollution. / 〜を取り除く remove pollution [pollutants]; decontaminate《an area, the soil》.
¶ **汚染管理区域** ⇨管理（管理区域） **汚染源** a source of pollution [contamination]; a pollution source. ★⇨飲用 **汚染原因** a cause of pollution. **汚染原因企業** a polluting company; a polluter. **汚染国**〔特定の伝染病などの発生国〕an infected country. / BSE［口蹄疫］〜国 a BSE-infected [an FMD-infected] country. **汚染者** a polluter. / 〜者負担原則 the Polluter Pays Principle（略：PPP）. **汚染除去** decontamination. **汚染対策** antipollution measures; measures against [to deal with] pollution. **汚染地域** a polluted area [region]. **汚染度**〔汚染されている度合〕a level [degree] of pollution [contamination]; a contamination

level;〔汚染する度合〕a polluting potential. / 〜度の高い水質 badly contaminated water; highly polluted water; water with a high degree of pollution [contamination]. / 〜度の低い水質 slightly polluted [contaminated] water; water that has not been badly contaminated; water with a low degree of pollution [contamination]. **汚染濃度** a [the] degree of pollution; pollution [contamination] concentrations. **汚染物質** a pollutant; a contaminant. **汚染率** a rate of contamination; a contamination rate. **汚染力** polluting potential. / 〜力の高い[強い]化学物質 a (chemical) substance with a high polluting potential; a major pollutant. **汚染レベル** a level of pollution; a pollution level. **油汚染** oil pollution [contamination]. **化学汚染** chemical pollution. **環境汚染** environmental pollution; pollution of the environment. **金属汚染** metal pollution. **景観汚染** scenic destruction; destruction of (the) scenery; scenic pollution. **視覚汚染** visual pollution. **重金属汚染** heavy metal pollution. **水質汚染** water pollution [contamination]; pollution [contamination] of (the) water. **騒音汚染** noise pollution. **大気汚染** atmospheric [air] pollution; pollution [contamination] of the atmosphere. / 大気〜の原因となる物質 substances that (can) cause atmospheric [air] pollution; atmospheric pollutants. **土壌汚染** soil contamination; contamination [pollution] of the soil; ground pollution. **熱汚染** thermal pollution. **放射能汚染** radioactive contamination.

オゾン ozone. ▶〜の ozonous; ozonic. / 〜を含む ozonous; ozonic; ozoniferous. ¶ **オゾン化物** an ozonide. **オゾン計** an ozonometer. **オゾン殺菌** ozone disinfection [sterilization]. **オゾン試験紙** (an) ozone paper. **オゾン処理[化]** ozonization; ozonation; ozonidation. **オゾン処理する** ozonize. **オゾン水** ozone water. **オゾン層** an ozone layer; the ozonosphere. / 〜層破壊 ozone (layer) depletion [destruction]; the depletion of the ozone layer. / 〜層破壊係数 an ozone depletion potential (略: ODP). / 〜層破壊物質 ozone-depleting substances. / 〜層保護 (the) protection of the ozone layer. / 〜層保護条約, ウィーン条

約〔1985年の〕the Vienna Convention for the Protection of the Ozone Layer. / 〜層保護法 the Ozone Layer Protection Law. ★正式名称は，特定物質の規制等によるオゾン層の保護に関する法律 (the Law concerning the Protection of the Ozone Layer through the Control of Specified Substances and Other Measures). **オゾン脱臭機[器，装置]** an ozone deodorizer. **オゾン破壊** ozone depletion [destruction]. **オゾン発生器[装置]** an ozonizer; an ozonator; an ozone apparatus. **オゾン反応** 〔オゾン酸化〕ozonolysis. **オゾン・ホール** ⇨英和 ozone hole

汚濁 taint; blight; pollution. ▶海水[湖]の〜 contamination of the sea [lakes]. ★⇨水質汚濁 / 河川の〜がひどい. The contamination of the rivers is appalling. / 湾の〜が進んでいる. Pollution of the bay is growing worse. ¶**汚濁水** polluted water. **汚濁物質** a water contaminant [pollutant].

汚泥(おでい) sludge; slime. ¶**汚泥処理** sludge treatment [disposal]. ★⇨活性(活性汚泥) ▶〜処理施設 a sludge treatment [disposal] facility. **汚泥乾燥処理** sludge drying (treatment) 〜乾燥処理施設[装置] a sludge drying facility [unit]. **汚泥減容化技術[減量化技術]** sludge reduction technology. **汚泥燃料** (a) sewage sludge fuel. **有毒汚泥** toxic sludge.

汚物 something unclean [unsanitary];〔糞便〕excrement; excreta. ▶下水管が破裂して〜が路上に流れ出した. A sewage [drain] pipe cracked and sewage spilled out onto the street. ¶**汚物入れ** 〔トイレなどの〕a sanitary napkin disposal bin [receptacle];〔病院などでの〕a fecal specimen receptacle. **汚物処理施設** a sewage disposal facility; a sanitation facility.

オルタ(ー)ナティブ 〔既存のものに代わる〕alternative. ¶**オルタ(ー)ナティブ・ツーリズム** 〔自然環境等に配慮した観光の在り方〕alternative tourism. **オルタ(ー)ナティブ・テクノロジー** 〔代替技術〕alternative technology. **オルタ(ー)ナティブ・メディシン** 〔代替医療〕alternative medicine.

音圧 sound pressure. ¶**音圧レベル** sound pressure level. ★単位は デシベル (decibel). ⇨英和 perceived noise decibel

温室 a greenhouse; a hothouse; a conservatory; a glasshouse; a stove;〔促成栽培の〕a forcing house;〔暖房装置のついていない〕a cold frame. ▶～で咲かせる force《a plant》to bloom in a hothouse. / ～で栽培する cultivate《a plant》in a greenhouse [hothouse]. / ～メロン a hothouse(-grown) [greenhouse(-grown)] melon.

¶ **温室効果** ⇨英和 greenhouse effect / 増加した二酸化炭素の～効果で地球が温暖化している．The Earth is warming up as a result of the greenhouse effect from increased amounts of carbon dioxide. / ～効果をもたらす．produce the greenhouse effect.
温室効果ガス ⇨英和 greenhouse gas / ～効果ガスによる地球の温暖化 global warming resulting from greenhouse gases. / ～効果ガスの排出を抑制する curb [check, control] emissions of greenhouse gases. / ～効果ガスを発する generate greenhouse gases. / ～効果ガスを削減する reduce greenhouse gases. / 経済の失速を恐れる国々は，一方的な～効果ガスの排出削減要求に反発している．Countries afraid of economic slowdowns are resisting unilateral demands that they cut their greenhouse gas emissions. / ～効果ガス排出権《trade》the right to emit greenhouse gases. / ～効果ガス発生源 a source of greenhouse gases. / ～効果ガス観測技術衛星〔宇宙航空研究開発機構（JAXA）の〕the Greenhouse Gas Observing Satellite（略：GOSAT）．★⇨二酸化炭素　**温室栽培** greenhouse [hothouse] (horti)culture.　**温室植物** a hothouse [greenhouse] plant.　**非温室植物** a hardy plant.

温泉 a hot [thermal] spring (spa [bath, facility, resort, inn]). ▶この辺はどこを掘っても～がわく［わき出る］．In this area you can dig anywhere and a hot spring will bubble up [gush out]. / ～が間欠的に噴き出す．A geyser shoots up intermittently [at intervals]. / ～に入る take a hot spring bath; take a bath in a hot spring. / この～には硫黄分が含まれている．This hot spring contains sulfur. / ～の効能 the effects [benefits] of a hot spring [spring's waters]. / ～の熱 a hot spring's heat. / ～の湯を飲む drink warm water from a thermal spring. / ～を引く have hot-spring water installed《in one's home [inn, etc.]》; lay on hot-spring water. / ～を掘る［探り当てる］dig into [strike, locate] a hot spring. / この～はリウマチによく効く．This hot spring [hot spring's water] is very effective for rheumatism. ¶ 温泉華〔温泉沈殿物〕travertine. 温泉学 balneology. 温泉掘削業者 a hot spring drilling company. 温泉権 the right to make use of hot spring waters. 温泉源 a hot [thermal] spring source. 温泉水 hot [thermal] spring water(s). 温泉生物 a thermal [hot] spring organism. 温泉分析書 a hot spring analysis table. 温泉法 the Hot Spring Law. 温泉脈 a hydrothermal vein; a vein of hot water. 温泉余土 solfataric clay. 温泉療法 balneotherapy; hot spring therapy [treatment]; a hot spring cure. 温泉療法医 a balneologist 温泉療法士 a (certified) balneotherapist. 温泉療養 convalescence at a hot spring; balneotherapy. 掘削[くみ上げ]温泉 a hot spring resulting from boring [drilling]; an *onsen* using subterranean water. 単純温泉 a simple hot [thermal] spring. 地下温泉 an underground hot spring. 天然温泉 a natural hot spring. / 天然〜表示マーク the symbol for a natural hot spring; the natural hot spring symbol. (天然)自噴温泉 a naturally occurring hot spring; a hot spring which comes up [breaks through, has erupted] naturally.

温帯 the [a] temperate zone; the temperate [variable] zones; a Temperate [Variable] Zone. ▶〜の temperate; temperate-zone. / 日本は〜に位置する. Japan is situated in the (North) Temperate Zone. / 広葉樹林は〜から熱帯にかけて分布している. Broadleaf forests extend from the temperate to the tropical zones. ¶ 温帯夏雨気候 a temperate climate with a rainy summer (season). 温帯気候 a temperate [mesotherm] climate. / 暖[冷]〜気候 a warm [cool] temperate climate. 温帯魚 a temperate zone fish. 温帯湖 a temperate lake. 温帯植物 a mesotherm; a temperate zone plant; 〔温帯全域の〕temperate zone flora. 温帯草原 temperate grassland. 温帯多雨気候 (a) temperate rainy climate; (a) humid temperate climate. 温帯多雨林 (a) temperate rain forest. 温帯地方 the temperate regions [latitudes]. 温帯低気圧 an extratropical cyclone [low, storm]. 温帯北部 the North Temperate Zone. 温帯林 a temperate forest.

温暖化 warming. ★⇨地球(地球温暖化), 気候(気候変動) ▶地球が〜する the earth [globe] warms up [grows warmer] 〜は地球全体の問題だ. Global warming is a problem that affects the world as a whole. | Global warming is a problem for the whole world. / 地球の〜が進むと, この辺りは海に沈んでしまう. If global warming continues, this area will sink [disappear] below the sea. / 地球〜の科学的合理性にはもはや疑問の余地がない. The scientific basis of [for] global warming can no longer be doubted. ¶**温暖化ガス** global warming gases. / 〜ガス放出量 global warming gas emissions. ★⇨温室(温室効果ガス) **温暖化対策税** a global warming tax. **温暖化ビジネス** 〔温室効果ガス削減の必要から生まれた新事業〕global warming businesses. **脱温暖化** prevention of [taking measures against] global warming. / 脱〜社会 a society [world] which combats [takes measures against, faces up to] global warming.

温度 (a) temperature. ▶〜の変化 a temperature change. / 〜の激変 a sudden [an extreme, a violent] change in (the) temperature. / 〜の差 a difference in temperature. / 〜の上昇[低下] a rise [fall, drop] in temperature. / ペルー沖の海水〜の異常な上昇はエルニーニョ現象と呼ばれる. An abnormal rise in ocean temperature off the coast of Peru is referred to as the El Niño phenomenon. / 〜を調節する adjust [control, regulate] the 《room》temperature. / 〜を一定に保つ maintain a constant [an even] temperature. / 〜を上げる[下げる] raise [lower] the temperature. / (エアコンの冷[暖]房)〜を26度に設定する (with the air conditioner on 'cool [warm]',) set the temperature to [at] 26 degrees. ¶**温度感覚** temperature sense; thermal [thermic] sense. **温度管理[制御]** temperature control. **温度効率比** thermal-efficiency ratio. **温度差発電** electricity generation by temperature difference. **温度周期性** 〔植物の〕thermoperiodicity. **温度上昇比** the temperature rise ratio; the rate of temperature rise. **温度性決定** 〔生物の〕temperature-dependent sex determination (略: TSD). **温度設定** 〔エアコンなどの〕the temperature setting 《of an air-conditioning [an air-cooling, a heating] apparatus》. / ～設定を1度上げる[下げる] raise [lower] the temperature setting 《of an air-conditioning [an air-cooling, a heating] apparatus》 by one degree. **温度センサー** a temperature sensor; a thermosensor. **温度知覚** sense of temperature; thermesthesia. **温度調節** temperature adjustment. / ～調節機能 the thermoregulation mechanism. / ～調節装置 a thermostat; a thermoregulator. **温度ヒューズ** a thermal fuse. **温度風** thermal wind. **温度分布** a temperature distribution. **快適温度** a comfortable temperature. **設定温度** 〔エアコンなどの〕the temperature setting 《of an air-conditioning [an air-cooling, a heating] apparatus》. **体感温度** ⇨体感 **表面温度** ⇨緑化

温排水 〔火力発電所などからの〕a thermal discharge; thermal effluent. ¶**温排水公害** thermal pollution. ★⇨排熱, 熱公害

か

カー・シェアリング, カープール 〔自動車の共同使用〕carpooling; car-pooling; car-sharing; ride-sharing; lift-sharing. ▶カー・シェアリングする organize a carpool. ★⇨相乗り, ライドシェア

カーフリー・デー 〔都市中心部での自家用車の使用を制限し,自動車と環境について考える日〕a car-free day.

カーボン ⇨英和 carbon, carbon offset, carbon neutral, carbon negative, Carbon Disclosure Project

ガイア仮説 ⇨英和 Gaia hypothesis [theory].

外因性内分泌攪乱化学物質 ⇨環境(環境ホルモン)

開花 flowering; blossoming; blooming; coming into flower [bloom, blossom]; efflorescence; inflorescence; florescence; anthesis. ▶〜する flower; bloom; blossom; come into flower [bloom]; effloresce. / 〜している be in flower [bloom, blossom]; be efflorescent; be inflorescent. / 〜しない蕾もある. Some buds don't open. / 雨季に入るとこの草原の草花がいっせいに〜する. When the rainy season sets in, the plants on these grasslands all burst suddenly into flower. / 今年は桜の〜が遅れる[早まる]だろう. This year the cherry trees will come into bloom later [earlier] than usual. | The cherry blossom will be late [early] this year. / 暖地では11月ごろから水仙の〜が始まる. In warmer areas the daffodils (come into) bloom as early as November. | In temperate districts daffodils are in bloom from early in November. / 人工的に〜を早める bring flowering forward artificially. / 短日処理を施して〜を調整する use short-day treatment to regulate flowering. / 春先の一時的な気温低下がこの花の〜を促す. A temporary temperature drop in early spring sets off [stimulates, activates] flowering in this plant. ¶**開花期** a flowering season;〔咲き始める時期〕the period when 《a plant》 flowers [blooms, comes into bloom];〔咲いている期間〕the period when 《a plant》 is in flower [bloom]; anthesis. /東京あたりの桜の〜期は4月上旬です. In the Tokyo area the cherry blossoms come [are] out early in April. | In the Tokyo area the cherry trees flower [blossom] in early April. **開花情報** information on 《cherry tree》 flowering [blossoming] times. **開花宣言** 〔主に桜の〕a declaration that 《cherry trees》 are in bloom; a declaration of 《cherry》 blossoms. **開花前線** a "front" (for the flowering of a plant species). /梅の〜前線は現在関東地方を北上している. The front line for *ume* blossom [The "*ume* blossom front"] is now moving north over the Kanto region. **開花調節** regulation of flowering; flowering regulation. **開花日** the date [day] when 《a plant》 flowers [blooms, blossoms, comes into flower]. **開花ホルモン** a flowering hormone; florigen. **開花予想** a forecast of 《cherry blossom》 flowering times. /気象庁が今年の桜の〜予想を発表した. The Meteorological Agency has announced its forecast of when the cherry trees (in different parts of Japan) will come into bloom [a forecast of cherry blossoming times]. **開花予想日** the day 《cherry trees》 are expected to (come into) bloom [flower, blossom].

海岸 the seashore; the (sea) coast; [seaside]; the waterfront; 〔渚〕the beach; the strand. ▶〜に on the shore [beach]; by the sea; at the seaside. /〜に打ち上げられる be washed [cast] ashore. /〜の coastal; seashore; seaside; 〔海岸に生息する〕limicoline 《birds》; littoral. ¶ **海岸管理者** the coast authority. **海岸気候** a maritime [coastal] climate. **海岸工学** coastal engineering. **海岸砂丘** a coastal sand dune. **海岸砂防** beach preservation [erosion prevention]. **海岸植生** coast [maritime] vegetation. **海岸植物** a beach plant. **海岸浸食** coastal [beach] erosion. **海岸図** a coast chart. **海岸線** the shoreline; the coastline; the strandline. /複雑な〜線 an irregular coastline. **海岸段丘** a coastal terrace. **海岸地形** a coastal landform. **海岸漂着物** things that have been washed [come] ashore; flotsam 《on the beach》; beached objects [material, items]. **海岸平野** a coastal plain. **海岸保安林** a coastal

回帰性 〔動物の帰巣能力〕(a) homing ability;〔帰巣本能〕a homing instinct.

回帰率 a rate of return. ▶サケの〜が低下した. The proportion of salmon returning has fallen.

解禁 〔猟・漁などの〕the opening of the hunting [fishing] season. ▶〜する〔猟・漁を〕open the hunting [fishing] season. / 鮎は来週〜となる. The *ayu* season opens [begins] next week. ¶**解禁期** 〔狩猟などの〕the hunting season; open season 《on…》. **解禁日** the first day 《of the *ayu* fishing season》.

回収 (a) collection; (a) withdrawal; recovery; retrieval;〔売った品物の〕recall; a calling in 《of…》;〔人工衛星・宇宙船の〕recovery. ★製品[廃品, 熱, 古紙, CO_2]回収などは⇨製品, 熱, 古紙, 廃品, CO_2(和英篇末尾)の各項目を参照. ▶〜する withdraw 《coins, books》 from circulation; collect; call in; retire; recycle 《disused things》;〔人工衛星・宇宙船を〕recover. / 〜可能な recoverable [retrievable] 《spaceship》; returnable 《bottles》. ★⇨リターナブル, デポジット / 流出重油を〜する recover spilled [leaked] fuel oil. / 欠陥車[欠陥品]を〜する recall defective vehicles [items, goods]. / 使用済みペットボトルは小売店で〜している. Empty plastic bottles are collected at retail shops. / その工場では使用済み核燃料からプルトニウムを〜している. That factory retrieves [is retrieving] plutonium from spent nuclear fuel. / 寿命の尽きた人工衛星がスペースシャトルによって〜された. A man-made satellite, its life expectancy over, was retrieved by a [the] space shuttle. / 金属類〜運動 a campaign to collect scrap metal. / 牛乳パックの〜ボックス a milk carton collection box [bin]; a milk carton recycle [recycling] bin [box]. ¶**回収型カプセル衛星** a satellite with a recovery capsule. **回収予定地** 〔宇宙船の〕a recovery area. **回収率** 〔アンケート用紙などの〕a return rate on 《a questionnaire》;〔再資源化のための〕a recovery rate. **無料回収車** 〔市街地を回って不要のテレビ・パソコン・バイクなどをただで引き取るトラック〕a free [no-charge] collection truck.

海獣 a marine [sea] animal.

海上濃霧警報 《issue》a dense (marine) fog warning.

海上風力発電所 an offshore wind power plant [station].

海上油田 an offshore oil field.

海水 seawater; brine. ▶〜から作った真水 desalinated seawater. / この藻は〜に生える．These algae grow in seawater. / 防潮堤が〜の浸入を防いだ．The seawall held back the seawater. / 浜風は〜を含んでいる．A beach breeze contains saltwater particles. ¶**海水魚** a saltwater fish. **海水蒸溜工場** a seawater distillation plant. **海水淡水化** desalination. ★⇨淡水化 / 〜淡水化装置 desalination equipment. / 〜淡水化プラント a desalination plant. / 〜淡水化膜 a desalination membrane. **海水温** (an) ocean temperature. ★⇨温度 **海水面** sea level. ⇨海面 / 〜面上昇 a rise in sea level. **海水浴場** a (swimming [bathing]) beach; a seaside swimming area; a beach [seaside] resort; a public beach for swimming [bathing]. **人工海水** artificial seawater.

海生 ¶**海生植物** a marine plant; submarine vegetation. **海生植物相** a marine flora. **海生生物** marine [sea] life. **海生動物** a marine [sea] animal. **海生動物相** a marine fauna. **海生哺乳類** marine [sea] mammals; mammals of the sea.

解体 〔建物などの〕taking [tearing] down; wrecking; demolition; disassembling;〔機械・車などの〕taking [pulling] apart [to pieces]; breaking up; wrecking; knocking down; disassembling; dismantling; dismantlement;〔死体・肉牛・クジラなどの〕dissection; dismemberment. ▶〜する〔建物を〕tear [pull]《a building》down; dismantle; demolish; wreck;〔機械・車などを〕take [pull]《a machine》apart [to pieces]; break《a car》up; knock down; disassemble; dismantle;〔動物の体を〕dissect; dismember. / 古い木造建築を〜して移築する take apart an old wooden building and reassemble it elsewhere. **解体移築**〔建物などの〕taking apart and reassembling in a new location. **解体核(兵器)** a dismantled nuclear weapon. **解体業者**〔建物の〕a demolition contractor; a building demolisher; a wrecker;〔機械・車などの〕a dismantler; an auto [a machine] parts recycler; an auto wrecker. **解体工事** demolition work. **解体工場** a《car》dismantling plant; a parts recycling shop; a chop shop. **解体ごみ** demolition debris; debris from demolition. **解体作業** demolition work; wrecking; dismantlement. **解体作業員** 〔建物の〕a demolition worker; a wrecker;〔機械・車などの〕a parts-recycling mechanic. **解体修理** taking apart and repairing; repairs that involve prior dismantling. / 古い建物の一部を〜修理する take apart [tear down] a portion of a house and repair it **解体撤去** demolition and removal《of buildings, etc.》. / 〜撤去費 the cost of demolition and removal. **解体費(用)**〔出費〕demolition costs; the cost of demolition;〔解体料金〕a demolition charge; the charge for demolition. **解体復元(工事)** dismantling and reconstruction [rebuilding]. / その建物は別の場所に〜復元された. The building was dismantled [taken down] and rebuilt [reconstructed, put up again] on a new site.

海中〔海面下〕the subsurface part of an ocean [a sea]. ▶〜に投棄する throw *sth* into the sea;〔船から〕throw《goods》overboard;〔海難の際に積み荷などを〕jettison. / 〜の生物 marine [sea] life. ¶ **海中居住実験計画** the Seatopia Project. **海中公園** an underwater park. **海中実験室** a wet lab [laboratory]. **海中林** a sea jungle.

害虫 a noxious [harmful] insect; an insect pest; a pest; vermin ★複数扱い. ▶リンゴ園に〜が発生した. There was an outbreak of insect pests in the apple orchard. / 畑に〜がはびこっている. The vegetable field is [fields are] swarming [infested] with vermin. / 〜の被害 damage done by insect pests. / 〜を駆除する exterminate insect pests [noxious insects, vermin]. / 〜駆除業者[会社] an exterminator (company).
¶ **害虫学** pestology.

海底 the bottom of the sea [ocean]; the sea floor; the seabed; the ocean floor [bed]; ground. ¶ **海底遺跡** seabed remains; submarine ruins; submerged ruins. **海底崖** a submarine [underwater] escarpment [cliff]. **海底(地形)測量** seabed surveying;〔1回の〕a seabed survey. **海底開発** exploitation [development] of the seabed. **海底火山** a submarine [an undersea, a seabed] volcano. **海底岩盤** the seabed (rock); the ocean floor (rock). **海底掘削船** a drillship; a drilling ship [vessel].

海底採鉱 seabed mining; deep-sea mining. **海底資源** seabed [sea floor, ocean floor, undersea] resources; the resources of the seabed. **海底地震** a submarine [an undersea] earthquake. ▶深さ40キロの〜地震が起きた. There was an earthquake at a depth of forty kilometers under the seabed. **海底地震計** an ocean-bottom seismograph [seismometer] (略: OBS). / ケーブル式〜地震計 an ocean-bottom cable seismograph [seismometer]. / 自己浮上式[ポップアップ型]〜地震計 a pop-up (type) ocean-bottom seismograph. **海底石油採掘** offshore oil drilling. **海底堆積物** (a) marine sediment. **海底探査[調査]** seabed investigation [surveying]; 〔1回の〕an investigation [a survey] of the seabed. **海底探査機** a submarine [an underwater] probe. **海底炭田** a submarine coal field [mine]. **海底地形** submarine topography [geomorphology]; physical contours [features] of the seabed. **海底地形学** submarine geomorphology. **海底地形図[地図]** a bathymetric chart; a map [chart] of the seafloor. **海底地質学** submarine geology. **海底トンネル** an undersea [a submarine] tunnel; a tunnel under the sea [ocean]. **海底熱水鉱床** a sea-floor [submarine] hydrothermal deposit. **海底パイプライン** a submarine [an underwater, an undersea] pipeline. **海底噴火** a submarine eruption; an eruption in [on] the seabed. **海底油田** a submarine [an undersea, an offshore] oil field. **海底林** an underwater forest.

改訂管理制度 〔国際捕鯨委員会 (IWC) の捕鯨監視制度〕the Revised Management Scheme (略: RMS).

改訂管理方式 〔国際捕鯨委員会 (IWC) の鯨類管理方式〕the Revised Management Procedure (略: RMP).

ガイドウェイ・バス a "guideway bus"; a rapid bus service running on rails.

開発 〔土地・資源・製品などの〕development; exploitation 《of resources》; growth 《of new industries》; reclamation 《of wasteland》. ▶〜する develop 《backward regions, an industry》; open up 《a country, a mine, an industry》; exploit 《resources》; tap 《resources》. / 持続可能な〜 sustainable development. ★⇨持続可能 / 未開地域

の〜 development of backward regions. / 大規模な〜 large-scale development. / むちゃな〜 insane [crazy] development [exploitation]．★⇨乱開発 / むやみな〜がもたらした環境破壊 environmental destruction [damage] caused by excessive [indiscriminate] development. / その地域は〜が進んでいる[遅れている]．The area is being developed [is behind in development]. / ゴルフ場〜にからむ汚職事件 a golf-course development corruption scandal [case]. / そののどかな村に〜の波が押し寄せた．The quiet village was overwhelmed by growth and development. | A wave of development swept over the peaceful village. / 〜を規制する control [regulate] development. / 野球場建設予定地で遺跡が発見され，〜か保存かでもめている．Archeological remains were discovered on the site of the proposed baseball stadium, and there is controversy about whether to develop or to preserve the site. / 〜の名のもとに目を覆うような自然破壊が進められている．Outrageous environmental destruction is going on under the name of development. | Development is being used as a pretext for destruction of the environment. / 新製品を〜する develop a new product. / 新しく〜した工程[技術] a newly-developed process [technique, technology]. / 〜中である be being developed; be under [undergoing, in the course of] development．★⇨開発段階 / 新薬の〜に携わる[取り組む] take part in [tackle] developing [the development of] new medicines [drugs].

¶ **開発援助** development aid [assistance]; help [assistance] with development. **開発援助委員会** 〔OECD の〕the Development Assistance Committee (略: DAC). **開発学** development studies. **開発格差** 〔国家間・地域間などの〕a development gap. **開発基金** a development fund. **開発教育** 〔国際社会における〕developmental education; education designed to foster an understanding of relations between rich and poor countries. **開発業者** a (commercial) developer; a development company [firm]. **開発競争** development competition; competition [a race] to develop 《new technologies》. / 新製品の〜競争 competition [a race] to develop new products. / 核兵器～競争 a nuclear arms race.　**開発許可**　a development permit.　**開発許可制度**　a development permit system.　**開発銀行**　a development bank.　**開発計画**　a development project [program]《in Africa》.　**開発経済学**　development economics; the economics of development.　**開発権**　exploitation [development] rights.　**開発コスト**　development costs; the cost of development. / ～コストを削減する reduce development cost(s).　**開発コミットメント指数**　〔米国のシンクタンク，世界開発センターが，先進国の政策に関して行う順位付け〕the Commitment to Development Index (略：CDI).　**開発三角地帯**　〔ラオス・カンボジア・ベトナム三国国境をまたぐ貧困地域〕the (Cambodia-Laos-Vietnam) Development Triangle.　**開発事業**　development (work).　**開発資金[資本]**　a development fund [capital].　**開発指導要綱**　〔宅地開発指導要綱〕housing development guidelines; guidelines for the development of housing estates; restrictions on housing development.　**開発段階**　《be at》a [the] development stage. / ～段階にある製品 a product (which is) at the development stage [under development] / 彼はこの新車には～段階から関わっている．He was involved with the new model from the development stage. | He was involved with developing the new car from the start.　**開発独裁体制**　(a) developmental dictatorship.　**開発費(用)[コスト]**　development costs.　**開発ブーム**　a development boom.　**開発法学**　law and development studies.　**開発モデル**　a development model.　**開発予算**　a development budget.　**宇宙開発**　space development.　**エネルギー開発**　energy development.　**海洋開発**　⇨海洋　**共同開発**　joint development.　**経済開発**　economic development.　**研究開発**　《carry out》research and development (略：R&D).　**産業開発**　industrial development [growth]; industrialization; development of industry.　**資源開発**　development [exploitation] of (natural) resources; resource(s) development.　**製品開発**　product development.　**石油[天然ガス]開発**　exploitation of oil [natural gas] resources.　**宅地開発**　opening up [development of] land for housing; development of housing land.　**地域開発**　community development. 都市開発 urban development. 土地開発 land development; development [exploitation] of land. 農業開発 agricultural development; development of agriculture; the growth of agriculture. 不動産開発 real estate [property] development. 未開発 undeveloped; unexploited; unused; untapped.

皆伐(かいばつ) large-scale felling (of trees).

外部不経済 external diseconomies. ★⇨英和 externality

海面 the surface of the sea;〔標準海面〕sea level. ▶～の上昇［低下］a rise [fall] in sea level. / ～上昇によって，沿岸部では台風の被害が年々深刻になっている．Rising sea levels are resulting in more coastal damage each year due to typhoons. ¶**海面温度** the temperature of the sea surface [at the surface of the sea]; (a) sea-surface temperature. **海面気圧** sea-level (atmospheric [air]) pressure; (the) atmospheric pressure at sea-level. **海面漁業** sea-water fishing. **海面更正**〔気圧の〕the reduction (of pressure) to (mean) sea level. **海面上昇** a rise in (the) sea level《due to global warming》. **海面変動** a change in sea level. **海面養殖** sea-water culture. **廃棄物海面処分場** ⇨処分場. **平均海面** a mean sea level. / 地球の平均～水位は 2030 年までには約 20 cm 上昇すると言われている．It is said that average world sea levels will rise [have risen] 20 cm by the year 2030.

界面活性剤 ⇨活性 (活性剤)

回遊魚 a migratory (species of) fish. ¶**季節回遊魚** a (seasonal) migratory fish. **降河性回遊魚** a catadromous fish. **昇河性回遊魚** an anadromous fish. **通し回遊魚** a diadromous fish.

海洋 the sea(s); the ocean. ▶世界の主な～ the main oceans of the world. / ～上で発達した低気圧 low pressure that developed over the ocean. / 地球の面積のうち～が占める割合 the proportion of the earth's (surface) area occupied [taken up] by the sea(s) [ocean(s)]. ¶**海洋安全保障** maritime [marine] security. **海洋エネルギー** ocean energy; energy from the ocean. **海洋汚染** marine pollution. / ～汚染物質 a marine pollutant. **海洋汚染防止条約** the International Convention for the Prevention of Pollution from Ships. ★ 1973 年に制定後，78 年に改定．現在の正式名称

は「1973年の船舶による汚染の防止のための国際条約に関する1978年の議定書」(the International Convention for the Prevention of Pollution from Ships, 1973, as modified by the protocol of 1978 relating thereto (略: MARPOL 73/78)). また略称からマルポール[MARPOL]条約 (the MARPOL Convention) とも言う． **海洋汚染防止法** the Marine Pollution Prevention Law. **海洋音響トモグラフィー** 〔ソナー技術による音響断層撮影法〕oceanic acoustic tomography. **海洋温度差発電** ocean thermal energy conversion. ★⇨英和 OTEC **海洋開発** ocean development; development of the ocean. **海洋開発研究機構** the Japan Agency for Marine-Earth Science and Technology (略: JAMSTEC). **海洋科学** marine science. **海洋科学調査** oceanographic [marine science] research; an oceanographic research project [program]. **海洋化学** marine chemistry; thalassochemistry. **海洋学** oceanography; oceanology. ★⇨日本(日本海洋学会), ユネスコ(ユネスコ政府間海洋学委員会) **海洋学者** an oceanographer; an oceanologist. **海洋隔離** 〔二酸化炭素の〕⇨英和 ocean sequestration. **海洋環境** a marine [an oceanic] environment. **海洋環境学** marine environmental studies [science]. **海洋監視衛星** an ocean-surveillance satellite. **海洋観測** oceanographic observation. **海洋観測衛星** a marine observation satellite. **海洋観測船** a marine research ship [vessel]. **海洋管理協議会** ⇨英和 Marine Stewardship Council. **海洋気象** marine weather; (the) weather at sea. / ～気象情報 a marine (weather) forecast. **海洋気象学** marine meteorology. **海洋気象台** a marine meteorological observatory. **海洋気団** a maritime air mass. **海洋基本法** the Basic Law on the Oceans. **海洋基本計画** 〔海洋基本法に基づく〕the Basic Plan on Ocean Policy. **海洋掘削船** a drillship; a drilling ship [vessel]. **海洋権益** 《the country's》maritime interests《in the East China Sea》. **海洋研究開発機構** the Japan Agency for Marine-Earth Science and Technology (略: JAMSTEC). **海洋工学** ocean [marine, maritime] engineering. **海洋航行不法行為防止条約** the Convention for the Suppression of Unlawful Acts Against the Safety of

Maritime Navigation; the SUA Convention. ★正式名称は「海洋航行の安全に対する不法な行為の防止に関する条約」. **海洋国(家)** a marine [maritime] country [nation, state];〔大国〕a sea power. **海洋資源** 《abundant》marine [ocean] resources; the resources of the sea. **海洋哨戒機** a maritime patrol aircraft. **海洋深層水** deep ocean [sea] water. **海洋図** an ocean chart; a map of the sea. **海洋性気候** a maritime climate. **海洋政策研究財団** the Ocean Policy Research Foundation (略: OPRF). **海洋生物** a marine organism;〈集合的に〉oceanic life. **海洋生物学** marine biology [bioscience]. **海洋生物学者** a marine biologist. **海洋生物地理区** a marine biogeographic region. **海洋戦略**〔一国の〕(a) maritime strategy. **海洋大気局**〔米国商務省の〕the National Oceanic and Atmospheric Administration (略: NOAA). **海洋大循環** ocean(ic) general circulation. **海洋探検** ocean [marine] exploration. **海洋地球物理学** oceanic earth physics. **海洋地質学** marine [ocean] geology. **海洋調査船** an oceanographic [an ocean, a marine] research vessel [ship]. **海洋底** the bottom of the sea; the ocean bottom. **海洋底拡大説** the seafloor spreading theory [hypothesis]. **海洋度** oceanity. **海洋投棄** 《regulate, stop》ocean [sea] dumping [disposal]《of nuclear waste》;《outlaw the》dumping at sea《of radioactive waste》. **海洋投棄規制条約** the Convention on the Prevention of Marine Pollution by Dumping of Wastes and Other Matter. **海洋の自由**〔国際法上の〕freedom of the seas. **海洋バイオテクノロジー** marine biotechnology. **海洋博覧会** an ocean(ic) exposition. **海洋物理学** physical oceanography. **海洋プレート** an oceanic plate. **海洋法** the law of the sea. ★⇨国連(国連海洋法会議[条約]) **海洋放牧** marine ranching. **海洋牧場** a marine [sea] farm [ranch]; a marifarm. **海洋保護区**〔海洋生態系を保護するための〕a marine protected area (略: MPA). **海洋療法** thalassotherapy. **日本海洋少年団** ⇨日本

外来 ¶ **外来河川** an exotic river [stream]. **外来魚** a non-native fish; a fish from abroad; an exotic;〔導入種〕an introduced species;〔定着種〕a colonist. **外

来種 〔動植物など〕an alien [a non-native] species; a foreign species; a species from abroad; an exotic;〔導入種〕an introduced species;〔定着種〕a colonist. **外来種被害防止法, 外来生物法** the Invasive Alien Species Law. 正式名称は「特定外来生物による生態系等に係る被害の防止に関する法律」. **外来生物** an alien [a non-native] organism. ▶特定〜生物 an invasive alien species（略: IAS）. / 未判定〜生物 an uncategorized alien species（略: UAS）. / 要注意〜生物 a potentially (dangerous) invasive species. **外来動物[植物]** an alien [a non-native] species of animal [plant]; an animal [a plant] from abroad; an exotic;〔導入種〕an introduced species;〔定着種〕a colonist.

海流 an ocean current; a (marine) current. ▶対馬[千島]〜 the Tsushima [Chishima] Current. / 日本〜⇨日本 **海流図** a current chart. **海流発電** ocean current (power) generation. **海流瓶** a drift bottle; a floater.

街路樹 a roadside [boulevard] tree; trees lining a street. ▶〜が色づき始めるころ the time when the roadside trees begin to change color. / 排気ガスで〜が枯れた. Some roadside trees have died from exhaust fumes. / 〜のある通り a tree-lined street. / 〜の剪定(せんてい)作業 pruning the roadside trees. / 私の街では〜としてイチョウを植えている. Where I live, ginkgo trees are planted by the roadside.

科学 science. ▶〜で説明[証明]できないこと something that cannot be explained [proved] by science. / 〜の目で見る see with a scientific eye. / 〜の進歩が即文明の進歩とはかぎらない. The progress of science does not necessarily equate with the progress of civilization. / 〜を応用する apply science. / 〜的な scientific. / 〜的な物の考え方 a scientific way of thinking; scientific thinking. / 〜的に考える[説明する] think [explain] scientifically. / それは〜的にありえないことだ. That's scientifically impossible. / 〜的観察 scientific observation. / 〜的管理[経営]法 scientific management. / 〜的研究 scientific research [study]. / 〜的研究方法 a scientific method of research. / 〜的見地から from a [the] scientific viewpoint. / 〜的根拠[裏付け] scientific grounds [support, backing]. / 〜的実験精神 the spirit of scientific experimentation. / 〜的証拠 scientific evidence [proof]. / 〜的真理 scientific truth. / 〜的正確さをもって with scientific exactitude [precision]. / 〜的説明 a scientific explanation. / 〜的大発見をする make a big [major, significant] scientific discovery. / 〜的農業 scientific farming. / 〜的方法 the scientific method. / 〜的合理性 scientific logic [sense]; a scientific basis; science. / 事業に〜的合理性があるかを確認する make sure that a project makes scientific sense. ★⇨温暖化

¶**科学映画** a science film. **科学衛星** a scientific satellite. **科学界** the world of science; the scientific world. **科学革命** a scientific revolution. **科学館** a science museum. **科学観測衛星** a scientific research satellite. **科学技術** science and technology; scientific technique(s); modern technology. / 最先端の〜技術 the most advanced [state-of-the-art] science and technology [modern technology, scientific technique(s)]. / 〜技術の進歩は人類に幸福をもたらすか？ Will the progress of modern technology bring happiness to the human race? / 〜技術者 a scientific technician. / 〜技術政策 a science and technology policy. / 〜技術立国 a national commitment to science and technology; national development based on science and technology. / 〜技術関係経費 science and technology-related expenditures. / 〜技術研究費 expenditure for scientific and technical research; science and technology research costs. / 〜技術振興調整費 special coordination funds for promoting science and technology. / 〜技術振興費 science and technology promotion expenditures. **科学技術白書** a white paper on science and technology;〔文部科学省の〕the White Paper on Science and Technology《2004》. **科学技術基本計画** 〔日本政府が5年に1度策定する〕the《Third》Science and Technology Basic Plan. **科学技術基本法** the Science and Technology Basic Law. **科学技術教育協会** the Foundation for Education of Science and Technology（略：FEST）. **科学教育** science education. **科学計量学** scientometrics. **科学研究費** scientific research expenses. **科学研究費**

補助金 〔文部科学省の助成金〕a government subsidy [government funding, a grant-in-aid] for scientific research. **科学コミュニケーター** a science communicator. **科学雑誌** a science magazine. **科学史** the history of science. **科学至上[万能]主義** scientism. **科学思想** scientific thought. **科学者** a scientist. **科学書** a scientific book. **科学振興** scientific advancement; the advancement [promotion] of science. **科学振興会** a society for the promotion of science. **科学性** scientificness. **科学知識** scientific knowledge. **科学哲学** (the) philosophy of science. **科学博物館** a science museum. **科学番組** a science program《on the radio, on TV》. **科学批判** scientific criticism. **科学用語** a scientific term;〈集合的に〉scientific terminology. **科学リテラシー** science [scientific] literacy. **科学理論** (a) scientific theory. **応用科学** applied science. **疑似[エセ]科学** (a) pseudoscience. 形 pseudoscientific. / 擬似[エセ]〜者 a pseudoscientist. **基礎科学** fundamental [basic] science. **自然科学** natural [physical] science. **社会科学** social science. **人文科学** the humanities. **数理科学** mathematical science.

化学 chemistry. 形 chemical. ▶〜的な chemical. / 〜的な刺激を加える apply a chemical stimulus. / 〜的に chemically. / 〜(的に)処理する dispose of *sth* chemically; treat *sth* chemically. / 〜的に変質する change chemically. / 〜(的)変化を起こす undergo a chemical change. / 〜的酸素要求量 ⇨英和 chemical oxygen demand / 水質サンプルの〜的酸素要求量を計測する test the chemical oxygen demand of a sample of water. / 〜的性質 a chemical property. / 〜的突然変異誘導 chemical mutagenesis. / 〜的突然変異誘導物質 a chemical mutagen. / 〜的風化 decomposition. / 〜的分類(学) chemotaxonomy. 形 chemotaxonomic; 〔化学的分類学者〕chemotaxonomist. / 〜的水処理〔排水の〕chemical waste-water treatment.

¶ **化学イオン化質量分析法** chemical ionization mass spectrometry. **化学遺伝学** chemical genetics. **化学エネルギー** chemical energy. **化学化石** a chemical fossil. **化学岩** chemical rock. **化学感覚** chemical sense. **化学機械** chemical machinery. **化学記号**

化学

a chemical symbol. **化学記号法** chemical notation. **化学気相成長法** chemical vapor deposition（略：CVD）. **化学吸着** chemisorption; chemical adsorption. **化学クロマトグラフィー** chemichromatography. **化学結合** a chemical bond. **化学元素** a chemical element. **化学研磨** chemical polishing. **化学工学** chemical engineering. **化学工業** the chemical industry. **化学工芸** chemical technology. **化学攻撃** a chemical (weapons) attack. **化学構造式** a chemical structural formula. **化学工場** a chemical factory [plant, works]. **化学合成** 〔人工的な〕chemical synthesis;〔生物の〕chemosynthesis. **化学合成細菌** chemosynthetic bacteria. **化学合成独立栄養** chemoautotrophy. 形 chemoautotrophic. **化学合成独立栄養生物** a chemoautotroph. **化学合成生物** 〔生物の〕a chemotroph. **化学構造** chemical structure; chemical constitution. **化学剤** a chemical agent; a chemical. **化学作用** a chemical action. / ～作用で chemically. / ～作用を起こさない inert; lacking a chemical action. **化学史** the history of chemistry. **化学式** a chemical formula. / 酢酸を～式で表す express acetic acid by a chemical formula. / 水の～式は H_2O である. The (chemical) formula for water is H_2O. **化学式量** chemical formula weight. **化学試験**〔再処理施設の〕《carry out》a chemical test. **化学実験** a chemical experiment. ▶～実験器具 a chemical apparatus. / ～実験室 a chemistry laboratory. **化学シナプス**〔神経伝達の〕chemical synapse. **化学シフト** a chemical shift. **化学者** a chemist. **化学試薬** a chemical reagent. **化学種** a chemical species. **化学修飾** chemical modification. **化学受容器[受容体]** a chemo(re)ceptor. **化学受容器引き金帯** a chemoreceptor trigger zone（略：CTZ）. **化学消火剤** fire-extinguishing chemicals. **化学蒸着(法)** chemical vapor deposition（略：CVD）. **化学消防車** a chemical fire engine. **化学消防隊** a chemical fire brigade [company]. **化学処理** a chemical treatment. ★⇨用例を参照 **化学処理工場** a chemical treatment plant. **化学飼料** chemical feed. **化学進化**〔化学組成上の進化〕chemical evolution. **化学浸透圧説** the chemiosmotic theory [hypothesis]; chemiosmosis. **化学親和力** chemical affinity. **化学性食中毒** chemical food poisoning. **化学清掃** 〔船舶の〕chemical cleaning. **化学生態学** chemical ecology. **化学製品** a chemical product; chemical goods [products]. **化学線** an actinic ray;〔特に紫外線〕an ultraviolet ray. **化学線学** actinology. **化学線作用** actinism. **化学線透過性** diactinism. 形 diactinic. **化学線量計** a chemical dosimeter. **化学戦** chemical warfare. **化学繊維** a synthetic [chemical] fiber. **化学繊維工業** the synthetic [chemical] fiber industry. **化学繊維紙** chemical fiber paper. **化学センサー** a chemical sensor. **化学増感** chemical sensitization. **化学増感剤** a chemical sensitizer. **化学操作**[処理] 《by》chemical manipulation [treatment]. **化学増幅** chemical amplification. **化学組成** chemical composition. **化学弾** 〔弾薬〕chemical ammunition;〔爆弾〕a chemical bomb. **化学脱水器** 〔船舶の〕a chemical dehydrator. **化学探鉱**[探査] geochemical prospecting. **化学テロ** chemical terrorism. **化学伝達** chemical transmission. **化学伝達物質** a chemical transmitter [mediator]. **化学電池** 〔化学反応による電池〕an electrochemical [a chemical] cell; a chemical battery. **化学天秤**(びん) a chemical [an analytical] balance. **化学動力学** chemical dynamics. **化学当量** a chemical equivalent. **化学毒性** chemical toxicity. **化学熱傷** a chemical burn. **化学熱力学** chemical thermodynamics. **化学発光** chemiluminescence. **化学反応** a chemical reaction. **化学反応式** 〔化学方程式〕a chemical equation; a reaction formula. **化学肥料** (a) chemical [commercial] fertilizer. / 〜肥料を使う use chemical fertilizer(s). / 〜肥料を減らす cut down on [reduce] (the amount of) chemical fertilizer. / 〜肥料工業 the chemical fertilizer industry. / 減〜肥料栽培 cultivation using reduced amounts of chemical fertilizer. / 減〜肥料農作物[米, 野菜] farm products [rice, vegetables] grown with reduced amounts of chemical fertilizer. **化学品の分類及び表示に関する世界調和システム** 〔2003年に国連が定めた有害物質の危険性表示の国際基準〕the Globally Harmonized System of Classification and Labelling of Chemicals (略: GHS). **化学物質** a chemical substance; a

化学

chemical agent; a chemical. **化学物質アドバイザー** a chemical risk assessor; an advisor on chemical substances (and their effects on people and the environment). **化学物質過敏症** multiple chemical sensitivity. **化学物質等安全データシート** ⇨製品(製品安全データシート). **化学物質排出把握管理促進法** ⇨PRTR法(和英篇末尾) **化学物質審査規制法** 〔化学物質の審査及び製造等の規制に関する法律 の通称〕the Law Concerning the Examination and Regulation of Manufacture, etc., of Chemical Substances. **化学物理学** chemical physics. **化学分析** a chemical analysis. **化学分類学** chemotaxonomy. **化学兵器** a chemical weapon. ★⇨弾頭/〜兵器の使用を禁止する prohibit [ban] the use of chemical weapons. **化学兵器テロ** chemical terrorism. **化学兵器物質** a chemical warfare agent (略: CWA). **化学平衡** a chemical equilibrium. **化学変化** chemical change. **化学変換** chemical transformation. **化学防護隊** a chemical protection unit. **化学防護服** a chemical (protection) suit. /陽圧式〜防護服 a chemical protection suit and positive-pressure breathing apparatus; an SCBA suit (SCBA は self-contained breathing apparatus の略). **化学放射線療法** chemoradiotherapy. **化学方程式** a chemical equation. **化学ポテンシャル** a chemical potential. **化学メッキ** chemical [electroless] plating. **化学(木材)パルプ** chemical (wood) pulp. **化学薬品** chemicals. **化学用語** a chemical term;〈集合的に〉chemical terminology. **化学予防** chemoprophylaxis. **化学量数** a stoichiometric number. **化学療法** chemotherapy. **化学療法医** a chemotherapist. **化学療法薬** a chemotherapeutic agent [drug]. **化学量論** stoichiometry. 形 stoichiometric. **化学レーザー** 【物】a chemical laser. **化学ロケット** a chemical rocket. **応用化学** applied chemistry. **工業化学** industrial chemistry. **実験化学** experimental chemistry. **純粋[理論]化学** pure [theoretical] chemistry. **精密化学** fine chemistry. **総合化学** synthetic chemistry. **超微量化学** ultramicrochemistry. **農芸[冶金]化学** agricultural [metallurgic] chemistry. **農産化学** chemurgy. 形 chemurgic. **微量[顕微]化学** microchemistry.

分析化学 analytical chemistry. **有機[無機]化学** organic [inorganic] chemistry.

核 〔細胞の核または原子核〕a nucleus《複数形 nuclei, nucleuses》. ★⇨核酸, 原子 ▶核を２つ持つ細胞 a binucleate cell. / 核による全人類の破滅 nuclear annihilation [holocaust]. / 核による報復 nuclear retaliation. / 核の nuclear. / 核の冬 (a) nuclear winter. ★英和 nuclear winter, nuclear autumn / 米国の核の傘の下で under the shelter of the US nuclear umbrella. / 核の均衡 nuclear parity《between two countries》. / 核の平和利用 (the) peaceful use of nuclear technology [power, energy]; using [the use of] nuclear technology for peaceful purposes. / 核の持ち込み the bringing in [introduction] of nuclear weapons《into Japan by other countries》. / 核(の)ボタン《push》the nuclear button. / 核の闇市場 a nuclear black market. / 核を取り外す remove a nuclear warhead; disarm《a nuclear missile》. ¶ **核アレルギー** Japan's "nuclear allergy"; a hypersensitivity induced by the atom bombs on Hiroshima and Nagasaki that has issued in a refusal to develop nuclear weapons or to allow their storage on its soil. **核医学** nuclear medicine. **核移植** nuclear transplantation. **核エネルギー** nuclear energy. ★⇨原子 **核オプション** 〔状況によっては核兵器の保持または使用が可能であるとする考え方〕the nuclear option. **核カード** 《play》the nuclear card. / 北朝鮮は核カードを巧みに使いながら日米韓に揺さぶりをかけているように見える．North Korea seems to be using its nuclear card cleverly to put pressure on Japan, the US and South Korea. **核開発** nuclear development; the development of nuclear capacity. / 核開発の国際監視 the international oversight of nuclear development. / 核開発競争に歯止めをかける put a stop to the race in nuclear development. / 核開発計画 a nuclear program. **核化学** nuclear chemistry. **核拡散** nuclear proliferation; proliferation of nuclear weapons. / 拡散防止条約 the Nuclear Nonproliferation [Non-Proliferation] Treaty (略: NPT). **核管理体制** the NPT Regime; the Nonproliferation Regime. **核関連物質** nuclear-related materials. **核危機** a nuclear crisis. **核技術** nuclear technology.　**核疑惑**　《be under》the suspicion of having nuclear weapons.　**核疑惑国**　a country suspected of having nuclear weapons; an undeclared nuclear country.　**核クラブ**　〔核兵器所有国〕《join》the nuclear [atomic] club.　**核軍縮**　nuclear disarmament; nuclear arms reduction.　**核軍備**　nuclear armament(s) [arms].　**核軍備縮小会議**　a nuclear disarmament conference.　**核原料物質**　(a) nuclear source material. / 核原料物質, 核燃料物質及び原子炉の規制に関する法律⇨原子(原子炉等規制法)　**核攻撃**　a nuclear attack [strike].　**核攻撃目標解除**　nuclear detargeting.　**核災害**　a nuclear disaster.　**核査察**　an inspection of nuclear facilities; a nuclear inspection.　**核シェルター**　a nuclear shelter.　**核事故**　a nuclear accident.　**核施設**　a nuclear facility [installation, complex].　**核時代**　the nuclear age.　**核実験**　nuclear testing; 〔1回の〕a nuclear test; a nuclear weapon(s) [arms, device] test. / 核実験がもたらす影響は完全に予測できるものではない. It is impossible to predict all the effects of nuclear testing. / 核実験に反対する oppose nuclear testing [tests]. / その国が核実験に成功したという発表に世界は驚愕した. The world was stunned by the declaration that the country had conducted a successful nuclear test [successfully tested a nuclear weapon]. / 核実験の禁止 the prohibition [banning, outlawing] of nuclear testing; a ban on nuclear tests; a nuclear test ban. / 核実験の自発的停止 a voluntary suspension of nuclear testing; 〔一定期間の〕a self-imposed nuclear test moratorium. / 核実験の中止を申し入れたが聞き入れられなかった. We demanded [requested] an end to nuclear testing, but we were ignored. | We made an appeal for nuclear tests to be stopped but they wouldn't listen to us. / 日本国会は核実験の即時停止を求める決議を採択した. The Japanese Diet adopted a resolution calling for an immediate ban on [halt to] nuclear testing. / 核実験を行う do [conduct, carry out] nuclear testing; 〔1回の〕do [conduct, carry out] a nuclear test; test a nuclear weapon [bomb, device]. / 核実験を中止する stop [discontinue, terminate, give up] nuclear testing; stop testing nuclear weapons. / 核実験を再開する resume nuclear weapons tests. / 核実験を繰り返す do repeated nuclear tests; do one nuclear test after another; reiterate nuclear testing; carry on conducting nuclear tests. / 核実験を凍結する (temporarily) halt [freeze, put a freeze on] nuclear testing. / 核実験を禁止する ban [forbid, prohibit, proscribe] nuclear testing. / その国は国民や周辺諸国の反発を無視して核実験を強行した．The country went ahead with the (nuclear) test in defiance of (the protests of) its own citizens and neighboring countries. / 高空核実験 high-altitude nuclear testing;〔1回の〕a high-altitude nuclear test. / 水中核実験 underwater nuclear testing;〔1回の〕an underwater nuclear test. / 大気圏外核実験 outer space nuclear testing;〔1回の〕an outer space nuclear test. / 大気圏内[空中]核実験 atmospheric nuclear testing;〔1回の〕an atmospheric nuclear test. / 地下核実験 underground nuclear testing;〔1回の〕an underground nuclear test. / 臨界前[未臨界]核実験 subcritical nuclear testing;〔1回の〕《conduct, carry out》a subcritical (nuclear) test. / 核実験禁止協定 a nuclear test ban agreement. / 核実験場 a nuclear test(ing) ground. / 核実験停止《break, observe》a moratorium on nuclear testing [nuclear testing moratorium]. / 核実験監視施設 a nuclear test monitoring facility.　**核ジャック**　the hijacking of nuclear material. / 核ジャック防止条約〔国際原子力機関の核物質防護条約の通称〕the Convention on the Physical Protection of Nuclear Material.　**核使用**　the use of nuclear weapons. / 米大統領は地域紛争での核使用に含みを残した．The US president did not rule out using nuclear weapons [the nuclear option] in regional conflicts.　**核小体**　a nucleolus《複数形 -li》.　**核戦争**　a nuclear war; nuclear warfare. / 核戦争勃発の危機 the risk of nuclear war breaking out. / 人類を核戦争による絶滅から救う save humanity from nuclear extinction [annihilation]. / 全面核戦争 an all-out nuclear war; total nuclear war.　**核戦争防止協定**〔1973年米ソが調印〕the Agreement on the Prevention of Nuclear War.　**核戦争防止国際医師会議**　the International Physicians for the Prevention of Nuclear War (略: IPPNW).　**核戦略**　nuclear strategy.　**核戦力**

核

nuclear capacity [capability, potential]. / 核戦力上の優位 nuclear superiority. / 短距離核戦力 short-range nuclear forces (略: SNF). / 短射程中距離核戦力 short-range intermediate nuclear forces (略: SRINF). / 長射程中距離核戦力 long-range intermediate nuclear forces (略: LRINF). **核装備** nuclear equipment; nuclear capability. / 核装備する nuclearize《a submarine》. / 核装備可能な nuclear-capable《bombers, missiles, countries》. **核大国** a nuclear power. **核弾頭** a nuclear warhead; an atomic warhead. ★⇨弾頭 / 核弾頭を装備した nuclear-tipped《missiles, torpedoes》. / 戦略核弾頭 a strategic nuclear warhead. / 核弾頭ミサイル a nuclear missile. / 核弾頭搭載可能ミサイル nuclear-capable missiles. **核たんぱく質** nucleoprotein. **核超大国** a nuclear superpower. **核テロ** (an act of) nuclear terrorism; a nuclear terrorist act. **核電荷** a nuclear charge. **核転換** (a) nuclear transmutation. **核時計** 〔核戦争による地球破滅までの残り時間を表す時計〕a doomsday clock. **核都市** 〔核兵器の開発・製造のための都市〕a nuclear city. **核都市イニシアチブ** 〔ロシアの核都市で働いていた科学者・技術者の平和的雇用を確保するための米・ロ共同プロジェクト〕the Nuclear City Initiative (略: NCI). **核燃料** nuclear fuel. / 使用済み核燃料《the reprocessing of》spent nuclear fuel. / 使用済み核燃料再処理工場 a spent nuclear fuel reprocessing plant. / 核燃料加工会社 a nuclear [uranium] fuel processing plant. / 核燃料サイクル the nuclear fuel cycle. / 核燃料サイクル基地 a nuclear fuel cycle center. / 核燃料サイクル施設 a nuclear fuel cycle facility [complex, plant]. / 核燃料再処理 nuclear fuel reprocessing. / 核燃料再処理工場 a nuclear fuel reprocessing plant. / 核燃料税 (a) tax on nuclear fuel; (a) nuclear fuel tax. / 核燃料物質 (a) nuclear fuel material. / 核燃料棒 a nuclear fuel rod. **核濃縮** pyknosis; pycnosis. **核廃棄物** nuclear waste (materials); 〔放射性廃棄物〕radioactive waste;《俗》nuke puke. / 高[低]レベル核廃棄物 high-level [low-level] nuclear waste. / 核廃棄物汚染 nuclear waste contamination. / 核廃棄物処分場 a nuclear waste dump [site, dumpsite]. / 核廃棄物処理工場 a nuclear waste processing plant; a plant for processing nuclear waste. / 核廃棄物貯蔵所［地］a《permanent underground》nuclear waste repository［site］. **核廃絶** the abolition［banning］of nuclear weapons; nuclear abolition. **核爆弾** a nuclear bomb. **核爆発** a nuclear explosion［blast, detonation］. / 核爆発が起こる a nuclear explosion occurs［takes place］. / 核爆発装置 a nuclear device. **核反応** (a) nuclear reaction. **核不拡散** nuclear nonproliferation. / 核不拡散条約 ⇨核拡散防止条約. **核武装** nuclear［atomic］armament(s). / 〜する arm《the air force》with nuclear weapons; arm oneself with nuclear weapons;《口》go nuclear. / 核武装している国 a nuclear-armed country; a nuclear power. / 核武装しうる nuclear-capable《aircraft, country》. / 核武装の撤廃 nuclear disarmament; a nuclear ban. / 核武装を禁止する ban nuclear weapons; denuclearize《a country》. / その国の核武装を阻止する stop the country arming itself with nuclear weapons. / 非核武装地帯 a denuclearized［an atom-free, a nuclear-free］zone. **核物質** nuclear material(s). / 兵器に転用可能な核物質 weapon-usable nuclear material. / 核物質の軍事利用を阻止する prevent the military use of nuclear materials. / 兵器級核物質 weapons-grade nuclear material. / 核物質輸送 shipment of nuclear materials. **核物理学**［化学］ nuclear physics［chemistry］. **核分裂**〔原子核の〕(nuclear, atomic) fission;〔細胞核の〕division of a cell nucleus; nuclear division;（無糸）amitosis;（有糸）endomitosis. / 〜する〔原子核が〕(undergo) fission. / ウラニウムに核分裂を起こさせる fission uranium. / 細胞質分裂に先行して核分裂が起こる. Nuclear division occurs［The cell nucleus divides］before division of the cytoplasm. / 間接核分裂 karyokinesis; indirect nuclear division;（有糸分裂）mitosis. / 直接核分裂 karyostenosis. / 核分裂性の fissile; fissionable. / 核分裂性核種 a fissile nuclide. / 核分裂生成物 a fission product. / 核分裂(性)物質 fissile［fissionable］materials; fissionables. / 核分裂中性子 a fission neutron. / 核分裂連鎖反応 a fission chain reaction. **核兵器** a nuclear［an atomic］weapon; nuclear arms. / 核兵器による破壊 nuclear destruction［devastation］. / 核兵器によらぬ

報復 non-nuclear retaliation. / 核兵器の廃絶 the (total) abolition [elimination] of nuclear weapons. / 核兵器の撤廃 the (total) elimination [abolition] of nuclear weapons. / 核兵器の究極的撤廃 the ultimate elimination of nuclear weapons. / 核兵器の撤去 denuclearization. / 核兵器の解体 the dismantling of nuclear weapons. / 核兵器の開発競争 a nuclear arms race. / 核兵器の開発凍結 a freeze on nuclear arms development. / 核兵器の大幅削減に合意する agree on a large-scale reduction in nuclear weapons. / 核兵器の貯蔵 a nuclear buildup; nuclear stockpiling. / 核兵器の貯蔵量 a nuclear arsenal [stockpile]. / 核兵器の先制不使用 no first use of nuclear weapons. / 核兵器を廃絶する abolish [eliminate] nuclear weapons. / 小型核兵器 a small-scale [low-power] nuclear weapon. / 戦術核兵器 a tactical nuclear weapon. / 戦場核兵器 a battlefield nuclear weapon. / 戦略核兵器 a strategic nuclear weapon. ▶中距離核兵器 a middle-range nuclear weapon. / 核兵器開発疑惑 suspected nuclear arms development. / 核兵器関連物資 nuclear weapons-related material(s). / 核兵器禁止区域 a nuclear-weapon-free zone (略: NWFZ); a nuclear-free zone. / (非)核兵器保有国 a (non-)nuclear power [nation]. ¶**核変換**【理】(a) nuclear transformation. **核防衛** nuclear defenses. **核防衛力** a nuclear defense capability. **核崩壊** 〔原子核の〕nuclear decay [disintegration]; 〔細胞の〕karyorrhexis. **核(兵器)保有宣言** 《North Korea's》declaration [announcement] that《it》has nuclear weapons;《North Korea's》declaration of possession of nuclear weapons. **核閉鎖都市** 〔周辺地域から遮断された核都市〕a closed nuclear city. **核放棄** relinquishing [getting rid of] nuclear capabilities; giving up [dropping] a nuclear program. / 完全[段階的]核放棄 the complete [gradual] abandonment of nuclear programs. / 完全で検証可能かつ後戻りできない核放棄 (a) complete, verifiable and irreversible dismantlement of《North Korea's》nuclear programs (略: CVID). **核[非核]保有国** a nuclear [non-nuclear] power [nation]. / 核保有国になる become a nuclear power;《口》go nuclear. **核保有能力** nuclear capability. **核ミサイル** a nuclear missile. **核**

融解 〔細胞の〕karyolysis; caryolysis. **核融合** 〔原子核の〕nuclear fusion;〔細胞の〕karyogamy. / 原子核融合が起きる. Fusion occurs [takes place]. / 常温[低温]核融合 tabletop [cold] (nuclear) fusion. / 熱核融合 thermonuclear fusion. / 核融合装置 a fusion device. / 核融合爆弾 a fusion bomb. / 核融合反応 a fusion reaction. / 核融合炉 a (nuclear) fusion reactor. / 核融合発電 (nuclear) fusion power generation. **核様体, 核様物質** a nucleoid. **核抑止** nuclear deterrence. **核抑止力** a nuclear deterrent. **細胞核** a cell nucleus. **戦術[戦略]核** tactical [strategic] nuclear weapons [arms]. **複合核** 〔原子の〕a compound nucleus.

核酸 a nucleic acid. ¶ **核酸塩基** a nucleic acid base. **核酸発酵** nucleic acid fermentation. **核酸分解酵素** nuclease. **核酸増幅検査** a nucleic acid amplification test (略: NAT). **デオキシリボ核酸** deoxyribonucleic acid (略: DNA). **リボ核酸** ribonucleic acid (略: RNA).

拡大生産者責任 ⇨生産

隔膜法 〔食塩電解法〕the diaphragm process.

学名 a scientific name; a technical name;〔植物学上の〕a botanical name;〔動物学上の〕a zoological name. ★生物学上の分類は上位から kingdom (界), phylum (門; 植物は通例 division), class (綱), order (目), family (科), genus (属), species (種(しゅ))の順になる. 学名はラテン・ギリシャ語をベースにした属名–種名の順に記すことが多く, これを二名法(binomial nomenclature)と呼ぶ. 例えばヒトの学名 Homo sapiens では Homo が属名, sapiens が種名. ⇨種, 属 ▶〜の命名法 nomenclature. / 〜の命名者 a nomenclator; a person who assigns scientific [biological] names. / …に〜をつける give [assign, allot] a scientific [biological, botanical, zoological] name to… / …という〜をもつ have [bear] the scientific [biological, botanical, zoological] name…; be scientifically [biologically, botanically, zoologically] named…. / 日本ザルの〜は何というの？ What is the scientific [biological, zoological] name for the Japanese monkey?

河口 the mouth of a river; a river's mouth;〔海に出る川の場合〕an estuary. ▶信濃川〜 the Shinano estuary; the estuary of the Shinano River. / 〜近くの

downriver. / 〜にできた三角州 a delta. / 新潟市は信濃川の〜にある． Niigata City is situated at the mouth of the Shinano River. / 〜の浅瀬 a swash. ¶**河口域** the area around the mouth of a river;〔海に出る川の場合〕an estuarine area; an estuary. **河口改良工事** estuary improvement (works). **河口港** an estuary harbor [port]. **河口水域** estuarine waters. **河口堰**(ぜき) an estuary weir [barrage]. / 利根川〜堰 the Tone River weir. **河口堆積物** an estuarine deposit; estuarine sediment.

加重等価平均感覚騒音レベル〔航空機騒音の程度を示す国際単位〕(a) weighted equivalent continuous perceived noise level. ★⇨英和 WECPNL

霞網(かすみあみ) a very fine net (for catching small birds); a "mist net."

風 a wind; a current of air;(そよ風) a breeze;(すき間風) a draft [draught];(扇風機などの) a breeze; a current (of air). ▶エアコンからの風 the wind [breeze] from an air conditioner. / 風速25メートルの風 a 25-meter-per-second wind; a 90-kph [90 kilometers-an-hour] wind; a wind (packing a speed) of 25 meters per second [90 kilometers an hour]. / 風で電気を起こす generate electricity using the wind. / 風の通り道 the path of the wind. / 風の道〔ヒートアイランド現象を防ぐための風の通り道〕an urban ventilation path. ★⇨ビル風, 風力

化石 a fossil; fossil(ized) remains. ▶生きている[生きた]〜 a living fossil. / 動物の〜 a fossil animal; a zoolite. / 植物の〜 a fossil plant; a phytolite. / 樹木の〜 a fossil tree; a dendrolite. / 魚の〜 a fossil fish; an ichthyolite. / 魚の歯の〜 an ichthyodont. / 鳥の〜 an ornitholite. / 骨[貝殻]の〜 a fossil [fossilized] bone [shell]. / 卵の化石 a fossil egg. / きば[歯]の〜 a fossil tusk [tooth]. / 葉の〜 a fossil leaf; a petrified [(per)mineralized] leaf. / マンモス[アンモナイト]の〜 a fossilized mammoth [ammonite]; a fossil of a mammoth [an ammonite]. / 類人猿の〜 an anthropoid fossil. / 恐竜の〜が見つかった． A dinosaur fossil was found. / 〜による地層同定の法則 the law of strata identified by fossils. / 〜の年代を決める date a fossil. / 〜の標本 a fossil sample. / 〜の採集 a fossil collection. / 〜を含む岩[地層] fossiliferous rocks [strata]. / 動植物の〜を含む contain zoic fossils. / その岩［地層］は〜を含んでいるに違いない．The rock [stratum] is bound to contain [bear] fossils. / 〜を発掘する dig up [excavate] a fossil; dig for fossils. / 〜から当時の気候や生態系がわかる．The climate and conditions of the time can be discovered from fossils. | Fossils can tell us the climate and conditions of the time. ¶ **化石アイボリー** fossil ivory. **化石エネルギー** fossil energy. / 非〜エネルギー nonfossil energy. ★⇨ 再生（再生可能エネルギー） **化石化（作用）** fossilization. / 〜化する fossilize. **化石樹脂** fossil resin [gum]. **化石賞** 〔温暖化防止に消極的な発言をおこなう国に NGO のネットワークである気候変動アクション・ネットワーク（the Climate Action Network, CAN）が贈る賞〕a "Fossil of the Day" award. **化石植物[動物，昆虫]** a fossil [fossilized] plant [animal, insect]. **化石水** fossil water. **化石層** a fossiliferous stratum; a fossil bed. **化石帯** a fossil zone. **化石土** fossil soil. **化石燃料** (a) fossil fuel. **化石燃料埋蔵量** fossil fuel reserves. **化石木** petrified wood. **化石林** a fossil forest.

仮設トイレ a portable toilet [latrine].

河川 a river; rivers (in general); a watercourse. ▶〜の増水に注意してください．Beware of flooding. / 〜の汚濁が進んでいる．River pollution is on the increase. ¶ **河川改修** river improvement. **河川学** potamology; fluviology. **河川環境管理財団** the Foundation of River & Watershed Environment Management. **河川管理** river management. **河川管理者** a river authority. **河川行政** river [riparian] administration. **河川漁業** river fishery. **河川局** the River Bureau. **河川計画** river planning. **河川工学** river [riparian] engineering. **河川航行** river [fluvial] navigation. **河川工事** river conservation work. **河川敷**（じき） a (dry) riverbed; a riverside area. / 〜敷公園 a riverside park. **河川事業** a river [watercourse] (improvement) project. **河川舟運[水運]** river transport(ation). **河川浄化** river purification. **河川審議会** 〔国交省の〕the River Council. **河川水** river water. **河川水量記録計** a fluviometer; a device for measuring [recording] river levels. **河川整備計画** a river development project. 河川総合開発 integrated river development. 河川測量 river surveying. 河川通行標識 a river traffic sign. 河川(の)争奪 〔隣接河川の合流による水系変化〕stream capture; stream piracy. 河川法 the River Law. 河川流域 a river basin. 河川流量 streamflow. /〜維持流量 river maintenance flow. 一級[二級]河川 a class-A [class-B] river. 感潮河川 a tidal river. 国際河川 an international river [watercourse]. 準用河川 a river, other than class-A and class-B rivers, and designated by mayor, to which parts of the River Law apply. 中小河川 small or medium-sized rivers. 適用河川 a river to which the River Law applies. 普通河川 an ordinary river to which the River Law doesn't apply.

可塑剤(かそざい) a plasticizer; a plasticizing agent. ▶非フタル酸系〜は使っていません．Non-phthalic plasticizers have not been used (in the manufacture of this product). | This product contains no non-phthalic plasticizers.

ガソリン gasoline;《口》gas; petrol. ▶〜が切れかかっている．The gasoline is running out [short]. | I am [My car is] running out [short] of gasoline. /〜が切れた．The gasoline has run out. | I have [My car has] run out of gasoline. /〜で走る run on gasoline. /〜を入れる[補給する] put in some gasoline; refuel (with gasoline). /〜を満タンにする fill 《*one's*》 car up (with gasoline). /中古車は一般に〜を食う．Used [Secondhand] cars usually burn a lot of [guzzle] gas. /現在〜は1リッターいくらですか．How much is (it for) a liter of gas now? /ハイオク〜 high-octane gasoline. /無鉛[有鉛]〜 unleaded [leaded] gasoline. /レギュラー〜 regular gasoline. ¶ガソリン・エンジン[機関] a gasoline engine. ガソリン・カー 〔ガソリンエンジンで走る気動車〕a gasoline car. ガソリン価格 《a hike in》the price of gasoline; gasoline prices. ガソリン切れ running out of gasoline. ガソリン(自動)車 a gasoline-engined [gasoline-fueled] car. ガソリンスタンド a gasoline (service) station; a filling [gas, petrol] station; a garage. /サービスのよい〜スタンド a good [friendly, helpful] gasoline station; a gasoline station that gives [offers] good service. /セルフサー

ビス(方式)の～スタンド a self-service gasoline station. / ～スタンドで洗車もしてもらった. I had my car washed [got a carwash] at the gasoline station. **ガソリン税** gasoline tax. **ガソリン代** a gasoline bill; the cost of gasoline; expenditure on gasoline. **ガソリン・タンク** 〔車などの〕a gasoline tank; a petrol tank;《口》a gas tank. **ガソリン添加剤** a gasoline additive. **ガソリン販売** the sale of gasoline; gasoline sales. **ガソリン不足** 〔社会における〕a gasoline shortage; a shortage of gasoline. **ガソリン補給** a fill-up. **ガソリン・ポンプ** 〔ガソリンスタンドの〕a gasoline pump.

家畜 a domestic animal;〈集合的に〉livestock; (ウシ科の) cattle. ▶～の改良 livestock improvement. / ～の群れ a herd of livestock [cattle];〔ぞろぞろ動く牛〕a drove of cattle. / ～を飼育する raise livestock; breed domestic animals. ¶**家畜市** a livestock [cattle] market. **家畜改良センター** the National Livestock Breeding Center (略: NLBC). **家畜改良増殖法** the Law for Improvement and Increased Production of Livestock; the Livestock Improvement and Increased Production Law. **家畜小屋[舎]** a barn; a stable; a livestock shed; a cattle shed [stall]. **家畜飼育** livestock husbandry [rearing]. **家畜商** a drover; a cattle dealer. **家畜飼料** livestock feed. **家畜人工授精所** a livestock artificial insemination station. **家畜人工授精師** a livestock artificial inseminator. **家畜審査** (a) livestock inspection. **家畜単位** an animal [a livestock] unit. **家畜伝染病** an infectious livestock disease; a livestock epidemic. **家畜登録** livestock registration. **家畜排泄物(処理)法** the Livestock Waste (Disposal) Law. 正式名称は「家畜排泄物の管理の適正化及び利用の促進に関する法律」(the Law Concerning Appropriate Management and Promotion of the Use of Livestock Waste). **家畜病** (a) livestock disease. **家畜病院** a veterinary hospital. **家畜防疫** control of communicable livestock disease(s); livestock epidemic prevention. **家畜防疫員** an animal quarantine officer (略: AQO). **家畜防疫官** 〔動物検疫所の〕an animal quarantine officer (略: AQO); an animal quarantine inspector. **家畜保険** livestock insurance. **家畜保健**

衛生所　a livestock hygiene service center.　家畜輸送車　a livestock [cattle] car [van, truck]; 〔鉄道の〕a stock car.　家畜輸送船　a livestock [cattle] ship [carrier].

課徴金　a surcharge.　¶輸入[輸出]課徴金　an import [export] surcharge; a surcharge on imports [exports].

渇水　a drought; (a) shortage of water.　★⇨節水, 取水制限, 給水制限　¶異常渇水　a serious drought.　渇水位　a drought water level; water levels during a drought.　渇水時　a (period of) drought [water shortage].　渇水対策　〔夏の〕water saving measures; measures for dealing with water shortages [drought(s)].　渇水調整協議会　a committee for water shortage [drought] control.　渇水年　a dry [drought] year; a year of extremely low rainfall [precipitation].　渇水被害　drought damage; damage from drought.　渇水(流)量　water flow during a drought.

活性　activity.　▶～のある active; activated / ～のない inert.　★⇨不活性 / 酵素～を高める[低下させる] enhance [reduce] enzyme activity.　¶活性汚泥　activated sludge.　活性汚泥法　〔下水処理法〕the activated sludge process [method].　活性化　〔反応性を高めること〕activation. / ～化する be activated; become active. / 界面を～化させる enhance surface activity.　★⇨活性剤　活性化吸着　activated adsorption.　活性型ビタミンD　the active metabolite of vitamin D.　活性剤　an activator. / 界面[表面]～剤 a surface-active agent; a surfactant.　活性錯体[錯合体]　an activated complex.　活性酸素[水素, 窒素]　active oxygen [hydrogen, nitrogen]. / ～酸素を除去する remove [eliminate] active oxygen. / ～酸素除去酵素〔スーパーオキシド・ジスムターゼ〕superoxide dismutase（略：SOD）. / ～酸素除去作用 action [effectiveness] in removing active oxygen. / ～酸素除去能力 (an) ability to remove active oxygen. / ～水素水 water containing active hydrogen.　活性代謝物　an active metabolite.　活性炭　activated charcoal; activated carbon. / ～炭シート an activated carbon sheet. / ～炭処理 activated charcoal treatment.　活性中心　an active center.　活性土　active earth.　活性白土　activated clay.　活性ビタミン剤　an activated vitamin preparation.　**活性部位** an active site.　**活性物質**　an active substance [material]. / 光学〜物質 an optically active substance [material]. / 生物[生理]〜物質 a biologically [physiologically] active substance.

褐虫藻(かっちゅうそう)　〔サンゴなどの体内に共生する単細胞藻類の総称〕a zooxanthella《複数形 -lae》. ★⇨白化(はっか)現象

活動　▶〜する〔動物などが〕be active;《口》be on the go;〔機能する〕function; work;〔組織や人が〕be active; play [take] an active part《in...》;《口》be on the go;〔奔走する〕canvass; campaign. ★⇨行動 / 夜〜する動物 a nocturnal [nocturnally active] animal. ★⇨夜行性 / 盛んに〜している be extremely active; be full of activity. / 浅間山は今なお〜していて，いつ噴火するかもしれない．Mt. Asama is still active, so there is no telling when it might erupt. / 〜中である be acting;（働いている）be at work; be in operation;〔火山が〕be active. / 最近，精子の数が少なかったり〜が鈍かったりする男性が増えた．In recent times there has been an increase in the number of men with a lower sperm count or lower sperm activity. / 梅雨前線の〜が活発になって九州地方は大雨になるでしょう．The seasonal rain front will pick up in activity and will bring heavy rain to the Kyushu area. / 〜を始める become active; begin to move about;〔火山が〕burst into activity. / 熊は春になると冬眠から覚め〜を開始する．In spring bears awaken from their hibernation and begin to move about. / 暖かくなって虫たちが〜的になる前に庭木の消毒をしましょう．Disinfect the trees in the garden before the weather becomes warm and the bugs become active. / 〜をやめる stop (all) activity [activities]. / 〜をやめた[休止している] extinct [dormant]《volcano》. / 鳥たちが一日の〜を終えて眠りにつくころ the hour at which the birds cease their day's activities and go to asleep. / 彼女は市民オンブズマンとして大いに〜しています．She is being very active [energetic] in her role of civic ombudsman. / アフリカの飢餓救済〜に参加する take part in relief activities on behalf of starving Africans. / 〜の記録[軌跡] a record [track] of one's activities. / 昨年度の〜報告 a report on the preceding year's activities [operations]. / ～方針 an action policy [program]. / 君のボランティアクラブではどんな～をしているの？ What sort of activities do you do in your volunteer club? / 私たちは川をきれいにするための～をしています. We conduct activities aimed at cleaning up rivers. ¶ 活動家 an activist. / 市民～家 a civic activist. 活動範囲 〔組織・動物などの〕the scope [range, sphere] of activity. 活動周期 〔火山・地震帯などの〕an active cycle; a cycle of activity;〔太陽の〕a [an active] solar cycle; a (solar) cycle of activity. 環境保護活動 ⇨環境(環境保護) クラブ[サークル]活動 club activities. 募金活動 fund-raising activity [activities]. ボランティア[奉仕]活動 volunteer work [activities].

合併処理 〔台所排水と屎尿の〕combined treatment (of kitchen waste liquids and toilet sewage). ¶ 合併(処理)浄化槽 〔台所の排水と屎尿の両方を浄化する〕a combined kitchen waste liquids and toilet sewage treatment tank; a combined septic tank.

家庭 ¶ 家庭ごみ household waste(s) [refuse, garbage, rubbish]. 家庭菜園 a private vegetable garden; a kitchen garden. 家庭雑排水 miscellaneous domestic liquid waste. 家庭排水[廃水] 〔台所・浴室などからの〕domestic drainage [waste water];〔水洗便所の〕domestic sewage. 家庭用水 water for domestic use.

家電リサイクル法 the Household Appliance Recycling Law. ★「特定家庭用機器再商品化法」(the Law for Recycling of Specified Kinds of Consumer Electric Goods) の通称.

カドミウム cadmium. ¶ カドミウム汚染 cadmium pollution. カドミウム中毒 cadmium poisoning.

カネミ油症(事件) 〔1968年の〕the Kanemi rice oil disease (incident).

可燃 ¶ 可燃ゴミ burnable trash [rubbish]; burnables. ★⇨ごみ 可燃材料 a combustible [an inflammable] material. 可燃物 a combustible; an inflammable; something that will burn.

カバークロップ[プランツ] ⇨グラウンドカバー・プランツ.

花粉 pollen. 形 pollinic(al). ▶ ～が飛散する pollen flies around [fills the air] / きょうは～がたくさん飛んでいる. There's a lot of pollen flying around [in the air] today. / 〜の季節が始まった．The pollen season has set in. / 〜を有する[生じる，運ぶ] polliniferous; pollinigerous. / めしべに〜をつける pollinate. / ハチが花から花へ〜を運んでいる．Bees carry pollen from flower to flower. / スギ[ブタクサ]〜 cedar [ragweed] pollen. ¶**花粉アレルギー** 《have, develop》a pollen allergy. **花粉花** a pollen flower. **花粉荷**(か) 〔ミツバチが花粉を花蜜で固めたもの〕bee pollen. **花粉塊** a pollinium 《複数形 -nia》. **花粉学** palynology. 形 palynologic(al). / 〜学者 palynologist. **花粉かご** 〔ミツバチの〕a pollen basket; a corbicula 《複数形 -ae》. **花粉管** a pollen tube. **花粉管核** a [the] pollen-tube nucleus. **花粉櫛**(し) 〔ミツバチの〕a pollen comb. **花粉計数器** a pollen counter. **花粉室** a pollen chamber. **花粉四分子** a pollen tetrad. **花粉症** hay fever; (a) pollen allergy; pollinosis; pollenosis. / 〜症になる develop hay fever. / 〜症に苦しむ[悩まされる] suffer from hay fever. / 杉〜症 an allergy to Japanese cedar pollen; Japanese cedar pollinosis. / 〜症患者 a hay fever sufferer. / 〜症緩和米 (genetically modified) rice that prevents hay fever. / 〜症対策 measures to combat [alleviate] hay fever. / 〜症対策グッズ hay fever [anti-hay fever] goods; hay-fever treatments and associated goods. / 〜症予防 hay fever prevention. **花粉情報** 〔飛散量の〕a pollen report. **花粉数** a 《high, low》pollen count. **花粉嚢**(のう) a pollen sac; a theca 《複数形 -cae》. **花粉培養** pollen culture. **花粉飛散量** an atmospheric pollen count; the amount of pollen in the air. **花粉分析** 〔堆積物中の〕pollen analysis. **花粉母細胞** a pollen mother cell. **花粉粒** a pollen grain; a microspore.

紙 paper. ▶トイレに紙がない．There's no paper in the toilet. / 牛乳パックから紙を作る manufacture paper from milk cartons; recycle milk cartons into paper. ¶**紙屑** wastepaper; a scrap of paper; a paper scrap; paper scraps. / 紙屑を拾う pick up a scrap of paper; collect wastepaper. / 紙屑をまるめる roll [crumple] a paper scrap into a ball. / 紙屑かご a wastebasket; a wastepaper basket. **紙コップ** a paper cup. **紙皿** a paper plate. **紙タオル** a paper towel. **紙パック** 〔飲料などの〕

a paper carton;〔掃除機の〕a paper (dust) bag.　**紙パックマーク**〔環境ラベルの一種〕a paper carton (recycling) mark.　**紙パルプ**　paper pulp.　**紙パンツ**〔使い捨て下着の〕disposable paper underwear;〔紙おむつ〕disposable paper diapers [nappies].　**紙フィルター**〔コーヒー用の〕a paper filter.　★紙製容器包装⇨容器包装

ガラス　glass;〔板状の〕plate glass;〔窓ガラス〕a pane (of glass). ▶〜が飛散するのを防ぐフィルム a shatter-resistant film (on glass). /〜の破片〔1個の〕a piece [bit, fragment] of (broken) glass; a shard of glass;〔粉々になったもの〕broken glass; pieces [bits, fragments] of broken glass　¶**ガラス器**　a glass vessel [container];〈集合的に〉glassware; glass.　**ガラス器具**〔化学実験用の〕scientific glassware.　**ガラス屑**　refuse glass.　**ガラス固化**〔放射性廃棄物の〕vitrification. /〜固化状態 a glassy state. /〔高レベル廃棄物の〕〜固化体《canisters of》vitrified high-level (radioactive) waste（略: VHLW）.　**ガラスびん**　a glass bottle;〔薬などを入れる小型の〕a phial; a vial.　**網(入り)ガラス**　reinforced glass; glass reinforced with wire netting.　**安全ガラス**　safety glass; nonshatterable glass.　**耐熱ガラス**　heat-resistant glass.

夏緑樹林（かりょくじゅりん）　a deciduous forest; a forest of deciduous trees.

火力発電　thermal power generation. ▶〜による電力 electricity generated by thermal power generation. /〜所 a thermal power plant [station].

カルキ〔さらし粉〕bleaching powder; chlorinated lime. ▶〜臭い水道水 tap water smelling of chlorine [bleach]. /(水道水などから)〜を抜く dechlorinate.

カルタヘナ議定書〔遺伝子組み換え生物などの国際取引についての〕the Cartagena Protocol on Biosafety.　★⇨英和　Biosafety Protocol

カルタヘナ法　the Cartagena Law.　★正式名称は「遺伝子組み換え生物等の規制による生物の多様性の確保に関する法律」(the Japanese Law Concerning the Conservation and Sustainable Use of Biological Diversity through Regulations on the Use of Living Modified Organisms).

瓦礫（がれき）〔建造物の破片〕(demolition) debris; rubble; wreckage. ▶〜の山 a heap of rubble. /いくつものビルが地震で〜

の山と化した．The earthquake reduced a number of buildings to rubble.

カレット 〔溶解用のガラスくず〕cullet.

簡易包装 simplified packaging; (a) simplified wrapping.

感覚公害 〔騒音・悪臭など〕sensory pollution.

灌漑(かんがい) ▶～する irrigate; water. ／この豊富な湧水は古くから～に利用されてきた．This abundant spring has long been used for irrigation. ¶**灌漑計画** an irrigation scheme. **灌漑工事** irrigation works. **灌漑水量** an irrigation amount [volume]. **灌漑地** irrigated land; land under irrigation. **灌漑池** an irrigation pond [reservoir]. **灌漑農業** irrigation farming. **灌漑用水** water for irrigation; irrigation water. **灌漑用水路[用井戸, 用堀割]** an irrigation canal [well, ditch]. **点滴灌漑** drip irrigation; trickle irrigation.

干害, 旱害(かんがい) damage from [caused by] a drought; a drought disaster. ★⇒旱魃(かんばつ) ▶～を受ける suffer (damage) from a drought; be (badly) affected by a drought. ¶**干害防備保安林** ⇒保安林

雁鴨(ガンカモ)**調査** a waterfowl census.

換気 ventilation; 〔1回の〕a change of air. ▶～がよい[悪い] be well [badly, ill] ventilated. ／～が悪い部屋にはカビが生えやすい．Mold develops easily in badly ventilated rooms [rooms with poor ventilation]. ／～のために窓を開ける open a window to let fresh air in [air a room]. ／こまめに部屋の～をする air [ventilate] a room frequently. ¶**換気回数** (a) ventilation frequency; the number of changes of air. **換気管** a ventilation [ventilating] pipe. **換気機能** 〔肺の〕ventilatory function. **換気孔[口]** a vent; an air vent; a ventilation [ventilating] hole [aperture, opening]. **換気坑** a ventilation [ventilating] shaft. **換気扇** a ventilation [ventilating] fan; an extractor fan. ／～扇を回す turn on [start] an extractor fan. ／～扇を止める turn off [stop] an extractor fan. **換気装置** a ventilator; a ventilating device; ventilation facilities [equipment]. **換気筒** 〔屋根の上などにある〕a ventilator; a ventilation pipe [duct, chimney]. **換気塔** 〔トンネルや地下街の〕a ventilation tower. **換気窓** a ventilation window; a window vent. **換気予備量** breathing reserve. **換気量** 〔肺の〕ventilatory volume; the amount of ventilation;〔建築物の〕the amount of ventilation. **強制[機械]換気** 〔部屋の〕forced [artificial, mechanical] ventilation. **自然換気** natural ventilation. **人工換気** 〔口移しまたは機械による〕artificial ventilation;〔機械による〕mechanical ventilation.

環境 〔特に自然の環境〕(an) environment;〔状況・条件〕surroundings; circumstances; a situation; conditions. ★⇨自然(自然環境), 体内(体内環境), 住環境, 地球(地球環境) ▶日本の地理的〜 Japan's geographical environment [situation]; the geographic setting of Japan. / 健康的[快適]な〜 a healthy [pleasant] environment; healthy [pleasant] surroundings. / 緑の多い〜 a green [《文》verdant] environment. / 緑の多い〜を生かして公園を設計する make a design for a park that puts to good use the abundant greenery already present; make a park without cutting down the plants that are already there. / 新しい建物のデザインは周囲の〜との一体性を持つものにしたい．I'd like the design of the new building to form an integrated whole with its surroundings. / 現在の地球を取り巻く〜 (the) earth's present environment; the present environment on earth; the current condition of the earth. / 酸性雨の〜への影響 the environmental effect(s) of acid rain. / (近所の)〜がいい[悪い] be in a decent [an undesirable] neighborhood. / この地域の健全な教育〜を守ろう．Let us protect the healthy educational environment [environment for education]. / 子供たちを取り巻く社会〜 children's social circumstances [environment]; the society in which children live [around children]. / 交通量の増加と共に都心の〜が悪化している．As traffic increases, the urban environment deteriorates. | The more traffic there is, the worse the environment becomes in central areas of cities. / 子育てには〜が大切だ．Environment is important in the education of children. | A child's upbringing is greatly influenced by its home environment. / そのホテルではペットを連れての宿泊のための〜が整っている．That hotel has everything needed [is fully

equipped] for staying with a pet. / 〜と調和する come [get] into harmony with an [the] environment; adjust to an [the] environment. / 〜に左右される be influenced [affected] by one's environment; be at the mercy of circumstance(s). / 〜に恵まれる have [enjoy] a good environment. / 〜に配慮する take the needs of the environment into consideration. / 〜配慮型の製品 an environmentally friendly [eco-friendly] product. / 〜への配慮 care [concern] for the environment. / 〜配慮(型)住宅 an environmentally friendly [eco-friendly] house. / 〜配慮(型)融資 environmentally friendly financing [credit]. / 〜に適応する adapt (to the environment). / 〜に対するこの事故の影響は計り知れない. It is difficult to imagine the enormous effect this accident will have on the environment. / 〜に敏感な市民 citizens who are sensitive to the needs of the environment. / 〜に熱心な行政 ecologically oriented administration. / 〜に無害な environmentally harmless 《detergents》. / 〜にやさしい environment(ally) friendly 《products, technologies, cars》; green 《technology》; environmentally benign; not harmful to the environment. / 〜にやさしくない harmful [damaging] to the environment; not environment-friendly; environmentally unfriendly. / 〜の影響 the influence of an [the] environment 《on ...》. / 〜の悪い場所 a place with [in] a bad [an unhealthy] environment; an environmentally poor location; an unhealthy situation. / 〜を維持[保全]する preserve [conserve, maintain] the environment; keep [leave] the environment as it is [untouched]. / 〜を美化する make the surroundings more attractive [less unpleasant] by tidying it up; clean up [tidy up] an area and make it more attractive. / 〜の美化 improving the appearance of the environment; making the environment more attractive; creating a more attractive environment. / この地の植物は厳しい〜を克服して現代に生き続けている. The plants here have mastered a harsh environment and continue to propagate today.
¶ 環境アセスメント　environmental (impact) assessment; 〔個々の〕 an environmental review; an environmental assess-

ment. / 政府の行う〜アセスメントは開発の免罪符にすぎない場合が多い. Often, environmental assessments by the government are no more than a prior justification for development. **環境悪化** deterioration of the environment; environmental deterioration [degradation]. **環境アドバイザー** an environmental adviser. **環境医学** environmental medicine; geomedicine. **環境意識** awareness of the environment; environmental awareness [consciousness]; being aware of [thinking about] the environment. 形 environmentally aware [conscious]. / 〜意識の高い主婦が増えてきた. More housewives are conscious of the environment now. / 〜(保護の)意識が高い be highly aware [conscious] of the need to protect the environment. **環境因子** an environmental factor. **環境影響調査[評価]** ⇨環境アセスメント **環境影響評価審査会[審議会]** an environment assessment council [committee]. **環境影響評価法** the Environment Impact Assessment Law. **環境衛生** environmental hygiene [health, sanitation]. **環境衛生監視員** an environmental sanitation inspector; an environmental health officer. **環境衛生指導員** an environmental sanitation instructor. **環境衛生学** environmental health [hygiene] (studies). **環境NGO** 〔環境保護に関わる非政府組織〕an environmental NGO. **環境エンリッチメント** 〔飼育動物の生活環境を豊かで充実したものにするための取り組み〕environmental enrichment. **環境ODA** environmental ODA; ODA for environmental protection. **環境汚染** environmental pollution [contamination]; pollution [contamination] of the environment. **環境汚染物質** an environmental contaminant [pollutant]; a pollutant; a substance that pollutes [contaminates] the environment. **環境汚染物質排出移動登録制度** ⇨排出(排出移動登録制度) **環境音楽** 〔インテリアの一部としての〕ambient music. **環境音響学** environmental acoustics. **環境会計** environmental accounting; accounting that takes environmental costs into account. **環境開発サミット** ⇨環境サミット **環境改変技術使用禁止条約** the Environmental Modification Convention; 〔正式名称〕the Convention on the Prohibition of Military or Any Other Hostile Use of Environmental Modification Techniques.　**環境科学**　environmental science.　**環境科学者**　an environmental scientist; an environmentalist.　**環境学**　environmental science; environmentology.　**環境格付け**　(an) eco-rating.　**環境格付け融資**　financing at preferential terms for a company with a good eco-rating.　**環境家計簿**　an environmental household account book;〔それをつけること〕environmental accounting.　**環境確保条例**　〔東京都の〕the Ordinance on Environmental Preservation. ★「都民の健康と安全を確保する環境に関する条例 (the Ordinance on Environmental Preservation to Secure the Health and Safety of Tokyoites)」の略称.　**環境価値**　(an) environmental value; an environmental benefit.　**環境価値取引**　environmental value [benefit] trading.　**環境活動家**　an environmental activist; an environmentalist.　**環境監査**　an environmental audit.　**環境監視**　environment(al) monitoring; monitoring (of) the environment.　**環境観測技術衛星**　the Advanced Earth Observing Satellite (略: ADEOS).　**環境管理**　environment(al) management.　**環境管理システム**　an environmental management system (略: EMS).　**環境関連技術**　an environment-related technology.　**環境関連商品**　green products.　**環境関連ビジネス**　an environment-related business.　**環境規格**　an environmental standard. / 国際〜規格〔ISO14001 (⇨和英篇末尾)など〕an international environmental standard.　**環境危機**　an environmental crisis; a crisis in [of] the environment.　**環境危機時計**　〔環境悪化による人類滅亡までの残り時間を表す時計〕the environmental doomsday clock.　**環境技術**　(an) environmental technology.　**環境基準**　an environmental standard. / PCB の〜基準をクリアする meet the environmental standard for PCB. / 〜基準に適合する自動車 an automobile that conforms to [meets] environmental standards. / 〜基準を設定する create [set, enforce] environmental standards. / 〜基準値 an environmental standard value.　**環境規制**　(an) environmental regulation.　**環境偽装**　〔企業などが環境問題への貢献を偽称すること〕environmental deception; phony environmentalism. ★⇨英和 greenwash　**環境基本計画**〔環境基本法に基づいて策定された〕the Basic Environment Plan.　**環境教育** environmental education. /〜教育者 environmental educator.　**環境行政** environmental administration.　**環境共生(型)社会** an environment(ally)-friendly society.　**環境共生住宅** environmentally friendly housing.　**環境共生住宅部品** environmentally-friendly housing components.　**環境共生都市**〔エコシティ〕an eco(-)city [Eco(-)city].　**環境経営** environmental management; green management.　**環境経営学会** the Sustainable Management Forum of Japan (略: SMF).　**環境経営格付機構** the Sustainable Management Rating Institute (略: SMRI).　**環境経営促進法** the Green Management Promotion Law.　**環境計画学** environmental planning (studies).　**環境経済学** environmental economics. /〜経済学者 environmental economist.　**環境芸術**（an) environmental art. /〜芸術家 environmental artist.　**環境権** environmental rights; the right to (live in) a good [decent, healthy] environment. /〜権訴訟 an environmental suit [case].　**環境工学** environmental engineering.　**環境考古学** environmental archaeology.　**環境後進国[先進国]** an environmentally backward [advanced] country; a country with a poor [good] record on environmental issues [protection].　**環境効率** eco-efficiency. /製品〜効率 product environmental efficiency; the environmental efficiency of a product.　**環境コスト** environmental costs; an environmental cost; the cost to the environment《of industrialization》.　**環境災害** an environmental disaster.　**環境サミット** an environment(al) summit;〔持続可能な開発に関する世界首脳会議(2002)〕the World Summit on Sustainable Development (略: WSSD).　**環境産業** environment-related industries.　**環境サンプル**〔環境汚染調査のための採取物質〕an environmental sample. /〜サンプル調査 environmental sampling.　**環境GNP**〔グリーンGNP〕the green GNP.　**環境試験** environmental testing; an environmental test.　**環境JIS** Environmental JIS.　**「環境・持続社会」研究センター**〔NPO法人〕the Japan Center for a Sustainable Environment and Society (略: JACSES). **環境自主行動計画** 〔経団連が1997年に発表〕the Keidanren Voluntary Action Plan on the Environment. **環境自治体** 〔環境保全に配慮する自治体〕environmentally minded [aware, conscious] municipalities. **環境自治体会議環境政策研究所** the Research Institute for Local Initiative of Environmental Policies (略: RELIEP). **環境指標** an environment(al) indicator. ★⇨指標種 **環境車** ⇨環境対応[対策]車. **環境社会学** environmental sociology. **環境社会検定試験** 〔東京商工会議所などが実施する〕the Certification Test for Environmental Specialists; "the Eco Test." **環境収容力[容量]** a carrying capacity; an environmental capacity. **環境省** 〔2001年, 環境庁を改組〕the Ministry of the Environment (略: MOE); the Environment Ministry. ★大臣および主な部局は以下の通り.

環境相, 環境大臣 the Minister of [for] the Environment
廃棄物・リサイクル対策部 Waste Management and Recycling Department
総合環境政策局 Environmental Policy Bureau
環境保健部 Environmental Health Department
地球環境局 Global Environment Bureau
水・大気環境局(旧称 環境管理局) Environmental Management Bureau
自然環境局 Nature Conservation Bureau
地方環境事務所〔2005年全国に7つ設置〕Regional Environment Office

¶**環境浄化** (an) environmental cleanup. **環境条件** environmental conditions. **環境譲与税** 〔国が地方公共団体の地球温暖化対策に充てる税〕environment tax income transferred (from the Central Government) to local government(s). **環境・食糧・地域振興省** Department for Environment, Food and Rural Affairs ★英国で2001年に成立. 省の長である環境・食糧・地域振興大臣は Secretary of State for Environment, Food and Rural Affairs と言う. **環境心理学** environmental psychology. / ～心理学者 environmental psychologist. **環境人類学** environmental anthropology. / ～人類学者 environmental anthropologist. **環境ストレス** (an) environmental stress; cost [damage, unfriendliness] to the environment.　**環境(ストレス)耐性**　environmental (stress) resistance.　**環境スワップ**　⇨英和 debt-for-nature swap　**環境税**　an environment(al) tax; an eco(logical) tax. /〜税を課す implement an environmental tax.　★⇨環境譲与税, 英和 carbon tax　**環境制御**　environmental control.　**環境制御工学**　environmental control engineering.　**環境制御装置**　〔障害者が身辺のものを操作するための〕an environmental control system（略: ECS).　**環境政策**　(an) environmental policy; (a) policy on [for] the environment. / 当 NPO は〜政策の提言に中心的役割を果たしている. This NPO plays a central role in proposing environmental policies.　**環境性能**　environmental performance.　**環境生物学**　environmental biology. /〜生物学者 environmental biologist.　**環境生命医学**　environmental life science(s).　**環境生理学**　environmental physiology.　**環境測定**　(an) ambient measurement.　**環境対応技術**　(an) environmentally friendly [eco-friendly, environment-friendly] technology.　**環境対応[対策]車**　a green vehicle [car]; an environmentally friendly vehicle [car]; a low-pollution car; an ecocar. /〜対策の切り札 a surefire environmental countermeasure. / わが社は〜対策の充実で化学工業界トップにランクされている. Our company ranks [is ranked] top [number one] in the chemical industry in terms of the completeness of our environmental measures. /〜対策に重点的に予算を配分する apportion funds predominantly [a preponderance of funds] to environmental measures. /〜対策の遅れがちな日本と対照的なのがドイツである. In stark contrast to Japan, with its dilatory approach to environmental policy, is Germany. | Germany is at the opposite extreme from Japan's dilatory approach to environmental policies. /〜対策で現地調査の占める比重は大きい. On-site surveys are a major component of environmental protection measures.　**環境団体**　⇨環境保護団体　**環境地図**　an environment map.　**環境中毒学**　ecotoxicology; environmental toxicology; (medical) ecotoxicology.　**環境彫刻**　(an) environmental sculpture.　**環境調査**　an environmental survey. / 事前〜調査[〜事前調査]

を実施する conduct a preliminary environmental survey. **環境調査衛星** an environmental survey satellite. **環境調整** 〔行政の〕environmental coordination. **環境調整済国内純生産** ⇨グリーン(グリーン GDP) **環境適応** (environmental) adaptation; adaptation to an [the] environment. / 〜適応能(力) environmental adaptability. **環境的公正[正義]** environmental justice. **環境デザイン** (an) environmental design. / 〜デザイナー environmental designer. **環境テロ** ecological terrorism; eco-terrorism; ecoterrorism. ★⇨英和 ecoterrorism **環境トイレ** 〔環境に負荷をかけないトイレ〕an ecological toilet; an environment(ally) friendly toilet. **環境と開発に関する世界委員会** 〔国連の〕the World Committee on Environment and Development (略: WCED). **環境と開発に関する国連会議, 環境開発会議** ⇨国連開発環境会議 **環境難民** an environmental refugee. **環境の日** ⇨英和 World Environment Day **環境破壊** environmental destruction; destruction of an [the] environment; 〔大規模な〕environmental devastation. ★⇨自然(自然破壊) / 〜破壊は猛烈なスピードで進んでいるが, 今ならまだ間に合う. Environmental destruction is advancing at a furious pace, but if we act now, there's still time. **環境白書** a white paper on the environment; 〔環境省の〕the Annual Report on the Environment in Japan《2006》. **環境発がん** environmental carcinogenesis. **環境パフォーマンス** 〔環境に関する自治体・企業などの活動とその実績〕environmental performance. **環境パフォーマンス評価** (an) environmental performance evaluation (略: EPE). **環境犯罪** 〔環境破壊につながる犯罪〕(an) environmental crime. **環境被害** environmental damage; damage [harm] to the environment. **環境ビジネス** eco-business; ecobusiness; an eco-business [ecobusiness]. **環境部** 〔役所の〕an environmental department [division]; a department for the environment. **環境ファンド** ⇨エコ(エコファンド) **環境負荷** (an) environmental load [burden]; the load [burden] on the environment《of pesticides》. ★⇨英和 ecological footprint / 低〜負荷型社会の構築 the creation of a type of society which imposes little damage on the environment [causes little damage to the environment]. ★⇨低環境負荷 / ～負荷を低減することが企業の社会的義務である．Lowering the burden on the environment is a corporate social responsibility. / ～負荷物質 an environmentally damaging [burdensome] substance; a substance that imposes a burden on the environment. **環境付加価値** environmental value added（略：EnVA）; environmental added value. **環境分析** an environmental assessment [analysis]（略：EA）. **環境変異** (an) environmental variation. **環境変化** (a)《global》environmental change; change(s) to the《global》environment. **環境法** environmental law. **環境報告書** 〔自治体・企業などが出す環境対策報告書〕an environmental report. **環境防災林** a forest for environmental preservation and disaster prevention. **環境放射線** environmental radiation. **環境保健基準** 〔世界保健機関（WHO）による，電磁波の〕(the) Environmental Health Criteria（略：EHC）. **環境保護** environmental protection; protection of an [the] environment. / ～保護のキャンペーンを張る mount an environmental protection campaign. / ～保護における先進的な試み a go-ahead attempt to protect the environment. / ～保護運動[活動] an environmental (protection) movement; an ecology movement. / ～保護運動家 an environmentalist; an environmental activist; an ecologist. / ～保護主義 environmentalism; ecology. / ～保護主義者 an environmentalist; an environmental protectionist; an ecologist. / ～保護団体 an environmental (protection) organization [group]; an environmentalist organization. ★⇨ダム **環境保護局** 〔米国の〕the Environmental Protection Agency（略：EPA）. **環境保全** environmental preservation [conservation]; preserving the environment. / ～保全意識の高まり rising awareness of environmental protection [preservation]. / ～保全運動 an environmental (preservation) movement; a campaign to preserve the environment. / ～保全主義者 a conservationist; an environmentalist. / ～保全対策 environmental (preservation) measures; measures to preserve the environment. / ～保全義務 the duty [obligation] to protect the environment. **環境保全活動・環境教育推進法, 環境教育推進法** the Law for the Promotion of [to Promote] Environmental Education. ★正式名は「環境の保全のための意欲の増進及び環境教育の推進に関する法律」. **環境舗装** environmentally friendly pavement [paving]. **環境ホルモン** 〔内分泌攪乱物質〕an endocrine-disrupting chemical; an endocrine-disrupting substance; an environmental estrogen; a hormone [an endocrine] disrupter. /～ホルモンの人体に対する作用機序 the mechanism by which environmental hormones work upon the human body. **環境未来都市** an environmentally advanced city. **環境モニター** an environment monitor. **環境モニタリング** environmental monitoring;〔原子力施設周辺の放射線量率などを調査する〕environmental monitoring near nuclear facilities. **環境問題** an environmental problem [issue, matter]; environmental affairs. /最近大手企業では～問題に積極的に取り組んでいることをアピールする動きが見られる. There is a recent trend for large enterprises to publicize the fact that they are doing something about [taking positive measures to deal with] environmental problems. /地球の～問題を気にはしているのだが, 何をどう具体的にすればいいのかわからない. I am concerned about global environmental problems, but I don't know in practical terms what I should actually do. /エネルギー問題と～問題の一体的解決 integrated solution of [an integrated solution to] energy and environmental problems. /～問題専門家 an environmental expert [specialist]; an expert on the environment; an environmentalist. **環境要因** ⇨環境因子 **環境予防医学** environmental (and) preventive medicine. **環境ラベル** 〔環境にやさしい商品であることを示すラベル表示〕environmental labeling;〔そのラベル〕an environmental label. ★⇨エコ(エコラベル) **環境リスク** an environmental risk. **環境リスク評価** (an) environmental risk assessment. **環境立国** a national commitment to the environment. **環境立法** environmental legislation. **環境リテラシー** 〔環境に配慮した行動を取れる能力〕environmental literacy. **環境倫理学** environmental ethics. **環境療法** environment therapy; milieu therapy. **音環境** a sound environment; an acoustic environment; an aural environment. / 音〜デザイナー a sound environment designer. / 私たちは静けさを含めた心地よい音〜作りを心がけています．We strive to create environments of soothing sounds, including silence. **職場[労働]環境** a [the] working environment; job conditions; the environment at work. / 劣悪な労働〜 squalid working conditions; an appalling working environment. **生育環境** a rearing environment. / 生育〜の違いで樹形は驚くほど変わる．The shape of a tree will differ astonishingly depending on the environment in which it is grown. **生活環境** *one's* environment [circumstances]; the circumstances [environment] in which *one* lives; *one's* situation [conditions]; *one's* life. **生息環境** a habitat. **内部[外部]環境** an [the] internal [external] environment. **日本環境学会など** ⇨ 日本(ほん)

緩衝(かんしょう)緑地 a green buffer zone; a green belt.

乾性沈着 〔汚染物質などの〕dry deposition.

感染 (an) infection;〔空気感染〕aerial [airborne] infection;〔接触感染〕contact [contagious] infection. ▶〜する get [become] infected 《with cholera》; catch 《influenza》;《口》pick up 《a virus》. / 〜性の〔空気などによる〕infectious;〔接触による〕contagious; catching. / そのウイルスに〜した人 a person infected by [with] the virus. / ネズミを介して〜する be infected via [through the medium of] rats;《a disease》communicated [transmitted] to humans by rats. / その病気はペットのオウムから〜することがある．The disease can be transmitted to humans by pet parrots. | Pet parrots can communicate the disease to humans [infect humans with the disease]. / この病気は〜してもすぐには発症しない．One does not develop symptoms immediately upon infection with the disease. / (免疫があって)〜しない be immune《to...》; have immunity 《to...》. / ヒトには〜しない病気 a disease that does not infect humans. / エイズ・ウイルスに〜した血清から製造した非加熱製剤 untreated blood products obtained from blood serum infected with the AIDS virus. / サルモネラ菌〜の恐れ[危険]があるのでこれらの魚の生食は避けること. These fish may be infected with salmonella and should not be eaten raw. / インフルエンザ・ウイルスの〜を防ぐワクチン influenza vaccine; a vaccine to prevent the spread of [infection by] influenza. / 乳児や老人が院内〜しやすい. Babies and the elderly are susceptible to nosocomial infection. / 発病した5人はMRSAによる院内〜であると判明した. The five patients who came down with the disease were identified as having an MRSA-derived nosocomial infection. / 次々に耐性菌が現れて院内〜を引き起こしている. One resistant strain of bacteria after another has turned up, each bringing on nosocomial infection. ¶**感染牛** an infected cow [bull]; 〈集合的に〉infected cattle. **感染経路[ルート]** a route of infection; an infection route. **感染源** a source of infection. **感染者** a carrier 《of a virus》; an infected person; a person with an infection. **感染性廃棄物** infectious waste.

乾燥 ▶〜する〔乾く〕dry (up, out); become [get] dry [arid, parched]; dehydrate; lose 《its》 moisture. /〜した dry; dried (-up[-out]); parched 《soil》; 〔土地・気候などが〕arid;〔水分を除いた〕dehydrated;〔木材が〕seasoned. /〜した大地を緑化する plant trees on [afforest] arid land. /(土地が)耕作には〜しすぎている be too dry [arid, parched] for cultivation. / 空気が非常に〜している. The air is extremely dry. | Humidity [Air humidity] is extremely low. / 雨が降らないために地面の〜が進んでいる. The ground is losing moisture due to lack of rain. / サボテンは〜に強い. Cactuses can cope with arid conditions. /〜を好む〔生物が〕xerophilous. ¶**乾燥季, 乾燥期**〔乾季〕a [the] dry season;〔長期間の乾燥期〕a dry period (in the history of the earth). **乾燥気候** an arid climate. / 半〜気候の semiarid. **乾燥休眠** drought dormancy. **乾燥限界**〔乾燥気候と湿潤気候の境界〕an arid boundary. **乾燥指数** an aridity index. **乾燥断熱減率** a dry adiabatic lapse rate. **乾燥地** arid land; dry ground. **乾燥地形** arid topography; an arid landform. **乾燥地帯** an arid zone. **乾燥注意報** a dry air advisory.

観測〔観察〕observation;〔測量〕a survey; surveying. ▶〜する

observe; carry out [conduct] observations 《on [of]...》; survey; make [carry out, conduct] a survey 《of [on]...》; record; watch. / そのとき神戸ではマグニチュード7.2を〜した．In Kobe the earthquake registered a magnitude of 7.2 on the Richter scale. | In Kobe a magnitude of 7.2 on the Richter scale was recorded. / 〜史上まれに見る大型の台風 a typhoon of a size rarely witnessed in the history of meteorological observation. / 〜史上の最高気温を記録した．We recorded the highest temperature on record. | It was the hottest day on record. / 今日東京では5月としては〜が始まって以来の暑さとなった．Today was the hottest day recorded for Tokyo since records began. / 太陽[日食]の〜を行う observe [make an observation of] the sun [a solar eclipse]. / 気象〜を行う make meteorological observations. / 火山活動の〜を強化する strengthen observation of volcanic activity; watch volcanic activity more carefully.

干拓 land reclamation by drainage. ★⇨埋め立てる ¶**干拓工事** reclamation work. **干拓事業** reclamation works. ▶国営諫早湾〜事業 the Isahaya Bay reclamation project. **干拓地** reclaimed land.

官能検査[試験] 〔食品・製品などに対する〕a sensory test [evaluation].

間伐(かんばつ) 〔森林の〕thinning. ▶森を〜する thin out a forest. ¶**間伐材** timber from forest-thinning [trees that have been weeded out]. **間伐(材)紙** paper made from forest thinnings.

旱魃(かんばつ) (a) drought; (a long period of) dry weather. ★⇨渇水 ▶〜で困っている be suffering because of drought [lack of rain] / 〜で作物が駄目になるだろう．The drought will ruin the crops. / 〜の被害 damage from drought; drought damage.

含有 ▶〜する contain; have 《in...》. / 豊富なミネラルを〜する自然塩 natural salt containing plenty of [rich in] minerals; a mineral-rich natural salt. ¶**含有成分** a constituent; a component; an ingredient.

含有率 (the) percentage of a constituent. **含有量** the quantity of a constituent;《vitamin C》content. / 健康飲料のカルシウム〜量 the amount of calcium (contained) in a health drink. / 鉱石の銀〜量 the silver content of an ore. / アル

コールの含有量が多い contain a high percentage of alcohol; have a high alcohol content.

管理 〔統制・支配・運営管轄〕administration; management; control; supervision; regulation; superintendence；〔保全・維持〕maintenance; preservation;〔修理〕repairs. ★⇨改訂管理制度 ▶〜する administer; manage; control; superintend; supervise; police; run；〔保全・維持する〕maintain; preserve; keep in a good condition [state of repair]; look after. / コンピューターでハウス内の温度を〜する control greenhouse temperature with a computer. / 学校の〜下で生じたけが an injury received while under school supervision [the school is responsible for 《a child》]. / この道路は市が〜している．It is the city (government) that maintains this road. / プールの〜 maintenance [upkeep] of a swimming pool. / 国立公園の〜 maintenance of [preservation of, care for, looking after] a national park. / 公園の〜を命じられる be put [placed] in charge of park maintenance [maintenance of a park, keeping parks in good condition]. / 厚生施設を〜運営する maintain and run a welfare organization. /（…の）管理地．〔掲示で〕Managed Property; Property Managed (by …). ¶**管理基準** a management [control] standard. / 危険物〜基準 a management standard for hazardous materials. **管理区域** 〔放射線などによる被爆から防護するための〕a controlled area. / 汚染〜区域 a contamination controlled area. / 放射線〜区域 a radiation controlled area. **管理使用** 〔危険物質などの，厳重な管理の下での使用〕controlled use《of asbestos》; (strictly) controlled [supervised] use; use under control [supervision, surveillance]. **管理責任** responsibility for management [administration, running]. **管理責任者** the [a] person responsible for management [administration]. **管理釣り場** a managed site for fishing; a《local-gorvernment》administered place for fishing.

寒冷 ¶**寒冷期** 〔地球の〕a cold period;〔1年のうちの〕a [the] cold season. **寒冷渦**(かんれいうず)(かんれいか) a cold vortex. **寒冷高気圧** a cold anticyclone. **寒冷荒原** a cold desert. **寒冷前線** a cold front. ▶〜前線が南下した．A [The] cold front moved south.

寒冷地　a cold district.　**寒冷地砂漠**　a cold desert.　**寒冷地仕様**〔住宅・自動車などの〕《a car equipped to meet》cold-area [cold-region] specifications.　**寒冷地農業**　cold-district farming [agriculture].　**寒冷低気圧**　a cold cyclone.

き

気圧 atmospheric [air] pressure. ▶~の関係で owing to atmospheric conditions. / 昨日の~は980ヘクトパスカルを示した。The barometer registered [stood at] 980 hectopascals yesterday. / 中心~940ヘクトパスカルの台風 a typhoon with a central pressure of 940 hectopascals. / ~の谷 a (low pressure) trough. ¶**気圧配置** a pressure pattern.

キーストーン種(しゅ) 〔個体数は少ないがその種が属する生物群や生態系に大きな影響を及ぼす種〕a keystone species.

気温 (atmospheric) temperature. ★⇨温度 ▶~セ氏18度 a temperature of eighteen degrees Celsius [18°C]. / ~20度の部屋 a room whose temperature is twenty degrees Celsius. / ~が上がる[下がる]. The temperature rises [falls]. / ~が急激に上がる[下がる] the temperature shoots up [plummets, plunges]. / ~が30度に上昇した。The temperature rose [went up, climbed] to 30°C. / ~が30度を超した。The temperature went [climbed] above thirty. / ~がセ氏10度に低下した。The temperature went down [fell, dropped] to ten degrees Celsius. / 昼になっても~が0度を下回ったままだった。Even the daytime temperature stayed below zero. / 上空の~がマイナス30度以下になると雨ではなく雪が降る。When the upper-atmosphere temperature falls lower than minus thirty it snows rather than rains. / ~が高い[低い]. The temperature's high [low]. / 当地の夏は~は高いが湿度が低いのでからっとしている。In summer the temperatures here are high but the humidity is low, so the air doesn't feel muggy. / ~が高くなると雪崩(なだれ)が発生しやすい。High temperatures are conducive to avalanches. / 山頂とふもとでは~が5度違う。There's a five-degree difference in temperature between the top of the mountain and the foot of the mountain. / 春先は~の変化が激しい。Early spring is a time of acute [drastic] changes in (atmospheric) temperature. / 砂漠では日中と夜の〜の差が激しい．In deserts the differences between daytime and nighttime temperatures are extreme. / 〜の上昇にともなって along with a rise [climb] in temperature. / 〜の急上昇 a sudden rise in temperature. / 〜の急低下 a sudden drop [fall] in temperature; a cold snap. / 〜の高い日が続いています．We are having a succession of warm [hot] days. / 〔天気概況で〕〜16度．The temperature is sixteen degrees (Celsius, Centigrade). / 明日〜は少し下がる見込みである．Tomorrow the temperature is expected to fall slightly [be slightly lower]. / 今日は大変暖かく，3月下旬ごろの〜です．It's very warm today, on a level with average temperatures for late March. / 最高[最低]〜 the maximum [minimum] temperature; 〔天気予報で〕today's high [low]. ¶**平均気温** an average temperature. **気温逆転** temperature inversion. **気温減率** a temperature lapse rate.

帰化種 a naturalized species.
帰化植物[動物] a naturalized plant [animal].

危機 a crisis 《複数形 crises》; a critical moment [point, juncture, hour, situation, stage]. ▶核戦争の〜 a nuclear war crisis. / 〜にある[瀕している] be in a state of crisis; be (teetering) on the verge [brink] of a crisis. / トキは絶滅の〜に瀕している．The *toki* [crested ibis] is faced with the threat [on the verge] of extinction. / エネルギー〜 an energy crisis. ¶**危機遺産** 〔世界遺産のうち緊急な保護が必要なもの〕an endangered (World Heritage) site. **危機遺産リスト** 〔緊急の保全策が必要な世界遺産のリスト〕the List of World Heritage in Danger; the World Heritage in Danger list. / ユネスコはガラパゴス諸島を〜遺産リストに登録することに決した．UNESCO has decided to include the Galapagos on its List of World Heritage in Danger.

危急種(ききゅうしゅ) ⇨絶滅危惧種
企業 an enterprise; a business; a business corporation [firm]; a company; a corporation; a firm. ▶欠陥商品による事故は〜の責任だ．Accidents caused by defective merchandise are the responsibility of the enterprise [business, corporation, company] (that manufactured

them)． / リサイクル費用を〜側が負担する the enterprise side bears the burden of recycling costs． ¶**企業責任** 《promote》 corporate responsibility（略：CR）．★⇨英和 corporate social responsibility　**企業平均燃費**〔米国で販売される自動車の，メーカーごとの平均燃費規制〕corporate average fuel economy（略：CAFE）． / 〜平均燃費基準［規制］(the) Corporate Average Fuel Economy standards［regulations］．

奇形　deformation; (a) malformation; (a) deformity; (a) monstrosity;〔形態学上の〕abnormality．▶〜の deformed; malformed; monstrous．¶**奇形魚[動物]**　a deformed fish［animal］．

危険　(a) danger; dangerousness; (a) peril; jeopardy; (a) risk; a hazard．▶〜な dangerous; perilous; risky; hazardous; unsafe． / 〜な物質 a hazardous［dangerous］substance． / 〜につき入るべからず．〔掲示〕Danger: Do not enter! | Danger: Enter at (your) own risk． / 〜：高圧電流．〔掲示〕Warning: high voltage． / 副作用の〜(性)が高い薬 a medicine［drug］that has a high risk of side effects． / 原子力より〜性が低いエネルギー源 an energy source that is less risky［dangerous］than nuclear power． / 堤防が決壊する〜性が増してきた．The danger of the dike's breaking［embankment's collapsing］has increased． / 環境ホルモンの〜性を認識する be aware of［recognize］the danger of endocrine-disrupting substances． / 絶滅の〜がある ⇨絶滅危惧種　¶**危険責任**　responsibility for danger．**危険速度**　critical speed［velocity］．　**危険地帯[区域]**　a danger spot［zone, area］．　**危険地点**　a danger point．　**危険手当**　danger money; hazard pay．　**危険半円**〔台風区域の右半円〕the dangerous semicircle．　**危険物**　a dangerous article; a hazardous material; hazardous materials;〔爆発・可燃物〕explosives and combustibles;〔航海・輸送上の〕dangerous cargo［goods］． / 〜物の処理[管理] the disposal［management, control］of hazardous materials［dangerous goods］．★⇨管理 / 〜物の輸送 the transportation of hazardous materials［dangerous goods］． / 機内への〜物の持ち込みを禁止する．Dangerous goods［Hazardous materials］are not allowed (to be carried) onto the plane． / 〜物の入ったドラム

缶が空き地に放置されている．Drum cans containing hazardous materials are left in a vacant lot [field]． / その工場は～物を取り扱うので，住宅街から離れたところにある．Because the factory handles hazardous materials, it is situated away from residential areas． / ～物持ち込み厳禁．〔掲示〕Dangerous goods forbidden《to be carried aboard》． / ～物貯蔵庫 a dangerous goods [hazardous materials] storehouse [warehouse]． / ～物取扱者 a hazardous materials engineer． / ～物取扱主任者 a hazardous materials chief engineer．　**危険防止**　risk prevention．　**危険薬物**　dangerous drugs．　**危険有害要因**　〔職場や作業などの〕a hazard factor．

気候　climate;〔天候〕weather;〔時候〕a season．★⇨気象 ▶よい[温和な]～ a fine [mild, temperate] climate． / 不順な～ unseasonable weather． / 変わりやすい～ a variable climate; changeable [unsettled] weather． / 健康によい～ a healthy [wholesome] climate． / 当地の～はおだやかで健康によい．The climate here is mild [temperate] and healthy． / ～がよければ under favorable climatic conditions; provided the climate is good． / ～の変化 a climatic change; variations in climate． / そこは～の変化が激しい．The place is subject to extreme [violent] climatic changes． / ～の影響 a climatic influence． / 海洋[大陸，島嶼(とうしょ)]性～ a(n) maritime [continental, insular] climate．¶**気候学**　climatology．　**気候学者**　a climatologist．　**気候型**　a climatic type．　**気候感度**　〔大気中の二酸化炭素の上昇に応じた気温上昇の程度〕climate sensitivity．　**気候区**　a climatic region [division]．　**気候区分**　a climatic classification．　**気候指数**　a climatic index．　**気候ジャンプ**　〔急激な気候レジームシフト〕a climate [climatic] jump．★⇨レジームシフト　**気候順化[順応]**　acclimatization．　**気候政策**　a climate policy．　**気候帯**　a climate [climatic] zone．　**気候地形学**　climatic geomorphology．　**気候投資基金**　the Climate Investment Fund（略：CIF）．　**気候取引所**　〔温室効果ガス排出権取引所〕a climate exchange．★⇨排出権 / 欧州～取引所 the European Climate Exchange（略：ECX）/ シカゴ～取引所 Chicago Climate Exchange（略：CCX）．　**気候表**　a climatic table．　**気候変動**　climate change. / 地球温暖化が様々な〜変動を引き起こしている．Global warming is causing many kinds of climate change. **気候変動に関する政府間パネル** ⇨英和 IPCC **気候変動枠組み条約** 〔国連の〕the (United Nations) Framework Convention on Climate Change (略: (UN) FCCC). ★「地球温暖化防止条約」(the Global Warming Convention) の正式名称．★⇨京都議定書 **気候変動枠組み条約締約国会議** the Conference of Parties to the UN Framework Convention on Climate Change. ★⇨英和 COP **気候変動税** 〔英国の〕a climate change levy (略: CCL). **気候モデル** a climate model. / 〜モデル計算 a climate model calculation.

旗国(きこく) 〔船籍国〕a flag state. ★⇨英和 flag state control ¶**旗国主義** the flag state principle. **旗国法** the law of the flag state.

基準 a standard; a criterion《複数形 criteria, criterions》; a basis《複数形 bases》; a yardstick; a benchmark; a gauge; a point of reference. ▶甘い［厳しい］〜 loose [strict] standards. / 検査の〜があいまいだ．The testing standards are vague. / 〜に合っている［合っていない］be in accord [not be in accord] with established standards. / 〜に達する meet standards. / …を〜にして決める set [fix] on the basis of… / 先進国を〜にして途上国を判断するな．Don't judge developing countries by the standards of advanced countries. / この問題では欧米を〜にして考える必要はない．In considering this problem, there's no need to use the West as the point of reference. / その病院の設備は〜を満たしていると認定されている．The hospital facilities have been accredited as fulfilling established standards. / 合格の〜を引き上げる［下げる］raise [lower] the passing grade. / 〜を緩和する loosen standards. / …のための〜を設ける set criteria [standards] for… / …の明確な基準を設定する establish clear standards for. / 1990 年を〜としてその収穫量を 100 とすると 1995 年は 103 である．Using 1990 as the base year in which the harvest is reckoned at 100, that for 1995 is 103. ★⇨基準年 ¶**基準年** 〔統計などの〕a base [benchmark] year. ★⇨英和 base year **基準・認証制度** 〔安全確保や経済取引の適正化のため，製品や施設に基準を設け，それが満たされているかを確認・

証明するシステム〕a standardization and certification system.　**環境基準**　environmental (quality) standards. ★⇨環境, 規制(規制基準)　**建築基準**　architectural standards [requirements].　**国際基準**　《by》international standards. / 国際的な統一～ a uniform international standard.　**設置基準**　〔施設などの〕standards [requirements] for establishing《a university, a nursing home, a network of highway signs》.　**耐震基準**　quakeproofing standards.　**適用基準**　application standards [criteria]; standards [criteria] for the application《of…》.　**二重基準**　a double standard. ★⇨排ガス(排ガス基準)

希少(きしょう)　▶～な scarce; rare. ¶**希少金属**[鉱物]　rare metals [minerals].　**希少資源**　scarce resources.　**希少種**　〔生物の〕a rare species.　**希少動植物**[野生生物]　rare [scarce] fauna and flora [wildlife].

気象　weather; weather [atmospheric] conditions. ▶～の変動 meteorological change; climatic fluctuation. ★⇨異常(異常気象) ¶**気象衛星**　a meteorological [weather] satellite; a weather eye.　**気象概況**　general [overall] weather conditions.　**気象学**　meteorology (圏 meteorological); climatology;〔高層の〕aerology. / 巨～学 macrometeorology. / 総観～学 synoptic meteorology. / 微～学 micrometeorology. / 理論～学 theoretical meteorology.　**気象学者**　a meteorologist; a climatologist; an aerologist.　**気象観測**　(a) meteorological [weather] observation; a weather survey. / 気象～をする make meteorological observations.　**気象観測衛星**　a weather observation satellite; a weather satellite.　**気象観測機**　a weather research craft.　**気象観測所**　a meteorological (observing) station; a weather station.　**気象観測船**　a weather ship; a meteorological observation vessel.　**気象観測飛行**　a weather survey flight; weather flying.　**気象観測ロケット**　a sounding [meteorological] rocket.　**気象記号**　a weather symbol.　**気象業務**　weather [meteorological] service.　**気象業務支援センター**　the Japan Meteorological Business Support Center (略: JMBSC).　**気象業務法**　the Weather [Meteorological] Service Law.　**気象警報**　a weather warning.　**気象現象**　atmospheric [meteorological, weather] phenomena. **気象災害** a weather [meteorological] disaster; climatic damage. **気象写真** a weather (satellite) photograph. **気象条件[状況]** weather conditions. **気象情報[通報]** weather news [information]; a weather [meteorological] report [bulletin]. **気象情報サービス** (a) weather information service. **気象図**〔天気図〕a weather chart [map]. **気象台** a meteorological observatory. /沖縄〜台 the Okinawa Meteorological Observatory. /海洋〜台〔4つある〕a marine meteorological observatory. /管区〜台〔沖縄以外の全国に5つある〕a district meteorological observatory. /地方〜台〔管区気象台の下部組織〕a local meteorological observatory. /航空地方〜台 an aviation weather service center. /〜台員 a meteorologist; a weatherperson. **気象注意報** a weather alert [advisory]. **気象庁** the Meteorological Agency. /〜庁長官 the Director-General of the Meteorological Agency. **気象潮** a meteorological tide. **気象調節** weather modification [control]. **気象データ** weather [meteorological] data. **気象統計** meteorological statistics. **気象用語** meteorological terms [vocabulary]. **気象予報**〔業務〕weather forecasting;〔1回の〕a weather forecast. **気象予報官** a weather forecaster; a weatherperson; a weatherman. **気象予報士** a certified meteorologist [weather forecaster]. **気象予報士試験** a weather forecaster test; a certification test for weather forecasters. **気象力学** dynamic meteorology. **気象レーダー** (a) weather radar. **気象情報会社** a weather [meteorological] information company. **自記気象計** a meteorograph; an aerograph.

汽水〔淡水と海水がまじりあう河口部などの水〕brackish water. ¶ **汽水域** brackish waters. **汽水湖** a brackish lake. **汽水成層[堆積物]** a brackish-water sediment.

寄生 ▶〜する be parasitic《on…》; be a parasite《on [to]…》; live (parasitically)《upon [with, in, within]…》. /ハチに〜されたハエの幼虫 a fly larva parasitized by a wasp. ¶ **寄生体** a parasite. **寄生虫** a parasite; a parasitic worm [insect];〈集合的に〉vermin. /植物の〜虫 an insect that lives parasitically on a plant;〔樹木

の〕an insect parasite that infests trees. / 〜虫がわく〈宿主が主語〉get (parasitic) worms; verminate. / 腸内〜虫 an intestinal worm;〔蠕虫(ぜんちゅう)〕a helminth. ★⇨外部寄生, 内部寄生 **寄生虫学** parasitology;〔蠕虫学〕helminthology. **寄生虫駆除剤**〔駆虫薬・虫下し〕a parasiticide. **寄生虫血症** parasitemia. **寄生虫性肝硬変** parasitic cirrhosis. **寄生虫塞栓** parasitic embolism. **寄生虫病** parasitic disease; vermination; helminthiasis. **寄生虫症** (a) disease caused by internal parasites; (a) parasitosis; (a) verminosis. **寄生動物** a parasite; a guest; a parasitic animal. / 外部〜動物〈総称〉ectozoa《単数形 ectozoon》. / 体表〜動物 an epizoon《複数形 -zoa》. **寄生微生物** a microparasite. **異種寄生** heteroecism. **一時寄生** temporary parasitism. **外部寄生** ectoparasitism; external parasitism. / 外部〜虫 an ectoparasite; an external parasite. **過寄生** superparasitism. **完全[不完全]寄生** complete [incomplete] parasitism. **共寄生** multiple parasitism; multiparasitism; symparasitism. **社会寄生** social parasitism. **条件的[任意]寄生** facultative parasitism. **絶対寄生** obligate [obligatory] parasitism. **全寄生** holoparasitism. **重複寄生** hyperparasitism; superparasitism. **定留寄生** stationary parasitism. **内部寄生** endoparasitism; internal parasitism. / 内部〜虫 an endoparasite; an internal parasite. **半寄生** hemiparasitism; semiparasitism. **寄生火山** a parasitic cone [volcano]. **寄生去勢** parasitic castration. **寄生根** a parasitic root. **寄生状態** parasitism. **寄生植物** a parasitic plant; a parasite; a guest. / 外部〜植物 an ectophyte. / 内部〜植物 an endophyte. / 腐性〜植物 a saprophyte.

規制 regulation; control. ▶〜する regulate; control; impose restraints. / この薬は法的〜の対象となっている[対象外である]. This medicine [drug] is [is not] subject to legal restrictions. ¶ **規制影響分析** (a) regulatory impact analysis [assessment](略: RIA). ★環境行政などの規制を導入するに当たりその影響と費用対効果などを客観的数値として算出する評価手法. **規制緩和** the easing [relaxation] of restraints or regulations;〔規制撤廃〕deregulation. **規制基準**〔公害などの〕

a regulatory standard. **規制強化** (a) tightening of regulations. **規制権限** (the) authority to control《over...》. / ～権限(の)不行使 failure to exercise the authority to control. **規制値** a regulated [regulatory, regulation] value; a limit value; a limit.

季節 〔気候区分〕a season; a time of the year. ▶雨の[乾燥した]～ the rainy [dry] season. / 桜の～ the cherry season. / 花粉症の～ the hay fever [pollinosis] season. / 新じゃがの出回る～ the time of year when new potatoes start appearing. / ～遅れの台風 a post-season typhoon. / ～はずれの out of season; unseasonable; off-season. / ～はずれの雪 an out-of-season snowfall. / 今年は～が遅れていてまだ寒い. Spring is late this year and it's still chilly. / ナスを植えるには～が遅すぎる．The season is too far advanced to plant eggplants [aubergines]. / ～の seasonal; in season. / ～の花 flowers of the season. / ～のものを食べる eat foods in season. / ～の変わり目 a time of seasonal change; a seasonal turning point. / ～の移り変わり a turn of season; a seasonal transition. / この～はいつも風が強い. It generally blows hard at this time of (the) year. / 松茸は今が～です．Matsutake(s) are now in season. / 温室栽培の野菜が一年中出回るようになり，食卓から～感が薄れている．Because greenhouse vegetables are available the year round, dining tables are losing their sense of season. ¶**季節移動** seasonal migration. **季節遷移** seasonal succession. **季節帯** the seasonal zone. **季節風** a periodic wind; a seasonal wind;〔インド洋の〕a monsoon. / 日本海側の冬は～風のため降雪量が多い．In winter the prevailing winds bring heavy snows to the Japan Sea side. / 反対～風 an antimonsoon. **季節風気候** a monsoon climate. **季節変異** seasonal variation. **季節予報** a seasonal weather forecast. **季節別時間帯別電力** a season- and time-specific power service. / 業務用～別時間帯別電力 a season- and time-specific commercial power service.

季相(きそう) a seasonal aspect.

帰巣(きそう) homing. ¶**帰巣本能** a homing instinct. **帰巣性** ⇨回帰性

偽装 camouflage; disguise. ▶～する disguise; camouflage; hide by camouflage; fake; dissemble. / 産卵の日付を偽装して卵を売る sell eggs with the dates of laying falsely labeled. ¶**偽装表示** 〔食品の産地や商品の内容などについての〕false labeling.

牛肉偽装事件 〔2001年, BSE対策事業の一環としての国産牛肉買い取り事業を悪用し, 食肉卸業者が輸入牛肉を国産牛肉と偽り補助金を詐取した詐欺事件〕the fraudulent beef-labeling case.

期限偽装 falsifying [falsification of] an expiry date. ★⇨産地(産地偽装)

擬態 (biological) mimicry. ★特に捕食者の嫌う他の動物に似る標識的擬態を言う. ▶〜する mimic. / 〜の imitative; mimetic. ¶**擬態色** mimetic coloration; a mimetic color. **擬態生物** mimicry animals. **隠蔽的擬態** 〔周囲の植物などにまぎれる擬態〕mimesis. **保護的擬態** protective mimicry.

喫煙 smoking (tobacco). ★⇨たばこ, 禁煙, 嫌煙 ▶〜する smoke《a cigarette, a pipe》; have a smoke. / 〜のため体をこわす smoke *oneself* sick. / 〜は本人だけでなく周囲の人の健康もむしばむ. Smoking damages the health not only of the smoker but also of those around him. / ここでは〜は禁じられている. Smoking is not allowed [prohibited] here. | No smoking is permitted here. / 未成年者の〜は法律で禁じられている. Minors are prohibited by law from smoking. ¶**喫煙室** a smoking [smoke] room. **喫煙者** a smoker. / 非〜者 a nonsmoker. **喫煙車(両)** a smoking car; a smoking carriage [coach]; a smoker. **喫煙習慣** the habit of smoking. **喫煙所** a smoking area [corner]. **喫煙人口** a [the] smoking population. **喫煙席** 〔列車・レストランなどの〕a smoking section [area]. **喫煙対策** an antismoking measure. / 職場における〜対策 measures (taken) against smoking in the workplace. ★⇨分煙 **喫煙マナー** smoking manners. **喫煙率** the smoking rate; the percentage of smokers. / 女性の〜率 the percentage of women who smoke. **間接喫煙** inhaling secondary [sidestream] smoke. **受動喫煙** passive smoking. **直接喫煙** inhaling primary [mainstream] smoke. **路上喫煙** smoking on [in] the street. / 路上〜禁止条例 an ordinance against smoking on [in] the street. ★⇨たばこ(歩きたばこ)

揮発(きはつ) volatilization. ▶〜する volatilize. / 揮発性の, 揮発し

やすい volatile《liquid》. ¶**揮発性**[度] volatility. **揮発性物質** a volatile; a volatile substance [compound]. **揮発性有機化合物** volatile organic compounds（略: VOC）. **揮発性有機塩素化合物** a volatile organochlorine compound. **揮発性溶剤** a volatile solvent. **揮発性ワニス** spirit varnish. **揮発物**[分] volatile matter; a volatile. **揮発油** a volatile oil; naphtha; gasoline. **揮発油税** (a) gasoline tax.

逆浸透法 〔浄水法の一つ〕reverse osmosis.

逆転 〔温度の〕an inversion. ¶**逆転層** an inversion layer.

キャスビー 〔建築物総合環境性能評価システム〕CASBEE; the Comprehensive Assessment System for Building Environmental Efficiency.

キャッチ・アンド・リリース 〔釣りあげた魚をすぐ放すこと〕catch and release (fishing).

ギャップ ⇨英和 GAP

キャップ・アンド・トレード ⇨英和 cap and trade

キャリング・キャパシティ ⇨英和 carrying capacity

吸音 sound [acoustic] absorption. ¶**吸音タイル** an acoustic(al) tile. **吸音パネル** an acoustic(al) panel. **吸音板** a sounding board; a sound board. **吸音率** acoustic absorptivity. **吸音力**[材] acoustic [sound-absorbing] power [materials]. **吸音天井** an acoustical [acoustic] ceiling.

給気 air supply; 〔液体への〕aeration; 〔通風〕ventilation. ★⇨給排気 ¶**給気管** an air duct. **給気口** an air inlet. **給気装置** 〔液体への〕an aerator; 〔通風装置〕a ventilator.

急傾斜地 a steep slope; steep(ly sloping) terrain. ¶**急傾斜地崩壊** (a) slope failure. ▶〜崩壊危険箇所 a location where there is a risk of slope failure; a (potential) slope failure area [zone].

吸血 ¶**吸血昆虫**[害虫] a bloodsucking insect. **吸血者** 〔特に昆虫〕a hematophagus. **吸血性の** hematophagous. **吸血動物** a bloodsucking animal; a bloodsucker.

休耕 ▶〜する do not cultivate《a field》; leave《a field》fallow [idle]. / 〜中である〔農地が〕lie fallow [idle]. **休耕地** land not cultivated [lying fallow, lying idle]. **休耕田** a fallow [an idle] rice field [paddy]; a rice field lying fallow [left uncultivated]《for a season》; a set-aside field.

吸湿 moisture absorption. ¶ **吸湿効果** moisture absorption efficacy. **吸湿剤** a moisture absorbent; a desiccant. **吸湿試験** a moisture absorption test. **吸湿性** (moisture) absorbency; hygroscopicity; a hygroscopic property. ▶〜性の hygroscopic. / 〜性が高い be highly [very] hygroscopic. **吸湿率** 〔含水率〕the percentage of water content. **吸湿力** (moisture) absorbency.

吸収 absorption;〔同化〕assimilation. ▶〜する absorb;〔同化する〕assimilate. / 音[光]を〜する absorb sound [light]. ★⇨吸音 / 皮膚を通じて体内に〜される be absorbed into the system through the skin. ¶ **吸収源**〔CO_2 を吸収する森林や海など〕a carbon sink. ★⇨シンク,森林(森林吸収源),緑化,英和 carbon sink **吸収口** a suctorial mouth. **吸収剤** an absorbent. **吸収作用** (a process of) absorption. **吸収性** an absorptive property; absorptiveness; absorbency. / 〜性の absorptive; absorbent. / 〜性のある with absorbency; absorptive; absorbent. / 〜性の高い with high absorbency; highly absorptive [absorbent]. **吸収線量**〔放射線の〕an absorbed dose. **吸収組織** an absorptive tissue. **吸収率** absorptivity; absorptance. **吸収力** absorbability; power to absorb; absorbency; absorptiveness; absorptive power. / 〜力のある able to absorb; absorbent; absorptive. **吸収効率**〔栄養分などの,体内での〕absorption efficiency.

給水 provision [supply] of water; water supply [service]; water feeding;〔供給された水〕supplied water; feedwater. ▶〜する supply《a town》with water; feed water《to a boiler》; feed《a boiler》with water. / 〜がない have no water supply《for the day》. / 〜が止まった. The water supply (to the area) has been stopped. / 〜を断つ cut off the supply of water. / 〜(量)を 15% カットする reduce [cut, cut back] water supply by fifteen percent. / 時間〜する restrict water supply to certain hours. ¶ **給水加熱器** a feedwater heater. **給水管** a water pipe;〔建物外部の〕a main;〔建物内部の〕a service pipe; a feedwater pipe《on a boiler》. **給水器** a waterer; a drinker. **給水区域** a water supply district; a water service area. **給水施設**

a water(-supply) system 《for [in] a city》. **給水車** a water(-supply) truck; a water wagon [cart]. **給水制限** 〔取水制限より進んで利用者への給水を制限する〕water (supply) restriction(s). ★⇨取水制限 / ～制限の緩和 relaxing [easing] of water (supply) restrictions. **給水設備** water-supply facilities; waterworks. **給水栓** a water tap; a (water) faucet;〔街路の〕a hydrant;〔ボイラーなどの〕a feed cock. **給水船** a water boat [tender]. **給水装置** water-supply equipment;〔ボイラーなどの〕the feed system. **給水タンク** a water tank. **給水調整器** a feed water regulator. **給水塔** a water tower; a standpipe;〔駅の〕a water column. **給水弁** a feed valve. **給水本管** a water main. **給水ポンプ** a feed pump. **給水量** the amount of water supplied. **給水濾過器** a feed water filter. **減圧給水** 〔給水を制限する方法の1つ〕water pressure reduction; reducing [reduction of] water pressure.

急性 ¶**急性吸入毒性** acute inhalation toxicity. **急性経口毒性** acute oral toxicity. **急性障害** 〔被曝による〕acute sickness.

急速濾過(ろか)**(法)** rapid filtration.

吸着 adsorption. ▶～する adsorb. ¶**吸着器** an adsorber. **吸着クロマトグラフィー** adsorption chromatography. **吸着剤** an adsorbent. **吸着水** adsorption water. **吸着性の** adsorbent; adsorptive. **吸着等温線** an adsorption isotherm. **吸着熱** heat of adsorption. **吸着媒** adsorbent. **吸着平衡** adsorption equilibrium. **吸着マット** an absorbent mat; absorbent matting. **吸着力** adsorbability; adsorptivity; adsorption [adsorptive] power. / ～力が強い[弱い] be highly [weakly] adsorptive. / ～力のある adsorptive; capable of adsorbing *sth*. **活性化吸着** activated adsorption. **静電吸着** electrostatic adsorption [adhesion]. **選択的吸着** 〔微粒子物質などの〕selective adsorption《of...》. **物理吸着** physical adsorption; physisorption.

給電 the supplying [supply] of electricity; electric [power] supply. ¶**給電回路** a feeder circuit. **給電所** a load-dispatching office. **給電線** a service wire; a feeder. **給電盤** a load-dispatching board; a feeder panel. **給電量** the amount [volume] of electricity sup-

牛乳 milk; cow's milk. ▶殺菌した〜 pasteurized milk. /〜アレルギーの子供 a child with a milk allergy [sensitivity]. ¶**牛乳パック** a milk carton. ★⇨紙 **還元牛乳**〔脱脂粉乳からの〕recombined milk;〔全脂粉乳からの〕reconstituted milk. **均質[ホモ]牛乳** homogenized milk. **成分無調整牛乳** nonhomogenized [unhomogenized] milk. **成分調整牛乳** homogenized milk. **低脂肪牛乳** low-fat milk. **濃縮牛乳** concentrated milk;〔糖分無添加の〕evaporated milk;〔糖分添加の〕condensed milk. **(紙)パック牛乳** a carton of milk; milk in a carton.

給排気 ventilation. ¶**給排気設備** ventilation equipment. **給排気筒** a ventilation duct. **自然給排気式**〔暖房器具などの〕balanced flue.

休眠 dormancy;〔蚕の〕quiescence; quiescency. ▶〜中の dormant; resting; diapausing《larvae》. /〜から覚める〔植物が〕break dormancy. ¶**休眠芽[胞子]** a dormant [resting] bud [spore]. **休眠期** a period of dormancy; a resting stage; diapause. **休眠打破**〔休眠状態を終わらせること〕dormancy breaking. **休眠誘導**〔休眠状態に入らせること〕dormancy induction.

休猟(きゅうりょう) suspension of hunting (activities). ▶〜する suspend hunting. ¶**休猟区** an area [a zone] in which hunting is temporarily banned. **休猟日** a day off from hunting (activities).

休漁(きゅうりょう) suspension of fishing (activities). ▶〜する suspend fishing (activities). ¶**休漁区**《set up》a zone where fishing is temporarily discontinued. **休漁日** a day off from fishing. **一斉休漁日** a day off from all fishing activity whatsoever.

供給 supply;〔電気・ガス・水道の〕service. ★⇨給水, 給電, 原子(原子力供給国グループ) ▶〜する supply; provide; serve《a town with water》. /原料を〜する supply《a factory》with material; feed《a machine》with material. /首都圏に野菜を〜している農家 farmers who supply vegetables to a metropolitan area. /送電線が切れて電力の〜ができなくなった. The lines were down and the power supply was cut off. /その町は利根川から水の〜を受けている. That town is supplied with water from the Tone River.

¶ **供給価格** a supply price. **供給過剰[過多]** an excessive supply; an oversupply;〔商品の〕a glut. **供給管[本管]** a supply pipe [main]. **供給業者** a supplier; a service provider. **供給曲線** a supply curve. **供給区域** a service area. **供給源** a source of supply. / たんぱくの〜源 a source of albumin supply. **供給コスト** (a) supply cost. **供給サイド** supply-side. **供給地** a region [an area, a district, a country] that supplies; a source of supply; a source of 《crude oil》. / 木炭の〜地 an area that supplies charcoal. / 中東は日本の主な石油の〜地である. The Middle East is the chief source of Japan's oil supply. **供給電圧** service voltage. **供給電源** a source of electric power supply. **供給熱量自給率**〔食料自給率の〕a ratio of self-sufficiency in supplying calories. ★⇨地域(地域熱供給) **供給不足** a short supply; an undersupply. **供給量** an amount supplied. **供給路** a channel of supply; a supply route; a lifeline. / 〜路を断つ cut a supply route.

狂牛病 mad cow disease. ★正式名はウシ海綿状脳症(bovine spongiform encephalopathy; 略: BSE).

凝集沈殿装置 coagulating sedimentation equipment.

共生 coexistence; association; symbiosis; commensalism;〔菌根による〕mycotrophy. ▶〜する coexist; live together [in symbiosis]. / 〜的な symbiotic(al); commensal. ¶ **共生関係** a symbiotic relationship. / 鳥と植物の〜関係 a symbiotic relationship between birds and plants. **共生体[生物]** a symbiont; a symbion《複数形 -bia》; a symbiote. **共生発芽** symbiotic germination. **共生発光** symbiotic luminescence. **共生(細)菌** a symbiotic bacterium《複数形 -ria》. **共生者** a symbiont;(片利共生の) a commensal. **住み込み共生** inquilinity. 形 inquilinous. **相利共生** symbiosis; mutualism. **片利共生** commensalism.

京都議定書 ⇨英和 Kyoto Protocol ▶〜への参加を拒否する refuse to join the Kyoto Protocol. / 気候変動問題に関する国際的枠組みとして〜が 1997 年に合意された. The Kyoto Protocol was agreed to in 1997 as an international framework on the issue of climate change. ¶ **京都議定書目標達成計画** the Kyoto Protocol Target Achievement Plan.

京都メカニズム ⇨英和 Kyoto mechanism

共同 ¶**共同漁業権** common fishing rights. **共同購入** cooperative [communal] buying [purchasing]; group purchasing. ▶～購入で無農薬の野菜を買う buy organic vegetables as a group. / ～購入をする buy things together as a group; make a communal purchase 《of...》. / ～購入を利用する make use of cooperative buying. / ～購入を申し込む apply to join a cooperative buying program. **共同使用** communal [joint, public] use; use by a community. / ～使用する use *sth* communally [in common, as a community]. **共同体** a community; a communal society. / 村落～体 a rural [village] community. **共同体主義** communitarianism. **共同地** 〔共有地〕a common; (a piece of) common land. **共同農場** a collective [cooperative] farm. **共同漁業水域** ⇨漁業 **共同実施** 〔京都メカニズムの〕⇨英和 JI **共同実施活動** activities implemented jointly (略: AIJ).

共有 joint [common] ownership; co-ownership; community;〔合有・総有に対し, 共同所有の一形〕joint ownership (in which the owners can dispose of their individual shares without restrictions). ▶～する own jointly; hold in common; share (ownership of) ... with *sb*. / ～の common; joint; jointly [communally] owned. ¶**共有化**〔土地・財産などの〕《asset》pooling [sharing];〔知識・情報などの〕《information》sharing. / 土地の大規模な～化を図る plan a large scale land sharing [pooling] project. / インターネットで情報を～化する share information over the Internet. **共有財産** common [jointly owned, public, community] property; property owned in common. **共有者** a joint owner; a co-owner. **共有地** a common; public [common] land. / 村の～地 a village common. 「**共有地の悲劇**」〔米国の生物学者 G・ハーディンが 1986 年『サイエンス』誌に発表した論文; 共有の資源は濫用されがちで, その濫用が資源を枯渇させてしまうと説いている〕The Tragedy of Commons. / 反～地の悲劇〔共有されるべき資源なのに, 所有権者が他者に使わせないことによって生じる悲劇〕the tragedy of the anticommons. **共有物** common property. / この机は～物だ.

This desk is for common use. / これはわれわれの〜物だ. It belongs to us all.　**共有部分[スペース]**　〔マンションなどの〕(a) common space.　**共有林**　a jointly owned [co-owned] forest; a forest owned in common.

許可　〔公的機関の承認〕(official) permission; (an) approval; sanction;〔認可〕authorization. ▶〜する〔承認する〕permit; sanction; approve (of)...;〔認可する〕authorize. / スーパーマーケットに盲導犬の同伴を〜してほしい. I wish supermarkets would admit guide dogs [allow guide dogs in, let guide dogs be brought in]. / 公園内での物品の販売は〜されておりません. Trading is not permitted in the park. |〔掲示〕Trading Forbidden. | No Trading. / その木を切る〜が下りた. Permission to cut down the tree was granted. / その土地の開発が早く〜になればいい. If only development of the land was approved quickly. | If only permission would come through quickly to develop the land. / 風致地区の高層建築は絶対〜にならないはずだ. High-rise construction is bound not to be approved in an area of scenic beauty. / 〜を申請する apply for [request] a license [permission, approval]. / 〜を得る[取る, もらう] get [gain, obtain, receive, be given, be granted] permission [a permit]. / 〜を得て営業する do business under license. / 〜を与える[出す] give permission; issue approval [a license, a permit]. / 〜を取り消す cancel permission; rescind [revoke, withdraw] approval. / 無〜で, 〜無しで without (official) permission [approval, authority]; unapproved.　¶**許可漁業**　licensed fishery.　**許可書[証]**　a license; a warrant; an authorization; (a) written permission; a permit (card). ★⇨ 漁業 / 建築〜証 a construction permit. / 採鉱〜証 a mining license. / (通関手続き上の)輸出[入]〜証 an export [import] permit. / 〜証を発行する issue [give, grant] a license [permit]《to...》. / 〜証を取る get [obtain, receive] a license [permit]《from...》. / 〜証を取り上げる rescind [revoke, withdraw] a license.　**許可申請**　application [a request] for permission [a permit].　**許可制**　a license [licensing] system. / 〜制になっている be on a license system [basis].　**許可手続**　an authorization procedure.　**許可料**　a license fee; a fee for permission. **営業許可** (official) permission to operate (a business); a license. / 営業〜が下りなかった. No license was granted [given] for the business. / 営業〜の下りている商売 a licensed [an authorized] business. / 営業〜の申請をする apply for a business license [permit]. / 営業〜を与える authorize *sb* to carry on a business. / 営業〜を受ける secure a business license [permit]. / 営業〜証 a business license [permit]. / 営業〜願い application for a business license [permit]. **建築許可** permission to build; a construction license. ★⇨許可証, 建設(建設業) **使用許可** permission to use *sth*; a license for use of *sth*. ★⇨水利(水利使用許可) **上陸許可** shore leave. **通行許可** permission to pass; admittance. **登山許可** 《be given》 permission to climb 《Mt. Everest》.

漁獲 fishery; fishing. ▶昨シーズンは〜が多かった. We got [made] a good catch last season. | Last season the catches were large. / この沖ではニシンの〜が多い. There are good catches of herrings off this coast. ¶**漁獲可能量, 漁獲枠** a total allowable catch (略: TAC). **漁獲高[量]** a haul [catch] (of fish); a fish catch; a take. ★⇨生物(生物学的許容漁獲量) **漁獲割当** a fishing [catch] quota. / 鮭鱒(さけます)〜(割当)量 the salmon catch quota. **漁獲規制** a fishing [catch] restriction.

漁業 fishery; fishing; the fishing industry. ★⇨沿岸[遠洋, 河川, 近海]漁業 ▶淡水[沖合い]〜 freshwater [offshore] fishery. ¶**漁業会社** a fishery; a fishery [fishing] company [corporation]; 〈集合的に〉fishery interests. **漁業管轄権** fisheries [fishery] jurisdiction; jurisdiction of fishing《for crab》. **漁業気象** weather forecasting for fishermen. **漁業共済保険** Fishery Mutual Aid Insurance. **漁業協定** a fisheries agreement [pact]. / 双務〜協定 a bilateral fisheries agreement [pact]. / 日米〜協定 the Japan-US Fisheries Agreement. ★⇨国連(国連公海漁業協定) **漁業協同組合** a fishermen's cooperative (association). **漁業許可証** a fishing license. **漁業組合** a fishermen's union [association]. **漁業権** a fishing [fishery] right. / 共同〜権〔他人の漁区内での〕a common of fishery [piscary]. 漁業交渉 fishery talks; fishing negotiations 《with Russia》. 漁業資源 fishing resources. 漁業情報サービスセンター 〔社団法人〕the Japan Fisheries Information Service Center. 漁業条約 a fisheries [fishing] treaty. 漁業振興 fishery promotion [development]. / ～振興費 the cost of promoting [developing] fishing; fishery promotion costs. 漁業水域 fishery waters; a fishery [fishing] ground [zone]. / 共同～水域 a joint fishing [fishery] zone; a common fishing zone. 漁業専管水域 exclusive fishing waters; an exclusive fishery [fishing] ground [zone]. 漁業調整 fisheries coordination. ¶漁業調整規則 rules [regulations] for fisheries coordination. ▶～調整規則ライン〔北海道近海の日ロの〕a line defined by rules [regulations] for fisheries coordination. / ～調整事務所 a fisheries coordination office. 漁業取締船[監視船] 〔密漁などを取り締まる〕a fisheries patrol boat [inspection vessel]. 漁業白書 ⇨水産(水産白書) 漁業被害 damage to fishery [fishing]. 漁業補償 compensation to fishermen [the 《local》 fishing industry]; fisheries compensation. / ～補償交渉 negotiations on compensation to fishermen. / ～補償費 money for [the cost of, expenses for] compensation to fishermen. 漁業保存水域 《establish》a fishery conservation zone. 漁業無線局 〔漁船に設置されている〕a ship radio station.

極 〔地球・磁石・電池・細胞などの極〕a pole. 形 polar. ¶極移動 polar wandering. 極渦 the (Antarctic [Arctic]) polar vortex. 極循環 〔大気大循環の一種〕the Polar cell. 極成層圏雲 a polar stratospheric cloud. 極地(方) the polar regions. ▶極地植物 an arctic [a polar] plant. / 極地付近の polar; circumpolar 《ocean》.

極相 a climax. ¶亜極相 a subclimax. 後極相 a postclimax. 前極相 a preclimax. 極相群落 a climax community. 極相林 a climax forest.

局地 ▶明日は～的に激しい雨が降るでしょう. Tomorrow there will be heavy rain in places [isolated showers of heavy rain]. ¶局地気候 a local climate. 局地気象 local weather; local meteorology. 局地激甚災害 a devastating localized [local] disaster; a devastating disaster in a restricted locality. **局地(的)豪雨** torrential rain(s) in a restricted area; torrential local(ized) rain. **局地地震** a local earthquake. **局地風** a local wind. **局地予報** 〔天気の〕 a spot [localized, high-resolution] weather forecast.

魚礁(ぎょしょう) a rocky place under the water where fish tend to gather; a breeding ground [reef] for fish. ¶**浮き魚礁** a fish aggregating device (略: FAD). **人工魚礁** an artificial fish reef. ▶廃船を海底に沈めて人工〜にする scuttle a decommissioned ship to create a fish reef on the ocean floor.

漁場(ぎょじょう) a fishing ground; fishing banks; a fishing place; a fishery. ¶**漁場標識** a fishery [fishing ground] marker. **好漁場** a good fishing place [spot]. ▶ワタリガニの好〜 a good place to catch swimming crab. **区画漁場** a demarcated fishery [fishing ground]. ★⇨区画(区画漁業) **定置(網)漁場** a fixed [set, stationary] net fishery [fishing ground].

去勢 〔家畜などの〕castration; gelding; 〔ペットの〕neutering. ★⇨不妊 ▶〜する castrate; geld; neuter. /〜した動物 a castrated animal; a neuter. ¶**去勢雄鶏** a capon. **去勢牛** a bullock. **去勢馬** a gelding; a castrated [cut] horse.

魚道 the regular course traveled by schools of fish (of a given species); 〔滝・ダムなどに設けた〕a fishway; (魚梯) a fish ladder.

許容 permission; allowance; acceptance; approval; tolerance. ★⇨許可 ▶〜する permit; allow; approve; tolerate. /〜しうる permissible; allowable. /〜できる誤差 an allowable [a permissible] error. /〜(される)範囲内で within permissible limits. /その数値は通常の〜範囲内にある[を超えている]. That figure [number, value] is within [over] the normal tolerance. ¶**許容限度** a tolerance limit; the maximum permissible limit. **許容摂取量** ⇨摂取量 **許容線量** 〔放射線の〕the (maximal [maximum]) permissible dose; a tolerance dose; the maximum safe dosage; a permissible [tolerable] level. **許容法規** a permissive regulation.

キラー海藻 ⇨英和 killer seaweed

禁煙 〔掲示で〕No Smoking. ★⇨分煙, 喫煙 ▶館内では〜です.

Smoking is prohibited inside the building. / 完全～．〔掲示で〕Absolutely No Smoking. | Smoking Completely Prohibited. |（ビル中禁煙）No Smoking in the Building(s)．/ 博物館内は全面～です．Smoking is not permitted anywhere in the museum. | There are no smoking areas in the museum. / 全面[完全]～列車 a completely nonsmoking train. / ～する give up [quit, abstain from] smoking. / 私は目下～中です．I'm off cigarettes at the moment. / 今日は(個人的に決めた)～デーだ．Today is my no-smoking day. / 世界～デー〔5月31日；世界保健機関(WHO)が定めた〕World No-Tobacco Day. ¶**禁煙運動** an antismoking [anti-tobacco] campaign. **禁煙エリア[コーナー]** a no-smoking [nonsmoking] area [corner]．**禁煙外来** smoking cessation outpatient services [treatment]．**禁煙ガム** stop-smoking gum; quit-smoking gum;〔ニコチンを含む〕nicotine gum. **禁煙週間** a no-smoking week. **禁煙先進国** an advanced country in terms of its non-smoking policies; an advanced anti-smoking nation. **禁煙サポート** smoking cessation support; support for quitting (smoking)．**禁煙サポート・サイト** 〔インターネット上の〕a smoking cessation support site. **禁煙車** 〔列車の〕a no-smoking [nonsmoking] car;（車室）a nosmoking [nonsmoking] compartment. **禁煙席** a no-smoking [nonsmoking] seat. **禁煙タクシー** a no-smoking taxi. **禁煙パイプ** a dummy cigarette. **禁煙表示** a no-smoking sign. **禁煙法**〔法律〕a no-smoking [an antismoking] law;〔たばこのやめ方〕a smoking cessation method; a method for quitting smoking. **禁煙補助具** a smoking cessation aid. **禁煙補助剤**[薬] a smoking cessation aid [drug]; an aid to stop smoking; an antismoking product. **禁煙療法** a smoking cessation treatment method.

近縁係数 a coefficient of relationship.

近縁種 a closely-related species.

近海 adjacent [neighboring] seas; coastal [home] waters. ¶**近海漁業** coastal [inshore] fishery [fishing]．**近海もの**[魚] a shore-fish.

近郊農業 agriculture in suburban areas.

近所迷惑 a nuisance [an embarrassment, an annoyance] to

(people in) the neighborhood; a source of inconvenience to (people in) the neighborhood. ★⇨近隣騒音 ▶～な causing a nuisance [annoyance, inconvenience] to (people in) the neighborhood. / ～近所迷惑になる become a nuisance [source of inconvenience, source of annoyance] to (people in) the neighborhood.

金属 (a) metal. 形 metallic. ★⇨アレルギー(金属アレルギー), 汚染(金属汚染), くず(金属くず) ¶**金属ごみ** metal waste. **金属スクラップ** metal scrap.

禁猟 prohibition of shooting [hunting]. ▶～の鳥獣 prohibited game. ¶**禁猟期** the closed [close] season. **禁猟区 [地]** a (game [hunting]) preserve; a no hunting zone;〔神苑などの〕a (wildlife [bird]) sanctuary.

禁漁 prohibition of fishing. ¶**禁漁期** the closed [close] season (for fishing). **禁漁区** an area closed to fishing; a no-fishing [no-take] area [zone]; a marine preserve.

近隣 ¶**近隣騒音** the din and bustle of the neighborhood. ★⇨騒音 **近隣型ショッピングセンター** a neighborhood shopping center. **近隣住区**〔人口に応じた生活関連施設を備えた，まとまりある住宅地域〕a residential area with a full range of social services; an integrated residential neighborhood. **近隣住民** neighboring residents《of a US army base》. **近隣商業地域**〔都市計画上の〕a neighborhood commercial district (to which certain building restrictions apply).

菌類 fungi. ¶**菌類学** mycology; fungology.

く

空気 air; the atmosphere. ▶汚れた〜 impure [foul, polluted] air / 〜が汚れている．The air [atmosphere] is contaminated [polluted]．/ 山の〜はうまい．The mountain air is bracing. ¶**空気清浄フィルター** an air (purifying) filter. **空気汚染** (indoor) air contamination [pollution]．**空気清浄機**[器，装置] an air cleaner [purifier].

空港 an airport. ¶**空港環境整備協会** the Airport Environment Improvement Foundation (略：AEIF)．**空港気象ドップラー・レーダー** airport doppler radar. **空港公害** environmental pollution from [caused by, due to] an airport. **空港周辺整備機構** the Osaka Airport Periphery Redevelopment Organization. **空港整備法** the Airport Development Law.

空中散布 ⇨散布

空調 air conditioning. ★⇨エアコン ¶**空調病** 〔空調から排泄される塵埃アレルギー〕air conditioner disease.

クーラー an air conditioner; an air-conditioning unit. ★⇨エアコン ▶〜がききすぎている．The air-conditioner is set too low. / 〜の冷気より自然の風のほうが心地よい．A natural breeze feels more pleasant than the cold air of air-conditioning. / 〜を弱める turn down the air conditioner.

クールビズ 〔環境省の夏の軽装キャンペーン〕"Cool Biz"; the Japanese government's program to encourage people to wear light clothes and use less air conditioning during the summer;〔ノーネクタイのビジネス服〕no-necktie office wear [clothes].

区画 〔市街などの〕a block;〔土地の〕a lot [plot, parcel] (of land)．▶〜する divide; partition; demarcate; draw a line (of demarcation)《between...》; delimit; mark off [out]《land》. ¶**区画漁業** demarcated fishery. ★⇨漁場 **区画漁業権** demarcated fishery rights. **区画整理** land readjustment [rezoning]; 〔都市の〕rezoning; replanning of streets; readjustment of town lots. / 〜する rezone (a site); readjust (ward boundaries). / 農地の〜整理 agricultural land readjustment. / 〜整理案 a zoning proposal [blueprint]. / 〜整理図 a rezoning map. **区画分譲地** a site of (housing) lots for sale; a rezoned site.

区系 〔動物相・植物相などの区〕a region. ¶ **植物区系** a floral region. **動物区系** a faunal [zoogeographic] region.

草刈り ▶〜する mow (the grass);〔鎌で〕cut the grass [weeds] with a sickle; make hay. ¶ **草刈り機** a mowing machine; a mower; a lawnmower; a grass cutter.

草地(くさち) ⇨草地(そうち)

草の根運動 a grassroots movement.

草の根交流 grassroots exchange; grassroots contact(s).

草むしり ▶庭の〜をする weed a garden; pull (out) weeds.

駆除 extermination; eradication; stamping out. ▶害虫を〜する exterminate [eradicate, stamp out] an insect pest. ★ ⇨害虫 / ネズミを〜する get rid of rats; rid 《a house》 of vermin. / ネズミ〜業者 a rat exterminator; a rat control company. ¶ **駆除剤** an expellant;〔害虫の〕an insecticide; a pesticide; an [a pest] eliminator; an eradicator. **駆除法** 〔シラミの〕a delousing method;〔シロアリの〕a method for exterminating [ridding 《a house of》] termites. **駆除対象動物** harmful animals [beasts] for killing.

苦情 a complaint; a grievance; an objection. ★⇨クレーム, 抗議 ▶〜が出てから慌てて対応する react hastily following a complaint. /〔担当部局が〕ごみ収集に関する〜をたくさん受けている receive a lot of complaints about garbage collection. / ほうぼうから〜が殺到した. We had a rush of complaints from everywhere. | Objections poured in from all quarters. / 正式に〜を申し入れる put in a formal complaint;〔当局に〕file a complaint with the authorities;〔裁判所・警察に〕lodge a complaint with the court [police];〔書面で〕register a (written) complaint. / 〜を処理する process [address, manage] a complaint. ★⇨すぐやる課 / 〜を無視する ignore a complaint. / 〜相談室〔警察などの〕a consultation section for citizens with problems. / 〜やお困りのことがあれば最寄りの相談室にお尋ねください. For complaints or help with problems, please visit your nearest consultation office. ¶**苦情承り係[所]** a complaint(s) department [section];〔担当者〕a complaint(s) officer. **苦情記入用紙** a complaint [grievance] form [slip]. **苦情請願**〔行政の怠慢などによって被った権利侵害などの訴え；参議院の行政監視委員会が審査する〕a petition on an administrative complaint. **苦情処理委員会**〔社内などの〕a grievance committee;〔全国的組織などの〕a complaints commission. **苦情処理係** a customer relations officer; a person responsible for [in charge of] handling customers' complaints; a troubleshooter. **苦情処理機関** an [a complaint] arbitration body;【労】a grievance (settlement) machinery. **苦情処理部** a complaints [customer relations] section [department]. **苦情電話[メール]** a complaining phone call [e-mail].

鯨(くじら) a whale. ★⇨捕鯨, 座礁死 ▶鯨が潮を吹く. A whale blows. | A whale spouts (water). / 鯨が水面におどり出た. A whale breached [surfaced]. / 鯨の子 a whale calf. / 鯨の群れ a gam [herd, pod, school] of whales. / 鯨の肉 whale meat. / 鯨の油 whale oil. / 鯨のひげ whalebone; baleen.

くず, 屑〔ごみ〕rubbish; refuse; trash; junk; scrap;（野菜の）waste; refuse; garbage;（繊維の）flock;（亜麻などの）hards [hurds];（パンの）crumbs;（木の）chips;（かんなくず）shavings;（おがくず）sawdust. ★⇨紙（紙屑）, ごみ ¶**くず入れ**〔駅や公園の〕a trash basket [can, box]; a dustbin. **くずかご** a wastebasket; a wastepaper basket [bin]. ▶～かごに入れる put《crumpled notes》into a wastebasket. **くず屋** ⇨廃品（廃品回収業者） **金属くず** scrap metal. **鉄くず, くず鉄** scrap iron; iron scraps;（やすりくず）iron filings. / ～鉄にする convert《a machine》into scrap iron; scrap《a machine made of metal》.

薬〔薬剤〕(a) medicine; a drug. ★⇨農薬, 殺虫剤 ¶**薬アレルギー** a drug allergy; an allergy to medicines. ▶今までに薬アレルギーを起こしたことがありますか. Have you ever had an allergic reaction to any drugs [medicine]? **薬漬け**〔患者への薬の過剰投与〕"pickling in medicine"; a doctor's excessive prescription of medicine; overprescription; overmedication. / 薬漬け医療を改善する reduce the overmedication of medical treatment.

国別登録簿システム 〔京都議定書に基づく，温室効果ガスの〕a national registry system.

組み換え ⇨遺伝子

グラウンド・カバー 〔地表を覆う地被植物〕ground cover. ¶ グラウンドカバー・プランツ〔グラウンドカバー用の園芸植物〕a ground cover plant.

グラウンドワーク 〔英国起源の市民・行政・企業が連携した地域環境の整備活動〕Groundwork.

クリアリングハウスメカニズム ⇨英和 Clearing-House Mechanism

クリーン ¶ クリーン・エネルギー〔環境汚染源とならないエネルギー〕clean energy; environmentally friendly energy (sources). ▶太陽エネルギーは〜エネルギーと言われる．The energy the sun produces [Solar energy] is said to be clean energy. / 〜エネルギーを産出する produce clean energy. **クリーン・カー**〔低公害車〕a low-emission vehicle; a "clean car." **クリーン開発と気候に関するアジア太平洋パートナーシップ** the Asia-Pacific Partnership on Clean Development and Climate (略: APP). **クリーン開発メカニズム** ★⇨英和 CDM **クリーン・コール・テクノロジー**〔環境低負荷型の石炭利用技術〕clean coal technology (略: CCT). **クリーン・ディーゼルエンジン** a clean diesel engine. **クリーン燃料** (a) clean fuel. **クリーン・ヒーター**〔室内の空気を汚さない暖房機〕a heater with a vent to the outside of the house; 【商標】a Clean Heater. **クリーン・ベンチ**〔微粒子や微生物を制御した実験台〕a clean bench. **クリーン・ルーム**〔半導体製造工場などの無塵室・病院の無菌室〕a clean room; a bio-clean room.

グリーン〔環境に配慮した；環境保護の〕green. ★⇨エコ **グリーン・インベスター**〔環境対策に積極的に取り組む企業に優先的に投資する投資家〕a green investor. **グリーン・エイド・プラン**〔途上国の公害防止のためにエネルギー環境技術の支援などを行う協力プログラム〕the Green Aid Plan (略: GAP). **グリーンエネルギー**〔太陽エネルギー・地熱・風力・水力など，汚染物質を出さないエネルギー〕green energy. **グリーン・カー**〔環境配慮型自動車〕a green car. **グリーン購入ネットワーク** the Green Purchasing Network (略: GPN). **グリーン購入法** the Green Purchasing

Law. ★正式名称は「国等による環境物品等の調達の推進等に関する法律」(the Law Concerning the Promotion of Procurement of Eco-Friendly Goods and Services by the State and Other Entities). ★⇨グリーン調達　**グリーン・コーディネーター**〔観葉植物などを使って室内空間を演出する人〕an interior foliage designer.　**グリーン・コンシューマー**〔環境に優しい商品を選択する消費者〕a green consumer; a consumer in favor of environmentally friendly products.　**グリーン・コンシューマリズム**〔環境に優しい商品を選ぼうとする消費者運動・意識〕green consumerism.　**グリーン・サービサイジングモデル事業**〔経済産業省が推進する，環境にやさしいビジネスを発掘支援する事業〕a Green Servicizing Model Business.　**グリーン産業**〔環境配慮型産業〕(a) green industry.　**グリーン証書**〔再生可能なエネルギー源による電力に政府が発行する証明書〕a green certificate; (正式名称) a tradable renewable energy certificate (略: TREC). ★⇨英和 green certificate　**グリーン・ジョブ**〔環境負荷の削減に貢献する活動〕a green (collar) job.　**グリーン税**〔自動車の〕a green tax.　**グリーン税制**　green taxation.　**グリーン調達**〔環境に配慮した部品や材料を調達すること〕green purchasing [procurement]; environmentally friendly [responsible] purchasing [procurement]; environmentally preferable purchasing [procurement] (略: EPP).　**グリーン調達調査共通化協議会**　the Japan Green Procurement Survey Standardization Initiative (略: JGPSSI).　**グリーン・ツーリズム**〔農村などで休暇を過ごすこと〕green tourism.　**グリーン・テクノロジー**〔環境配慮型技術〕green technology.　**グリーン電力**〔太陽光・風力など環境に悪影響を与えないエネルギーによる電力〕green power.　**グリーン電力証書(システム)**〔太陽光・風力など自然エネルギーにより発電された電力を証書化して取引するシステム〕a green power certificate (system).　**グリーン・パートナーシップ**〔製造業者が仕入れ業者などと提携して環境負荷の低減をめざす活動〕a green partnership.　**グリーンハウス**〔温室〕a greenhouse.　**グリーンハウス・イフェクト**　⇨英和 greenhouse effect　**グリーンピース**〔国際的環境保護団体〕⇨英和 Greenpeace　**グリーン・ビジネス**　indoor plant hire and sales.

グリーン物流 〔環境に優しい物流システム〕green logistics. **グリーン物流総合プログラム** 〔グリーン物流の構築を目指す施策〕the Green Logistics Comprehensive Program(s). **グリーン・プラ** 〔生分解性プラスチック〕(a) biodegradable plastic; (a) "green" plastic. **グリーンベルト** 〔都市周辺の緑地帯〕a greenbelt. **グリーンベルト運動** ⇨英和 Green Belt Movement **グリーン包装** 〔環境に配慮した包装のしかた〕green packaging. **グリーンマーク** 〔環境ラベル；古紙利用製品に表示される〕a "green mark"; a recycled-paper mark [logo]. **グリーン GDP** 〔環境汚染被害・自然資源の消耗などによる経済的損失を考慮して算出する GDP〕the green GDP. **グリーン GNP** 〔環境汚染被害・自然資源の消耗などによる経済的損失を勘案して求める GNP〕the green GNP. **グリーン IT** 推進協議会 the Green IT Promotion Council.

車 〔自動車〕a motorcar; an auto(mobile); a car. ★⇨自動車 ¶**車公害** 〔排気ガスによる〕auto(mobile) [motor-vehicle, vehicle] exhaust pollution. **車社会** an automobile society; a society dependent on cars; an automobile-[a car-]dominated society. **車離れ** (a) movement away from [loss of interest in] automobiles [cars, driving]. ▶若者の車離れにガソリン高騰が追い討ちをかけて自動車メーカーは軒並み大幅減益となった．The decreased use of cars by young people, combined with a sharp rise in gasoline prices, resulted in large across-the-board declines in automaker profits.

車椅子(いす) a wheelchair. ▶〜用のスロープ《build, install》a wheelchair ramp. / 〜で入れる浴室 wheelchair-accessible bathroom. / 〜バスケットボール wheelchair basketball. / 電動〜で走行する drive an electric wheelchair. / 足こぎ〜 a pedal-driven [foot-driven] wheelchair; a cycling chair. / 〜移動車 a wheelchair accessible vehicle (略: WAV); a wheelchair accessible car. / 〜用トイレ a wheelchair (accessible) toilet. / 〜ランナー a wheelchair competitor《in a marathon》.

クレーム 〔苦情・抗議・不満〕a complaint; an objection《to...》；〔損害賠償の請求〕a claim《for damages》．★⇨苦情, 抗議 ▶顧客[消費者]からの〜 a customer [consumer] complaint. / その番組に動物愛護団体から〜がついた

[来た]. A complaint was lodged against the TV show by an animal rights group. / ～の電話 a phone call from a person with a complaint. / 商品に～をつける complain [make a complaint] about a commodity. / 損害賠償の～を拒む reject a claim for damages. ¶**クレーム隠し** covering up [hiding] complaints《about defective vehicles》. / A社は会社ぐるみで～隠しを行っていた. Company A implemented a company-wide cover-up of the complaints. | Company A as a whole covered up the complaints. **クレーム処理** settlement of complaints. **クレーム処理係** a complaints clerk. **クレーム処理部門** the complaints department.

グローバル ▶～な〔世界的な・地球的な〕global《views》; worldwide. / ～な市場 a global market. / ～な見方をする have a global perspective [point of view]. / これからは地球温暖化問題のように, ますます～な発想が要求される. From now on global thinking will be more and more demanded, as happened with the problem of global warming. ¶**グローバル化** globalization. / 経済の～化 globalization of the (world) economy **グローバル・ガバナンス**〔国家を超えた問題解決への取り組み〕global governance. **グローバル企業** a global corporation. **グローバル・ギャップ**〔世界的な適正農業規範〕⇨英和 GLOBALGAP **グローバル・コンパクト**〔国連の〕the (United Nations) Global Compact. **グローバル作物多様性トラスト** the Global Crop Diversity Trust. **グローバル・スタンダード**〔世界的な標準〕a global standard;〔国際基準〕an international standard. **グローバル・パートナーシップ**〔世界的な問題解決のための国家間の協力関係〕(a) global partnership. **グローバル・ビレッジ**〔通信手段の発達で狭くなって1つの村のようになった世界; 地球村〕a global village.

クローン a clone. ▶羊の～をつくる clone a sheep. ¶**クローン技術** cloning technology. **クローン（技術）規制法** the Human Cloning Techniques Regulation Law. ★正式名称は「ヒトに関するクローン技術等の規制に関する法律」(the Law Concerning Regulation Relating to Human Cloning Techniques and Other Similar Techniques). **クローン実験** a cloning experiment. **クローン動物** a cloned animal. **クローン人間** a human clone; a cloned human being. クローン胚 a cloned embryo. / ヒト〜胚 a cloned human embryo. クローン羊[牛] a cloned sheep [cow]. クローン・ペット 〔クローンの愛玩動物〕a pet clone; a cloned pet. クローンES細胞 a cloned embryonic stem cell. 再クローン a reclone; recloned 《cattle》.

黒潮(くろしお) the Japan [Kuroshio] Current. ¶黒潮系 the Kuroshio System. 黒潮続流 the Kuroshio Extension.

クロルデン 〔殺虫液〕chlordan(e).

クロルピリホス 〔農薬〕chlorpyrifos.

クロロフィル 〔葉緑素〕chlorophyll.

クロロフルオロカーボン a chlorofluorocarbon（略: CFC）. ★⇨フロン

群居 gregariousness; the tendency to herd together; aggregation. ★⇨群れ ▶〜する herd together; live gregariously. / 〜を好む鳥類 birds that flock together; social birds. ¶**群居性** gregariousness; sociability. / 〜性の gregarious; social. / 〜性が強い be highly gregarious. 群居単位 a social unit. 群居動物 a gregarious animal. 群居本能 the herd instinct; (inherent) gregariousness.

群集 〔生物の〕a (biotic) community. ★⇨群れ ¶**群集生態学** synecology. 競争群集 〔植物の〕competitive association. すみ分け群集 〔植物の〕complementary association.

燻蒸(くんじょう) 〔いぶして消毒すること〕fumigation. ▶〜する fumigate; smoke (out). / 硫黄で〜(消毒)する sulfur; sulfurize; fumigate [smoke out]《a room》with sulfur. ¶**燻蒸剤** 〔殺虫剤〕a fumigant. 土壌燻蒸剤 a soil fumigant. 燻蒸消毒器 a fumigator.

群生 ▶〜する grow gregariously;〔植物が〕grow in profusion; (牛・豚が) herd together; (鳥・羊が) flock together; live in flocks; (ハチが) hive. / (蟻が巣を作って)〜する live together in colonies. ¶**群生植物[動物]** gregarious [social] plants [animals].

群落 a (plant) community. ▶〜を作る live in community. ¶**群落生態学** ⇨群集生態学 群落適合 ⇨適合

け

警戒 ▶天候の急変を〜する watch out [be on the alert, be on the lookout] for a sudden change of weather. / 河川の氾濫を〜する watch out [be on the alert] for river flooding. / 台風のもたらす大雨に〜が必要だ．Precautions need to be taken against heavy rainfall brought on by the typhoon. / 徹夜で火山噴火活動の〜に当たる stand watch all through the night for any volcanic eruption(s). / 津波の恐れあり．沿岸部は〜を要する．There is a threat of a tsunami. Coastal areas must be on their guard. / 台風の来襲に備えて24時間態勢の〜を敷いている．We are on 24-hour alert awaiting the typhoon's assault. ¶ **警戒色** aposematic [warning] coloration; 〔特定の色〕a warning [an aposematic] color. **警戒水位** the warning level of a river; a warning water level. / 特別〜水位 a special river level warning; a special warning water level. / 川の水が〜水位を越えた．The waters of the river passed [went over, rose above] the warning level. **警戒声** 〔鳥の発する〕an alarm note [call]. **警戒宣言** (a) proclamation [(an) issuance] of a warning [an alert]; a warning. / 気象庁から大地震発生の兆候の連絡を受けると, 総理大臣は直ちに〜宣言を発することになっている．When the Prime Minister is informed by the Meteorological Agency of signs of a major earthquake, he immediately issues a warning to the public. **警戒値** 〔医療・防災・保安等に関連する〕an alarm [a warning] trigger level; a warning [an alarm] threshold. / 〜値に達する reach a level high enough to trigger a warning; be high enough to set off an alarm. **警戒フェロモン** 〔仲間に危険を伝える〕an alarm pheromone. ★⇨警報

計画 a plan; a project; a scheme. ▶〜する plan; make a plan. / 降って沸いたようなダム建設の〜 a plan to build [construct] a dam that comes [from] out of the blue. / 〜中の新校舎[事業] a new school building [enterprise] under consideration [in the works]. / 〜が中止になった. The plan was cancelled [called off]. / 工場の移転〜が発表された. A plan to move the factory was [has been] announced. / 当初の〜ではここにダムを作る予定だった. According to the original plan, a dam was to be built here. / 都市再建〜を立てる draw up a plan for rebuilding [reconstructing] a city. / 町はここにゴミ処理場を建設する〜だったが, 住民の反対にあって断念した. The town(ship) had planned to construct a waste disposal plant here, but the plan was abandoned because of opposition from residents. / ダム建設はまだ〜段階にある. The dam construction is still in the planning stage. / ゴルフ場の開発〜 a golf course development project. / ビルの建築〜 the plans for constructing buildings [building constructions]; building construction plans. / 発電所の誘致〜 a plan to attract a power plant [station]《to this prefecture》. ¶**計画高水流量** design flood discharge. **計画停止** 〔発電所などでの〕a planned shutdown.

景観 a landscape; a scene; scenery; a scenic spot [view]; a sight; a view; a spectacular sight [view]. ▶都市の〜 a cityscape / 開発の名目で〜を破壊する destroy a scenic spot [landscape] under the pretext of development. / そのような建物は自然の〜を損なうことになろう. Such buildings would spoil the natural landscape [scenery]. ¶**景観アセスメント** (a) landscape [scenery] assessment. **景観汚染** ⇨汚染 **景観規制** aesthetic zoning (regulations). **景観工学** landscape engineering. **景観権** 〔生活空間の中で良好な景観的利益を享受する権利〕the right to a view. **景観重要建造物** 〔景観法に基づいて指定される〕an object or structure of scenic importance. **景観条例** a landscape ordinance. **景観税** ⇨英和 view tax **景観地区** an area of outstanding beauty; a scenic zone. **景観(の)利益** the benefits of a view. **景観破壊** scenic destruction; destruction of (the) scenery. **景観法** the Landscape Law. ★関連する二つの法をあわせて「景観緑(みどり)三法」the three laws concerning the conservation of green space and landscapes と総称される. **景観保全[保護]** scenic [landscape] preservation; preservation of scenic spots [the landscape]. **景観まちづくり条例** an urban landscape ordinance.

警告表示 〔家電製品の〕a warning label;〔たばこの箱などの〕⇨健康(健康警告表示).

経済 ▶〜至上主義 the belief that the economy should come first [is more important than anything else]; an [the] "economy-first" principle. / 自然保護より〜効率を優先する[追求する] give priority to [demand] economic efficiency over ecological conservation. / 〜効率だけで農業は語れない．One cannot speak of agriculture in terms of economic efficiency alone. | Agriculture is not an issue that can be dealt with simply by looking at the economics (of it).

警報 a warning (signal); an alarm (signal); an alert. ★⇨警戒 ▶今後の〜に十分注意してください．Please be fully alert for any future warnings. / 〜を伝える[鳴らす] raise [sound] an alarm《for a fire》; give warning;〔大声で〕shout an alarm;〔ベルを鳴らして〕ring an alarm bell. / 〜を解除する give [sound] the all clear. / 〜を発令する issue [post] a《flood》warning; issue [post] an alert. / 〜発令中．An alert is now in effect [has been issued, has been posted]. ¶ **警報装置** an alarm device [system]; a warning device. **警報物質**[フェロモン]【生態】an alarm substance [pheromone]. **大雨[大雪]警報**《issue》a heavy rain [snow] warning. **火災警報** a fire alarm. **空襲警報** an air-raid alarm [warning]. **非常警報** an emergency alarm (signal). **暴風雪[洪水]警報** a snowstorm [flood] warning. / 神奈川県全域にただいま洪水〜が発令中です．A flood warning is now in effect [has been issued, has been posted] for all of Kanagawa Prefecture.

軽油 〔ディーゼル燃料〕diesel oil [fuel];〔軽質油〕light oil;〔石油軽油・ガス油〕gas oil. ★⇨低硫黄(ていいおう) ¶ **軽油代替燃料** (an) alternative diesel fuel. **軽油発電機** a diesel generator. **軽油引取税** (a) diesel oil delivery tax. **不正軽油** adulterated diesel oil.

ケース・スタディ 〔事例[症例]研究〕a case study. ▶環境保護のための１つの〜になる運動 a movement that will be a case study for environmental protection. / 〜を行う conduct a case study.

毛皮 〔服飾用の〕(a) fur;〔生皮〕a skin; a pelt; a hide; a fell;〔犬・猫などの〕a coat;〔羊などの〕a jacket; a fleece. ▶〜反対運動〔動物愛護団体などの〕the anti-fur movement. / 〜をとるために狐を狩る hunt foxes for their skins.

下水 〔汚水〕black [foul, waste] water; sewage;〔下水設備・下水道〕sewerage; drainage; the drains; a sewer system;〔下水管〕a sewer; a drain;〔溝〕a ditch;〔人道と車道の間の〕a gutter;〔道路・鉄道の下などを横切る〕a culvert. ▶〜が詰まった. The drain is blocked [obstructed]. / 雨水を〜に流す run rainwater (off) into [with] the sewage. / 〜を掃除する clean [scour] a drain. / 〜を作る lay [dig] a drain [sewer]. ¶ **下水汚泥** sewage sludge. / 〜汚泥リサイクル sewage sludge recycling. **下水管** a drainpipe; a sewer pipe; a sewer; a wastepipe;〔排水管〕a cesspipe. **下水管渠** a sewer. **下水溝** a sewer; a drainage ditch [channel]. **下水口** a sewage [waste] outlet; an outfall; a gully hole. **下水工事** drainage [sewerage] works. **下水工事人** a drain [sewer] digger. **下水浄化** sewage purification. **下水消化ガス** sewage [wastewater] digester gas. **下水処理** sewage treatment [disposal, processing]. / 〜処理施設 a sewage treatment facility. / 〜処理場 a sewage treatment plant; a sewage works. / 〜処理能力を上回る雨量 an amount of rain beyond the capacity of the drains; rainfall that the drains cannot cope with. **下水溜め** a cesspool; a cesspit; a sink(hole). **下水道** a drain; the drains; drainage; a sewer; the sewers; sewerage; a sewer system. / 〜道の整備 installing [laying (out)] a sewerage system. / 〜道普及率 a sewerage coverage ratio. / 〜道完備〔掲示〕With full mains sewerage. / 〜道管 a sewer; a sewer [sewage] pipe. **下水道法** the Sewerage Law. **私設下水** a house sewer.

欠陥 a defect; a fault; a flaw; a shortcoming; an imperfection; a deformity. ▶その車はエンジンに〜がある. There is a fault in the car's engine. | The car has a defective engine. ¶ **欠陥隠し** 〔メーカーなどによる〕concealment of [concealing] defects. **欠陥建築(物)** a defective [《口》jerry-built] building [construction, structure]. 欠

陥車 a defective [faulty] car;《口》a lemon. **欠陥住宅** a defective house;〈集合的に〉defective housing. / 〜住宅保険 construction-[building-]defect(s) insurance (for private housing). **欠陥商品**《recall》a defective [faulty] product;《口》a lemon;〈集合的に〉defective merchandise [goods].

ケミカル 〔化学的な〕chemical. ¶ **ケミカル・タンカー** 〔化学薬品を運ぶ〕a chemical tanker. **ケミカル・リサイクル** 〔化学原料としての再利用〕chemical recycling.

煙 smoke; fumes. ▶たばこの煙 cigarette [cigar] smoke; tobacco smoke ★⇨喫煙 / 他人の吸うたばこの煙 smoke from a cigar [cigarette] somebody else is smoking;〔副流煙〕sidestream smoke. / たばこの煙が部屋にこもっていた. A cloud of tobacco [cigarette, cigar] smoke hung in the room. / たばこの煙を吐く blow cigarette [cigar] smoke. / ダンプカーが黒い煙をまき散らしながら走り去った. A dump truck sped away, spewing a trail of black smoke. ¶ **煙感知器[探知器]** a smoke detector;〔警報器〕a smoke alarm. **煙公害** ⇨煙害(えんがい) **煙フィルター** a smoke(-moderating) filter.

ゲリラ豪雨 a sudden heavy downpour [cloudburst] causing big damage; a "guer(r)illa downpour."

ゲリラ雪 a sudden heavy snowfall restricted to a small area (somewhat like a guer(r)illa attack).

原因 a cause《of...》;〔要因〕a factor《in...》;〔根源〕the origin [source, root]《of...》. ★⇨因果関係 ▶地球温暖化の〜 the origins [causes] of global warming. / なんらかの〜でその薬剤が食品に混入した可能性がある. It's possible that the chemical somehow found its way into the food. / 〜を調査する investigate the cause《of...》. / 〜を突き止める[特定する, 明らかにする] trace the origin(s)《of...》; trace ... (back) to its origin(s); establish [determine, identify, clear up] the cause《of...》. / これがこの事件の〜だ. This is at the bottom of the affair. / 〜不明の病気 a disease without a known cause; a disease of unknown etiology. ¶ **原因企業** 〔公害の〕a company that causes《pollution》; a company to blame《for emitting pollution》; a source《of pollution》. **原因物質** a causative agent; 〔アレルギーの〕an allergen.

検疫 quarantine; (a) quarantine inspection. ▶～する quarantine; put in quarantine; inspect for contagious diseases [pests]. / ～中である be in quarantine. / 海外からの動植物の持ち込みは空港で～が必要となる．Animals and plants brought in from overseas require a quarantine inspection at the airport. / ～のため上陸を禁止される be held [detained] in quarantine. / 厳重な～を受ける be subjected to a rigorous quarantine inspection. / ～を開始する institute [start] quarantine 《on a vessel》. / コレラ多発地帯からの入国者に～を行う carry out a quarantine inspection of arrivals from cholera-stricken areas. / ～済み．〔掲示〕Passed Quarantine Inspection. ¶**検疫官** a quarantine officer. **検疫感染症**[**伝染病**] 〔検疫の対象となる〕a quarantinable infectious disease. **検疫旗** a quarantine [yellow] flag. **検疫期間** a quarantine period. **検疫規則** quarantine regulations. **検疫(探知)犬** a customs [quarantine] dog. **検疫港**[**泊地**] a quarantine port [anchorage]. **検疫所** a quarantine station; a lazaretto. **検疫証明書** a quarantine certificate. / 植物～証明書 a phytosanitary certificate. **検疫信号** a quarantine signal. **検疫船** a quarantine ship; a lazaretto《複数形 -s》. **海上**[**陸上**]**検疫** maritime [land] quarantine inspection. **機内検疫** (an) onboard quarantine inspection《of a flight》(before disembarkation after landing). **空港検疫** (a) quarantine inspection at an airport. **国境検疫** (a) quarantine inspection at a frontier. **船内検疫** (an) onboard inspection carried out by quarantine officials. **輸入**[**輸出**]**検疫** an import [export] quarantine.

嫌煙 a nonsmoker's antipathy to tobacco smoke; a nonsmoker's dislike of smoking [tobacco smoke]. ¶**嫌煙運動** an antismoking campaign. **嫌煙家** a (militant) nonsmoker; an anti-smoker. ▶～家と愛煙家の反目 antagonism [animosity] between smokers and nonsmokers. **嫌煙権** nonsmokers' rights; the rights of nonsmokers.

限界削減費用 〔環境汚染物質の〕(a) marginal abatement cost (略: MAC). ★⇨ごみ(ごみ処理)

限外濾過(げんがいろか) ultrafiltration. ¶**限外濾過膜** an ultrafiltration membrane.

嫌気性 ▶〜の anaerobic. /通性[偏性]〜性のバクテリア facultative anaerobic [strictly anaerobic, obligatory anaerobic] bacteria. /〜(細)菌 anaerobic bacteria; anaerobes.

元凶 the main [chief] culprit; the main cause [source]. ▶大気汚染の〜を断つ cut off the main source of air pollution. /産業廃棄物が環境破壊の〜となった. Industrial wastes have become the main culprit behind [cause of] environmental destruction.

健康 health; fitness;〔健全さ〕healthiness; wholesomeness. ★⇨喫煙 ▶〜な healthy; sound; well; fit. /〜によい[悪い] be good [bad] for the health; be healthy [unhealthy]. /〜に有害な物質 a substance detrimental to health. /住民の〜を守る guard [protect] the health of the population. ¶**健康影響調査**〔公害などの〕《conduct》a health effects survey. **健康影響評価** a health impact assessment (略：HIA). **健康管理士** a certified health care manager. **健康管理手帳**〔危険有害業務に従事した労働者に交付される〕a health care handbook. **健康強調表示**〔食品の〕a health claim. **健康グッズ[用品]** health(-related) products [goods]. **健康警告表示**〔たばこの箱の〕a《cigarette》health warning (message). **健康教育** health education. **健康権** the [a] right to health. **健康産業** the health industry; the health care industry. **健康サンダル** health sandals. **健康志向** health consciousness; health awareness. /最近，年齢・性別を問わず〜志向が高まっている. Recently health awareness has been increasing among both sexes and all age groups. /〜志向商品[食品] health-oriented products [foods]. **健康指標** a health index. **健康弱者** the (physically) vulnerable; (physically) vulnerable people. **健康週間** Good Health Week. **健康住宅** healthy housing. **健康寿命** (a) healthy life expectancy. /平均〜寿命 (an) average healthy life expectancy《of 70 years for children born in 1999》. **健康食** a healthy meal; a healthy diet; healthy food. /〜食を取っている be on a healthy diet; be eating healthy food. **健康食品** (a) health food; healthy [wholesome] food. / 〜食品店 a health food store [shop]. **健康診査** a health [medical] check-up [examination]. **健康診断** a medical [physical, health] examination [checkup]. / 会社[職場]の〜診断 a medical examination arranged by *one's* company [conducted at *one's* workplace]. / 児童生徒のために年一度の〜診断が義務づけられている. Conducting an annual medical examination for all pupils and students is compulsory. / 〜診断でひっかかる have something found wrong with *one* in a medical examination / 〜診断を受ける undergo a medical examination [a physical examination] / 〜診断をする conduct a health examination; check (up) on *sb's* health. / 定期〜診断 a regular [periodic(al)] medical examination [checkup, physical]. / 〜診断結果 the results of a medical examination. / 〜診断書 a certificate of health. **健康水準** a health standard. **健康増進** promotion of health. **健康増進センター** a health-promotion center. **健康茶** health tea. **健康調査** a health survey 《of first-graders》. **健康度** health status. / 〜度評価 (a) health status assessment [evaluation]; (an) assessment [evaluation] of *sb's* health status. **健康配慮義務**〔被雇用者に対する雇用者の〕the obligation [duty] to ensure worker health; the responsibility to maintain a healthful workplace. **健康配慮義務違反** a breach of the obligation to maintain a healthful workplace. **健康保持** health maintenance; the maintenance [preservation] of (good) health. / 〜保持に必要な条件 the requisites for good health. / 〜保持用摂取品 a health (maintenance) supplement. **健康被害** health damage 《from pollution》. **健康被害救済制度** a system to assist people whose health has been damaged 《by pollution》. **健康余命** healthy life expectancy (略: HLE).

検査 (an) inspection; (an) examination; a test; a checkup; a check;〔機械・船舶などの〕an overhaul;〔繊維の〕conditioning;〔器具・物品の〕verification. ▶〜する inspect; examine; make an examination of…; check (up on…); test; overhaul《a machine》. / 目で〜する visually check [inspect, examine]《the condition of a cable》. / 計量器を〜する check scales and measures. /〜の結果…であることが判明した．As a result of the test it was proven that.... | The examination showed .... /〜を受ける undergo [go through] an examination; be examined; be inspected; be subjected to inspection; present《*one's* automobile》for inspection; get tested《for hepatitis B》/〜を[に]通る，〜に合格する pass an examination [an inspection, a test];〔規格に合う〕come [measure] up to the standards; be okayed. /〜済み．〔表示〕Examined [Inspected]. | OK. /〜中で under examination [inspection]; being examined [inspected, tested, checked]; under review. ¶検査穴[ゲージ, ピット] an inspection hole [gauge, pit]. **検査員** an inspector; a surveyor. **検査技師** a testing [laboratory] technician;〔医療関係の〕a clinical examination technician. **検査技術** testing technique(s). **検査キット**〔病気などの〕a diagnosis kit. **検査合格証** a certificate of passing inspection. **検査所** an inspecting office [station]. **検査証** a test certificate. **検査体制** an《meat》inspection system. **検査報告書** an inspection [examination] report. **検査忌避**〔監督官庁の検査に対する非協力や妨害〕evasion of inspection; evading an inspection. **検査済証**〔竣工時の〕a certificate of completion. **検査データ** inspection data. **検査被曝** diagnostic radiation exposure. **検査妨害** obstruction of [interference with] inspection; obstructing [interfering with] an inspection. **検査ライン**〔製造工場などでの〕an inspection line. **検査率**〔輸入食品などに対する〕the inspection rate《for imported beef》. **抜き取り検査** sampling; a sampling inspection; a spot check. **品質検査** quality inspection. **目視検査**《do, make》a visual inspection [examination, check]《of the products》.

原材料 ⇨原料

原産 ▶熱帯アジア〜の鳥 a bird native [indigenous] to tropical Asia. /ユーカリはオーストラリア〜である．The eucalyptus is native to [originally comes from] Australia. ¶**原産国** the country of origin. **原産国表示**〔表示すること〕country-of-origin labeling (略：COL, COOL);〔その表示〕a country-of-origin label. **原産地** the place [country] of origin; the provenance; the home; 〔動植物の〕the habitat. ★⇨産地 / コーヒーの〜地 the (original) home of the coffee plant. / 〜地不明の動植物 plants and animals of unknown origin. / ジャガイモは南アメリカが〜地といわれている. South America is said to be the country of origin [place of origin, home] of the potato.

原産地規則 the rule of origin; the local content rule. **原産地証明書** a certificate of origin [provenance]. **原産地表示** 〔食品などの〕an indication of (the place of) origin. **原料原産地表示** 〔加工食品の〕an indication of (the place of) origin of ingredients. ★⇨原料

原子 an atom. ★原子核(実験[反応, 分裂, 融合など])⇨核 ▶〜の atomic. / 〜を分裂させる split the atom. ¶**原子雲** an atomic cloud; a mushroom cloud. **原子エネルギー** atomic [nuclear] energy. **原子核エネルギー** nuclear energy. ★⇨原子力, 原子炉 **原子核工学** nucleonics. **原子核物理学** nuclear physics. **原子物理学者** nuclear physicist. **原子核模型** a nuclear model. **原子化熱** heat of atomization. **原子爆弾** an atomic [atom] bomb; an A-bomb. ★⇨核(核爆弾, 核兵器) **原子放射線の影響に関する国連科学委員会** the United Nations Scientific Committee on the Effects of Atomic Radiation (略: UNSCEAR). **原子力** atomic energy; nuclear power. / 〜力で動く atomic-powered; nuclear-powered 《submarines》. / 〜力を平和目的に役立てる[平和利用する] harness the atom for peaceful purposes. / 〜力の平和利用 the peaceful use of atomic [nuclear] energy; the use of atomic [nuclear] energy for peaceful purposes. / 〜力平和利用計画 an atoms-for-peace program. / 〜力平和利用三原則 the three principles on the peaceful use of atomic energy (in Japan). ★⇨原子力三原則 **原子力安全委員会** the Nuclear Safety Commission (略: NSC). **原子力安全白書** 〔内閣府の〕the《2004》White Paper on Nuclear Safety. **原子力安全・保安院** 〔資源エネルギー庁の〕the Nuclear and Industrial Safety Agency (略: NISA). **原子力安全基盤機構** 〔独立行政法人〕the Japan Nuclear Energy Safety Organization (略: JNES). **原子力委員会** 〔日本の〕the Atomic Energy Commission. ★米国の the Atomic Energy Commission (略: AEC)は1974年廃止さ

れ，エネルギー研究開発局 the Energy Research and Development Administration（略：ERDA; 1975-77）と原子力規制委員会 the Nuclear Regulatory Commission（略：NRC; 1975-）とに改組．**原子力委員会及び原子力安全委員会設置法** the Law for Establishment of the Atomic Energy Commission and the Nuclear Safety Commission. **原子力衛星** a nuclear-powered satellite. **原子力エネルギー** atomic energy. **原子力エネルギー協会**〔米国の〕the Nuclear Energy Institute（略：NEI）. **原子力エンジン** an atomic engine. **原子力開発利用長期計画** a long-term program for research, development and utilization of nuclear energy. **原子力管理** atomic energy control. **原子力基本法** the Atomic Energy Basic Law. **原子力供給国グループ** the Nuclear Suppliers Group（略：NSG）. **原子力協定** an atomic energy agreement. **原子力計画** an atomic power plan [program]. **原子力公害** nuclear pollution. **原子力工学** nuclear engineering. **原子力航空母艦[空母]** a nuclear (aircraft) carrier. **原子力災害対策特別措置法** the Law on Special Measures Concerning Nuclear Emergency Preparedness. **原子力産業** the nuclear industry. **原子力三原則** the three fundamental rules of atomic energy（: autonomous management; democratic operation; freedom of information）. **原子力時代** the (post) atomic age; the nuclear age. **原子力推進**〔潜水艦などの〕atomic [nuclear] propulsion. /〜力推進の nuclear-propelled. **原子力政策大綱**〔原子力委員会が2005年に策定〕the Framework for Nuclear Energy Policy. **原子力設備利用率**〔電力会社などの〕a facility utilization rate at a nuclear power facility. **原子力船** a nuclear-powered vessel [ship]. **原子力潜水艦** a nuclear(-powered) submarine; an N-sub(marine). **原子力戦争** an atomic war. **原子力損害賠償法** the Atomic Energy Damage Compensation Law. **原子力損害賠償補償契約法** the Atomic Energy Damage Compensation Indemnification Contract Law. **原子力電池** a nuclear battery; a radioisotope [radioisotopic] battery. **原子力廃止措置機関**〔英国の〕the Nuclear Decommissioning Authority（略：NDA）. **原子力白書**〔内閣府の〕the White Paper on

Nuclear Energy《2004》. **原子力発電** atomic power generation. / ～力発電のウエートを高めることによって火力発電による環境破壊を抑える contain the environmental destruction caused by thermal power generation by putting more emphasis on nuclear (electric) power generation. **原子力発電環境整備機構** the Nuclear Waste Management Organization of Japan（略: NUMO）. **原子力発電機** an atomic power generator. **原子力発電技術機構** the Nuclear Power Engineering Corporation（略: NUPEC）. **原子力発電所** an atomic [a nuclear] power station [plant]. ★⇨原子炉, 原発, 建設 **原子力飛行機** a nuclear(-powered) plane [aircraft]. **原子力保険** nuclear [atomic] energy insurance. **原子力ロケット** a nuclear-powered rocket. **原子力防災センター** 〔原発などの緊急事態応急対策拠点施設〕an off-site center. ★⇨耐震 **原子炉** a (nuclear [fission chain]) reactor; a (nuclear) pile. / ～炉を廃止[廃炉に]する decommission a nuclear reactor. / 発電用～炉 a nuclear power reactor. / ブリーダー型～炉 a breeder reactor. / 湯沸かし[コールダーホール, 水泳プール]型～炉 a water boiler [Calder Hall, swimming-pool] type reactor. **原子炉衛星** a nuclear-powered satellite. **原子炉事故** a nuclear reactor accident; an accident in a nuclear reactor. **原子炉格納容器** a (nuclear) reactor containment vessel. **原子炉隔離冷却系** 〔原子炉の給水系停止時に炉心に給水を行う設備〕a reactor core isolation cooling system（略: RCIC）. **原子炉工学** (nuclear) reactor engineering. **原子炉出力** reactor power. **原子炉等規制法** the Nuclear Reactor Regulation Law. ★正式には「核原料物質, 核燃料物質及び原子炉の規制に関する法律 the Law for the Regulation of Nuclear Source Material, Nuclear Fuel Material and Reactors」という.

検出 ▶砒素の痕跡を～する detect a trace of arsenic. / 化学分析で～可能な detectable by chemical analysis. ¶ **検出器[装置]** a detector. **検出限界** the detection limit; the limit of detection. **検出濃度** (a) measured [detected] concentration《of dioxin》. **検出力** the ability to detect ….

原始林, 原生林 a virgin [primeval] forest.

建設 construction; building; erection. ★⇨建築 ▶〜する construct; build; put up; erect; raise; lay out 《a baseball field》; lay down 《a highway》; make. / 空港［ビル］を〜する construct an airport [a building]. /（建物が）〜中だ be under construction; be going up. / 橋の〜が急ピッチで進んでいる. Construction of [on] the bridge is progressing rapidly. /（工場などの）〜を認可する approve [authorize] (the) construction 《of a factory》. ¶**建設・運転一体認可**〔米国における原子力発電所建設についての原子力規制委員会（NRC）の認可方法〕a Combined Construction and Operating License（略: COL）; a COL. **建設基準法** the Building Standards Law（略: BSL）. **建設業許可**〔建設業法に基づく〕a construction license. / 一般〜業許可 an ordinary construction license. / 特定〜業許可 a special construction license. **建設工事紛争審査会**〔各都道府県に1つずつ設置〕a [the] Committee for Investigation of Construction Disputes. / 中央〜工事紛争審査会〔国土交通省に設置〕the Central Committee for Investigation of Construction Disputes. / 都道府県〜工事紛争審査会 Prefectural Committees for Investigation of Construction Disputes. **建設残土** surplus soil from a construction site. **建設廃棄物［廃材］**《a dumping ground for》construction debris. **建設副産物**〔建設工事に伴い発生する再生資源と廃棄物〕construction by-products. **建設（資材）リサイクル法** the Construction Recycling Law. ★正式名称は，「建設工事に係る資材の再資源化等に関する法律」(the Law concerning Recycling, etc. of Materials from Construction Work).

健全 ⇨健康 ▶〜な食生活 healthy eating habits. / 〜な生育 healthy growth. / 心身の〜な発達 (the) sound development of body and mind. ¶**健全性指標**〔河川・土壌などの〕an environmental health index. **健全性評価制度**〔原子力発電設備の〕an integrity evaluation system for nuclear power plants.

建造物 a building; a structure; an edifice. ¶**建造物損壊罪** the crime of causing destruction to buildings and structures. **建造物緑化** making buildings green; the greening of buildings; decorating buildings with greenery. **歴史的建造物** a historic structure.

建築 〔建てること〕construction; building; erection;〔建築物〕a building; a structure;〈集合的に〉architecture. ★⇨建設 ▶~する build; construct; erect; put up. ¶建築確認 building confirmation. / ~確認申請 a building-confirmation application. / ~確認制度 the building-confirmation system. / ~確認審査[検査] an inspection to confirm that a building conforms to building standards. 建築環境・省エネルギー機構 〔財団法人〕the Institute for Building Environment and Energy Conservation (略: IBEC). 建築基準 building standards. 建築基準法 the Building Standards Law. / ~基準法に違反する violate the Building Standards Law. / ~基準法の改正[緩和] revision [relaxation] of the Building Standards Law. 建築規制 building regulations. 建築業 the building [construction] industry; the building trade. 建築業者 a builder; a constructor. 建築協定 〔建築基準法で定める〕a building agreement. 建築許可 a building [construction] permit. 建築限界 the construction gauge; the clearance limit; the track clearance. 建築限界図 a clearance diagram. 建築現場 a construction site. 建築工学 architectural engineering. 建築工事 construction work. 建築工事請負契約 a contract (agreement) for construction work. 建築構造 building construction. 建築コンペ 〔設計競技〕a (building-)design competition. 建築材料 building [construction] material(s). 建築士 a registered [licensed] architect; an authorized architect and builder. / 一級[二級]~士 a first-class [second-class] registered architect. 建築条令 a building-restriction ordinance. 建築税 a tax on construction; a building tax. 建築制限 building restrictions. 建築統制 building control [regulation]. 建築主 an owner; a client. 建築物環境衛生管理基準 〔ビル衛生管理法に定める〕standards for the environmentally hygienic management of buildings. 建築物緑化 making buildings green; the greening of buildings. 建築物総合環境性能評価システム ⇨キャスビー 建築法規 the building code. 建築面積 building area. 建築模型 an architectural model. 建築用地 a building [construction] site;〔住宅の〕a housing lot.　建築残材　waste (products) from construction; building waste.　建築廃材[廃棄物]　《illegal dumping of》construction debris.　建築防火工学　fire protection engineering.　建築防災学　disaster prevention engineering.

減農薬　¶減農薬栽培　cultivation using reduced amounts of agricultural chemicals.　減農薬農作物[米, 野菜]　farm products [rice, vegetables] grown with reduced amounts of agricultural chemicals.

原発　〔原子力発電所〕a nuclear power plant [station].　★⇨原子　▶～の建設に反対する oppose (the) construction of a nuclear power plant. / ～を誘致する invite 《an electric power company》to build a nuclear power plant《in one's prefecture》; try to have a nuclear power plant built. / 反～活動家[運動] an antinuclear activist [movement].　原発銀座　a string of nuclear power plants.　原発事故　an accident at a nuclear power plant; a nuclear accident.　原発住民投票　a (local) referendum on building a nuclear power plant; a (local) referendum on nuclear power.　原発アレルギー　an aversion to nuclear power.　原発ジプシー　〔原子力発電所を渡り歩く労働者〕a nuclear gypsy [nomad]; a nuclear power worker who moves from job to job.　原発耐震指針　⇨耐震　脱原発　denuclearization. / 脱～の動き a movement away from (reliance on) nuclear power.

原油　crude oil [petroleum]; crude.　★⇨石油, 重油, 軽油　¶原油価格　《raise》the price of crude oil; crude-oil prices.　★⇨国際(国際価格)　原油生産国　an oil-producing country [nation].　原油タンカー　an oil [a crude-oil] tanker.　原油中継基地　a crude-oil transshipment station (略：CTS).　原油流出　an (accidental) oil spill.　★⇨流出, 沿岸　買い戻し原油　buy-back oil.　重質[中質, 軽質]原油　(a) heavy [medium, light] crude (oil); (a) heavy-[medium-, light-]gravity crude (oil).　サワー原油　〔硫黄化合物を多量に含む原油〕sour crude (oil).　スイート原油　〔硫黄分をほとんど含まない原油〕sweet crude (oil).　軽質スイート原油　light sweet crude (oil).

原料　raw [crude, rough] materials; material(s); an ingredient.　▶～の入手[割り当て] acquirement [allotment] of

原料

raw materials. / 〜を確保する secure [procure] raw materials. / チョコレートの〜はカカオ豆だ．Cacao seeds are used to make chocolate. / お酒には穀物を〜とするものが多い．Many alcoholic drinks are made from grain. / バーボンウイスキーはトウモロコシを主〜とする．Bourbon has corn as its main ingredient. / 日本のビールは，主な〜は麦芽だが，副〜として米やトウモロコシが用いられることが多い．In Japanese beer, the main ingredient is barley malt, but often such things as rice or corn can be used as secondary ingredients. / プルトニウムは核兵器の〜となる．Plutonium is used to make nuclear weapons.
¶ **原料表示** 〔食品の〕food product labeling; a food product label. ★⇨原産, アレルギー物質 **原料物質** 《nuclear》source material. **原料油** raw [stock] oil. **原料価格** raw material(s) prices; the price of raw materials. **原料調達ルート** a procurement [supply] route for raw materials; the route [channels] through which raw materials are procured.

広域 a wide [large] area. ¶ 広域応援 〔災害時の〕broad(er) regional aid. 広域基幹林道 an arterial forestry road. 広域行政 integrated administration of a large region. 広域下水道 wide-area sewerage. 広域ごみ処理 wide-area waste matter processing. 広域緊急援助隊 a regional emergency rescue team. 広域災害 a disaster covering a wide region; an [a geographically] extensive disaster. 広域災害救急医療情報システム a medical information system for use in geographically extensive disasters. 広域市町村圏 a greater city area (consisting of a major city and smaller surrounding municipalities). 広域停電 a wide area [an extensive] power outage [blackout, power cut]. 広域都市 an extensive city. 広域搬送 〔災害時の〕transportation 《of victims to medical facilities》in geographically extensive disasters. 広域被害 〔公害などの〕extensive [wide-area] damage; damage over a wide area. 広域連携 〔複数の自治体間などの〕cooperation (between neighboring municipalities) over a wide area. 広域連合 an association [a confederation] of public services set up to cover several local government areas.

降雨 rainfall; rain(s). ¶ 降雨強度 rainfall intensity. 降雨計 a rainfall meter. 降雨継続時間 duration of rainfall. 降雨図 a rainfall chart [map]. 降雨前線 a rain [wet] front. 降雨帯 a rain belt. 降雨量 (the amount of) rainfall. ▶この冬は〜量が少なかった．We haven't had much rain this winter. / 今度の台風で当地の〜量は30ミリであった．Thirty millimeters of rain fell here during the latest typhoon. | Precipitation here registered 30 millimeters during the typhoon. ★⇨雨量 年間[平均]降雨量 annual [average] rainfall. / この土地の年間〜量は300ミリだ．There are 300 millimeters of rain [rainfall] per annum here. 降雨林 a

豪雨　torrential rain(s); (a) torrential rainfall; torrents of rain; a torrential downpour; floods [a deluge] of rain; a cataract (of rain). ▶その地方に〜があった. There was torrential rain in the area. | The area was hit by torrential [devastating] rain. |《口》It bucketed down [came down in buckets] in the area. / 激しい〜だった. It rained torrentially. | The rain fell in sheets [came down in cataracts, was torrential]. ¶集中豪雨　a localized (torrential) downpour [cloudburst]; torrential rain falling over a limited area; a concentrated downpour. ★⇨ゲリラ豪雨

公益　public welfare [good, benefit, interest(s)];《in》the public interest. ★⇨公共　▶〜のために尽力する work for the public good [in the interests of the public]; serve the community. / 〜を図る work for [serve] the public (good). / 〜を害する be prejudicial [detrimental] to the public interest; be injurious to public welfare. ¶公益委員　a public [neutral] member《of a labor relations board》. 公益会社　a public service [utility] company. 公益学　public [community] service studies. 公益企業　a public utility service. 公益財団法人　a public interest incorporated foundation. 公益財団　a public utility foundation. 公益事業　public enterprise [works]; non profit-making activities; a public service; the public service industry; a public utility. 公益施設　a public facility [utility]. 公益社団法人　〔営利社団法人に対して〕a public service corporation. 公益信託　a charitable [public] trust. 公益性　public benefit [welfare, interest]. ★⇨ダム　公益代表　a representative of the public (interest); a person representing the (general) public. 公益団体　a public corporation. 公益通報者保護制度　a protection system for public-interest whistle-blowers. 公益通報者保護法　the Whistleblower Protection Law. 公益法人　a nonprofit foundation; a public utility [public-service] corporation. 公益法人白書　a white paper on nonprofit corporations;〔総務省の〕the Annual Report on Public Interest Corporations. 公益目的事業　〔公益法人認定の要件となる事業〕a business for public interest purposes.

公園 a (public) park; a public garden. ▶都市の〜は，人々の住環境を守ると同時に災害時の防火帯や避難場所となるよう設計されている．Urban parks are designed to protect people's living environment and at the same time to serve as firebreaks and places of refuge in times of disaster. / 都心に〜を造る make [create, lay out] a park in an inner city; provide a city center with a park. ¶**公園課** 〔市の〕the Parks Section. **公園管理事務所** a park administration office. **公園管理者** a park superintendent [manager]; a parks authority. **公園管理人** a (park) warden; a (park) ranger. **公園区域** a park district [area]. **公園使用規則** park regulations. **公園掃除人** a park cleaner. **公園道路** a road through a park; a parkway. **公園入場者** a visitor to a park. **公園墓地** a memorial park; a garden [park] cemetery; a cemetery in a park [surrounded by green]. **公園緑地** parks and other green areas. **公園緑地管理財団** the Parks and Recreation Foundation. **海中公園** a marine park. **記念公園** a memorial park; a park in memory of 《the world's fair》. **国定公園** a semi-national [quasi-national] park; a secondary protected area of outstanding natural beauty. **国立公園** a national park; a primary protected area of outstanding natural beauty. **自然公園** a (protected) wildlife area; a wilderness [nature] park; 〔サファリパーク〕a safari [an animal] park. **森林公園** a forest park; an area of woodland [forest] open to the public. **動物公園** a zoological park.

公海 the open sea; the high seas; international waters; a marine highway. ▶〜上で核実験を行う conduct nuclear tests over [in, on] international waters. ¶**公海漁業** fishery in international waters. **公海自由の原則** (the) freedom of the high seas. **公海条約** 〔1958年に国連の定めた〕the Geneva Conventions on the High Seas.

公害 〔環境汚染〕(environmental) pollution; pollution《of the atmosphere》; contamination; 〔環境破壊〕environmental destruction. ★⇨汚染，環境，低公害，無公害，感覚公害，振動公害，騒音公害，低周波公害，温排

水公害, 交通(交通公害) ▶もらい〜 pollution (which comes [derives, spreads]) from another area; pollution which affects areas other than its source. ★⇨越境公害 /〜に対する苦情 a complaint about (environmental) pollution; a pollution complaint. /〜のたれ流し irresponsible pollution《of the environment [a river]》; dumping pollutants [uncontrolled discharge of pollution] into the environment. ★⇨垂れ流す /〜を除去する[なくす] remove [eliminate, get rid of] a pollutant [contamination, pollution]. /〜を引き起こす cause [give rise to] pollution; pollute《the atmosphere》. /〜をまき散らす pollute a wide area; disperse pollution [a pollutant] over a wide area; spread pollution. / 典型七〜〔大気汚染, 水質汚染, 土壌汚染, 騒音, 振動, 悪臭, 地盤沈下〕the seven most common public nuisances (of air pollution, water pollution, soil contamination, noise, vibration, offensive odor, and ground subsidence). ¶**公害監視機関** a pollution-monitoring body. **公害企業** 《sue》an industrial polluter; a company which causes pollution. /〜企業に鉄槌を下す crack down on industrial polluters. ★⇨原因企業 **公害源** a source of environmental pollution;〔汚染物質〕a pollutant; a contaminant. **公害健康被害補償不服審査会** 〔環境省の〕the Pollution-Related Health Damage Compensation Grievance Board. **公害健康被害補償法** the Pollution-Related Health Damage Compensation Law. **公害罪** a pollution offense; an offense under environmental pollution legislation. **公害裁判[訴訟]** pollution litigation; a pollution suit. **公害除去設備** pollution-reduction equipment. ★⇨公害防止 **公害審査会** a pollution mediation board. **公害対策** antipollution [pollution control] measures; measures against [to control, to reduce] pollution. / 政府の〜対策はいつも後手後手だ. The government's antipollution measures are always too late [never in time]. | The government never deals with pollution until it's too late. **公害対策基本法** the Basic Law on Pollution Control [Measures to Prevent Pollution]; the Pollution Countermeasures Basic Law. **公害調査[監視]官** a pollution inspector [investigator]. **公害等調整委員会**〔総務省の外局〕the Environmental Disputes Coordination Commission. **公害都市** a (heavily) polluted city; a city with pollution problems. **公害発生型産業** a pollution-generating industry. **公害犯罪処罰法** the Environmental Pollution Offenses Law. **公害反対運動** a campaign against environmental pollution [disruption]; an anti-pollution campaign [movement]. **公害被害者救済制度** a relief system for pollution victims. **公害病** (an) illness caused by [arising from] (environmental) pollution; a pollution-caused disease;〔公害と関連のある〕a pollution-related disease; a pollution disease [illness]. /～病患者 a patient with a pollution-related disease. /～病認定 official registration [certification] of a patient as suffering from a pollution-related disease. /～病認定患者 a patient officially recognized [registered, certified] as suffering from a pollution-related illness [disease]; a certified pollution victim. ★⇨認定 **公害紛争** a pollution [an environmental pollution] dispute. **公害紛争処理法** the Law for the Settlement of Environmental Pollution Disputes. **公害防止** prevention of environmental pollution [disruption]; pollution prevention [control]. /～防止管理者 a manager in charge of pollution control; a pollution control supervisor. /～防止協定 an agreement on pollution control [prevention]; a pollution prevention [control] agreement. /～防止策 antipollution measures; an antipollution policy. /この工場の～防止設備は不完全だ. This factory doesn't have adequate [proper] antipollution facilities. | The factory is inadequately equipped to deal with pollution. /～防止産業 the environmental protection [pollution control] industry. /～防止条例 a pollution control [an antipollution] ordinance. **公害防止事業費事業者負担法** the Pollution Control Public Works Cost Allocation Law; the Law for Allocation of Pollution-Control Public Works Costs. **公害補償** compensation for pollution. **公害問題** a problem of environmental pollution; a pollution problem [issue]. **公害輸出** pollution export; export of pollution [contamination].

宇宙公害 the pollution of space.　**産業公害** industrial pollution.　**食品公害** food pollution [contamination]; contamination of food.　**排煙公害** exhaust pollution; air pollution caused by exhaust fumes. ★⇨煙害　**煤煙公害** soot pollution.

鉱害 pollution [environmental disruption] caused by mining. ★⇨鉱毒

光害 〔天体観測に支障をきたす都市の明かり〕light pollution. ¶**光害防止条例** a light-pollution control [prevention] ordinance.

光化学 ¶**光化学オキシダント**〔光化学スモッグの原因となる物質〕a photochemical oxidant.　**光化学オキシダント注意報** 《issue》a photochemical oxidant warning.　**光化学汚染** photochemical pollution.　**光化学スモッグ** (a) photochemical smog.　**光化学スモッグ警報** a photochemical smog warning.　**光化学スモッグ注意報** a photochemical smog advisory.

降下煤塵(こうかばいじん)　(a) dustfall. ★⇨煤塵

抗議 a protest; protestation;〔弱い〕a remonstrance;〔反対〕an objection;〔異議〕an exception;〔陳情〕a representation;〔苦情〕a complaint;【法】a demur. ★⇨苦情　▶〜する (make a) protest; object; complain. / 〜のデモ行進をする join [take part in] a protest march / 住民たちは産業廃棄物の不法投棄に対する〜運動を開始した. The local residents launched a movement against the illegal dumping of industrial wastes. / …に対して〜行動を起こす stage a protest against…. / …に〜のFAX[メール]を送る send a complaint by fax [e-mail] to…. ¶**抗議集会** a protest meeting [rally]; an indignation meeting.　**抗議声明** 《deliver》a statement [message] of protest 《against a nuclear test》; a protest statement.　**抗議電話** a complaint call; a (tele)phone complaint.

公共 ▶〜の public; community; communal; common. / 〜の福祉 public welfare. / 私権は〜の福祉に従う. All private rights shall conform to the public welfare. / 〜の利益 the public interest [good]; public interests [benefits]; the common good. ★⇨公益 / 〜の利益を図る work in behalf of the public interest [in the interests of the public]; promote public good. ¶**公共機関** a public institution. ★⇨公共輸送機関, 交通(交通機関)　**公共企業体** a public corporation [enterprise]; a public service corporation; a government corporation.　**公共組合**　〔健康保険組合などの社団法人〕a public corporation.　**公共経済学**　public economics.　**公共芸術**　public art.　**公共下水道**　public sewerage.　**公共圏**　public domain.　**公共広告**　〔麻薬追放や未成年の喫煙防止などを呼びかける広告〕public service advertisement [advertising].　**公共工事**　public works.　**公共サービス**　public services.　**公共財**　public goods.　**公共(用)財産**　public property.　**公共事業**　a public (utility) enterprise; a public works project; (public) utilities; public services. / 〜事業の前倒し the "front-loading" of spending for public works. / 〜事業会社 a public utility company; a public service corporation; public utilities. / 〜事業(関係)費 public works(-related) expenditures. / 〜事業部門 the public works sector. / 〜事業補助金 public works subsidies. / 〜事業予算 appropriations [a budget] for public works.　**公共施設**　public [community, communal] facilities; public accommodations [utilities].　**公共車両優先システム**　a Public Transportation Priority System (略: PTPS).　**公共団体**　a public body [institution, organization]; a public entity. / 地方〜団体 a local [regional] public body [entity].　**公共投資**　public works spending [investment]; investment(s) in public utilities. / 〜投資基本計画 a basic plan for public investment. / 〜投資関係費 expenses related to public works.　**公共図書館**　a public library.　**公共部門**　the public sector.　**公共法人**　a public-service corporation.　**公共放送**　public [non-commercial] broadcasting; public service broadcasting.　**公共放送ラジオ[テレビ]**　public service radio [television].　**公共マナー**　public manners.　**公共輸送機関**　public transportation [transport]. / 〜輸送機関を利用する use [take] public transportation [transport].　**公共用水**　water for public use.　**公共用地**　land for public works projects.　**公共料金**　〔水道・ガス(・電気)代など〕a public utility charge; utility bills [rates]. / 〜料金の引き上げ[引き下げ] an increase [a decrease] in utility rates.

工業　(an) industry; the manufacturing industry; 〈総称〉the industries.　▶〜用の industrial;

for industrial use [purposes]. ¶ **工業暗化[黒化]** 〔工業都市において蛾などが暗色に変異する現象〕★⇨英和 industrial melanism.　**工業化** industrialization. / 〜化する industrialize《a country》. / まだ〜化されていない地域 a nonindustrialized area; an area that is not yet industrialized. / 中南米諸国の〜化の進み具合はまちまちである. The Latin American countries are industrializing at varying rates.　**工業化社会** an industrialized society. / 脱〜化社会 a postindustrial society.　**工業災害** an industrial accident.　**工業再配置計画** inducement(s) for industrial relocation.　**工業整備特別地域** a special area for industrial consolidation.　**工業団地** an industrial park [estate].　**工業地域** an industrial area [district].　**工業地帯** an industrial zone.　**工業中毒** industrial poisoning.　**工業廃水** 《the discharge of》industrial effluent(s) [waste water].　**工業用水** water for industrial use; industrial water; a water supply for industry.　**工業用水道** 〔事業〕industrial water supply [service];〔施設〕industrial waterworks.　**工業用水道事業法** the Industrial Water Supply Business Law.　**工業用水法** the Industrial Water Law.　**工業立地** location [siting] of industries.

抗菌 ▶〜性の antimicrobial; antibacterial;〔抗真菌性の〕antifungal. ¶ **抗菌活性** antimicrobial [antibacterial] activity.　**抗菌活性物質** an antimicrobially active substance.　**抗菌靴下** antibacterial socks.　**抗菌グッズ[商品]** antimicrobial [antibacterial, bacteria-resistant] products [merchandise].　**抗菌作用** antimicrobial [antibacterial, antifungal] action.　**抗菌処理[加工]** antimicrobial [antibacterial, antifungal] treatment.　**抗菌砂** 〔砂場用の〕antimicrobial sand; sand treated with antimicrobial agents.　**抗菌スペクトル** antimicrobial [antibacterial] spectrum.　**抗菌セラミック** antibacterial [antimicrobial] ceramics.　**抗菌繊維** (an) antimicrobial [antibacterial] fiber.　**抗菌タイル[マスク, まな板]** an antibacterial tile [mask, chopping board].　**抗菌塗装** antimicrobial [antibacterial] coating.　**抗菌物質[剤]** an antimicrobial [antibacterial, antifungal] substance [agent].　**抗菌薬**

an antimicrobial [antibacterial, antifungal] drug [agent].
抗菌力 antibacterial [antimicrobial] power.
航空 ¶**航空(機)事故** a plane [an airplane, an air] accident; (墜落事故) a plane [an airplane] accident [crash]. **航空機事故調査** an aircraft accident investigation; an investigation of an aircraft accident. **航空機衝突防止装置** an airborne collision avoidance system (略: ACAS). **航空(機)騒音** 《reduce》aircraft noise; the noise of an airplane. **航空・鉄道事故調査委員会** 〔国土交通省の〕the Aircraft and Railway Accidents Investigation Commission.
光合成 photosynthesis. 形 photosynthetic. ▶有機物を〜する photosynthesize organic matter. / 〜を行う photosynthesize. / 植物は光が当たると〜を行う．Photosynthesis occurs when light strikes plants. **光合成細菌** photosynthetic bacteria. **光合成色素** a photosynthetic pigment. **光合成比** a photosynthetic ratio. **光合成物** a photosynthate.
黄砂 yellow sand. ¶**黄砂対策** countermeasures against yellow sand. **黄砂被害** damage from yellow sand. ▶〜被害が深刻になってきている．Damage from yellow sand is becoming severe. **黄砂飛来情報** 〔環境省の〕dust and sandstorm [DSS] information;《provide》real-time information on the spread of dust and sandstorms [DSS]. **黄砂予報** yellow sand [DSS] forecasting.

鉱滓(こうさい) slag; scoria. ¶**鉱滓セメント** slag cement. **鉱滓道床** (a) slag ballast. **鉱滓綿** slag wool; mineral wool; silicate cotton.

耕作 farming; cultivation《of corn》; farm work. ▶〜する farm; cultivate《land》; till《a field》; plow《land》. / 〜に適している be arable; be suitable for [capable of] cultivation. / 〜不能の土地 barren land [soil]; land [soil] too poor for cultivation. ¶**耕作機(械)** farm [farming] tools; farm [farming, agricultural] equipment; agricultural implements [machinery]. / 動力[手動]〜機(械) a powered [manual] tiller [cultivator]. **耕作権** farming [cultivation] rights; the right to cultivate the land. **耕作限界** the margin of cultivation. **耕作限界地** marginal land. **耕作地** farm land; arable land; cultivated land; land under cultivation. **耕作面積** the area 《of a farm》 under cultivation [tillage]; the acreage of a farm (used for cultivation). **耕作放棄地** agricultural land left fallow for a year or more; uncultivated farmland.

交雑 〔動植物の〕crossing; hybridization. ¶ **交雑育種** crossbreeding. ▶～育種する cross 《two species, species A with species B》; crossbreed 《A with B, with...》;《two species》 cross [crossbreed]. **交雑種** a hybrid; a cross. **交雑不稔[不妊]** 〔交雑によって生じた子が不稔[不妊]になること〕cross-sterility; amixia. **交雑防止** the prevention of hybridization. **交雑防止基準** standards for preventing hybridization. **複交雑** a double cross. **戻し交雑** a backcross.

高山(ざん) a high mountain. ▶～の alpine; mountainous. ¶ **高山気候** a highland climate; a mountain [an alpine] climate. **高山植物** an alpine plant; an alpine;〔植物相〕an alpine flora. **高山植物園** an alpine garden. **高山草原** an alpine meadow. **高山帯** an alpine belt [zone]. **高山地帯** a mountainous [an alpine] region; the high reaches《of Tibet》. **高山鳥** an alpine bird. **高山蝶** an alpine butterfly. **高山動物** an alpine animal;〔動物相〕an alpine fauna.

鉱山 a mine. ▶～を掘る dig [excavate] a mine. /～を経営する run [operate] a mine. ¶ **鉱山開発** development of a mine; mine development. **鉱山監督署** a mining inspection [control] office. **鉱山災害** a mine [mining] disaster [accident]. **鉱山採掘権** mining rights; a mining concession. ★⇨利権 **鉱山保安法** the Mining Safety Law.

工事 construction; construction work(s); engineering work. ★⇨道路(道路工事) ▶～中である be under [in course of] construction; be being built [constructed, made]. /～中.〔掲示〕Under Construction. |〔道路が〕Road Works [Roadworks]. |〔人がいるという意味で〕Men at Work. |〔修理で〕Under Repair. | Repair Work [Repairs] in Progress. / この先～中.〔掲示〕Construction [Road Works] Ahead. /～が大いにはかどった. Good progress has been made with the work. | Construction went ahead fast. /～に取りかかる set to work; start work. /～の遅れ a construction delay; a delay in construction (work). /

その〜の完成には約7年を要した. It took about seven years to complete the work. / 〜を請け負う contract to undertake construction [undertake repairs, do a job]. / 〜を起こす start [commence] work. / 〜を施行する carry out [do, execute] construction work. / 〜を監督する supervise [direct] work [construction]. ¶**工事現場** a work [construction] site; a worksite. **工事監督(者)** a site [construction, works] manager; a superintendent of works; an overseer. **工事監理** construction work(s) supervision; works supervision. **工事監理者** a construction work(s) supervisor; a works supervisor. **工事監理報告書** a construction supervision report. **工事差し止め** ⇨差し止め **工事車両** a construction vehicle. / 〜車両出入口 〔掲示〕Gate for Construction Vehicles. **工事渋滞** a traffic jam [traffic congestion] caused by roadworks. **工事入札** a bid [bidding] for a construction project. **工事費(用)** the cost of construction; construction costs. **工事費内訳書** a (written) breakdown [an itemized list] of construction costs. **改修[修復]工事** repairs; restoration; repair [restoration] work. **拡張工事** 〔建物の〕extension work(s). **欠陥工事** faulty [defective] construction (work).

工場 a factory; a plant; a work(s); a mill; a workshop; a shop; a manufactory; an industrial plant. ▶〜を閉鎖する close [shut] down a factory. / 〜を誘致する (try to) attract industry [factories]. ¶**工場衛生** factory hygiene [sanitation]; industrial hygiene. **工場街** a factory area [district]《of Osaka》; an industrial area [district, quarter]《of Kawasaki》. **工場監督** a factory superintendent;〔現場の〕a foreman; a supervisor. **工場監督官[検査官]** a factory inspector. **工場管理** factory management [administration, control]. **工場管理制度** a factory management system. **工場事故[災害]** an accident [a disaster, a tragedy] in [at] a factory. **工場設備** 〔操業用の〕factory [plant, shop] equipment; the equipment in [of, for] a factory;〔従業員用の〕factory [plant, shop] facilities. **工場騒音** factory noise. **工場地域[地帯]** a factory [an industrial] district [area]. **工場排ガ

ス gas from a factory; factory exhaust(s). **工場廃棄物** industrial refuse. **工場排水[廃水]** industrial wastewater; liquid industrial waste; liquid waste(s) from a factory. **工場排水規制法** the Factory Waste Water Control Law. **工場誘致** attracting industry [factories]《to an area》. **工場用地** a site for a factory; factory land; a potential industrial site. **工場立地法** the Factory Location Law.

恒常性 homeostasis.

硬水(こうすい) hard water. ¶**硬水軟化剤** a (water) softener.

降水 precipitation; hydrometeor. ★⇨降雨, 雨量 ¶**降水域** an area of precipitation. **降水確率** a probability [likelihood, chance] of precipitation [rain, rainfall]; a precipitation percentage. ▶今夜の〜確率は30パーセントです. The probability of rainfall tonight is 30 percent. | There is a 30 percent likelihood [probability, chance] of rain [precipitation, rainfall] tonight. **降水確率予報** rainfall probability forecasting; forecasting of likely precipitation; 〔個々の〕a rainfall probability forecast; a forecast of likely precipitation. **降水予報** precipitation forecasting; 〔個々の〕a forecast of precipitation. **降水時間[密度, 強度]** precipitation duration [density, intensity]. **降水短時間予報** (very) short-range precipitation forecasting; 〔個々の〕a (very) short-range forecast of precipitation. **降水量** (a quantity [an amount] of) precipitation. **月間[年間]降水量** a monthly [(an) annual] precipitation.

更生 ⇨再生 ¶**更生タイヤ** a recapped tire; a recap. **更生廃棄物** processed scrap. **更生品** a reconditioned article; a make-over. **廃品更生** reconditioning (of discarded equipment).

合成 composition; synthesis. ▶〜する synthesize; compose; compound; combine《into a whole [one]》. / 〜の compound; mixed; composite; complex; combined; synthetic ¶**合成化学** synthetic chemistry. / 有機〜化学 organic synthetic chemistry. **合成化学工業** the synthetic chemical(s) industry. **合成化学薬品** synthetic pharmaceuticals; a synthesized drug. **合成ガス** synthesis gas. **合成甘味料** a synthetic [an artificial] sweetener; a synthetic sweetening agent. **合成**

金属 (a) synthetic metal. 合成香料 (a) synthetic scent [perfume]. 合成ゴム synthetic rubber. 合成紙 synthetic paper. 合成種 an additive species. 合成樹脂 a synthetic resin; a plastic. 合成樹脂工業 the plastics industry; plastics. 合成樹脂接着剤 (a) plastic glue; (a) synthetic resin adhesive. 合成樹脂繊維 a plastic [synthetic resin] fiber. 合成樹脂塗料 synthetic resin paint [coating (material)]. 合成樹脂シート a synthetic resin sheet; a plastic sheet; plastic sheeting. 合成繊維 (a) synthetic fiber; (a) man-made [(an) artificial] fiber; a synthetic fabric. 合成洗剤 a synthetic detergent; syndet. 合成染料 (a) synthetic dye. 合成着色料 a synthetic colorant [coloring agent]; synthetic coloring. 合成燃料 (a) synthetic fuel; synfuel. 合成皮革 synthetic [simulated, artificial, composition] leather. 合成肥料 (a) compound fertilizer. 合成品 synthetic goods; goods [an item] made with artificial materials; goods containing artificial substances [made with synthetic materials]. 合成品種 a synthetic variety. 合成物 a compound; a composite thing; a synthetic (product); a synthesized [synthetic] product. 合成物質 a synthetic [an artificial, a man-made] substance;〔繊維・生地〕a synthetic. 合成保存料 a synthetic preservative. 合成ホルモン a synthetic hormone.

洪積層(こうせきそう) a diluvium; a diluvial formation. ¶洪積層土壌 (a) diluvial soil [deposit].

酵素 an enzyme. 形 enzymatic, enzymic; a ferment. ▶～の活性中心 an active center of an enzyme. / ～の特異性 enzymatic specificity. /〔広告などで〕～パワーで汚れを分解（します）. Breaks down dirt through enzyme power. / この洗剤は～の働きで汚れを分解する. This detergent uses the action of enzymes to break down stains. ¶ 酵素化 zymogenesis. 酵素学 enzymology. 酵素学者 an enzymologist. 酵素活性 enzyme activity. 酵素機能 (an) enzyme function. 酵素工学 enzyme engineering. 酵素製剤 an enzyme preparation. 酵素性分解 enzymolysis. 酵素前駆体, 酵素原 〔プロ酵素〕a zymogen; a proenzyme; an enzyme precursor. 酵素洗剤 (an) enzyme detergent. 酵素

阻害剤 an enzyme inhibitor. 酵素病 enzymopathy. 酵素免疫測定法 enzyme immunoassay（略：EIA）. 酵素誘導 enzyme induction. 酵素療法 enzymotherapy. 黄色酵素 a yellow enzyme. 抗酵素 an antienzyme.／抗〜性の antienzyme. 酸化酵素 an oxidase; an oxidizing enzyme. 消化酵素 a digestive enzyme. 転化酵素 invertase. 糖化酵素 carbohydrase.

高層 ▶〜の multistory; multistoried; high-rise; high; tall.／中〜の建物 a medium high-rise [multistory] building; a mid-rise building. ¶高層階 an upper floor [story] of a high-rise building. 高層気象 meteorology of the upper atmosphere; aerology. 高層気象学 aerology. 高層気象観測 aerological observation; upper-air observation. 高層気象図 an aerological diagram. 高層気象台 〔気象庁の〕an aerological observatory. 高層気流 an upper air current. 高層建築（物） a multistory [multistoried] building; a high-rise (building); a skyscraper; a high [tall] building.／超〜建築物［ビル］a super high-rise building; a skyscraper; an extremely high [tall] building. 高層湿原 a high moor. 高層大気 the upper air [atmosphere]. 高層風 (a) high-altitude wind; (an) upper-level wind. 高層住居誘導地区 〔都市の〕an urban zone in which regulations are relaxed in order to encourage the construction of high-rise buildings; a high-rise building zone.

構造活性相関 structure-activity relationship（略：SAR）.

高速 ¶高速（増殖）炉 a fast breeder; a fast-breeder reactor（略：FBR）. ▶液体金属〜増殖炉 a liquid metal fast-breeder reactor（略：LMFBR）. 高速大量輸送システム a high speed mass transportation [transport] system. 高速中性子炉 a fast neutron reactor. 高速道路 a superhighway; an expressway; an express highway; a freeway; a thruway; a motorway;〔有料の〕a turnpike.／〜建設の推進 the promotion of highway construction. 高速道路網 an expressway system; a motorway system. 高速道路料金 an expressway [a motorway] toll [fee]; a toll for an expressway.

公聴会 a public [an open] hearing; (public) hearings. ▶〜を

開く hold public hearings《on...》. **地方公聴会**〔地方に委員を派遣して行う〕a local [regional] public hearing attended by《officials》from《the central government》. **中央公聴会**〔国会内で行う〕a public hearing (held) in the Diet.

交通〔人の行き来〕traffic; coming and going;〔運輸・通信〕transportation [transport];〔河川の〕river navigation. ▶空の～ air transportation [transport]; air [aerial] traffic; transportation by air. / 雪で麻痺した～ snowbound traffic. / すべての～が止まった. All transportation came to a standstill. / ～が一時とだえた所も少なくない.〔道路の〕In several places traffic was tied up [paralyzed, brought to a standstill] for a while. |〔輸送機関の〕In several places transportation [transport] was interrupted [suspended, paralysed] for a while. / この路線は年々～が激しくなるだろう. There will be more and more traffic [Traffic will get heavier and heavier] on this route every year. / ～の流れ (the flow of) traffic [people and vehicles]. / ～を整理する control [regulate] traffic. / ～を復旧する make roads and railways function again; restore communications《with the mainland》. / 過密化している都市～を改善する improve urban transportation [transport] by reducing congestion; make urban transportation [transport] less overcongested [crowded]. / 環状道路の開通でこの辺の～事情はかなり改善された. With the completion of the ring road the traffic situation around here has improved considerably.

¶ **交通アセスメント**〔事前に交通量などの影響を検討しその対策を立てる制度〕(a) traffic assessment; a transportation [transport] impact assessment. **交通安全** traffic [road] safety. / ～交通安全を図る encourage traffic [road] safety; try to make the roads safe. **交通安全運動** a traffic safety campaign [drive]. / 春の全国～安全運動 the National Spring Road Safety Campaign. **交通安全教育** traffic [road] safety education. **交通安全協会** a traffic safety association. **交通安全施設** a traffic safety facility. **交通安全週間** Traffic [Road] Safety Week. **交通安全環境研究所**〔国土交通省所轄の独立行政法人〕the National Traffic Safety and Environment Laboratory.　**交通違反** a traffic offense; a violation of traffic regulations. / 〜違反をする commit a traffic offense; violate traffic regulations. / 〜違反を取り締まる police traffic (to catch people violating the regulations).　**交通エコロジー・モビリティ財団** the Foundation for Promoting Personal Mobility and Ecological Transportation.　**交通課** 〔警察署の〕a traffic section.　**交通監視員** 〔通学児童の〕a school crossing warden;《口》a lollipop man [woman]; 〔誘導員〕a traffic controller.　**交通監視カメラ** a traffic(-monitoring) camera.　**交通監視システム** a traffic monitoring system.　**交通管制所** a traffic control station [post].　**交通機関** (a) means of transportation [travel, locomotion]; transportation [traffic] facilities; a transit system. / 渋滞緩和のため公共〜機関の利用を呼びかける urge [encourage] people to use public transportation to help ease [relieve] traffic congestion.　**交通規制** traffic restrictions [controls, regulations]. / 駅伝競走のため国道1号線には〜規制が敷かれていた．There were traffic restrictions (in place) on Route 1 because of a relay race. / 〜規制を行う impose traffic restrictions.　**交通機動隊** 〔警察の〕a mobile traffic unit.　**交通公害** 〔排気ガスなどによる〕traffic pollution; (environmental) pollution caused by traffic; 〔歩行者や沿線住民が被る迷惑〕problems caused by traffic; traffic problems.　**交通事故** a traffic accident; 〔自動車などの〕a car accident; 〔電車などの〕a train [railroad] accident. / 〜事故が多発する交差点 a crossroads where accidents are common; a bad crossroads for accidents. / 〜事故が年々増えている．Traffic accidents are increasing year by year. | There are more road accidents every year. / 〜事故で負傷する[死ぬ] get hurt [get killed, die] in a traffic accident. / 交差点を横断中に〜事故にあった．I got hit by a car while crossing the street at an intersection. / 〜事故に巻き込まれる get [be] involved in a traffic accident; get mixed up [caught up] in a traffic accident. / 〜事故を起こす cause a traffic accident. / 〜死亡事故 a fatal traffic accident; a traffic accident resulting in death. / 〜死亡事故多発警報 a warning about (frequent) road deaths [fatalities]. **交通事故死** dying [being killed, death] in a traffic accident;〔統計上〕a traffic death [fatality]. **交通事故死亡者** a victim of a fatal traffic accident; a person killed in a car accident. **交通事故死亡者数** (the number of) road deaths; the number of people killed in car accidents; road accident fatalities. **交通事故傷害保険** car accident insurance; traffic personal accident insurance. **交通事故遺族** the families of people killed in traffic accidents; people bereaved in traffic accidents. **交通事故多発地域** an accident blackspot;〔掲示〕Caution! Accident Black Spot. **交通事故発生マップ** a traffic accident map. **交通事故総合分析センター**〔財団法人〕the Institute for Traffic Research and Data Analysis (略: ITARDA). **交通地獄** commuter hell; hellish traffic congestion. **交通指導員** a police officer charged with promotion of traffic safety. **交通弱者** a vulnerable road user;〔高齢者・子供・障害者など〕a mobility-impaired person;〔交通の便に恵まれない人〕a transportation-disadvantaged person. **交通渋滞** a traffic jam; (traffic) congestion;〔完全な麻痺状態〕gridlock. ★⇨渋滞　**交通整理**　traffic control [regulation]. / 〜整理をしている警官 a policeman directing traffic [on traffic duty]. **交通整理員**　a traffic control guard; a person who guides pedestrians and traffic《at a building site》. **交通騒音**《reduce》traffic noise (pollution);《be kept awake by》the noise of (the) traffic. **交通道徳** traffic manners; good behavior on the roads. / 〜道徳を守る behave properly [well, decently] on the roads. **交通麻痺**　a traffic jam; traffic paralysis; a snarl-up. **交通問題**　a traffic problem;〔輸送関係の〕a transportation [transport] problem. **交通誘導**　〔工事現場付近などでの〕traffic control; directing traffic; flagging. / 〜誘導する direct traffic; flag. **交通誘導員** a traffic control person; a flagger. **交通誘導灯**　an illuminated traffic control baton. **交通量**　(a volume [an amount] of) traffic; (a) traffic volume [density]. / この通りは〜量が少ない．There's not much traffic on this street. | This street has little traffic. / このあたりは〜量が多いので道路の傷みがはやい．There's a lot of traffic here, so road deterioration is fast. / 〜量の多い道路 a road with a lot of traffic; a crowded [busy] road. / 〜量の最も多い時間 the heaviest traffic hours. / 〜量の少ない街路 a street with little [light, sparse] traffic. / 〜量の少ない時間帯 a period when there is little traffic. / 1号線の〜量は最近減ってきている．Traffic has decreased [There's been less traffic] on Route 1 recently. **交通量図** a traffic discharge map. **交通量調査** a traffic census [survey].

口蹄疫（こうてい）〔家畜伝染病〕foot-and-mouth [hoof-and-mouth] disease; aftosa; aphthous fever.

高度 (an) altitude; (a) height. ¶**高度地域** 〔建物の高さを制限する〕a《class 1, class 2》building height restriction district. **高度道路交通システム** an intelligent transport system（略：ITS）. **高度分布** ⇨分布 **高度利用地区** an intensive land utilization zone; an intensely used area (of land).

行動 ⇨活動，採餌行動，捕食行動 ▶芋洗い〜〔ニホンザルの〕sweet potato washing. ¶**行動遺伝学** behavioral genetics; the genetics of behavior. **行動観察** observation(s) (of behavior); behavioral observation. / 野生の象の〜観察を行う carry out observations on [of] the behavior of elephants in the wild. **行動圏** 〔定住性動物の〕a home range. **行動生態学** behavioral ecology. **行動生物学** ethology. 形 ethological. **行動生物学者** an ethologist. **行動特性** a behavioral trait [characteristic]. **行動展示** 〔動物園の〕exhibiting《animals》so that they display natural behavior; allowing《a species》to behave naturally. **行動範囲** a territory; a sphere of action; an [a geographical] area of activity [activities]; 〔動物の〕a habitat. / 〜範囲の広い[狭い]動物 animals with an extensive [a circumscribed] habitat; animals whose activities cover a large area [are restricted]. **生得的[獲得性]行動** innate [acquired] behavior; an innate [acquired] form of behavior.

鉱毒 mineral pollution [poisoning, contamination]; pollution by minerals. ★⇨足尾鉱毒事件 ¶**鉱毒被害** (damage from) mineral [poisonous metal] pollution. **鉱毒被害地** an area of [affected by, suffering from] mineral pollution [poisoning]. **鉱毒被害民** victims of mineral pollution; people suffering from metal poisoning. 鉱毒水 water polluted [contaminated] with poisonous minerals.

交配 mating;〔異種交配〕hybridization; crossbreeding (形 crossbred); crossing;〔植物の〕cross-fertilization. ▶〜する cross; crossbreed; hybridize; mate《one breed with another》;〔同血統と〕inbreed;〔植物を〕cross-fertilize. / ライオンとヒョウを〜して新種の動物をつくる produce a new kind of animal by crossing a lion and a leopard. ¶**交配現象** hybridism. **交配種** a hybrid; a crossbreed.

荒廃 destruction; desolation; devastation; dilapidation; waste; ruin. ★⇨荒れる ▶〜する decay;〔田んぼなどが〕be ruined; fall into decay; run down; revert to nature [the wild];〔建物などが〕go to ruin; fall apart [to pieces, into decay]; get run down; become dilapidated; mo(u)lder; crumble;〔手入れされずに〕be in bad [a bad state of] repair; be in disrepair / 〜させる〔土地などを〕destroy; wreck; lay waste; devastate《land》; ravage; ruin. ¶**荒廃地** a waste; a wasteland;〔戦争・噴火などによる〕a devastated region. **荒廃地区**〔復興計画上の〕《be designated》a blighted area. **森林荒廃** (total) destruction of the forests; devastation of forest land.

後発発展[開発]途上国〔国連の規定による〕the least developed countries (略: LDC).

合流式下水管[下水道]〔雨水と汚水を一本の下水管で処理場へ送る方式〕a [the] combined sewer system; a system which combines different sewers [drains] into one.

小型(自動)車〔法律上の分類で〕a compact automobile [car] (less than 4.7 m in length, 1.7 m in width, 2.0 m in height, and with a non-diesel engine with a total displacement of less than 2,000 cc);〔一般に，小型の自動車〕a small [compact] car; a minicar; a mini. ★⇨燃費

護岸 strengthening a river bank [shore, coast]; shore [bank] protection; a revetment. ¶**護岸工事** shore [coast, bank] protection work; embankment work; riparian work. ▶〜工事を施す carry out embankment work; carry out work to strengthen a shore [coast, bank]; revet. **護岸堤防**〔海岸の〕a seawall;〔河川の〕a river wall. **護岸コンクリート** a concrete sea wall.　**護岸整備**　〔造成〕embankment (construction);〔維持〕embankment maintenance; maintenance of embankments.

国際　▶〜間の international; world; global; universal; among [between] the nations.　¶**国際一般名**　〔WHO が定める医薬品の国際的な名称〕an international nonproprietary name (略: INN).　**国際運河**　an international canal.　**国際エネルギー機関**　the International Energy Agency (略: IEA).　**国際エネルギースタープログラム**　the International Energy Star Program.　**国際エネルギーフォーラム**　the International Energy Forum (略: IEF).　**国際会議**　an international conference [convention, congress].　**国際海峡**　an international strait.　**国際海事機関**　〔国連の機関〕the International Maritime Organization (略: IMO).　**国際海洋法**　international maritime law; the international law of the sea.　**国際海洋汚染防止条約**　⇨海洋(海洋汚染防止条約)　**国際価格**　international prices; prices on the international market;〔各国間で取り決めた価格〕an internationally agreed price. / 原油の〜価格の上昇 an increase [a rise] in international oil prices.　**国際河川**　an international river.　**国際環境技術センター**　〔国連環境計画(UNEP)の〕the International Environmental Technology Centre (略: IETC).　**国際環境法**　international environment law.　**国際監視制度**　〔包括的核実験禁止条約(CTBT)にもとづく〕the International Monitoring System (略: IMS).　**国際感染症**　an international infectious disease.　**国際慣行**[慣習, 慣例] international practice [usage]; a convention [practice] accepted internationally.　**国際慣習法**　customary international law.　**国際規格**　an international [a global] standard 《for ...》.　**国際機関**　an international body [organization].　**国際企業**　an international enterprise [company, corporation].　**国際希少野生動植物種**　International Endangered Species of Wild Fauna and Flora.　**国際協調**　international harmony; harmony between nations.　**国際協定**　an international convention.　**国際協力**　international cooperation.　**国際協力 NGO センター**　the Japan NGO Center for International Cooperation (略: JANIC).　国

際協力活動　international cooperation [collaboration] activities; international assistance.　**国際漁業協力**　international cooperation on fishing.　**国際クマ会議**　the《17th》International Conference on Bear Research and Management《2006》.　**国際再生可能エネルギー機関**　the International Renewable Energy Agency.　⇨英和 IRENA　**国際査察制度**　a system of international inspection; an international inspection system.　**国際湿地条約**　⇨ラムサール条約　**国際自然保護連合[連盟]**　the International Union for Conservation of Nature and Natural Resources.　⇨英和 IUCN　**国際疾病分類**　the International Classification of Diseases（略：ICD）.　**国際条約**　an international treaty [agreement, pact, convention].　**国際水域**　《in》international waters;《on》the high seas.　**国際生活機能分類**〔世界保健機関（WHO）で採択された〕the International Classification of Functioning, Disability and Health（略：ICF）.　**国際生態学センター**　the Japanese Center for International Studies in Ecology（略：JISE）.　**国際生物多様性の日**〔5月22日；国連制定〕the International Day for Biological Diversity（略：IBD）; World Biodiversity Day.　**国際世論**《appeal to》world [international] opinion.　**国際単位**　an international unit（略：IU）.　**国際地球観測年**〔国連制定；1957-58年〕International Geophysical Year（略：IGY）.　**国際的枠組み**　an international framework.　★⇨京都議定書　**国際伝染病**　an international infectious [epidemic] disease.　**国際熱帯木材機関**　the International Tropical Timber Organization（略：ITTO）.　**国際熱帯木材協定**　the International Tropical Timber Agreement（略：ITTA）.　**国際農業開発基金**　the International Fund for Agricultural Development（略：IFAD）.　**国際農業交流・食糧支援基金**　the Japan International Agricultural Council（略：JIAC）.　**国際バイオエネルギー・パートナーシップ**　the Global Bioenergy Partnership（略：GBEP）.　**国際排出量取引**〔温室効果ガスの〕international gas emissions trading.　**国際法**　(public) international law; the law of nations. / 〜法違反　a violation [an infringement, a breach, a contravention] of international law; an offense [a crime] under international law. / …することは〜法上認められている．…ing is recognized [accepted] under international law. **国際法規** international laws and regulations. **国際法廷** an international tribunal. **国際捕鯨委員会** ⇨英和 International Whaling Commission **国際捕鯨取締条約** the International Convention for the Regulation of Whaling. **国際保護鳥[動物]** ⇨保護 **国際命名規約** ⇨命名 **国際年[デー]** an international year [day]; the International Year [Day]《of Education》. **国際標準** an international [a global] standard《for …》. **国際標準規格** an international standard; (an) international standard specification. **国際油濁補償基金** The International Oil Pollution Compensation Funds; the IOPC Funds. **国際ルール** international rules《of [for] air transportation》. **国際連合** the United Nations; the UN. ★⇨国連

国産 〜の〔自国の〕domestically produced [made, grown, manufactured]; domestic;〔日本の〕produced [made, grown, manufactured] in Japan; Japanese-made; of Japanese make; Japan-made. ▶〜の野菜 domestically produced [grown] vegetables; home-grown vegetables. ¶**国産牛**〔雌〕a cow〔〔雄〕a steer,〈総称〉cattle〕reared in Japan;〔食肉〕Japanese-reared beef. **国産競走馬** a racehorse bred in Japan [the country where it races]. **国産奨励** encouragement of domestic [home] production [industries]. **国産大豆** domestically grown [home-grown] soybeans. **国産品**〔自国の〕domestically produced [made, grown, manufactured] products; domestic goods [products]; home-produced [-manufactured] articles; articles of domestic manufacture; products of domestic industry;〔日本の〕Japanese products; articles produced [made, grown, manufactured] in Japan; articles of Japanese make. / 〜品を優先的に買う[使用する] buy [use] domestic goods in preference to foreign [imported] ones. / 純〜品 100-percent Japanese [all-Japanese] products. / 〜品使用運動 a "Buy Japanese" campaign [movement].

国土 a territory; a realm; a domain; land. ▶広い[狭い]〜 a

large [small] territory. / 〜の保全 national [nationwide] land conservation. / 〜を開発する develop [improve] the land. ¶ **国土開発** (national) land development; national land planning. **国土基本図** 〔国土地理院作成の地図〕a national land map; a basic map of the whole country. **国土計画** national land planning;（案）a program for land development. **国土形成計画** the Plan for Sustainable National Land Development [the Sustainable Development of National Land]．★2005年，全国総合開発計画から改称．**国土形成計画法** the Law for Sustainable National Land Development [the Sustainable Development of National Land]．★2005年，国土総合開発法から改正．**国土交通省** the Ministry of Land, Infrastructure, Transport and Tourism（略：MLIT）．**国土交通相[大臣]** the Minister of Land, Infrastructure, Transport and Tourism. **国土交通白書** 〔国土交通省の〕the White Paper on Land, Infrastructure, Transport and Tourism in Japan《2008》．**国土総合開発計画** 〔国土交通省の〕Japan's [the National]《First, Second, New, etc.》Comprehensive National Development Plan. ★「全国総合開発計画」（全総）と呼ばれ，一全総（1962），二全総（1969），三全総（1977），四全総（1987），五全総（1998）がある．⇨国土形成計画 **国土調査** a national land survey. **国土利用計画** a national land-use plan; a national plan for the use of the land. **国土緑化** national (land) afforestation; greening the nation;《the campaign for》a greener nation. **国土緑化運動** the National Land Afforestation Campaign. **国土緑化推進機構** the National Land Afforestation Promotion Organization（略：NLAPO）．

国内 ▶〜の internal; domestic; home; inland. ¶ **国内環境政策法** ⇨英和 NEPA **国内希少野生動植物種** the National Endangered Species of Wild Fauna and Flora. **国内純生産** (a) net domestic product（略：NDP）. / 環境調整済〜純生産 ⇨グリーン（グリーン GDP）**国内材** 〔木材〕domestic [domestically produced] lumber [timber, building materials];〔石材〕domestic [domestically quarried] rock [stone]. **国内在庫** 〔石油などの〕domestic stocks《of oil》;《oil》held in stock within the country.

国民生活 people's lives《in Japan》; the life of the people [citizen]; the way people live《in the country》; people's livelihoods; national life. ▶〜に大いにかかわる be closely connected with citizens' livelihood(s). /〜の水準を上げる improve the (citizen's) standard of living; raise [elevate] living standards in the country. /〜を安定[向上]させる stabilize [improve] national life; make citizens' lives more stable [better]. ¶**国民生活センター** the National Consumer Affairs Center of Japan（略: NCAC）. **国民生活白書** a white paper on the national [people's] life [lifestyle];〔内閣府の〕the White Paper on the National Lifestyle《2007》. **国民生活様式** the national lifestyle [way of life]; citizens' lifestyles. **国民生活基礎調査**〔厚生労働省が行う〕the Comprehensive Survey of the Living Conditions, Health and Welfare of the People. **国民生活指標** the People's Life Indicators（略: PLI）. **国民生活選好度調査**〔内閣府が行う〕the National Survey on Lifestyle Preferences. **国民生活動向調査**〔国民生活センターが毎年実施する〕the Survey of Trends in People's Livelihood, an annual survey of consumer attitudes. **国民生活モニター調査**〔内閣府が行う〕a survey on national lifestyle.

穀物 cereals; grain; corn. ▶（貨車・船などで）〜を運ぶ carry grain [cereals, corn]. ¶**穀物運搬船** a grain carrier. **穀物価格** grain prices. **穀物乾燥機** a grain drier. **穀物在庫率** a grain stock-to-use ratio. **穀物自給率** a [the] grain self-sufficiency rate [ratio]. **穀物市場** the grain [corn] market. **穀物商**〔商人〕a dealer in grain [cereals, corn]; a grain [corn] merchant;〔商店〕a shop dealing in grain [corn]. **穀物飼料**〔家畜用の〕(a) cereal feed. **穀物倉庫** a granary; a grain elevator. **穀物相場** the market price of grain [cereals, corn]; grain prices. **穀物取引所** a grain [corn] exchange. **穀物問屋[仲買人]** a grain broker; a corn factor. **穀物農耕** grain growing; growing grain [cereals, corn]. **穀物畑**〔栽培用の〕a grainfield; a cornfield; a field for grain [cereals, corn];〔栽培されている〕a field of grain [corn]. **穀物肥育** grain feeding. / ～肥育牛[牛肉] grain-fed cattle [beef]. **穀物メジャー** a major multinational grain company. **食用穀物** [小麦・米など] (a) food grain. **粗粒穀物** [トウモロコシ・大麦など] (a) coarse grain.

国有林 a national [state] forest; national woodland. ▶～を払い下げる auction (off) (a piece of) national forest land. ¶**国有林野事業特別会計** the Special Account for the National Forest Service.

国立 ¶**国立環境研究所** the National Institute for Environmental Studies (略: NIES). **国立環境研究所 地球環境研究センター** the Center for Global Environmental Research (of the National Institute for Environmental Studies) (略: CGER). **国立公園** ⇨公園, 管理

国連 [国際連合] the United Nations; the UN. ▶～を脱退する leave [secede from, withdraw from] the United Nations. **国連オブザーバー** [オブザーバーとして国連活動に参加する非加盟国] a United Nations observer. **国連会議** the United Nations Congress. **国連海洋法会議** the United Nations Conference on the Law of the Sea (略: UNCLOS). **国連海洋法条約** the United Nations Convention on the Law of the Sea. **国連加盟** membership of [in] the United Nations; UN membership. **国連加盟国** a member of the United Nations; a UN member. **国連環境開発会議** the United Nations Convention on Environment and Development (略: UNCED). **国連環境計画** the United Nations Environment Program (略: UNEP). **国連環境計画金融イニシアティブ** the UNEP Finance Initiatives (略: UNEP FI). **国連監視団** a United Nations observer mission. **国連旗** the United Nations Flag; a UN flag. **国連機関** a United Nations organ. **国連気候変動事務局** the United Nations Climate Change Secretariat. ★⇨気候(気候変動) **国連緊急特別総会** an Emergency Special Session of the UN General Assembly. **国連決議** [総会の] a United Nations [UN] General Assembly resolution; [安全保障理事会の] a United Nations [UN] Security Council resolution. / ～決議 1441 号 UN resolution 1441. **国連公海漁業協定** the United Nations Agreement on Fishing on the High Seas.

国連システム the United Nations system; the UN system. **国連食糧農業機関** the Food and Agriculture Organization of the United Nations (略: FAO). **国連人口委員会** the United Nations Population Commission. **国連人口基金** the United Nations Population Fund (略: UNFPA). ★ UNFPA は旧称 国連人口活動基金 (the United Nations Fund for Population Activities) による. **国連制裁** 《impose, enforce》a United Nations sanction. **国連世界気象機関** the (United Nations) World Meteorological Organization (略: WMO). **国連専門機関** 〔ILO, UNESCO, IMF などの〕a specialized agency of the United Nations. **国連総会** the United Nations [UN] General Assembly (略: UNGA); the General Assembly of the United Nations. **国連大使** an ambassador to the United Nations. **国連代表** (a member of) the 《Japanese》delegation to the United Nations. **国連代表部** the Permanent Mission 《of Japan》to the United Nations. **国連中心外交** a United Nations-centered diplomacy. **国連統計** United Nations statistics. **国連人間環境会議** a United Nations Conference on the Human Environment (略: UNCHE). ★通称 ストックホルム会議 (the Stockholm Conference). **国連人間居住計画** the United Nations Human Settlements Programme; 〔通称〕UN-Habitat. **国連派遣軍** a United Nations expeditionary force. **国連分担金** 《Japan's》financial contribution to the United Nations; a《country's》UN assessment [(assessed) share of the budget of the United Nations]. **国連放送** UN [(the) United Nations] Radio. **国連理事国** a member of the United Nations 《Security, Economic and Social, Trusteeship, Human Rights》Council.

古紙 wastepaper; used paper. ▶〜のリサイクル[再利用] recycling [reuse] of wastepaper; wastepaper recycling. / 〜を回収する collect wastepaper (for recycling). ¶ **古紙回収(業)** (the business of) collecting wastepaper for recycling; wastepaper collection. **古紙回収業者** a dealer in wastepaper for recycling; a used paper recycler. **古紙配合[利用](比)率** the ratio of recycled paper used in 《printing paper》. / 〜利用率 70％の再生紙 recycled paper with 70% wastepaper content. **雑誌古紙** old magazine paper (略: OMG). **新聞古紙** old newspapers. **段ボール古紙** old corrugated cardboard (略: OCC). **ミックス古紙** mixed (waste) paper (for recycling); mixed used paper.

湖沼(こしょう) lakes and marshes; inland waters. ▶〜の群集 a lake (biotic) [len(i)tic] community. ★⇨群集 ¶**湖沼学** limnology. 形 limnologic(al). **湖沼学者** a limnologist. **湖沼型** a type of lake, pond or marsh; a lake type. **湖沼漁業** (a) lake fishery; fishing in (freshwater) lakes and marshes. **湖沼堆積物** a lacustrine sediment [deposit]. **湖沼水質保全特別措置法** the Law Concerning Special Measures for the Preservation of Lake Water Quality; the Law for Special Measures for Preserving Clean Lake and Marsh Waters. ★通称「湖沼法」 the Clean Lakes Law. **湖沼養殖** freshwater fish breeding; breeding fish in (freshwater) lakes and marshes.

個体 an individual. ¶**個体距離** individual distance. **個体群** a population. **個体群生態学**[生物学] population ecology [biology]. **個体群成長** the growth [increase, augmentation] of a population. **個体群動態論** population dynamics. **個体群密度** (a) population density. **個体差** an individual difference; individual specificity. **個体識別** individual recognition [identification]. **個体識別番号** 《a tag bearing》an individual identification number. **個体数** population size; (a) population. **個体数推定** estimation [an estimate] of population size; a population estimate **個体数ピラミッド** an ecological pyramid; a pyramid of numbers. **個体性** individuality. **個体淘汰** individual selection. **個体発生** ontogeny. **個体変異** individual variation.

国家環境政策法 ⇨英和 NEPA

固定価格買取制度 ⇨英和 feed-in tariff

古都保存法 the Ancient Capitals [Cities] Preservation Law.

庫内温度 〔冷蔵庫などの〕the internal [box, case] temperature; the temperature inside《a refrigerator》.

粉石鹸(こなせっけん) ⇨石鹸

コプラナーPCB 〔毒性が強いポリ塩化ビフェニール〕a coplanar PCB. ★⇨英和 coplanar PCB

ごみ，ゴミ 〔不要物〕garbage; rubbish; junk; waste; trash; litter;〔汚らしいもの〕dirt; filth;〔ちり・ほこり〕dust. ★⇨くず，廃棄物，解体(解体ごみ)，家庭ごみ，金属ごみ，混合ごみ，災害(災害ごみ)，事業(系)ごみ，生活(生活ごみ)，剪定(せんてい)ごみ，粗大ごみ，都市ごみ，浮遊ごみ，生ごみ，破砕ごみ ▶可燃[燃える]〜 combustible [burnable, flammable] garbage [trash]. / 不燃[燃えない]〜 nonburnable [nonflammable] garbage [trash]. / 燃える〜と燃えない〜を分ける[分別する] separate burnable from nonburnable trash; separate rubbish into burnable and nonburnable items. / 〜を分別する separate [segregate] garbage (according to type); separate garbage before disposal. ★⇨分別，ごみ袋 / 今日は燃える〜[燃えない〜，資源〜]の日だ．Today is burnable trash [nonburnable trash, recycling] day. ★⇨資源(資源ごみ) / 〜の(収集)日 a garbage [trash] collection day. / 燃えない〜の収集は何曜日ですか．On what days is nonburnable [noncombustible] trash collected? / 台所の〜 kitchen refuse; garbage. ★⇨一般 / 行楽地の〜 picnic-ground litter. / 〜だらけの部屋 a littered [filthy, messy] room. / うちでは1日に袋2つ分の〜が出る．Our house produces a couple of bags of garbage a day. / 排出される〜の量はこの地区だけで年間数百トンにも上る．The trash thrown away in this area alone amounts to several hundred tons a year. / 家庭から出る〜の量を減らそう．Let's reduce the amount of rubbish thrown out by each family [household]. / 〜が増えるから包装しないでください．Forget the wrapping—it just makes for more garbage. ★⇨容器包装 / この残飯は全部〜になってしまう．All these leftovers are just going to become garbage. / 富士山から〜をなくしたい．We want Mt. Fuji to be free of trash. / 〜の減量[削減] reducing (the amount of) garbage [waste]; a reduction in the amount of trash (produced); waste [garbage, trash] reduction. / 〜の量がどっと増えた．The volume of garbage suddenly increased. / 〜を捨てる throw out garbage. ★⇨投棄，捨てる / (収集日に)〜を出す take out the garbage [trash]. / 〜は夜出さないで，その日の朝に出してください．Don't put the garbage out in the evening; put it out early next day [in the morning]. / 〜の出し方が変わったことがまだ住民に徹底していない．Not all the residents were aware that the garbage collection procedure had changed. / 〜を掃き集める sweep up the dust. ★⇨ごみ拾い / 砂浜に〜を散らかす litter [strew garbage on] the beach. / カラスが〜を散らかす．Crows strew garbage all over. / 野良猫は〜を漁(あさ)って生きている．Stray cats survive by scavenging in the rubbish. / カラスが〜をあさりに来るので困っている．We are having a lot of difficulty with crows, which come and raid [pick over, peck through] the garbage. ★⇨ごみ漁り / ここに〜を捨てないでください．〔掲示〕No litter please. | No dumping. / 〜を捨てるにも費用がかかる．Getting rid of things [Throwing out rubbish] costs money too. / 〜はお持ち帰りください．〔掲示〕Please take your litter home with you. / 〜持ち帰り運動 a "take your litter home" campaign. / 〜は持ち帰るという方式を登山者たちに徹底させる drum into mountain climbers the practice of taking their trash back with them. ¶**ごみ漁り** digging through the trash [rubbish]; dumpster [skip] diving. / 〜漁りをする forage through the trash. ★⇨ごみ箱　**ごみ埋め立て地**　a refuse [garbage] landfill. ★⇨埋め立て地　**ごみ置き場**　〔回収用の〕a garbage [trash] collection [disposal] area; a place to leave rubbish [household waste]. ★⇨ごみ捨て場 / 〜置き場を利用している以上，掃除当番は当然だ．I consider the taking of turns to clean the rubbish disposal area as something obligatory for me, as a user (of the area). **ゴミ学**　garbology. **ごみ固形燃料**　refuse-derived fuel. ★⇨英和 RDF　**ごみ収集（作業）**　garbage [refuse, trash, waste] collection. / 〜収集作業員 a garbage collector; a sanitation man [worker]; a dustman. / 〜収集車 a garbage [sanitation] truck; a dust cart. / 〜収集車が町内を回っている．The garbage truck is making its rounds of the neighborhood. / (ごみの日の)〜収集所[集積所]　⇨ごみ置き場 / 〜収集カレンダー a garbage [rubbish, trash, refuse] (collection) calendar. / 〜収集を有料化する begin charging a fee for garbage collection. **ごみ焼却**　refuse [garbage] incineration. / 〜焼却施設[場] a refuse

ごみ, ゴミ

[garbage] incineration plant. / 近隣住民の要請に応じて〜焼却場の移転が議会で決まった. The Assembly decided, in response to the demand of nearby residents, to have the waste incineration plant moved. / 〜焼却炉 a garbage incinerator. / 〜焼却灰 refuse incineration ash.　**ごみ処理[処分]**　refuse [garbage, waste] disposal. / 〜処理施設[処分場] a garbage treatment [disposal] plant; a refuse [garbage] dump. / この〜処理施設の素晴らしい点はまったく環境に悪影響を及ぼさないということです. The beauty of this garbage disposal facility is that it does not harm the environment in any way. / 当面〜処分場の建設を凍結する put [impose] a temporary freeze on the construction of garbage-treatment plants. / 〜処理場の建設が間に合わない. The building of waste-disposal facilities isn't keeping up with our needs. / 〜処分場がじき満杯になる. The garbage dump will soon be filled to the brim. / 実際問題としてすでに各地の〜処分場は満杯になりつつある. In practice, landfill sites up and down the country are already approaching capacity. / 大都市の〜処理能力は限界に達している. The ability of large cities to dispose of waste [rubbish] has reached its limits. ★⇒計画　**ごみ捨て場**　a garbage disposal area; a (garbage) dump [dumping ground]; a garbage pit.　**ごみゼロ**　zero waste [garbage, rubbish]. / 〜ゼロ地区 zero-waste communities. / 〜ゼロ運動 a zero-waste campaign. / 〜ゼロ社会を実現させる bring a trashless society into being.　**ごみ溜め**　a garbage dump; a dumpsite; a scrap heap; a dust hole.　**ごみ燃料**　＝ごみ固形燃料.　**ごみ箱**　a trash [garbage] can [pail]; a garbage box; a litter box [pail]; a dustbin. / 〜箱を漁る forage through a trash pail.　**ごみ挟み**　garbage [refuse, rubbish] tongs.　**ごみ発電**　〔ごみ焼却廃熱を利用した発電〕electric power generation by refuse incineration. ★⇒都市(都市ごみ)　**ごみ拾い**　garbage [rubbish, refuse, waste, litter] collecting; garbage collection; picking up garbage; collection of garbage;〔清掃〕cleaning up; a clean-up. / 〜拾いをする collect garbage; pick up litter [garbage]; clean up.　**ごみ袋**　a garbage [trash] bag. / 〜は市が指定する〜袋を使って出してく

ださい．Put out [Bin, Dispose of] your garbage using the bags specified by the city. / ～の分別収集を徹底させるというねらいで透明な～袋の使用が始まった．The use of clear plastic garbage bags was initiated with the aim of all-out enforcement of itemized garbage collection. / 半透明の～袋 a semitransparent [half-transparent] garbage bag. **ごみ問題** the garbage (disposal) problem. **ごみ屋** ⇨ごみ収集作業員, 廃品回収業者 **宇宙ごみ** space debris. / 働きを終えた人工衛星はそのまま宇宙～になるの？ Do satellites that have completed their tasks just stay up there as space trash? **散在ごみ** (public) litter. **粗大ごみ** oversized [large, bulky] refuse [trash]. ⇨粗大ごみ **日常ごみ** household garbage [rubbish, trash, refuse]; domestic garbage; everyday garbage. **漂着ごみ** beach waste [rubbish, garbage]; waste washed ashore [on shore].

コミュニティー 〔共同体〕a community; 〔地域社会〕a (local) community. ¶**コミュニティー活動** community activities. 图 community activist. **コミュニティー・ケア** 〔在宅介護〕community care. **コミュニティー・センター** 〔公民館〕a community center. **コミュニティー放送** community broadcasting. **コミュニティー・ホール** a community hall. **コミュニティーFM** 〔局〕a community FM radio station;〔放送〕community FM (broadcasting). **コミュニティー・オーガニゼーション** 〔地域組織活動〕community organization. **コミュニティー・ガーデン** 〔地域の公共用地や空き地などにつくり出された緑地空間〕a community garden. **コミュニティー・スクール** 〔地域社会が運営に参画する学校〕a community school. **コミュニティー道路** a community road. **コミュニティー・バス** 〔地域社会の需要に合わせて運行されるバス路線〕(a) community bus service;〔1台のバス〕a community bus. **コミュニティー・ビジネス** 〔地域社会の要求に応え，地域の利益を重視する事業〕a community business. **コミュニティー・ホテル** 〔結婚式場など地域社会の様々な需要を満たすホテル〕a community hotel. **コミュニティー・ワーク** 〔社会福祉の援助方法の1つ；地域援助技術〕community work.

固有 ▶～の proper; inherent; endemic; characteristic; intrinsic. ★⇨原産 / 日本〜の動植物 plants and animals indigenous [native] to Japan. / 本能は動物〜のものだ. Instinct is innate [an innate propensity] in animals. ¶ **固有種** 〔特定の地域に限って分布する生物〕an endemic [indigenous] species. **固有植物[動物]** indigenous flora [fauna].

ゴルフ場農薬 pesticides used on golf courses.

混獲 〔流し網漁などでの目的魚種以外の漁獲〕incidental catch 《of dolphins》.

混合 ¶ **混合飼料** 〔家畜用の〕(a) mixed feed. **混合ごみ** mixed refuse [garbage]. **混合水域** 〔黒潮(暖水塊)と親潮(冷水塊)が混ざって存在する場所〕the Mixed Water Region (略: MWR). **混合農業** mixed farming. **混合肥料** (a) compound manure; (a) mixed fertilizer; (a) compost. **混合[混交]林** 〔複数の優占樹木をもつ森林〕(a) mixed forest.

昆虫 an insect. ▶〜食の動物 entomophagous [insectivorous] animal. ¶ **昆虫アレルギー** an insect allergy. **昆虫学** entomology; insectology. / 〜学者 an entomologist. **昆虫忌避剤[薬]** an insect repellent. **昆虫恐怖症** entomophobia; insect phobia; a fear of insects. **昆虫採集** insect collecting; entomologizing; 《口》bug hunting. / 〜採集をする collect insects; go hunting for insects. / 〜採集家 an insect collector; 《口》a bug-hunter. **昆虫刺傷[咬傷]** an insect bite. **昆虫媒介感染症** an insect-borne infectious disease. **昆虫ホルモン** an insect hormone.

コンバインド・サイクル発電 〔複合発電〕combined cycle power generation. ★⇨英和 combined heat and power

コンパクト・シティー 〔都市・商業機能を中心区域に集中させた都市〕the [a] compact city.

コンポスター 〔コンポスト容器〕a composter; a compost bin.

コンポスト ⇨英和 compost ¶ **コンポスト・トイレ** a compost toilet.

さ

サーベイ・メーター　〔放射線の携帯用測定器〕a portable (Geiger) survey meter.

サーマル・リサイクル　〔燃焼材料としての再利用〕thermal recycling.

サーモカルスト　⇨英和 thermokarst

菜園　a vegetable [kitchen] garden;〔市場向け野菜の〕a truck farm [garden]; a market garden.　¶**菜園経営**　truck farming; market gardening.　**菜園経営者**　a truck farmer; a market gardener.　**菜園作り**　vegetable gardening.　**家庭菜園**　a household vegetable garden; a family garden; an allotment.　★⇨収穫　**ベランダ菜園**　a veranda garden; a (little) garden on the veranda.

災害ごみ　disaster garbage; garbage generated by a disaster.

災害廃棄物　disaster waste; waste generated by a disaster.

催奇形性(さいきけいせい)　teratogenicity; teratogeny.　▶～の teratogenic; teratogenetic.　¶**催奇(形)性作用**　teratogenic effect.　**催奇(形)性試験**　a teratogenicity test.　**催奇(形)性物質**　a teratogen.

採餌(さいじ)　〔動物の〕feeding; food getting.　¶**採餌行動**　feeding behavior.　**採餌場**　a feeding area; feeding grounds.　**採餌方法**　a feeding method; a method of feeding.

再資源化　recycling; resource recovery.　★⇨資源, リサイクル　▶～する recycle《cellphones》; compost《kitchen waste》.　¶**再資源化率**　a recycling rate.

採集　collection; collecting; gathering.　▶～する collect; gather.　★⇨昆虫, 植物 / その虫は箱根以東ではまだ～されたことがない. That (species of) insect has never been collected in Hakone or parts east of Hakone.　¶**採集網**　an insect net;〔チョウの〕a butterfly net.　**採集家**　a collector.　**採集地点**　〔動植物が発見された〕a station.

最終処分場　a final disposal site.　★⇨処分場, 処理場

最終沈殿池　〔水道の〕a final settling tank.

再循環　⇨循環

再使用　reuse; reusage; further [repeated] use [usage].　★⇨使

用, 利用, 再利用 ▶～する reuse; use《cans》again; make further [repeated] use of《a container》; recycle《materials, cans》. / ～可能な reusable; recyclable.

最小殺菌濃度 the minimum bactericidal concentration (略: MBC).

最小致死量 the minimum lethal dose.

最小治療量 the minimum therapeutic dose.

最小(発育)抑制濃度 〔微生物の〕the minimum inhibitory concentration (略: MIC).

最小有効量 〔薬物の〕the minimum effective dose; the minimum therapeutic dose.

再商品化 〔廃品などの〕recycling materials into saleable products [commodities]. ▶～する turn recycled materials into saleable products [commodities].

菜食 a vegetable [plant] diet. ▶～する live on vegetables. ¶**菜食主義** vegetarianism. **菜食主義者** a vegetarian. **完全菜食主義者** a vegan. **乳菜食主義者** 〔乳製品は食べる菜食主義者〕a lacto-vegetarian; a lactarian. **卵菜食主義者** 〔卵は食べる菜食主義者〕an ovo-vegetarian. **乳卵菜食主義者** 〔乳製品と卵は食べる菜食主義者〕a lacto-ovo-vegetarian; an ovo-lactarian.

再処理 〔再利用のための〕reprocessing. ▶～する reprocess. ¶**再処理工場[施設]** a reprocessing plant [facility]. / 使用済み核燃料[プルトニウム, ごみ]～工場 a spent nuclear fuel [plutonium, waste] reprocessing plant [facility].

再生 〔廃品からの再生産〕regeneration; recovery; reclamation; (the) recycling《of...》. ★⇨リサイクル ▶汚れた川を～させる restore a polluted river. / ～する reclaim; regenerate; recycle. / ～できる素材 recyclable material. / ペットボトルは～されて衣料品になる. PET containers are recycled and become articles of clothing. ★⇨再生樹脂 ¶**再生可能エネルギー** ⇨英和 renewable energy **再生原料** 〔再生資源から作られる原料〕recycled (raw) materials. **再生ゴム** reclaimed [regenerated] rubber. **再生剤** 〔イオン交換樹脂の〕regenerant. **再生産業** the recycling industry. **再生紙** recycled paper. / この名刺は～紙を利用しています. This business card is made of recycled paper. / このトイレットペーパーは100パーセント～紙です. This toilet paper is 100-percent recycled paper.　**再生紙使用マーク** the recycled paper logo.　**再生紙製品** a recycled-paper product.　**再生資源** recycled material.　**再生資源利用促進法** the Law for Promotion of Utilization of Recyclable Resources.　**再生樹脂** reprocessed resin. / ペットボトル〜樹脂 recycled PET (bottle) resin.　**再生水**〔下水を高度に浄化処理した水〕recycled [reclaimed] water.　**再生セルロース[繊維素]** regenerated cellulose.　**再生繊維** regenerated fiber.　**再生タイヤ** a recap(ped tire); a retread.　**再生品** a recycled [reclaimed, salvaged] item.　**再生プラスチック** recycled [reprocessed] plastic.　**再生油** reclaimed oil; recovered oil.　**再生用古紙** old [used] paper for recycling.　**再生羊毛** reworked wool.　**再生率**〔廃棄物のリサイクル率〕a recycling rate.　**再生利用** recycling; reclamation. /〜する reclaim; recycle; regenerate.　★⇨再利用　**再生利用業者** a scrap dealer.

最大許容線量 the maximum [maximal] permissible dose (略: MPD).

最大許容濃度 the maximum permissible concentration.

最大経済生産量 the maximum economic yield (略: MEY).

最大持続生産量 the [a] maximum sustainable yield (略: MSY).

最大耐(薬)量 the maximum [maximal] tolerated dose.

最大無作用量 the maximum no-effect level.

最大有効量〔薬物の〕the maximum [maximal] effective dose.

栽培 cultivation; culture; raising; growing; growth.　★⇨促成栽培，減農薬，化学(化学肥料)，特別栽培，自家栽培　▶〜する cultivate; grow; raise; rear. / トマトを〜栽培している土地 tomato-growing land; land cultivated with [under] tomatoes. / トマトを〜している農家 a tomato (-growing) farm. / この地方は果樹の〜が盛んである．Fruit trees are widely cultivated in this region. / キノコの〜キット a mushroom growing kit.　¶**栽培技術** cultivation techniques.　**栽培漁業** fish farming.　**栽培限界** a cultivation limit.　**栽培者** a grower; a cultivator.　**栽培種** agricultural [cultured] species.　**栽培植物** a domesticated [cultivated] plant.　**栽培品種** a cultivar;〔農作物〕an agrotype.　**栽培法** a cultivation method. / 菊の〜法 a method of chrysanthemum culture;

how to raise [cultivate, grow] chrysanthemums.　**栽培面積**　the area of (the) land under cultivation 《with tomatoes》.　**栽培物**　〔天然物の野菜に対して〕cultivated [farm, garden] vegetables. / このキノコは～ものです．This is a cultivated mushroom.　**周年栽培**　year-round cultivation.　**人工栽培**　artificial cultivation 《of edible mushrooms》.

細胞　a cell.　★⇨核, 生殖　¶**細胞分裂**　cell division.　▶バクテリアはおよそ20分ごとに～分裂をする．Bacteria divide about once every 20 minutes.　**多[単]細胞生物**　a multicellular [unicellular] organism.

債務環境スワップ　a debt-for-nature swap.　★⇨英和

在来　¶**在来種**　〔外来種に対して〕a native variety [species]; a local [domestic] breed;〔改良種に対して〕a wild [natural, non-cultivated, non-modified] species.　★⇨外来, 帰化　▶～種のイチゴ a native strawberry. / 非～種〔外来種〕a non-native [an introduced] variety [species]; an imported breed;〔改良種〕a non-natural [a cultivated, an improved, a modified] species.　**在来生物**　a native [an indigenous] organism;〔種〕a native [an indigenous] species.

再利用　reuse; recycling.　★⇨リサイクル, 再生　▶～する reuse; recycle. / ～可能な reusable; recyclable.

砂丘　a (sand) dune; a sandhill; a sandbank.　¶**砂丘植物**　sand-dune plants.　**砂丘地農業**　sand dune farming.　**砂丘林**　a forest on a sand dune.

削減　▶～する reduce; cut (back); retrench; trim; scale back; pare (back [down]);〔人員を〕slim down.　★⇨排出(排出削減), 温室(温室効果ガス) / EUは，温室効果ガス排出量の一律大幅～を主張した．The EU argued for a large uniform reduction in the emission of greenhouse gases. / 二酸化炭素の排出量をもっと～しなければならない．We must cut our emissions of carbon dioxide back even further.

作物　crops; (farm [agricultural]) products.　★⇨農作物　▶ひょうが降ってホウレンソウやキャベツなどの～に被害が出た．Hail damaged our spinach and cabbage crops. / ～の病気 crop disease. / 今年は～の出来が早い[遅い]．The crops have ripened early [late] this year. / 気候は大きく～の出来に影響する．Weather has a large effect on crops. / イノシシに畑の～を荒ら

された．Wild boars [Feral pigs] ravaged the crops in our fields. / サツマイモはやせた土地でもできる〜だ．Sweet potatoes are a crop that can be grown even in barren [poor, depleted] soil. ¶**作物限界** crop limits. **園芸作物** garden products. **救荒作物** a famine relief crop. ★⇨減農薬，化学(化学肥料)，遺伝子(遺伝子組み換え)，除草

差し止め 〔禁止〕prohibition; a ban; 〔停止〕suspension; an injunction; a proscription; a veto. ▶差し止める〔禁止する〕prohibit; forbid; ban; embargo; lay [put, place] a ban [taboo] on…; place *sth* under a ban; lay an embargo on《exports》; 〔停止する〕suspend; suppress《a publication》. /埋め立て計画を差し止める suspend [call a halt to] plans for a landfill site. / 〜になる be banned; be forbidden; be prohibited; be proscribed. ¶**差し止め請求** a request for (an) injunction. **差し止め請求権** the right to request (an) injunction. **差し止め命令** (an) injunction. **工事差し止め** an order to stop construction; a construction ban. **工事差し止め仮処分** a provisional disposition [temporary restraining order] against construction.

座礁(しょう) ▶〜する hit [strike] a rock [reef]; 〔動けなくなる〕run aground [ashore]; be stranded. ¶**座礁死** 〔クジラ・イルカなどの〕a《whale》stranding death. / クジラがここで大量に〜死する原因はいまだ不明だ．It is still unknown why so many whales become stranded and die here.

サステ(イ)ナビリティー，サステ(イ)ナブル sustainability, sustainable ★⇨持続可能 ¶**サステイナビリティー学[サイエンス]** sustainability science [studies]. **サステイナビリティ学連携研究機構** the Integrated Research System for Sustainability Science (略: IR3S).

殺菌 disinfection; pasteurization; sterilization. ▶〜する sterilize; disinfect; pasteurize. / 〜性の disinfectant; germicidal. ¶**殺菌温度** a thermal death point (略: TDP). **殺菌機** a sterilizer. **殺菌牛乳** sterilized [pasteurized] milk. **殺菌効果** a disinfectant [disinfection] effect. / アルコール製剤のほうが石鹸をつかった手洗いより〜効果が高い．An alcohol-based hand rub is more effective than soap as a disinfectant [disinfects the hands more effectively than soap]. **殺菌剤** an antimicrobial; a bactericide; a germicide; a sterilizer; a disinfectant; a germicidal agent. **殺菌作用** bactericidal action. **殺菌試験** a bactericidal test. **殺菌素** bacteri(o)cidin;〔溶菌素〕bacteriolysin. **殺菌灯** a germicidal lamp. **殺菌法** a germ-killing process; pasteurism. **殺菌薬** bactericides; germicides. **殺菌力** sterilizing [germicidal] power. **加熱殺菌** heat sterilization. **低温殺菌** low-temperature pasteurization.

雑種 〔異種間に生まれた生物〕a mixed breed; a crossbred; a crossbreed; a half-breed; a cross《between...》; a mongrel; a hybrid; a bastard;〔つぎ木の〕a graft hybrid. ★⇨交雑, 交配 ▶〜の crossbred; half-bred [-breed, -blood(ed)]; mixed-breed; of mixed breed; hybrid; mongrel. /〜の馬 a crossbred horse. /〜の犬 a mongrel (dog);《口》a pooch; a mutt. / 純血種より〜の方が病気に強い. Crossbreeds are less susceptible to disease than thoroughbreds. /〜を作る cross two breeds [one breed with another]; interbreed; hybridize; produce a hybrid. ¶**雑種強勢** heterosis; hybrid vigor. **雑種細胞** a hybrid cell. **雑種性** hybridity. **雑種世代** a filial generation. **雑種第一代** the first filial generation (略: F1). **雑種胚** a mixture of human and animal cells. **雑種不稔性** hybrid sterility. **一代雑種** a first-filial-generation hybrid. **栄養雑種** a vegetative hybrid. **種間雑種** an interspecific hybrid. **単性[一遺伝子]雑種** a monohybrid. **両性[二遺伝子]雑種** a dihybrid.

殺処分(さつしょぶん) 〔動物の〕slaughter(ing); euthanasia. ▶〜する〔家畜を〕slaughter; destroy;〔ペットなどを〕euthanize;《口》put down; put to sleep. / 感染を防ぐために数万頭の牛が〜された. Many tens of thousands of cows were put down in order to avoid an epidemic.

雑草 a weed;〔イネ科以外の, 広葉の〕a forb. ★⇨スーパー雑草, 除草, 草むしり ▶イネ科の〜 a gramineous weed. / 一雨降るとすぐ〜が生えてくる. The weeds come up straight after a shower [rainfall]. / 庭一面に〜が生えている. Weeds have overrun the garden. | The garden is overgrown [overrun] with weeds. / 畑は手を抜くと〜がはびこる. If you don't take constant care of the fields, the weeds will take over. / 〜の生い茂った weedy [weed-infested, weed-grown, weed-ridden] 《garden》; overgrown with weeds. / 〜のたくましさ[生命力] the toughness [tenacity] of weeds. / 芝生の中の〜を抜き取る weed a lawn. ¶**雑草地** a weedy [weeded] area; a place overgrown with weeds. ★⇨草地(そう)

殺鼠剤(さっそざい) a rodenticide; a raticide; a rat poison; a ratsbane.

殺ダニ剤 a miticide; an acaricide.

殺虫 ¶**殺虫剤** an insecticide; a vermicide; a pesticide; a bug killer; 〔粉の〕an insect powder; 〔幼虫の〕a larvicide; 〔成虫の〕an adulticide. **殺虫剤噴霧器** an insecticide sprayer; a flitgun; an aerosol bomb. **殺虫灯** an insect(-killing) lamp.

雑排水(ざっぱいすい) ⇨家庭雑廃水, 生活雑排水

里海(さとうみ) a section of ocean that coexists with a nearby populated area.

里地 mountain village (region). ¶**里地里山保全再生モデル事業** a model project for regenerating mountain villages and the woodland around them.

里山 a seminatural area that coexists with a nearby populated area and is an important source of fuel, food, compost etc.; a *satoyama*. ¶**里山林** (a) *satoyama* woodland.

砂漠 a desert. ¶**砂漠化** desertification. ▶過度の放牧, 農業用水の無計画な取水もまた, 〜化の進行に拍車をかけている. Desertification is also being accelerated by excessive grazing and out-of-control water use for agriculture. **砂漠化対処条約**〔国連の〕the (United Nations) Convention to Combat Desertification (略: UNCCD, CCD). ★正式名称は「深刻な干ばつまたは砂漠化に直面する国(特にアフリカの国)において砂漠化に対処するための国際連合条約」(the United Nations Convention to Combat Desertification in Those Countries Experiencing Serious Drought and/or Desertification, Particularly in Africa). ⇨英和 COP **砂漠気候** a desert climate. **砂漠植物[動物]** a desert plant [animal]. **砂漠土** (the) desert soil. **砂漠(の)緑化** greening a desert; the greening of a desert. **国際砂漠・砂漠化年**〔2006年〕the International Year of Deserts and Desertification.

サバンナ a savanna(h). ¶サバンナ気候 a savanna climate.
サヘル ⇨英和 Sahel
砂防 erosion control; sand arrestation; sandbank fixing. ¶砂防工学 erosion control engineering. 砂防工事 erosion control work; antierosion work; sand arrestation work. 砂防造林 afforestation for erosion control. 砂防ダム a sand control [sand-trap] dam. 砂防林 tree-fences (to prevent sand movement).

サマー・タイム 〔夏時間〕daylight saving (time) (略: DST); 〔英国の〕British summer time (略: BST). ▶〜になる go on daylight saving time [British summer time]. / 太陽の恩恵を積極的に活用しようというのが〜の目的である. Making active use of the benefits of the sun is the purpose of daylight saving time.

3R(さんアール) ⇨ 3R(スリーアール)

産業 industry ¶産業安全 industrial safety. 産業医 an industrial physician. 産業医学 industrial medicine. 産業衛生 industrial hygiene. 産業公害 industrial pollution. 産業廃棄物 industrial waste [waste products]; waste from industry. ★⇨産廃, 残存 産業廃棄物運搬車(両) an industrial waste truck. 産業廃棄物汚染 industrial waste pollution. 産業廃棄物管理票 〔マニフェスト〕a manifest. 産業廃棄物処理場 ⇨処理

三元触媒(さんげんしょくばい) ⇨触媒

サンゴ, 珊瑚 coral. ★⇨白化現象 ▶〜状の coralliform. / 〜質の coralline. / 〜色の coral. ¶珊瑚石 corallite. 珊瑚環礁 an [a coral] atoll. 珊瑚採取 coral fishing. / 〜採取漁船 a coral boat. 珊瑚珠(さんごじゅ) coral beads. 珊瑚礁 a coral reef; (環礁) an atoll; (海上にわずかに頭を出しているもの) a cay [key]. 珊瑚石灰岩 coral limestone. サンゴ虫 a coral polyp [insect]. 珊瑚島 a coral island. 国際サンゴ礁年 〔1997年, 2008年〕the International Year of the Reef. 粒珊瑚 a seed coral.

酸性 acidity. ▶〜の acid(ic). / 〜になる, 〜化する become acid; acidify. ¶酸性雨 acid rain; 〔酸性降下物〕acid fallout. 酸性塩 acid salt; supersalt. 酸性岩 acidic rock. 酸性酸化物 an acidic oxide. 酸性紙 acid [acidic] paper. 酸性試験 an acidity test. 酸性食品 acid [acidic] food; acid residue food. 酸性せっけん acid soap. 酸性泉

an acid spring. 酸性染料 an acid dye. 酸性沈着 acid deposition. 酸性土壌 acid soil. 酸性土植物 an oxylophyte. 酸性粘土 acid clay. 酸性白土 (Japanese) acid clay [earth]. 酸性反応 (an) acid reaction. 酸性肥料 acid fertilizer. 酸性霧 acid mist [fog]. 酸性雪 acid snow.

残存 ▶～する survive; be extant; be still existent [in existence, alive]; be left; remain. / 産業廃棄物に～する有害化学物質 toxic chemical substances residual in industrial wastes. ★ ⇨残留 ¶**残存器官** a residual [rudimentary] organ. **残存種** a relict. **残存動物群[植物群]** relict fauna [flora].

山地 a mountainous [mountain] district [region]; a hilly district. ¶**山地草原** an upland meadow. **山地帯** a mountain zone. **山地多雨林** a montane rain forest. **山地氷河** a montane glacier.

産地 a producing center; a place [an area] of production [origin]; (馬などの) a breeding center; (植物の) a growing district; a locality. ★⇨原産地 ▶米の～ a rice-producing[-growing] district. / 肉の～を表示する label meat with (the name of) the place of origin. / ～を偽って販売する sell under (the name of) a false place of origin. / ～直送[産直]の direct from the producers. / ～直送[産直]の野菜 farm-fresh vegetables. / ～直送のカキ[ミカン] oysters [mandarines] sent directly from the place where they are grown. / ～直送サービス (a service) sending goods directly from the place of production. / ～直売のリンゴ apples sold directly by producers. ¶**産地偽装** fraudulently claiming that《produce, vegetables, fruit, etc.》comes from a particular area or country. **産地呼称** naming a product after the place where it is produced; giving a product an appellation of origin. **産地証明書** a certificate of origin [provenance]. **産地直結** a direct tie-up with the place of production. **産地廃棄** 〔豊作による値崩れを防ぐための，野菜などの〕throwing out produce [vegetables, fruit] (in order to keep up the price). **産地表示システム** a place-of-origin labeling system. **産地銘柄** a local brand (name).

三点比較式臭袋法 ⇨臭袋法(しゅうたいほう)

残土 surplus soil. ★⇨建設残土, ウラン残土 ¶**残土処理** removal of surplus soil.　**残土処理[処分]場**　a surplus soil disposal site; a surplus soil dump.

産廃　〔産業廃棄物〕industrial waste(s). ★⇨捨てる　¶**産廃運搬車(両)**　〔産業廃棄物運搬車両〕an industrial waste truck.　**産廃業者**　an industrial waste collector.　**産廃銀座**　〔産業廃棄物処理場などが密集している地域〕an area with many industrial waste processing and disposal sites; a haven [mecca] for industrial waste.　**産廃処分場[処理場]**　⇨処分場, 処理場

産廃特措法　the Law on Special Measures against Industrial Waste. ★正式名称は,「特定産業廃棄物に起因する支障の除去等に関する特別措置法」(the Law Concerning Special Measures for the Removal of Obstacles Caused by Specified Industrial Waste).

傘伐(さんばつ)作業　the shelter-wood system.　¶**傘伐林**　a shelter-wood forest.

散布　(a) scattering; (a) sprinkling; (a) spraying; dispersion; dispersal. ▶～する scatter; sprinkle; spray; dust; spread; (飛行機から) drop. / 床下に石灰を～する sprinkle lime under the floor.　¶**空中散布**　〔農薬・殺虫剤などの〕aerial (crop) spraying; crop dusting;〔肥料の〕aerial topdressing. / 農薬の空中～ aerial crop dusting. ★⇨農薬　**散布器**　a sprinkler.　**散布剤**　dusting powder; epipastic.　**散布装置**　a spraying [sprinkling] device; a sprayer; a sprinkler.　**散布薬**　dusting powder.

産卵　laying eggs; egg-laying; (魚貝の) spawning; (昆虫の) oviposition. ▶～する lay [deposit] eggs; spawn; shoot [deposit] spawn;〔昆虫が〕oviposit;〔ハエが〕blow《a wound》.　¶**産卵回遊**　spawning migration.　**産卵管**　〔昆虫の〕an ovipositor.　**産卵期**　〔魚の〕a spawning season.　**産卵場[場所]**　〔魚の〕a spawning ground.　**産卵礁**　a spawning reef.　**産卵地**　a spawning ground. / アカウミガメの～地 the spawning grounds of the loggerhead turtle. / 毎年,～の時期を迎えたウミガメが夜間この浜辺にやってくる. Every year turtles that have entered their egg-laying season come to this beach at night. ★⇨海(海亀), 偽装　**産卵能力**　〔ニワトリなどの〕egg-laying performance.　**産卵率**　《daily, monthly, annual》egg production (rate) [output].

残留　¶**残留塩素**　residual chlorine. **残留抗生物質** 〔食品としての畜産動物や養殖魚に残留する抗生物質〕(an) antibiotic residue; residual antibiotic agents [antibiotics]. **残留性** 〔有機汚染物などの〕persistency; persistence. ▶〜性の(ある) persistent; residual. **残留性有機汚染物質** 〔有害化学物質の総称〕persistent organic pollutants (略: POPs). ★⇨残存, 英和 Stockholm Convention **残留洗剤** residual detergent. **残留熱** 【原子力】afterheat. **残留濃度** 〔農薬などの〕a residual [residue] concentration. **残留農薬** agrochemical residues; residues from agricultural chemicals; pesticide [herbicide, insecticide, fertilizer] residues. **残留農薬基準** agrochemical residue standards. **残留噴霧** 〔残存性殺虫剤をあらかじめ噴霧する方法〕residual spray. **残留放射能** residual radiation [radioactivity].

山林 〔山と林〕mountains and forests; 〔山の林〕a forest on a mountain. ▶〜を造る[伐採する] afforest [deforest] a mountain. ¶**山林開拓** deforestation; disafforestation. **山林学** forestry; dendrology. **山林学者** a dendrologist. **山林業** the forestry industry. **山林所得** an income from forestry. **山林保護** forest conservation. **山林乱伐** reckless deforestation.

し

飼育 breeding [raising] 《of cattle》; rearing 《of silkworms》. ▶〜する breed; raise; rear; keep. / コオロギ〜キット a cricket breeding kit. ¶**飼育係** a person in charge of handling animals [birds]; a handler of animals [birds]. **飼育学** thremmatology. **飼育技術** handling and feeding technique. **飼育小屋**〔学校などの〕a pen 《for rabbits》; a small animal enclosure. **飼育(容)器**〔昆虫・爬虫類などの〕a rearing container. **飼育者** a raiser; a breeder; a rearer; a 《bird》 fancier. **飼育場** a breeding ground. **飼育箱**〔昆虫などの〕a cage for raising insects; a rearing box. **飼育びん** a bottle for raising insects. **屋外[室外]飼育** outdoor 《pet》 ownership; keeping 《a pet》 outdoors. **屋内[室内]飼育** indoor 《pet》 ownership;〔ペットの〕keeping 《a pet》 indoors. / 室内〜の猫 an indoor cat; a house cat. **促成飼育** forced breeding.

ジーン〔遺伝子〕a gene. ¶**ジーン・セラピー**〔遺伝子治療〕gene therapy. **ジーン・バンク**〔遺伝子銀行〕a gene bank.

ジエチルスチルベストロール〔合成女性ホルモン；環境ホルモン〕diethylstilbestrol.

四塩化(しえんか) ¶**四塩化鉛[炭素，チタン]** lead [carbon, titanium] tetrachloride. **四塩化炭素中毒** carbon tetrachloride poisoning.

潮(しお)〔海面の上昇下降〕the tide;〔潮流〕a current. ▶潮が上げて[さして，満ちてきて]いる．The tide is rising [coming in, flowing in]. / 潮が満ちている．The tide is at the full. / 潮が引いている．The tide is ebbing [going out, flowing out, on the ebb]. / 潮が引いた後にはいくつもの貝殻が散らばっていた．After the tide went out there were a lot of shells lying scattered about. / 潮の満ち干(ひ) the ebb and flow of the tide / クジラが潮を吹く．A whale blows [spouts] (water [air]). ¶**潮溜り** a tidal [tide] pool. **潮干潟** a tidal flat.

自家 ▶〜醸造の home-brewed. ¶**自家菜園** a home garden. ★⇨菜園 **自家栽培** home gardening. ★⇨自給/作物を〜栽培する grow [raise] *one's* (own) food./〜栽培の野菜 homegrown vegetables. **自家受精[受粉]** ⇨受精, 受粉 **自家消費** in-house [person] consumption [use]. **自家処理** 〔生ゴミの〕disposal of (their own) garbage by each household. **自家発電** independent [private] (electric) power generation [production]; in-house power generation./〜発電に切り替える switch over to an independent generator. **自家発電設備** power-generation facilities of *one's* own. **自家発電装置[機]** an independent (electric) power plant《of a hospital》; an in-house power generator. **自家用車** a private car [automobile]; a car for *one's* personal [private, own] use; an owner-driven car. ★⇨マイカー

市街 〔市内の通り〕the streets《of a city [town]》;〔市〕a city; a town;〔市街地〕an urban district [area]. ¶ **市街化** urbanization. **市街化区域** an urbanization (promotion) area [zone]; an urbanized area [zone]. **市街化区域内農地** agricultural land within an urbanization promotion area [zone]. **市街化調整区域** an urbanization control area [zone]. **市街植樹** the planting of roadside trees. **市街地開発** urban development. **市街地開発区域** an urban development area [zone]. **市街地開発事業** an urban development project. **市街地再開発事業** an urban renewal [redevelopment] project.

紫外線 ultraviolet rays. ★⇨英和 UV-A[B, C], sun protection factor ¶ **紫外線カットガラス[カットフィルム]**〔自動車などの〕UV-blocking[-filtering] glass [film]. **紫外線吸収[防止]ガラス** ultraviolet ray absorbing [intercepting] glass. **紫外線吸収剤** an ultraviolet absorber [absorbent]. **紫外線吸収物質** a UV-absorbing substance. **紫外線情報** an ultraviolet forecast.

シカゴ気候取引所 Chicago Climate Exchange ★⇨気候(気候取引所), 英和 CCX

時期〔時〕time; the times;〔季節〕a season; a time of (the) year. ★⇨産卵 ▶菊の〜 the chrysanthemum season./稲刈りの〜時期 the rice-harvesting season; harvesttime for the rice./旬(しゅん)の〜 the right season《for...》. ★⇨旬/種まきの〜はとうに過ぎた. It is long past the seeding time. / これは花の〜が長い植物です. This is a plant with a long flowering season [that blooms over a long period]. / 電灯を当てて開花の〜を調節する adjust the time of flowering with electric light. ★⇨開花 / もう紅葉の〜は終わった. The season for autumn leaves is over.

自給 self-support[-supply, -sustenance]. ▶〜する support [provide for] *oneself*; supply *one's* own needs; meet *one's* own demands; be self-supporting[-sustaining]. ¶**自給器**〔餌料・材料などの〕a self-feeder; an automatic feeder. **自給経済主義** autarky (形 autarkic(al)); (the principle of) economic self-sufficiency. **自給自足** self-sufficiency[-containment]. / 今後日本はこの面では〜自足していかねばならない. In future Japan must be self-sufficient in this area. / 経済上の〜自足を成しとげる achieve [attain] economic self-sufficiency. **自給自足経済** (an) autarky; a self-sufficient economy. **自給自足政策** a self-supporting and self-sufficient policy. **自給的農家** a subsistence farmer. **自給肥料** a self-supplied fertilizer; homemade manure. **自給用作物** a subsistence crop. **自給率** a rate of self-sufficiency 《in energy》; self-sufficiency. ★⇨食料(食料自給率), 穀物(穀物自給率), エネルギー(エネルギー自給率[国]), 供給(供給熱量自給率) **自給炉** a self-feeding furnace.

事業 an undertaking; an enterprise; an activity; an operation; a project; a scheme. ★⇨公共(公共事業) ▶政府の〜 a government undertaking. / 国の援助を受けた〜 a government-aided project. / その〜に関係している各省庁 the ministries involved in the undertaking. / 私は長い間原子力開発〜に携わってきた. I've been working on nuclear power development for a long time. ¶**事業改善命令** a business improvement order. **事業(系)ごみ**〔一般家庭に対して, 会社などの出す〕commercial waste. ★⇨家庭ごみ **事業停止命令** an order to stop [cease] operations [(doing) business]. **事業認可** project approval [authorization]. **事業認定**〔土地収用法に基づく〕project authorization.

ジクロロメタン〔溶剤〕dichloromethane

資源 resources;〔天然資源〕natural resources;〔原材料〕raw materials. ★⇨再資源化, 再生,

海底(海底資源), 海洋(海洋資源), 水産(水産資源), 石油(石油資源) ▶国全体[国家]の〜 national resources. / 未開発[手付かず]の〜 undeveloped [dormant] resources; untapped natural resources; the untouched store of natural resources. / 未発見〜量を推定する estimate how many resources are still undiscovered [how much of a resource remains undiscovered]; estimate undiscovered resources. / 生産に必要な基本的〜 a basic resource (required [needed]) for production. / 〜が豊かである be rich in natural resources; have abundant natural resources. ★⇨資源大国 / 〜が乏しい be poor in natural resources. / 日本は〜がないので,輸入に頼っている. Because Japan has no natural resources, it relies on imports. / ペットボトルや牛乳パックは回収されて新たな〜になる. PET bottles and milk cartons are collected and made into new materials. / 〜の保護[保全] protection [conservation] of resources; resource conservation. / 〜の有効活用[節約] effective use of [cutting down on the use of, economizing on] resources. / 〜節約型[省〜]社会 a resource-conserving [resource-saving] society. / 〜多消費型産業[社会] resource-guzzling industry [society]. / 天然〜を開発する exploit [develop, tap] natural resources. / 〜を保護する[節約する] conserve [economize on] natural resources. ★⇨地球 / 〜を利用する use (natural) resources. / 〜を浪費する be wasteful of resources. / 〜を枯渇させる deplete (natural) resources;〔枯渇する〕dry up; be exhausted. / 国内の〜を枯渇させる[使い果たす] drain *one's* country of its resources.

¶ **資源インフレ** rising prices of natural resources. **資源エネルギー庁**〔経済産業省の〕Agency of Natural Resources and Energy. **資源外交** resource diplomacy. **資源回収** resource recovery [retrieval]. **資源開発** resource development. **(再)資源化施設** a resource recovery plant. **資源株** a resources [resource] stock. **資源(国)カルテル**〔資源産出国によるカルテル; OPEC, OAPEC など〕a resource cartel. **資源管理** management of (natural) resources; resource management. **資源管理型漁業** resource management-type fishery [fish-

資源

ing〕. **資源関連株** a resource-related stock. **資源供給国** a resource-supplying country [nation]. **資源工学** resources [resource] engineering. **資源国通貨** 〔天然資源の輸出を主産業とする国の通貨〕a commodity currency. **資源ごみ** recyclable waste. / 〜ごみの回収日 a day on which recyclable waste is collected. / 月に一度〜ごみの回収がある. Recyclable materials are collected once a month. | There is a monthly collection of materials for recycling. / 〜ごみのリサイクル the recycling of recyclable waste《from other wastes》. / 〜ごみを分別する separate recyclable waste. / 当市はペットボトル以外のプラスチックは〜ごみとして回収しません. This city does not collect as recyclable waste any plastics except PET bottles. **資源産出国** a resource-producing country. **資源循環型** ⇨循環 **資源小国** a country poor in [with limited] natural resources; a resource-poor country. **資源戦争** a war for [over] natural resources. **資源大国** a country rich in natural resources; a country with abundant [ample] resources; a resource-rich nation. **資源探査** resource exploration. / 〜(探査)衛星 an earth resources observation satellite. **資源地質学** resource geology. **資源ナショナリズム** resource nationalism. **資源配分** allocation [distribution] of natural resources; resource allocation. **資源摩擦** friction over resources. **資源メジャー** a major resource company. **資源有限時代** an era [age] of limited natural resources. **資源有効利用促進法** the Law for Promotion of Effective Utilization of Resources. **資源輸入国** a resource-importing country [nation]. **資源リサイクル** recycling of resources; resource recycling. **資源リサイクルセンター** a resource recycling center. **資源略奪** resource plundering. **観光資源** tourism resources. **鉱物資源** mineral resources. / 豊富[貧弱]な〜資源 rich [poor] mineral resources. / 非常に鉱物〜の豊かな地方 a district rich in minerals [of great mineral wealth]. **枯渇資源** exhaustible [non-renewable] resources. / 非枯渇〜 inexhaustible [renewable] resources. **人的資源** manpower (resources); human resources. **地下資源** underground resources. **物的資源** material [tangible] resources. 非在来型[在来型]資源 unconventional [conventional] resources.

時差出[通]勤 〔フレックスタイム〕flextime;〔通勤時間をずらすこと〕commuting at staggered commuting hours. ▶時差出勤制度を採用する adopt staggered working hours. / 朝夕ラッシュ時の混雑緩和のため時差通勤にご協力ください. Please try to stagger commuting hours in order to ease congestion during the morning and evening rush hours.

自主 ▶環境や福祉・教育に関する市民の〜的な活動を市は応援しています. The city supports the independent activities [initiatives] of its residents in regard to the environment, social welfare, and education. ¶**自主回収**〔欠陥製品の〕(a) voluntary recall《of defective products》. ★ ⇨リコール, 欠陥 / 欠陥商品を〜回収する voluntarily recall defective goods. / 不純物が混入したパック入り牛乳を〜回収する (voluntarily) recall carton milk that was contaminated by foreign matter. **自主開発原油**[**石油**] independently developed (crude) oil. **自主開発石油比率** the percentage of independently developed (crude) oil. **自主開発油田** an independently developed oil field. **自主規制** self-imposed control [voluntary restraints, self-restraint]《on cotton textile exports》. / 〜規制する apply self-restraint; impose self-restraint [self-control] on…. **自主防災** locally managed disaster prevention [relief]. **自主防災組織**[**会**] a locally managed disaster prevention [relief] organization.

自浄 self-purification; self-cleansing; autopurification. ★ ⇨浄化 ¶**自浄作用** 〔土地などの〕autopurification;〔川や水道などの〕self-purification. ▶自然の〜作用 the self-cleansing action of nature; nature's self-cleansing functions; natural (self-)purification. **自浄能力** a [the] power of self-purification; (a) self-cleansing ability [capability, power]; an [the] ability to cleanse [purify] itself. / 自然は〜能力を備えているがそれも限度がある Although nature has self-purifying powers, there are limits to them. | Nature's ability to purify itself is not unlimited.

自生 spontaneous generation; autogenesis; autogeny; abiogenesis;〔野生〕wild [natural] growth. ▶～する grow (in the) wild [naturally]. / この地方にはアツモリソウが～している．Cypripedium grows naturally in this area. / ツバキは本州以南に～する．Camellias grow in the wild in Honsh and parts south. / 人間が食料にしている野菜はもともと山野に～していた植物である．The vegetables that people eat are plants that originally grew wild [were found in the wild] in the mountains and meadows. / ～の spontaneous; autogenous; abiogenetic;〔野生の〕native《plants》; aboriginal; of wild [natural] growth. ¶ **自生種** a natural species. **自生植物** native [wild, spontaneous] plants; volunteer plants. **自生地** a natural [wild] growth area. **自生林** a natural woods [grove]. / シャクナゲの～林 a natural rhododendron grove [thicket].

史跡〔場所〕a historic site [place, spot]; a place of great historical interest;〔建造物など〕historic relics [remains]. ★⇨ 世界遺産 ▶その建物は最近国の～に指定された．That building was recently designated a national historical monument. / ～を保存する preserve historic remains ¶ **史跡名勝天然記念物**〔文化財保護法で指定する〕a historic landmark, scenic spot, or natural monument.

自然 nature. ★⇨野生，自浄，ネイチャー，環境，地球 ▶雄大な～ majestic nature / ～の雄大さに感動した．I was deeply moved by the grandeur of nature. / 厳しい～(の力) the vast [harsh] forces of nature. ★⇨ 自然条件 / 大～，母なる～ Mother Nature; (Mighty) Nature. / 大～を舞台に野生動物たちが繰り広げるドラマ the drama unfolded by wild animals on the vast stage of nature. / アフリカの大～を背景にして against the vast backdrop [background] of nature in Africa. / ～の猛威[威力] the violence of nature ; the power of nature; natural [elemental] forces. / 大～の威力を見せつけられた思いがした．〔被災地での感想〕I felt as though I had witnessed a demonstration of nature's awesome force. / 人間は～の猛威に打ち勝つことはできない．Humankind cannot prevail against [is no match for] the violence of Nature. / ～の脅威を学ぶ learn to respect the power of nature. / ～との対話 communication [(a) dialogue, (a) conversation] with nature. / この海洋公園は「人間と〜の調和」というコンセプトを掲げて開設された．This marine park was established under the slogan "co-existence of people and nature" [with the idea of harmony between people and the natural world in mind]. / 〜と人間の共生 the coexistence [symbiosis] of man and nature. / その都市からは〜が日に日に失われてゆく．That city is losing more of its natural environment with each passing day. / こんな大都会にもまだところどころに〜が残っている．Even a large city like this still retains pockets of nature. / かつての東京にはもっと〜があった．In former times there was more wildlife in Tokyo. | Tokyo used to have more wild plants and animals. / 豊かな〜に恵まれている be abundantly blessed with nature [natural settings]．★⇨自然環境 / この島にはまだ豊かな〜が残っている．Nature still abounds on this island. / このあたりにはまだ〜が一杯残っている．There is still a lot of natural scenery left around here. / 今なお〜が残る地域 a district where nature still thrives [is still to be found]. / この貴重な〜をできる限り無傷で次世代に伝えていくのがわれわれの重要な役目だ．It is our serious duty to convey this priceless natural world with as little damage as possible to the next generation. / 手付かずの〜 untouched [pristine] nature. / 〜に返れ．Back to nature! ★⇨自然回帰 / その施設では人の手で育てた鶴を〜に帰す試みが今も続けられている．In that facility they are still trying to return to the wild a crane that has been raised by humans. / 死んだら〜に帰りたいという人の散骨が増えている．More and more, the ashes of people who "want to return to nature" are being scattered at sea or in the mountains. / この食器は廃棄後は〜に帰る素材からできている．These dishes and utensils are made of materials that decompose naturally after they are discarded [biodegradable materials]．★⇨生分解性 / 〜の驚異 the wonders of nature. / 〜の法則 the laws of nature. / 〜界 the natural [physical] world; (the realm of) nature. / 〜界のおきて the laws of nature. / 野生動物は〜のおきての中で生きている．Wild animals live according to Nature's laws. / 絶滅する生物が増え，〜界のバランスが崩

れてきた．Because more and more living things are becoming extinct, the balance of nature has been upset. / ～界には人間の手ではとうてい模倣できないものが多い．There are many things in nature that human ingenuity cannot (hope to) imitate [that defy imitation by human beings]. / 人間による～の征服 man's dominion over nature. / 人類による～の征服はとどのつまり環境破壊であった．Man's conquest of nature has ended up in destruction of the environment. / ～を破壊する destroy nature. / ～との戦い a battle against nature. / ～の摂理 the laws governing nature. / ～の営み the workings of nature. / ～の斉一性 the uniformity of nature. / ～の美しさ natural beauty; the beauty of nature; the beauty of the natural world. / ～の恵みを享受する enjoy the blessings of nature. ★⇨恵み / 彼らは～の中で遊ぶ楽しさを知らないのだ．They haven't experienced the delight [pleasure, joy] of playing in a natural setting. / ～のふところに抱かれて育つ grow up in the bosom [lap] of nature. / ～のふところではぐくまれるキタキツネの赤ちゃん a northern fox cub raised in the bosom of Mother Nature. / ～を愛する love nature. / ～を愛する人 a nature lover；〔自然観察者〕a naturalist. ★⇨自然観察 / 子供たちが～に触れる機会が減った．Opportunities for children to be in contact with nature have decreased. ★⇨触れあい，自然体験 / ～に親しむ get close to nature. / ～を友とする be a friend of nature. / ～探索に出かける go out on a nature hunt. / 子供たちが～に親しむ催し an event for children to get to know and love nature. / ～を歌った詩 a nature poem. / ～を大切に．Treat the environment with respect. / ～を守る preserve [protect] nature. / ～の［な］〔天然の〕natural; wild《plants》; native;〔人工的でない〕unartificial;〔生得の〕inborn; inherent. / ～のままの in a [*one's*] natural state. / ～のままの山河 mountains and streams as nature made them. / ～の色 a natural color. / ～の甘み natural sweetness. / 流木の～の形を生かした工芸品 handcrafts that use the natural shapes of driftwood. / ～の地形を活かした城 a castle that incorporates the natural features of the land. /〔夏，窓を開けて〕～の風のほうが

ずっといいねえ．Natural breezes are so much better. / この珍しい岩は波の浸食によって〜にできたものです．This unusual rock was formed naturally by the erosion of waves.

¶ **自然育雛** natural brooding 《of chicks》. **自然遺産** 《our》 natural heritage; an item [a site] of natural heritage; a natural heritage item [site]. ★⇨世界（世界自然遺産） **自然エネルギー** natural [alternative] energy. ★⇨自然力 **自然塩** natural salt. **自然海塩** natural sea salt. **自然海岸** natural coast lines. **自然回帰** going back [a return] to nature. / ここ数年, 若者の間で〜回帰の傾向が強まっている．In recent years there has been something of a return to nature among young people. **自然改造** the remodelling [reshaping] of nature. **自然観** a view of [an outlook on] nature. / 彼が書く紀行文には彼独特の〜観が織り込まれている．His travel writing is shot through with his own particular view of nature. | His characteristic view of nature is woven into his accounts of his travels. **自然換気** natural ventilation. **自然環境** a natural environment; the environment. / 厳しい〜環境 a harsh (natural) environment. / 私は豊かな〜環境で育った．I grew up in a rich natural environment [with nature all around me, surrounded by nature]. / 子孫のために〜環境を守る preserve the environment for our posterity. **自然環境学** natural environment(al) studies [science]; study of the natural environment. **自然環境保全基礎調査** 〔緑の国勢調査〕a Green Census. **自然環境保全地域** a nature conservation area. **自然観察** nature observation [study]. **自然観察会** a nature (study) gathering; an outing to look at nature; a [an amateur] naturalist gathering;〔その同好会〕a nature (study) club. **自然観察指導員** a nature observation [conservation] instructor. **自然乾燥** 〔木材の〕natural seasoning 《of wood》;〔髪などの〕natural drying. **自然休養村** a natural recreation village. **自然休養林** a natural recreation forest. **自然教育園** a park for nature study;〔国立科学博物館付属の施設〕the Institute for Nature Study. **自然享受権** 〔他人の私有地を自由に散策できる権利〕public access rights; a [the] right of public access.

自然

自然景観 a view of natural scenery; a nature scene. ★⇨景観 **自然権** 〔自然法に基づく人間の権利〕a natural right. **自然現象** a natural phenomenon. **自然光** natural light;〔白色光〕white light. **自然公園** a nature [natural] park. / 都道府県立〜公園 a prefectural nature [natural] park. **自然公園審議会** the Council on Natural Parks. **自然公園法** the Natural Parks Law. **自然交配** 〔動植物の〕natural hybridization [crossbreeding, cross-fertilization];〔動物の〕natural mating. / 大島桜と緋寒桜が〜交配して河津桜が生まれた．The Kawazu cherry is a natural hybrid of the Oshima and Taiwan cherries. **自然公物** natural public resources. **自然酵母** ⇨天然(天然酵母) **自然災害** a natural disaster [calamity]. **自然再生型公共事業** a natural restoration-type public works project. **自然再生推進法** the Nature Regeneration Promotion Law. **自然散策** a nature walk. / 〜散策する take a nature walk. **自然残留磁気** natural remnant magnetization. **自然史** natural history. / 〜史博物館 a natural history museum; a museum of natural history. **自然志向** a desire for nature [the natural]. / 都市部では〜志向が高まっている．In urban areas there is an increasing desire for the natural [people are increasingly nature-oriented]. **自然条件** natural features; an [a natural] environment; environmental conditions. / 彼らはやせた土地とわずかな降雨量という厳しい〜条件の中で生活している．They live in a harsh environment of poor soil and little rain. | Conditions are harsh and they subsist on poor soil with little precipitation. / 日本は米の栽培に適した〜条件を備えている．Japan has [possesses, is endowed with] the right conditions for rice cultivation. | Japan possesses a suitable environment for growing rice. **自然状態** 《in》a natural state;《under》natural conditions. **自然食(品)** 〔食品〕natural foods;〔行為〕eating natural foods. / 〜食品の店を探している．I'm looking for a shop that sells natural foods. **自然植生** 〔代償植生・人為植生に対して〕natural vegetation. **自然崇拝** nature worship; the cult of nature; naturism. **自然崇拝者** a nature worshipper. **自然生態系** a natural ecosystem. **自然石**

fieldstone.　**自然染色**　〔草木染め〕dyeing with natural dyes of vegetable origin. / ～染色の敷物 a vegetable-dyed rug.　**自然選択**　natural selection.　**自然選択説**　the theory of natural selection.　**自然葬**　a natural burial.　**自然増**　a natural increase 《in population》.　**自然素材**　(a) natural material;〔食料品の〕a natural ingredient.　**自然体験**　《children's》experience of nature.　**自然体験型観光**　nature (experience) tourism.　**自然地理(学)**　physical geography; physiography.　**自然堤防**　a natural bank [embankment, levee].　**自然淘汰**　⇨自然選択　**自然動物園**　a natural zoo.　**自然毒**　a naturally occurring poison.　**自然突然変異**　spontaneous mutation.　**自然農法**　organic [chemical-free] farming [agriculture]; macrobiotic farming.　**自然の権利訴訟**　〔動物など人間以外の自然を原告として起こす訴訟〕a nonhuman rights lawsuit.　**自然破壊**　the destruction of nature [natural environments].　**自然発火**　autoignition; spontaneous combustion [ignition]; self-ignition. / ～発火する combust spontaneously.　**自然発火防止剤**　an agent to prevent spontaneous combustion.　**自然発酵**　natural [spontaneous] fermentation. / ～発酵する ferment naturally [spontaneously]. / ～発酵ビール (a) naturally [spontaneously] fermented beer.　**自然発生**　a spontaneous [a natural, an unpremeditated] occurrence;〔生物の〕spontaneous generation; abiogenesis; autogenesis. / ～発生する occur spontaneously [naturally, without premeditation]; grow [develop] naturally;〔生物が〕generate [be generated] spontaneously; be produced by spontaneous generation.　**自然繁殖**　natural breeding [propagation, reproduction]; propagation in nature.　**自然孵化(ふか)**　natural incubation《of eggs》. / ～孵化した鮭の稚魚 naturally incubated salmon fry.　**自然物**　a natural object.　**自然分娩**　natural childbirth.　**自然分娩法**　natural delivery;〔自然無痛分娩法；ラマーズ法〕the Lamaze technique [method].　**自然分類**　〔生物学上の〕natural classification.　**自然放射性核種**　a natural radionuclide.　**自然放射線**　natural radiation.　**自然放射能**　natural radioactivity.　**自然木**　a native tree.　**自然保護**　(the) conservation [preservation] of nature; (the) protection of the natural environment. / 環境省とタイアップした〜保護キャンペーン a nature conservation campaign in collaboration with the Ministry of the Environment. ★⇨環境(環境保護) / 経済発展が第一で，〜保護は二の次というのがこれらの国々の現状だ．The situation for these countries is that environmental protection must take a back seat to economic development. / 〜保護の精神がその国に根を下ろすのはまだだいぶ先のことだろう．It will be yet some time before the spirit of conservation strikes root in that country. / 〜保護の観点から発言する speak from the standpoint of nature conservation. / 〜保護を訴える appeal for (the) conservation of nature. **自然保護運動** a (nature) conservation movement; a movement for nature conservation [preservation]. / 〜保護運動に参加する take part [participate] in a conservation campaign. **自然保護運動家** a (nature) conservationist. **自然保護官** a forest [(nature) conservation] ranger; a ranger《in a national park》; a (national) park ranger. **自然保護官補佐** ⇨アクティブレンジャー **自然保護監視員** a nature conservation guard [inspector]. **自然保護区** a nature reserve; a (nature) conservation area. **自然保護条例** a nature conservation ordinance; an ordinance for nature conservation. **自然保護スワップ** ⇨債務環境スワップ． **自然保護団体** a (nature) conservation group. **自然保護論者[主義者]** a conservationist. **自然歩道** a nature trail; a nature path. **自然療法** naturopathy; a nature cure. **自然力** 〔自然界の作用〕(a) natural agency;〔風力・水力など〕the forces [powers] of nature; elemental forces. / 〜力を利用する harness the forces of nature《for...》. ★⇨自浄 **自然緑地** a natural green tract of land. **自然林** a natural forest.

持続 ▶〜可能な開発[経済成長] sustainable development [economic growth] ★⇨サステナブル，GRI / 〜可能な社会 a sustainable society. / ...の〜可能な[持続的な]利用 the sustainable use of.... ¶ **持続可能性** sustainability. **持続可能性報告書** a sustainability report. **持続可能な開発のための教育の10年** 〔国連の〕the United Nations Decade of Education for Sustainable Development (略：UN-

DESD). ★2002年に決議された2005年から2014年までの10年計画. **持続可能な開発のための世界経済人会議** the WBCSD; the World Business Council for Sustainable Development. **持続的養殖生産確保法** the Law to Ensure Sustainable Aquaculture Production.

下刈り weeding; removing undergrowth. ▶森の〜をする[下草を刈る] remove undergrowth from a forest.

下草 weeds covering the floor of a forest [growing under the trees]; undergrowth. ★⇨下刈り

室外機 （エアコンの）outdoor [external] unit (of an air conditioner).

シック・ハウス 〔体調不良を引き起こす化学物質を建材に含んだ住宅〕a sick house. ¶**シック・ハウス症候群** 〔住宅建材中の化学物質により引き起こされる体調不良〕sick house syndrome（略: SHS). ★⇨英和 sick house syndrome **シック・ハウス対策** sick house countermeasures; measures to prevent sick house syndrome. **シック・ハウス問題** the sick house problem.

湿原 a wetland; a swamp; a marsh; a bog; marshy ground; marshland; a moor. ★⇨湿地 ▶釧路〜 Kushiro Marsh [Wetlands] (in Hokkaido). ¶**湿原植物** a marsh plant.

実験 experimentation;（実験室における操作）laboratory [experimental] work;〔1回ごとの〕an experiment; a test. ★⇨動物（動物実験，実験動物），核（核実験），実証 ▶〜する experiment 《on [in]...》; conduct [do] an experiment [a test]《on...》; carry out [do, conduct] laboratory work. / 新素材の強度を〜する test the strength of a new material. ¶**実験室** a laboratory; an experimental laboratory;《口》a lab. **実験台** a testing bench; an experiment stand; a laboratory table;〔実験対象〕the subject of an experiment;（人）a human guinea pig. ★⇨モルモット / ...を〜台に使う use *sb* [*sth*] as the subject of an experiment. / 彼らは〜台になろうと申し出た. They volunteered to be [serve as] human guinea pigs. **実験農場** an experimental farm; a pilot farm. **実験炉** 〔原発の〕an experimental (nuclear) reactor.

実施 enforcement; operation; execution; implementation. ▶〜する〔手続きを〕carry out [follow]《a procedure》;〔催しを〕

hold [stage, put on]《an event》;〔作業を〕conduct《an operation》;〔制度を〕bring [put]《a system》into operation [force];〔政策を〕implement; institute《a policy》; effectuate; effectivate. / 条例を〜する enforce a regulation. / その法律は今も〜されている．The law still remains in force [operation]. / 条約の〜 the enforcement of a treaty. / 省エネ[廃品回収]の〜 the implementation of energy conservation [waste article collection] measures. / 消火訓練〜にあたっての注意事項を申し上げます．I will now read out the precautions to be followed when carrying out fire-fighting drills. / 〜を取りやめる[延期する] cancel [postpone]《an event》; cancel [delay] (the) implementation《of...》.

実証 ▶〜する corroborate《a proof》; establish《a fact》; prove《a theory》; demonstrate《a fact》; bear out《a conclusion》; verify《a fact》; underpin《a method》; certify《evidence》. / では，私の考えを〜する事実を述べましょう．Now let me state some facts that support [back up, corroborate] my ideas. / この仮説は〜されるだろうか．Will this hypothesis be proved? / まだ〜があがっていない．The proof has not yet been obtained. / 〜を踏まえた議論 a debate based on positive evidence. / 〜済みの proved; proven. / この学説は〜済みだ．This theory has been proved [substantiated, demonstrated to be true]. / 再生水の〜プラント a water recycling demonstration plant. ¶**実証実験[試験]** a demonstration [verification] experiment [test]. **実証走行試験**〔新型車などの〕an evaluation [a verification] test drive; test driving《a green vehicle》(to ensure that it performs satisfactorily). **公道実証実験**〔燃料電池車など新種の車両や新しい交通システムなどの，公道を使っての実験〕a verification test for [verification testing of]《fuel cell electric vehicles》on public roads. **焼却実証試験**〔廃プラスチックなどの〕an incinerator test [incinerator testing] (for《toxic emissions》); testing and monitoring of waste incineration. **耐震実証試験** a seismic simulation test; a seismic vibration test; seismic vibration [simulation] testing.

湿生 ▶〜の hydrarch. ¶**湿生雑草** hydrarch weeds. **湿生植物**

a hygrophyte.　**湿生遷移**　hydrarch succession.

湿地　damp [marshy] ground; a bog; a swamp; a marsh; wetlands.　▶～にすむ生物 wetland creatures [fauna].　¶**湿地草原** a marshy meadow.　**湿地帯** wetlands; a marshy district; marshland; bogland.　**湿地保全** wetland(s) conservation; conservation of wetlands.　★⇨ラムサール条約, 湿原

湿度　humidity.　★⇨体感, 除湿　▶今日は～がとても高い. It is very humid [damp] today. / 当地は～が低くてからっとしているから30度あってもそんなに暑くない. It's a (pleasantly) dry, low-humidity area, so even when the temperature reaches 30°C it doesn't feel that hot. / 大気中の～を測定する measure the humidity of the atmosphere. / ～をコントロールする control the (level of) humidity. / ～は現在80です. The humidity is 80 (percent) now.　¶**湿度計** a hygrometer.

室内　¶**室内飼い**　⇨飼育(室内飼育)　**室内環境** an indoor environment.　**室内環境汚染** indoor environmental pollution.　**室内気候** an indoor climate.　**室内空気汚染** indoor air pollution.　**室内犬** an indoor dog; a house dog.

指定　appointment; designation; assignment; specification.　▶～する appoint; designate; assign; name; specify; earmark 《for a specific use》.　★⇨史跡, 天然(天然記念物)　¶**指定可燃物** designated combustibles [burnables].　**指定漁業**〔農林水産大臣の許可を必要とする漁業〕a licensed fishery.　**指定伝染病** a designated [specified] communicable diseases.　**指定(ごみ)袋** designated garbage bags; rubbish bags of a type specified 《by the local government》.　★⇨ごみ(ごみ袋)

自転車　a bycycle [bike]　★⇨放置自転車　¶**自転車専用道路** a cycling path.　**自転車通学[通勤]** bicycling to school [work]; commuting by bicycle.　▶～通学する go to school by bicycle.　**自転車ツーキニスト**〔自転車で通勤する人〕a person who commutes by bicycle; a bicycle [《口》bike] commuter.　**自転車利用環境整備モデル都市**〔国土交通省指定の〕a model bicycle-friendly city.

自動車　a car; a motorcar; a motor vehicle; an auto(mobile)　★⇨小型自動車, マイカー, 車, 電気(電気自動車), 使用(使用済み自動車)　▶～の安全性 car safety; the safety of a car. ¶**自動車アセスメント**〔自動車安全性能評価〕(米国の) the New Car Assessment Program (略: NCAP); (日本の) the Japan New Car Assessment Program (略: JNCAP) [⇨自動車安全情報]; (欧州の) the European New Car Assessment Programme (略: Euro NCAP). **自動車安全運転センター**〔警察庁所管の特殊法人〕the Japan Safe Driving Center. **自動車安全情報**〔衝突安全性能などのテスト結果を車種ごとに公表したもの〕Automobile Safety Information. 自動車アセスメント (New Car Assessment Japan) ともいう. **自動車公害** automobile pollution; (environmental) pollution produced by cars [automobiles]. **自動車ジーメン**〔東京都などの自動車公害監察員〕《Tokyo Metropolitan Government》diesel vehicle inspectors; automobile "G-men." **自動車事故** a car [an automobile] accident. **自動車事故対策機構** the National Agency for Automotive Safety & Victims' Aid (略: NASVA). 2003年自動車事故対策センターより移行. **自動車シュレッダー・ダスト** 自動車破砕屑[残渣] an automobile shredder residue (略: ASR). **自動車衝突防止装置** the Car Collision Avoidance System (略: CCAS). **自動車騒音** traffic [car] noise; the noise of traffic [cars, automobiles]. **自動車騒音対策** traffic noise (control) measures. **自動車乗り入れ規制** limitations [restrictions] on entry by car. **自動車排ガス規制** (car) exhaust emission control. ★⇨排ガス **自動車排ガス測定局** the automobile exhaust measurement office. **自動車排ガス対策** exhaust emission measures. **自動車フロン券**〔廃車のエアコンフロンの回収破壊費用払込済み証〕an automobile CFC coupon. **自動車保管場所標章** a parking (permit) sticker. **自動車リサイクル促進センター** the Japan Automobile Recycling Promotion Center (略: JARC). **自動車リサイクル法** the Automobile Recycling Law. ★正式名称は「使用済み自動車の再資源化等に関する法律」(the Law Concerning Recycling Measures for End-of-Life Vehicles). **自動車 NOx・PM 法**【法】〔改正自動車 NOx 法〕the Automobile NOx/PM Reduction Law. **貨物自動車** a truck; a lorry.

屎尿(にょう) raw sewage; human waste. ¶**屎尿汚泥** raw sewage

sludge. 屎尿処理 sewage disposal; treatment (and disposal) of raw sewage. 屎尿処理施設[処理場] a raw sewage treatment [disposal] facility [plant]. 屎尿溜(だ)め a cesspool.

芝生 a lawn; grass; turf; a grass plot; a patch of grass. ▶～に寝転ぶ lie [throw *oneself*] down on the grass. / ～に入るべからず.〔掲示〕Keep off the grass. / ～の手入れをする take care of a lawn [the grass, the turf]. / ～を刈る mow a lawn. ¶芝生散水器 a lawn sprinkler.

地盤 〔地面〕(the) ground;〔土台〕a base; a foundation. ▶この辺は～が緩いから地震の時は危険だ. The ground is not firm around here, so it's dangerous in earthquakes. / この家は～がしっかりしているから揺れない. Since this house has [is built on] firm foundations, it doesn't shake. / 雨で～が緩んでいるので土砂崩れにご注意ください. Rain has made the ground unstable, so beware of landslides. ¶地盤陥没 ground subsidence. 地盤基礎工学 geotechnical (and) foundation engineering; geotechnical engineering. 地盤工学 geotechnical engineering; geotechnics. 地盤調査 subsurface investigation [exploration]. 地盤沈下 subsidence (of the ground); land subsidence. / ～沈下する sink; subside. / 辺り一帯の～沈下 subsidence all over an area. / ～沈下がひどい. There is severe subsidence (in the ground). / ～沈下が進んでいる. Subsidence is proceeding [getting worse]. | There is increasing subsidence. / ～沈下を食い止める halt subsidence; hold subsidence in check. / 地下水の過度のくみ上げは～沈下を招くことがある. Pumping too much water from the ground can cause [increase the likelihood of] subsidence. 地盤補強 〔ビル・建物などの〕foundation reinforcement.

指標 an indicator; a barometer. ▶健康の～ a barometer of how healthy one is; a health index. ¶指標種 an indicator species. ★⇨英和 indicator species 指標植物 an indicator plant. 指標生物 〔自然環境の状態を調査するのに使われる生物〕an indicator (organism); a bioindicator. ★⇨生物(生物指標)

市民 a citizen; a resident; an inhabitant. ★⇨住民 ▶地元の～グループ a local citizens' group; a group of local residents. ¶市民運動 a citizens' [civic] campaign《for...》. **市民運動[活動]家** a citizen [civic] activist. **市民感覚[感情]** popular [local] feeling(s) [sentiment]. **市民教育** civic education. **市民参加型行政** participatory government [administration]. **市民生活** the life [livelihood] of the citizens; civic life. **市民大会** a citizens' [residents', civic] rally. **市民大学** a college [school] for adult citizens (run by a local government). **市民団体** a citizens' [civic, residents'] organization [group]. **市民農園** a community garden; an allotment (garden). **市民フォーラム** a citizens' forum. **市民福祉課** 〔役所の〕a citizens' welfare section. **市民ボランティア** a citizen volunteer.

地元 ▶～(住)民 a local (resident); the locals; local people [inhabitants]; people of the district. ★⇨地域 / この計画に対して～にかなりの反感がある. There is considerable feeling against the project locally [among the locals]. / その原発は～自治体の反対によって建設工事が中断している. Due to opposition by the local government, work on the atomic power station has been put on hold. / この地点に高速道路のインターチェンジを造ろうという計画が持ち上がったが，～調整がつかず頓挫した. There were plans to build an expressway interchange on this site, but negotiations at the local level were unsuccessful and the plan was aborted.

遮音(しゃおん) sound insulation. ★⇨防音, 消音 ¶**遮音効果** (a) sound insulation effect. **遮音材** a sound-insulating material. **遮音床** a soundproof floor. **遮音装置** a sound arrester. **遮音壁** a soundproof [sound-insulating] wall.

社会 society; the world; the public; 〔共同社会〕a community. ▶ミツバチ[ニホンザル]の～ the society of bees [Japanese monkeys]. / ～一般の利益 public interest [welfare, good]. ★⇨公共 / ～に貢献する[尽くす] contribute to society [the community, public welfare]; labor [work, exert *oneself*] for society [the public good]; serve society. / ～を変える[動かす] change [turn the wheels of] society. / 企業の～貢献 the social contribution of a business. ★⇨企業 / ～奉仕をする do social service; work for the public benefit. / ～奉仕家 a social worker. / ～奉仕事業 social service. ¶**社会生物**

学　sociobiology.

遮光(しゃこう)　▶〜する shield [shade] (*sth* from) the light. ¶**遮光カーテン**　a thick curtain to keep out light; a shading curtain;〔灯火管制用〕a blackout curtain.　**遮光ガラス**　(a) light-shielding [light-resistant] glass.　**遮光スクリーン**　an occulter.　**遮光装置**　shading.　**遮光幕**〔灯火管制用の〕(窓の) a blackout curtain; (灯火のまわりの) a shade;〔テレビカメラの〕a flag.

遮水(しゃすい)　water interception [insulation]; seepage control. ▶〜する intercept water; control seepage.　¶**遮水工**　seepage control work.　**遮水材**(料)　an [a water] impermeable material; (a) seepage control material.　**遮水シート**　a water-impermeable [seepage control, liner] sheet.　**遮水性**　(water) impermeability; imperviousness (to water).　**遮水性舗装**　(water) impervious paving.　**遮水壁**　an impermeable wall; a (watertight) bulkhead.　**遮水膜**　an impermeable membrane.

斜線制限〔建築基準法上の〕a setback restriction [regulation]. ▶北側〜 a north-side setback restriction [regulation] / 道路〜 a street-side [front] setback restriction [regulation].

砂利　gravel; (small) pebbles;〔小砂利〕fine gravel; (海岸などの) shingle;〔(軌道などの) 敷砂利〕ballast; (道路用の) road metal; hoggin. ▶道路には〜が敷いてある．The roads are graveled [covered with gravel]. / 〜を敷く spread gravel; gravel《a road》;〔軌道に〕ballast《a railroad》.　¶**砂利採取**　gravel digging [collecting].　**砂利採取業者**　a gravel digger [collector]; a gravel digging [collecting] firm　**砂利採取場**　a gravel pit.　**砂利トラック**　a gravel truck.　**砂利道**[**歩道**]　a gravel road [walk].

種(しゅ)　a species《複数形も species》;〔変種〕a variety.　★⇨学名, 種類, 雑種, 新種, 交配, 交雑　▶〜の specific. / 〜の多様性 ★⇨英和 biodiversity.　¶**種の保存法**〔絶滅のおそれのある野生動植物の種の保存に関する法律の通称〕the Species Preservation Law.

樹医(じゅい)　⇨樹木(樹木医)

雌雄(しゆう)　〔メスとオス〕male and female; the two sexes.　¶**雌雄異花**　diclinism. 形 diclinous.　**雌雄異形**　dioecism.　**雌雄異株**　dioecism. 形 dioecious.　**雌雄異熟**　dichogamy. 形 dichogamous, dichogamic.　**雌雄異熟花**

a dichogamous flower. **雌雄産み分け** sex choice. **雌雄鑑別** 〔ヒヨコの〕sexing (of a chick). ▶~鑑別する determine the sex 《of…》; sex《a chicken》. **雌雄鑑別士** 〔ヒヨコの〕a (chick) sexer. **雌雄選択** sexual selection. **雌雄同株** monoecism. 形 monoecious. **雌雄同熟** synacmy. **雌雄同熟花** an adichogamous flower. **雌雄淘汰** sexual selection. **雌雄同体** hermaphroditism. 形 hermaphroditic. **雌雄モザイク** a gynandromorph. **雌雄両花具有** androgyny. 形 androgynous.

収穫 〔農作物の取り入れ〕harvesting;〔取り入れた農作物〕a harvest; a crop; a yield. ▶~する gather (in) a harvest; harvest; reap; gather [take] in. / 大麦を~する harvest [gather in] barley [the barley crop]. / 畑で熟してから~されたトマト tomatoes that were harvested after ripening on the vine. / このハーブは秋口まで何度でも若芽を~できる. Young buds can be harvested from this herb any number of times until early autumn. / 秋の~ autumn harvesting; an autumn crop. / 米［野菜，果物］の~ rice [vegetable, fruit] harvesting; a rice [vegetable, fruit] crop. / 予想を上回る［下回る］~ a greater-[less-]than-expected harvest [crop, yield]. / 豊かな［ささやかな］~ an abundant [a meager] harvest. / ~が多い［少ない］have a good. [bad, poor] harvest [crop]. / 北海道でも米の~が増えてきている. Rice harvests in Hokkaido have been increasing [on the rise]. / 全国で小麦の~が減ってきた. Wheat harvests have been declining over the entire country. / バナナは年二度の~がある. We get two crops of bananas a year. / 今年のリンゴはかなりの~が望めそうだ. It looks like we can look forward to a pretty good apple crop this year. / 妻の家庭菜園から初めての~があがった. We had our first crop from my wife's home garden. / ~に感謝する be thankful [grateful] for the harvest. / ~の多い highly productive《farm》; high-yielding《wheat》. / ~率の高い［低い］米の品種 a high-[low-]yield(ing) variety of rice. / ~のない，無収穫の harvestless. / 入植当初はまったく~のない年もあった. When we first settled in there were years when we had no crop at all. / 今年は天候に恵まれて，大きな~をあげることができた. Thanks to this year's great weather, we had a bountiful harvest. / ～を祝う celebrate a (good) harvest. / この川は流域の村々に豊かな～をもたらしてきた．This river has produced abundant harvests for the villages along the river valley. / 今年の米の～は平年作以下の見積もりだ．This year's rice crop is estimated to be below the average. ¶ **収穫蟻** a harvester ant. **収穫期** the harvesting season; (the) harvesttime. **収穫祭** a harvest festival. **収穫高** the yield; the crop. / 予想～高 an estimated crop [yield]. **収穫年度** a crop year. **収穫予想** a crop [harvest, yield] estimate. **収穫予想高** an estimated yield [crop] for the year.

住環境 *one's* living environment; housing (conditions); *one's* housing situation. ★⇨公園 ▶市民の～の整備を第一の課題とします．My first priority will be to improve the living [housing] conditions [environment] of our citizens. / ～を重視する give importance to living [housing] conditions [environment].

臭気 an offensive smell; a bad [foul, nasty, fetid] odor; a stench; a stink. ★⇨悪臭, 異臭, 臭袋法(しゅうたいほう) ▶～のある bad-smelling; stinking; malodorous. / ～を放つ give off an offensive smell; emit a foul [bad] odor. / ～を消す destroy a bad odor [the bad odor 《of...》]; deodorize. ¶ **臭気止め** a deodorizer; a deodorant. **臭気判定士** an odor analyst.

重金属中毒 heavy metal poisoning. ★⇨汚染(重金属汚染)

充実不足 〔米・穀粒の〕incomplete grain filling.

終息 ▶～する cease; end; come to [be brought to] an end [a close]; be eradicated. / この地方のコレラは今もって～していない．In this district cholera has not yet been stamped out [eradicated]. / …の～宣言を出す issue an announcement proclaiming the eradication of ….

渋滞 a (traffic) jam [snarl, tie-up, backup]; (traffic) congestion; (市街地交差点での) (a) gridlock. ▶～する〔交通が〕be jammed up; snarl. / 中央道上りは約20キロの～です．Traffic into Tokyo on the Chuo Highway is backed up for 20 kilometers. / 東関東自動車道は宮野木料金所を頭に20キロの～です．On the Higashi-Kanto Expressway there is congestion stretching back (for) twenty kilometers from the Miyanogi tollgates. / この事故で上下線ともかなり〜しています．Due to the accident, traffic in both directions is badly backed up. / 大〜 a huge backup; a mammoth tie-up. / 朝［夕方］の〜 morning [evening] (traffic) delays. / 車の〜で約束の時間に遅れる be late for an appointment due to a traffic jam. / 〜に巻き込まれる be [get] caught in a (traffic) snarl [jam]. / 〜を抜ける escape from [get out of] a traffic jam. / この〜を抜けるには３時間かかるだろう．It'll probably take three hours to get out of this snarl-up. / 〜を避けて迂回する take a detour to avoid a traffic jam. / 〜を避けて抜け道を探す look for a way around backed-up traffic [a traffic jam]. / 交通〜を緩和する relieve [reduce, alleviate] traffic congestion; ease a traffic jam; make traffic flow more smoothly. ★⇨交通（交通機関） / 〜を緩和するためにバイパスを通す construct a bypass to relieve [reduce] traffic congestion. / このあたりは一日中〜が激しい．Traffic congestion is terrible all day around here. | There are terrible traffic jams all day around here. / 〜した道路 a (badly) congested road; a road where a traffic jam has developed; a gridlocked road. / 事故現場の見物〜 a traffic jam caused by rubbernecking drivers at the scene of an accident. ¶渋滞学 congestion studies. 渋滞情報 traffic jam information; information on congestion. 渋滞予測 traffic congestion prediction.

臭袋法(しゅうたいほう)〔におい測定法〕the odor bag method. ¶三点比較式臭袋法 the triangular odor bag method.

集団 a group; a mass. ★⇨群集 ▶〜を作る動物 a gregarious [herd] animal. ¶集団営巣地 a (nesting) colony;〔カラス・ペンギンなどの〕a rookery. 集団越冬 mass wintering. / 〜越冬する (over-)winter en masse [in large groups]. 集団生活 aggregation.

集中 ¶集中管理 centralized [integrated] control《of information》. ▶プルトニウムの〜管理 the integrated control system for handling plutonium. 集中豪雨 ⇨豪雨

充電 charge; charging; electrification. ▶〜する〔蓄電池に〕charge《an accumulator》(with electricity); give a charge of electricity to《a storage battery》. / 太陽電池を利用し

てニッカド電池を再〜する use a solar battery to recharge a nickel-cadmium battery. ¶**充電器** a (battery) charger. **充電スタンド** 〔電気自動車用の充電所〕a charging [charge] station.

住民 inhabitants; residents; dwellers;〔人口〕the population. ▶(意見を把握するため)〜にアンケートをとる do a survey of residents [residential survey]; carry out a questionnaire (survey) of residents (to find out what they feel). / 〜説明会を開く hold an explanatory meeting for local residents; organize a meeting to inform local people《about the projected expressway》. ¶**住民運動** a local residents' campaign; a neighborhood《protest》movement. / 原発反対の〜運動 a residents' campaign against a nuclear power plant. **住民監査請求** a citizens' petition for audit; a residents' audit request. **住民監視** 〔住民による〕surveillance by local residents《of illegal dumping》; a neighborhood《crime》watch;〔住民に対する〕surveillance of local residents. **住民訴訟** a citizens' lawsuit (against municipal authorities for malfeasance). **住民投票** a local referendum [plebiscite]. / …に関して〜投票を行う hold a referendum [plebiscite] on … **住民発議** a citizens' [residents'] initiative. **住民発議制度** a citizens' [residents'] initiative system.

重油 〔燃料油〕fuel oil;〔重質油〕heavy oil. ▶〜まみれの魚介類 sea creatures covered with thick oil. / 〜が流れ出した．There was a fuel oil spill. ★⇨油, 原油流出, オイルフェンス, 回収, 流出, 沿岸 / 〜に汚染された海岸 a seashore polluted with fuel oil. / 〜汚染 fuel oil pollution.

銃猟(じゅうりょう) gun-hunting; shooting. ▶〜に行く go hunting [shooting]. / 〜を禁ず．〔掲示〕No Gun-hunting. ¶**銃猟家** a gun-hunter. **銃猟禁止期** the closed [close] season for gun-hunting. **銃猟禁止区域** a no-hunting zone. **銃猟税** a gun-hunting [shooting] tax.

樹上 ▶〜の[に, で] on [in] a tree;《live》in the trees. ¶**樹上性の** arboreal; tree-dwelling. **樹上生活** arboreal life. / 〜生活に適している be suited for arboreal existence. **樹上生活者** a tree dweller; an arboreal [a tree-dwelling] animal.

取水 water intake; taking water from a river [lake] (for the water supply). ▶～する take [draw, draw off, use] water from 《a lake》. / 利根川からもっと～する必要がある. We'll have to take more water from the Tone River. | More water will have to be supplied from [of the water supply will have to come from] the Tone River. ¶**取水源** a water supply source; the source of water supply. **取水口** a sluice gate. **取水施設** a water intake facility. **取水制限** 〔河川などからの取水を制限すること；給水制限の前段階〕restrictions on the use of water (from 《a lake》). ★⇨給水(給水制限) / 30パーセントの～制限が必要だ. We'll have to reduce [cut down] the amount of water we take (from 《the river》) by thirty percent. / ～制限を緩和する ease restrictions on the amount of water taken 《from the reservoir》. **取水堰** ⇨堰(せき) **(川からの)取水量** the amount of the water taken 《from a river》.

受水槽(じゅすいそう) a water tank [receptacle].

受精, 授精 〔卵子が精子と受精すること〕fertilization; 〔人工的に授精すること〕fertilization; insemination; fecundation. ▶～する〔受精する〕be fertilized [inseminated, fecundated]; 〔授精する〕fertilize; inseminate; fecundate. ¶**受精毛** 〔紅藻類の〕a trichogyne. **受精嚢** a seminal receptacle; a spermatheca. **受精能力** fertility; 〔卵子の〕fertilizability. **授精能力** 〔精子の〕inseminating capacity [ability]. **受精胚** a fertilized embryo. **受精膜** a fertilization membrane. **受精[未受精]卵** a fertilized [an unfertilized] egg. **受精卵移植** (an) embryo transplant [transfer]; transplantation of a fertilized egg [ovum] **受精卵クローン** an embryo clone **受精卵診断** 〔着床前診断〕preimplantation diagnosis. **受精率** an insemination [a fertilization] rate. **自花[自家]受精** self-fertilization; autogamy. / 自家受精の self-fertilized; autogamous; autogamic. **重複受精** double fertilization. **人工授精** artificial insemination [fertilization]. / 人工授精で妊娠した女性 a woman impregnated with artificial insemination. / サケの人工授精 artificial fertilization of salmon. / 配偶者間人工授精 artificial insemination by husband (略: AIH). / 非配偶者間人工授精 artificial insemination by donor (略: AID); artificial insemination with donor's sperm [semen] (略: AID); heterologous artificial insemination. **体外受精** external fertilization;〔人工的な〕in vitro fertilization;《口》test-tube fertilization. **体内受精** internal fertilization;〔人工的な〕in vivo fertilization. **他家受精** cross-fertilization; allogamy.

受粉, 授粉 pollination. ★⇨花粉 ▶～する〔花に授粉する〕pollinate;〔花が受粉する〕be pollinated; receive pollen. / このハチはリンゴの受粉に一役買っている．This kind of bee helps to pollinate apples. /〔農家が行う〕イチゴの授粉作業 strawberry pollination; the pollination of strawberries; the process of pollinating strawberries. ¶ **受粉樹**〔他の樹に受粉するために使われる樹〕a pollinating tree; a pollinator tree. **自家[自花]受粉** self-pollination. / 自家受粉の self-pollinated. **人工受粉** artificial pollination. **他花受粉** cross-pollination. **虫媒受粉** pollination by insects; insect pollination. **風媒受粉** wind pollination; anemophily.

樹木 a tree; an arbor. ★⇨植樹 ▶～の arboreal. / ～の茂った山々 wooded [woody] mountains. / ～のない woodless; naked; bare (of trees); treeless. / ～を好む鳥[昆虫] dendrophilous birds [insects]. ★⇨樹上 ¶ **樹木医** a tree surgeon [doctor]. **樹木園** an arboretum. **樹木学** dendrology. **樹木学者** a dendrologist. **樹木限界線** a timberline. **樹木測定器** a dendrometer.

主要排出国会議[会合] ⇨英和 MEM

狩猟 ⇨猟 ¶ **狩猟解禁日** the first [opening] day of the hunting season. **狩猟期** the hunting [open] season. **狩猟許可証** a hunting [shooting] license. **狩猟禁止期** the closed [close] season. **狩猟地[場]** a hunting ground;〔ふだんは鳥獣を保護しておく〕a (game) preserve. **狩猟鳥獣**〈集合的に〉game. **狩猟法**〔法律全般〕the game laws;〔日本の法律名〕⇨鳥獣. **狩猟免許証** a hunting [shooting] license.

種類 a kind; a sort; a variety; a class;〔動植物の〕a species. ★⇨種(しゅ);〔型〕a type;〔性質〕nature. ▶植物[動物]の～ species of plants. [animals];

plant [animal] species. / この バラ園にはピンクのバラが数十〜植えられている. In this rose garden are planted several dozen varieties of pink roses. / 熱帯雨林は動植物の〜が豊富だ. In tropical rain forests there are abundant varieties of plants and animals. / 珍しい〜のキノコ an unusual variety of mushroom. / ありふれた〜の魚 the usual [common, ordinary] run of fish. / 未知の〜の昆虫 unknown varieties of insects. / この地方で取れる〜の海藻 the types of seaweed harvested in this region. / いろいろな〜の貝 different species of shells. / 調査隊はこの山域で希少植物20〜を確認した. The survey group confirmed the existence of 20 types of scarce plants in this mountainous area.

シュレッダー・ダスト 〔書類細断機から出るごみ〕shredder dust.

旬(しゅん) ▶旬の野菜は値段も安く栄養も豊富だ. Vegetables in season are low in price and rich in nourishment. ★⇨季節 / 旬のもの foods in season;〔食材〕seasonal ingredients at their peak. / 旬を外れた果物 fruit out of season. / 牡蠣(かき)は今が旬だ. Oysters are now in season. | It is the season for oysters now.

順化 acclimatization; naturalization; acclimation. ★⇨気候順化 ▶〜する acclimatize ; acclimate. / その植物はまだ日本の風土に〜していない. The plant is not yet acclimatized in Japan [has not yet acclimatized itself to Japan]. ¶**高所順化**[順応, 適応] adaptation to high altitude(s); altitude accommodation [adaptation]. **野生順化** adaptation [rehabilitation]《of a species》to the wild.

春化処理 〔植物の開花を促進させる処理〕vernalization. ★⇨開花 ▶〜する vernalize.

循環 circulation; rotation; cycle. ▶〜する circulate 《through…》; rotate; cycle; recur; repeat; move in a cycle; go in cycles [circles]. / この噴水の水は〜している. This fountain recycles its water. / 四季[年]の〜 the cycle [round] of the seasons [year]. / 空気[水]の〜 air [water] circulation. ¶**循環(型)産業** a recycling industry. **(資源)循環型社会** 〔廃棄より再利用を第一に考えた環境保全型社会〕a recycling-oriented society; a recycling community. **循環型社会形成推進基本法** the Basic Law for Establishing a Recycling-based

Society. **循環型社会白書**〔環境省の〕the White Paper on the Recycling-Based Society. **循環処理**〔温泉の〕safe recirculation《of hot-spring water》; a system which disinfects《hot-spring water》and renders it safe before recirculating it. **循環流**〔海洋の〕a circling drift. ★⇨深層(深層循環)，海洋(海洋大循環)，極(極循環)，大気(大気大循環) **循環利用**〔廃棄物などの〕recycling; recirculation. **循環利用率**〔資源全体の再利用量を，消費した天然資源量で割った値〕the ratio of recycled resources to natural resources used. **循環濾過** filtering circulation; circulation filtering. **原子炉再循環系** a primary loop recirculation system. **原子炉再循環系配管** primary loop recirculation system piping;〔その1基〕a primary loop recirculation system pipe. **再循環ポンプ**〔原子炉冷却材用の〕a recirculation pump (for the reactor core coolant). ★⇨排気(排気再循環)

浚渫(しゅんせつ) dredging. ▶〜する dredge《a channel, a harbor》. ¶ **浚渫汚泥** dredged sludge. ★⇨泥土 **浚渫機** a dredging machine; a dredger; a dredge. **浚渫作業** a dredging operation. **浚渫作業員** a dredger. **浚渫船** a dredge (boat); a dredger. **浚渫ポンプ** a dredge [dredging] pump.

順応(じゅんのう) ▶〜する adapt [adjust, accommodate, acclimatize, acclimate] (*oneself*)《to new circumstances》. ★⇨順化，適応 / 環境に〜する adapt (*oneself*) to the environment. / 人間は住む場所に容易に〜できるように作られている．Human beings are created in such a way that they can easily acclimatize [become accustomed] to living wherever they are. ¶ **順応力** adaptability; capacity for adaptation.

使用 ⇨利用，再使用 ▶〜済み製品回収 product return [collection]. ★⇨製品回収 / 〜済み容器の回収 the collection of used containers. ¶ **使用済み核燃料** ⇨核(核燃料) **使用済み核燃料税** (a) spent nuclear fuel tax. **使用済み核燃料棒** a spent nuclear fuel rod. **使用済み自動車**〔廃車〕an end-of-life vehicle (略: ELV).

飼養 breeding; raising; rearing. ★⇨飼育 ▶〜する breed; raise; rear; keep. / 家畜[豚]を〜する raise [rear] cattle [pigs]. ¶ **飼養者** a breeder; a rearer; a raiser. **飼養場** a《chicken》farm. **飼養法** a method of

省エネ(ルギー)

breeding [raising, rearing]; how to breed [raise, rear] 《pigs》. **人工飼養器** an artificial mother.

省エネ(ルギー) energy conservation [saving]; saving (of) energy; conservation of energy. ▶省エネが売りの家電製品 household electrical appliances whose selling point is that they save energy. ★⇨消費(消費電力) / 省エネ家電 an energy-saving home appliance. / 省エネタイプのエアコン an energy-saving type of air conditioner. / 省エネ運動に協力的な店 a shop that cooperates with an energy-saving campaign. / 省エネ(ルギー)の energy-efficient; energy-saving[-conserving]. ¶**省エネ(ルギー)機器** energy-saving equipment. **省エネ(ルギー)基準** an energy conservation standard. **省エネ(ルギー)基準達成率** an energy conservation standard achievement rate. **省エネ技術** energy-saving technologies. **省エネ(ルギー)車[船]** an energy-efficient car [ship]. **省エネ(ルギー)住宅** an energy-efficient [-saving] house. **省エネ(ルギー)製品** an energy-saving product. **省エネ効果** (an) energy-saving effect. **省エネ(ルギー)支援** 〔企業などに対する〕support [a subsidy] for energy conservation. **省エネ性(能)** (an) energy-saving efficiency [effectiveness]; efficiency in saving energy; the energy-saving performance《of a new model of solar water heater》;《assess》how efficient《a solar panel》is at reducing energy consumption. **省エネ表示** energy-efficiency labelling; an energy-efficiency label. **省エネ法** the Energy Saving Law. ★正式名は「エネルギーの使用の合理化に関する法律」(the Law Concerning the Rational Use of Energy). **省エネマーク[ラベル]** 〔環境ラベルの一種〕an energy conservation mark. **省エネラベリング制度** an energy conservation labeling system. **省エネ・リサイクル支援法** the Energy Conservation and Recycling Assistance Law. ★正式名称は「エネルギー等の使用の合理化及び再生資源の利用に関する事業活動に関する臨時措置法」. **省エネルギーセンター** 〔財団法人〕the Energy Conservation Center, Japan (略: ECCJ). **省エネルック** an energy-saving look.

消音 ¶**消音装置**, **消音器** a sound arrester;〔バイクなどの〕

a muffler; a silencer. **消音ピアノ** a silent piano.

常温 〔恒温〕a constant temperature;〔平常の温度〕the normal temperature. ▶この牛乳は〜で3か月保存がきく．This milk can be kept at normal room temperature for up to 3 months. ¶**常温核融合** ⇨核(核融合) **常温倉庫** a non-refrigerated [an unrefrigerated] warehouse. **常温輸送** non-refrigerated [unrefrigerated] transport.

浄化 purification; purgation; elutriation. ★⇨自浄, 炭(すみ), 汚水, 浄水 ▶〜する purify; cleanse; deterge; elutriate. / 水の〜 water purification. / 水質〜法 ⇨英和 Clean Water Act ¶**浄化設備** 〔下水の〕sewage disposal facilities; sanitation facilities. **浄化槽** 〔飲料水の〕a water-purification tank 《for drinking water》; a tank for purifying drinking water;〔下水の〕a septic tank 《for sewage》. **浄化槽汚泥** septic tank sludge. **浄化槽汚泥処理** septic tank sludge treatment. **浄化装置** a purifier; an apparatus for purifying. **浄化能力[作用]** purification capacity [capability].

障害 〔身体の〕(a) disorder; (a) disability; (a) dysfunction; trouble;〔組織機能の〕a lesion; a disorder;〔精神の〕derangement. ▶この物質はラットの肝臓に〜を引き起こすことがわかっている．It has been established that the substance will cause liver dysfunction in rats.

焼却 destruction by fire; incineration; cremation. ★⇨実証(焼却実証試験) ▶〜する incinerate; burn up; destroy *sth* by fire; cremate; reduce *sth* to ashes. / ごみ〜器 a waste [trash, rubbish] incinerator; a waste [trash] burner. ★⇨ごみ(ごみ焼却) ¶**焼却施設** an incineration plant [facility]. **焼却場** an incineration facility (for household and industrial waste). **焼却処分** disposal by burning [incineration]. / 〜処分(に)する dispose of 《diseased animals, confidential documents》 by burning; burn 《diseased animals》. **焼却炉** an incinerator.

商業捕鯨 ⇨捕鯨

衝撃波 a shock wave. ¶**衝撃波音** 〔超音速の航空機などの発する〕a sonic boom.

省資源 ⇨資源

消臭 deodorization; odor neutralization. ▶〜する deodorize. ¶**消臭ゲル** (a) deodorant gel. **消臭効果** a deodorant effect.

消臭剤　a deodorant; a deodorizer; a deodorizing agent; an air freshener; an odor eater.
消臭スプレー　a deodorant spray.　消臭繊維　a deodorant fabric.　消臭芳香剤　a scented deodorizer; an [a scented] air freshener.

上昇　⇨温度，海面　¶上昇気流　an ascending air current; an updraft.　▶〜気流に乗る ride on an updraft; ride a rising current of air.

浄水　¶浄水管　a pipe for clean water.　浄水器　a water purifier.　浄水施設　a water purification facility [plant].　浄水車　〔被災地・戦場などで用いる〕a water purification vehicle.　浄水場　a filtration plant; a water purification plant; a drinking water treatment plant.　浄水錠剤　a water(-purification, -purifying) tablet.　浄水装置　a water-purifying device; a water-purification system.　浄水池　a clean [pure] water reservoir.　浄水フィルター　a water(-purification [-purifying]) filter.

上水　▶玉川〜 the Tama aqueduct; the culvert that provided water to Edo from the Tama River.　¶上水道　〔設備〕a water (supply) pipe; waterworks.　★⇨水道

消費　consumption; spending.　▶〜する consume; spend;〔時間や労力を〕expend; use (up). /〜社会 (a) consumer [consumption] society.　★⇨大量(大量消費社会) / 私の家では夏より冬のほうが電力を余計に〜する．My home uses more electricity in winter than summer.　★⇨消費電力　¶消費期限　a "consume by" date.　★⇨賞味期限　消費者　a consumer;〔食物連鎖上の〕⇨英和 consumer. /〜者の反応 customer response. / 一般〜者〈集合的に〉general [ordinary] consumers; the consuming public.　消費者委員会　〔内閣府の消費者行政監視組織〕the Consumer Commission.　消費者基本法　the Consumer Fundamental Law.　消費者(保護)運動　a consumer (protection) movement. /〜者運動家 a consumer advocate [activist].　消費者教育　consumer education; the education of consumers.　消費者行政　consumer administration.　消費者月間　Consumer Month.　消費者信頼感指数　〔米国などの〕the Consumer Confidence Index.　消費者政策委員会　〔国際標準化機構(ISO)の〕the Consumer Policy Committee (略：COPOLCO).　消費者相談

室　a customer consultation center;〔自治体などの〕a consumer counselling center.　**消費者訴訟**　consumer litigation.　**消費者代表**　a consumer representative.　**消費者団体**　a consumer group [organization].　**消費者団体訴訟制度**　〔消費者団体に団体訴訟を認める制度〕a consumer-group class-action system.　**消費者庁**　〔内閣府の外局〕the Consumer Affairs Agency.　**消費者被害**　damage to consumers; consumer damage.　**消費者被害救済制度**　a consumer damage relief system.　**消費生活アドバイザー[相談員]**　a consumer adviser [advisor].　**消費生活センター**　a consumer service center.　**消費生活専門相談員**　〔国民生活センターが資格認定する〕a consumer counselor [advisor].　**消費生活用製品安全法**　the Consumer Products Safety Law.　**消費電力**　consumption of electricity; electricity [(electric) power] consumption;《the total of》energy consumed. / この新型エアコンは〜電力が少ない．This new air conditioner consumes little electricity [doesn't use much electricity]．★⇨省エネ / 〜電力の少ない[小さい]集積回路を開発する develop a low power integrated circuit. / 通産省は，家庭用電気冷蔵庫の消費電力を2000年度までに13％削減するようメーカーに求めた．MITI called on producers to reduce domestic refrigerator power [electricity] consumption by 13% by the year 2000. / 待機時〜電力 (a) standby current; (a) standby power consumption. / 低〜電力を達成[実現]する achieve low power consumption. / 低〜電力の機種 a low consumption [low power-consumption] model. / 定格〜電力〔電球などに表示された〕(a) rated wattage. / 〜電力量 a [the] number of kilowatt [watt] hours; (a) wattage per hour;《price per》kilowatt hour (略: kWh)．　**消費都市**　a consuming [consumer] city.　**消費量**　(the amount of) consumption．★⇨消費電力 / 夏になると冷房をつけっぱなしにするため電気の〜量がうなぎ上りとなる．In summer, with air conditioners on all day, electricity consumption rises sharply. / 年間〜量 (an) annual consumption.

情報　information; a report; news.　¶**情報アクセス権**　the right of free access to information.　**情報開示[公開]**　disclosure of [disclosing] information; information disclosure. ▶…に〜開示する reveal [disclose, let out] information《to…》. / 〜開示[公開]を請求する request the disclosure of《personal》information. / 〜公開条令 a freedom-[disclosure-]of-information ordinance; regulations governing [guaranteeing] disclosure of information. / 〜開示[公開]請求権 the right to request the disclosure of information. / 全国〜公開度ランキング〔各都道府県の〕a national information disclosure [transparency] ranking. / 〜公開法〔日本の〕the Freedom of Information Law;〔米国の〕the Freedom of Information Act（略：FOIA）. **情報隠し** concealing [hiding] information; information concealment; an information cover-up. ★⇨リコール（リコール隠し）/ 国交省では，その自動車会社により意図的な〜隠しが行われたと確認されれば刑事告発の対象になるとしている. The Ministry of Land, Infrastructure and Transport regards that auto maker as a target for criminal charges if the company is confirmed to have intentionally covered up information. **情報公害** information pollution.

賞味期限〔食品の〕a "best before" date; an expiry date; an expiration date; a use-by date. ★⇨消費（消費期限）▶このチーズは〜が切れている This cheese is past its sell-[use-]by date. / 〜期限を1日過ぎたからといって食べても問題はない. Food that is one day past its sell-by date will do you no harm. / 〜はいつまでですか When does the sell-[use-]by date expire? ｜ When is the use-by date? / 〜 2003 年 8 月 31 日〔表示〕Best [Consume] before Aug. 31, 2003.

静脈産業〔廃棄物処理産業〕a waste processing [recovery] industry.

静脈物流〔家電などの廃棄物回収のための物流〕waste recovery logistics.

条約 a treaty; a pact; a convention; an agreement. ★ treaty, pact は主に比較的限られた当事国同士の政治的・軍事的な条約に用い，環境保護など世界的取り組みの関連条約には convention を使う傾向があるが，通称として treaty を用いることもあって絶対的な区別ではない. ⇨ワシントン条約，ラムサール条約，生物（生物多様性条約），気候（気候変動枠組み条約）▶〜に調印する sign a treaty. / 〜の発効 (the) implementation of a treaty. / 〜の適

用 the application of a convention [treaty clause]. / 〜を批准する ratify a treaty. / 〜を守る [破る] observe [violate] a treaty. / 〜を廃棄する annul a treaty. ¶**条約参加[加盟]国** a signatory (country); a treaty power [participant]; (a country that is) a party to a treaty.

照葉樹林 〔常緑広葉樹林〕an evergreen broad-leaved forest. ¶**照葉樹林文化** a [the] culture of the evergreen broad-leaved forest(s) of East Asia.

常緑 ▶〜の evergreen; indeciduous. ¶**常緑広葉樹** an evergreen broad-leaved tree. **常緑広葉樹林** an evergreen broad-leaved forest. **常緑樹** an evergreen [indeciduous] tree; an evergreen;〈集合的に〉evergreens.

条例 regulations; rules;〔個々の〕an ordinance; a regulation; a law; a bylaw. ▶市〜に従い in accordance with city regulations [a city ordinance] / それは〜または規約によって規定されている. It is prescribed by rules and regulations. / 〜を発布する[取り消す] issue [revoke] regulations. ★⇨実施 ¶**条例違反** a violation of regulations. **条例制定** enactment of an ordinance. **市条例** a city [municipal] ordinance [regulation]. **新聞条例** a press law; press regulations. **都条例** a Metropolitan bylaw [ordinance].

食育 〔栄養教育〕nutrition education;〔正しい食生活を通じて健全な心身を育むこと〕diet [dietary] education. ★⇨食農教育 ¶**食育基本法** the Basic Law for Food Education. **食育推進会議** 〔食育の推進を図るために内閣府に設置された組織〕the (Cabinet Office) Council for Food Education Promotion. **食育推進基本計画** 〔食育推進会議が策定する〕the Basic Plan for the Promotion of Food Education.

食害 〔害虫・鳥獣による〕damage (to plants or crops) caused by vermin [insects, birds and animals]; vermin damage.

食事 ¶**食事性アレルギー** an alimentary allergy. ★⇨食物, アレルギー **食事摂取基準** 〔厚生労働省が5年ごとに改定〕the Dietary Reference Intakes (略: DRIs). **食事バランスガイド** 〔2005年, 農水省と厚労省が作成; 一日に何をどれだけ食べたらよいかをイラストで示したもの〕Japanese Food Guide; a guide to a balanced diet (produced by the Japanese Ministries of Agriculture and Health).

植樹 tree planting. ★⇨植林, 市街植樹, 街路樹 ▶～する plant trees. ¶植樹の日 〔アメリカ・カナダ・オーストラリアなどで〕Arbor Day. **植樹運動** a tree-planting campaign [drive]. **植樹祭** a tree-planting ceremony. **植樹造林** afforestation.

職住 ¶職住一体(型)住宅 a combined living and working housing unit. **職住近接型都市** a city with housing near workplaces.

植生 〔ある地域の全植物〕vegetation. ★⇨植物, 海岸(海岸植生), 自然(自然植生), 人為植生 ▶熱帯の～ tropical vegetation. ¶植生学 vegetation science. **植生管理** vegetation management. **植生自然度** the 《first, tenth》 degree of natural vegetation; the degree of naturalness of vegetation. **植生図** a vegetation map. **植生図作成** vegetation mapping. **植生帯** a zone of vegetation; a vegetation zone. **植生分布** (a [the]) distribution of vegetation; (a) vegetation distribution.

食生活 eating [dietary] habits; diet. ★⇨食育 ▶～を改善する improve *one's* diet [eating habits]. / 健康のために野菜や魚中心の和風の～を基本にしています. For my health I mainly eat Japanese-style meals centered on fish and vegetables. / 日本では～の欧米化が進んだ. In Japan Westernization of the dietary regime proceeded apace.

食虫 ▶～性の insectivorous; entomophagous. ¶食虫動物[植物] an insectivore; an insectivorous animal [plant]. **食虫類** Insectivora.

食農教育 dietary and agricultural education. ★⇨食育

触媒 a catalyzer; a catalyst; a catalytic (agent). ¶三元触媒 〔車の排ガス中の一酸化炭素・炭化水素・窒素酸化物を無害化する触媒〕a three-way catalyst. ▶(三元)～コンバーター a (three-way) catalytic converter. **選択還元触媒** 〔自動車が排出する窒素酸化物を減らす方式〕a [the] selective catalytic reduction (system)(略: SCR).

食品 food(s); food products; foodstuffs. ▶～の鮮度を保つ keep (a) food fresh; maintain a food's freshness. / 実際問題, 遺伝子組み替え～が人体に今後どう影響するか, だれにもわからない. The fact of the matter is that no one knows what future effect genetically modified [GM] foods will have on the human body. / ～の表示制度 a food labeling system. ¶**食品安全委員会**〔食品安全基本法に基づく〕the Food Safety Commission (略: FSC). **食品安全基本法** the Food Safety Basic Law. **食品安全担当大臣**〔内閣府の特命担当大臣〕the Minister of State for Food Safety. **食品安全部**〔厚生労働省の〕the Department of Food Safety. **食品売り場** the food section [department]. **食品汚染** food contamination; food pollution. **食品汚染物質** a food contaminant; a food pollutant. **食品会社** a food company. **食品化学** food chemistry. **食品科学** food science. **食品学** sitology. **食品環境検査協会** the Japan Inspection Association of Food and Food Industry Environment (略: JIAFE). **食品偽装** food fraud. ★⇨産地偽装 **食品群** a 《basic》food group. **食品公害** food contamination. ★⇨公害 **食品交換表** a food-exchange list. **食品工業** the food industry. **食品工場** a food(-processing) factory. **食品香料** (a) flavor. **食品混入物** a food contaminant. **食品サンプル**〔食堂の〕a《vinyl, wax》food sample. **食品事故** a "food accident"; an accident involving (contaminated) food; a food poisoning [contaminated food] incident. **食品照射**〔放射線による〕food irradiation. **食品スーパー(マーケット)** a food(-based) supermarket. **食品製造用剤** an agent for food processing; a food manufacturing agent. **食品成分試験法** food constituent analysis. **食品成分表** a food-composition table. **食品店** a grocery store [shop]; a grocery; a grocer's (shop). **食品添加物** a food additive; an additive. **食品添加物公定書** an official formulary of food additives. **食品取扱者** a food handler. **食品トレー**〔食料品店の〕a《Styrofoam》food tray. **食品廃棄物** commercial food waste. **食品廃棄率**〔賞味期限切れ,食べ残しなどによる〕the ratio [percentage] of food discarded. **食品表示** a food label;〈集合的に〉food labeling. **食品表示ウォッチャー**〔農林水産省の〕a food labeling watcher [monitor]. **食品分析** food analysis. **食品包装用ラップ** plastic wrap; clingfilm;〔商標〕Saran Wrap. **食品防腐剤[保存料]** (a) food preservative. **食品リコール**〔メーカーによる不良食品の自主回収〕《initiate》a food recall. **食品リコール(費用)保険** food recall insurance. **食品リサイクル法**

the Food Recycling Law. **食品履歴** a food product's history. **食品履歴制度** a food traceability system. **食品ロス** food waste. **食品ロス統計** 〔農水省の発表する〕statistics on food waste. **食品ロス率** a food waste ratio. **インスタント食品** instant foods. **加工食品** processed foods. **主要食品** staple foods; staples. **冷凍食品** frozen foods.

植物 a plant;〈集合的に〉plant life; vegetation;〔一時代・一地域の植物相〕(a) flora. ★⇨植生, 塩生植物, 寄生(寄生植物), 帰化植物, 指標植物 ▶〜を採集する collect plants; botanize; herborize. /台湾の〜(相)を研究する study the flora of Taiwan. /熱帯の〜を植える plant tropical plants. /〜質[性]の vegetable. /〜性バター vegetable butter.

¶**植物育種学** plant breeding. **植物遺伝学** plant genetics. **植物ウイルス** a plant virus. **植物ウイルス病** a plant-virus disease. **植物エキス** a plant extract. **植物エクジソン** a phytoecdysone. **植物園** a botanical garden; botanical gardens. /小石川〜園 Koishikawa Botanical Gardens. **植物塩基** a plant [vegetable] base. **植物界** the plant [vegetable] kingdom. **植物解剖学** phytotomy; plant anatomy. **植物化学** plant chemistry; phytochemistry. **植物観察ハンドブック** 〔軽便な図鑑など〕a field guide to plants. **植物記載学** descriptive botany. **植物季節観測** plant phenological observation. **植物極** 〔卵細胞のうち植物性器官を作る；動物極と対照〕a vegetal pole. **植物区系** a floral zone. **植物群系** a formation. **植物群落** a plant community. **植物形態学** plant morphology; morphological botany. **植物検疫** plant quarantine. **植物公園** a botanical garden; a park. **植物工場** 〔人工的に環境を調節して植物栽培を行う〕a plant factory. **植物細胞[繊維]** a vegetable cell [fiber]. **植物細胞学** plant cytology. **植物色素** 〔食品添加物〕(a) plant pigment. **植物質** vegetable matter. **植物社会学** phytosociology; plant sociology. **植物新品種保護国際同盟** ⇨UPOV(和英篇末尾) **植物図鑑** an illustrated [a pictorial] guide to《Japanese》flora. **植物ステロール** phytosterol. **植物ステロール・エステル** a plant sterol ester; a phytosterol ester. **植物性殺虫剤** a botanical insecticide. **植物性自然毒** vegetative natural poison; natural plant poison. **植物性峻下薬** a drastic plant cathartic. **植物性食品** plant [vegetable] foods. **植物性神経系** a vegetative nervous system. **植物性繊維** (a) vegetable fiber. **植物性染料** a vegetable dye. **植物性たんぱく質** (a) vegetable protein; (a) protein of vegetable origin. **植物性薬品** a botanical drug. **植物生態学[組織学]** plant ecology [histology]; ecological [structural] botany. **植物生長調節物質** a plant growth-regulating substance. **植物成長ホルモン** (a) plant growth hormone. **植物生長抑制剤** a plant growth retardant. **植物生理学** plant physiology; physiological botany. **植物生理学者** a plant physiologist. **植物組織培養** plant tissue culture. **植物帯** a floral [vegetation] zone; a zone of vegetation. **植物地理学** geographical botany; plant geography. **植物毒素** a phytotoxin. **植物標本** a botanical specimen; 〈集合的に〉a herbarium. **植物病理学** plant pathology; phytopathology. **植物プラスチック** ⇨バイオマス(バイオマス・プラスチック). **植物プランクトン** phytoplankton. **植物分布** plant distribution; the distribution of plants. **植物分類学** systematic botany; plant taxonomy. **植物分類学者** a plant taxonomist. **植物防疫法** the Plant Quarantine Law. **植物防疫所** Plant Protection Stations **植物ホルモン** (a) plant hormone; (a) phytohormone. **植物油** vegetable oil. /純〜油 100-percent [100%] vegetable oil. **植物油脂** vegetable oil and fat. **植物由来成分[物質]** a vegetable(-derived) ingredient [substance]. **隠花植物** a flowerless plant; a cryptogam. **陰生[陰地]植物** a shade plant. **海浜植物** a seaside plant. **顕花植物** a flowering plant; a phanerogam. **薬用植物** a medicinal plant [herb]. **陽生植物** a sun plant.

食物 food; foodstuffs; victuals; 〔一定の〕a diet; 〔糧食〕provisions. ★⇨食べ物, 食事, 食生活, 食品 ¶固形[流動]食物 (a) solid [liquid] food. **食物アレルギー** 《have》a food allergy. **食物アレルギー表示制度** a food allergy labeling system. **食物環** a food cycle. **食物源** a food source. **食物繊維** dietary fiber. ★⇨掃除(ぞう) **食物繊維飲料** a fiber drink. **食物網** a food web. **食物連鎖** a food chain. ★⇨英和 food chain

食用 ▶〜赤色2号〔食用色素〕

Food Red No.2; amaranth. /〜である[になる] be good to eat; be edible; be fit for food [the table]. ¶**食用油** cooking [salad] oil. **食用インク** edible ink. **食用花** an edible flower. **食用蛙**〔ウシガエル〕a bullfrog. **食用カタツムリ**〔エスカルゴ〕an edible snail; an escargot. **食用菊** an edible chrysanthemum. **食用牛**〈集合的に〉beef cattle. **食用魚** an edible fish; a food fish. **食用菌** an edible mushroom. **食用香料** food flavor. **食用作物** a food crop. **食用色素** a food dye; a food color. **食用色素タール**〔合成色素〕a food additive synthetic dye. **食用植物** an edible plant. **食用米** rice (fit) for human consumption. /非〜米 rice not (fit) for human consumption; rice for purposes other than human consumption. **食用油脂** edible oil and fat.

食料[糧]〔食べ物, 備蓄食糧〕food; foodstuffs;〔備蓄食糧〕provisions. ★⇨食品, 食物 ▶〜を仕入れる[買い込む] lay in provisions [(a supply of) food] /〜を確保する secure a supply of food [provisions]. /〜が欠乏[不足]してきた. Provisions are running low. | We are running low on [short of] provisions. /〜の備蓄が底をつきかけている. Food supplies are running low. /〜を蓄える lay in [up] provisions; stock [store up] provisions. /〜を確保する secure provisions [(a supply of) food]. /〜を供給する provide [supply] *sb* with food; provide food《for *sb*》; furnish food《to *sb*》; provision《a district》. /〜援助[支援]を再開する resume food assistance [aid]. ¶**食料安全保障** food security. /(世界)食糧安全保障委員会〔食糧農業機関(FAO)の〕the Committee on World Food Security. **食糧委員** a provision committee. **食糧危機** a food crisis. **食料自給** food self-sufficiency. **食料自給率** the food self-sufficiency rate [ratio]. ★⇨自給 **食料需給表** a food balance sheet (略: FBS). **食糧証券** a food financing bill. **食糧政策** (a) food policy. **食料生産** food production. **食糧戦争** a food war; a war for [over] food. **食糧庁** the Food Agency. ★⇨総合食料局 **食料貯蔵室** a larder; a pantry. **食糧統制** food control. **食糧難** the difficulty of obtaining food. **食糧農業機関**〔国連の〕the Food and Agriculture Organization (略: FAO). **食料・農業・農村基本計画**〔農林

水産省の〕the Basic Plan on Food, Agriculture and Rural Areas.　**食料・農業・農村基本法**　the Basic Law on Food, Agriculture and Rural Areas.　**食料・農業・農村白書**　〔農林水産省の〕the Annual Report on Food, Agriculture and Rural Areas in Japan《2004》.　**食料白書**　a white paper on food.　**食糧費**　〔地方自治体の〕food expenses.　**食糧備蓄**　food supplies [reserves]; reserves of provisions.　**食糧問題**　a food problem.

植林　〔樹を植えること〕tree planting; afforestation;（再植林）reforestation;〔植林地〕an afforestation area; a plantation.　★⇨植樹, 造林　▶〜する　afforest《a mountain》; plant trees;〔再び〕reforest《denuded land》.　¶ **植林計画**　a tree-planting [an afforestation] project.　**植林政策**　an afforestation [a reforestation] policy.　**植林ボランティア**　〔活動〕volunteer tree-planting (activities);〔人〕a tree-planting volunteer; a volunteer tree-planter.　**植林用地**　land [an area] for afforestation; an afforestation area [site].

除湿　dehumidification.　▶〜する　dehumidify.　¶ **除湿器**　a dehumidifier.　**除湿剤**　a dehumidifying agent.

除草　▶庭の〜をする　weed (out) a garden.　¶ **除草機**　a weeder.　**除草剤**　a herbicide; a weed killer; a weedicide.　**除草剤耐性**　herbicide tolerance.　**除草剤耐性作物**　a herbicide-tolerant crop.

除伐(じょ)　〔樹木の〕improvement [salvage] cutting; clean-cutting; extraction.

処分　disposal; disposition.　▶〜する　dispose of…; deal [do] with…; make a clearance of《unsold goods》;〔病気の家畜を〕⇨殺処分.／ごみの〜　garbage disposal.　★⇨処理　¶ **処分場**　a disposal plant.　**ごみ[産業廃棄物]処分場**　a waste [an industrial waste] disposal plant.　**廃棄物海面処分場**　〔1960年代の東京湾の夢の島など〕an offshore [a coastal] landfill [waste disposal] site.

処理　（薬品などを用いての）treatment.　★⇨再処理, ごみ(ごみ処理), 下水(下水処理), 屎尿(にょう)処理, 中間処理　▶〜する〔薬品などを用いて〕treat.／未〜の下水　raw [untreated] sewage.　¶ **処理場**　a treatment [disposal] plant.　★⇨埋め立て, ごみ／産廃[産業廃棄物]〜場　an industrial waste treatment [disposal] plant; an industrial waste processing facility. **処理水** treated water. **処理法** a treatment method; a method of disposal. **汚水処理** sewage treatment [disposal]. **簡易処理** 〔汚水などの〕simplified 《wastewater》 treatment. **殺菌処理** sterilization. **熱処理** heat treatment. **排水[廃水]処理** wastewater treatment.

人為 ¶**人為災害** ⇨人災. **人為植生** artificial vegetation. **人為選択[淘汰]** artificial selection. **人為突然変異** artificial mutation. **人為分類** 〔博物学の〕artificial classification.

新エネルギー ⇨代替(代替エネルギー) ¶**新エネルギー・産業技術総合開発機構** the New Energy and Industrial Technology Development Organization (略: NEDO). **新エネルギー利用(等に関する)特別措置法, 新エネ法** the Special Measures Law for Promoting the Use of New Energy. ★通称⇨ RPS 法(和英篇末尾)

進化 evolution; progress. ▶〜する evolve [develop]《from... into...》./ 人間は猿から〜したと主張する insist that humans evolved from apes. / 人類の〜 human evolution; the evolution of humankind. ¶**進化経済学** evolutionary economics. **進化工学** evolutionary engineering. **進化主義** evolutionism. **進化生態学** evolutionary ecology. **進化生物学** evolutionary biology. **進化発生生物学** evolutionary developmental biology; evolution of development; evo-devo. **進化分子工学** evolutionary molecular engineering. **進化論** the theory [doctrine] of evolution; evolutionary theory; evolutionism. **進化論者** an evolutionist. **収斂(しゅうれん)進化** convergent evolution. **漸変進化** anamorphosis. **定向進化** orthogenesis. **平行進化** parallel evolution.

深海 the depths of the sea; an abyss. ★⇨深層 ¶**深海救助潜水艇** a deep submergence rescue vehicle (略: DSRV). **深海魚** a deep-sea [an abyssal] fish. **深海漁業** fishing [fishery] for abyssal fish; deep-sea fishing. **深海散乱層** a deep scattering layer. **深海潜水艇** a deep submergence vehicle (略: DSV). **深海測深** bathymetry; deep-sea sounding. **深海測深器** a bathymeter; a depth-sounder. **深海ダイバー** a deep-sea diver. **深海探査** deep-sea exploration. **深海底** the deep seabed; the deep seafloor. **深海底鉱業** deep seabed mining. **深海底鉱**

物資源 deep seabed mineral resources. **深海底生物群集** a deep-sea biological community. **深海底帯** the abyssal zone. **深海トロール漁** deep-sea trawling. **深海波** an abyssal [a deep-sea] wave. **深海油田** a deep-sea oil field. **深海流** a deep-sea current.

シンク〔森林に吸収される温室効果ガス(の量)〕sink; absorption of greenhouse gases by forests. ★⇨森林(森林吸収源),英和 carbon sink

人口 a population; the number of inhabitants. ★⇨世界(世界人口会議) ▶～の少ない国 a thinly [sparsely] populated [inhabited] country. / ～の多い国 a densely [highly] populated [inhabited] country. / この町は～が増えた．The population of this town has grown. / その町は～がどんどん減っている．The population of the town is decreasing rapidly. / ～の都市集中化 the gravitation of people toward(s) cities. / ～の都市流入 a shift of population [population shift] to cities; an influx of people into cities. / 東京の～は1,000万以上である．The population of Tokyo is more than 10,000,000. ｜ Tokyo has a population of over ten million. / 高校卒業～ the high-school-graduate [school-leaving] population; the population of high-school graduates [school leavers]. / 十八歳～ the 18-year-old population; the population of 18-year-olds. ¶**人口圧**〔経済活動を圧する人口過剰の状態〕population pressure. **人口移動** demographic shift; population drift [mobility]. **人口学**〔人口統計学〕demography. **人口カバー率** a population coverage rate. **人口過少** underpopulation. **人口過剰** overpopulation. **人口過密国** a country with a high population density; an overcrowded country. **人口減少時代** a period of population decrease. **人口減(少)社会** a society with a decreasing population. **人口構成** population composition. **人口構造** population [demographic] structure [composition]. **人口高齢化** aging of the population (of a country). **人口重心** the population center of gravity 《of Yokohama》. **人口集積度** (the degree of) population density [concentration]. **人口集中** population concentration; concentration of the population 《in urban areas》. **人口集中地区** a

人口

densely inhabited district (略: DID).　**人口政策**　a population policy; a demographic policy.　**人口静態**　a static population.　**人口増加[減少]**　an increase [a decrease] in population; population growth [decrease].　**人口増加率**　a population growth rate.　**人口置換水準[置換率]**　a [the] population replacement level [rate].　**人口調査**　a census. / ～調査を行う take a census (of the population).　**人口調整**　population control.　**人口転換理論**　the demographic transition theory.　**人口統計**　demographic [population, vital] statistics.　**人口統計学**　demography.　**人口統計学者**　a demographer.　**人口動態**　a fluid [dynamic] population.　**人口動態調査**　〔厚生労働省の〕a demographic survey.　**人口動態統計**　vital statistics (of a population); demographics.　**人口動態統計特殊報告**　〔厚生労働省の〕a special report on vital statistics.　**人口爆発**　a population explosion.　**人口表**　a table of population.　**人口ピラミッド**　〔年齢別のグラフ〕a population pyramid.　**人口比率**　a [the] population ratio 《of urban to rural residents》; a [the] ratio 《of women to men》 in the population; a [the] percentage 《of women》 in the population.　**人口普及率**　a population penetration rate.　**人口扶養力**　〔土地の〕《the land's》 capacity to support the population; the population-carrying capacity 《of the land》.　**人口分布グラフ**　a population distribution graph.　**人口ボーナス**　〔全人口に占める生産年齢人口の割合が増大すること〕a population bonus; an increase in the working-age population.　**人口密集地**　a densely populated district [area].　**人口密度**　population density. / ～密度が高い[低い] be densely [sparsely] populated; be thickly [thinly] peopled. / その地方の～密度は1平方キロにつき100人である．The density of population in the region is 100 people to a square kilometer. | The per-square-kilometer population density of the region is 100.　**人口問題**　the population problem. / 世界の～問題が広く取りざたされている．The population problem of the world is widely talked about. / 現在の出生率だと，あの国にもまもなく～問題が起こるだろう．With the present birthrate that country also may soon have a population problem.　**人口問題研究所**

the Institute of Population Problems. **人口要件** population requirements. / 市になるための～要件を満たす meet the population requirements to become a city. **人口抑制** population (growth) control. **人口流出** population drain [flight]. **居住人口** a permanent [resident] population. **交流人口** the nonresident population. **就業人口** the working population; the population of the employed; the work force. **世界人口** world [the world, the world's] population. **総人口** the total population. **昼間[夜間]人口** the daytime [nighttime] population. **定住人口** a permanent [resident] population. **登録人口** the registered population. **浮動人口** a floating population. **幽霊人口** a "ghost" [bogus, spurious] population. **労働人口** the working [laboring] population.

人工 ●～の，～的な artificial; unnatural. ★⇨人為，合成，栽培(人工栽培) / ～的に雷を起こす，～雷を起こす (artificially) induce lightning / ～雷発生機 an artificial lightning generator. / ～的に合成されたダイヤモンド a synthesized diamond.
¶ **人工育雛** artificial brooding 《of chicks》. **人工池** an artificial [a man-made] pond. **人工遺伝子** an artificial gene. **人工雨(う)，人工降雨** 〔人為的に降らせる雨〕artificial rain;〔人為的降雨作業〕rainmaking. / ～降雨を起こすために雲にドライアイスをまく seed a cloud with dry ice to make rain. **人工オーロラ** an artificial aurora. **人工海岸** a man-made [an artificial] beach. **人工甘味料** an artificial sweetener; artificial sweetening. ★⇨合成 **人工観葉植物** an artificial leafy [decorative] plant. **人工魚礁** ⇨魚礁(ぎょしょう). **人工湖** an artificial [a man-made] lake. **人工降雪機** a snowmaker; a snowmaking machine. **人工交配** artificial crossing. **人工飼育** artificial feeding. **人工地震** an artificial [a man-made] earthquake. **人工受粉** artificial pollination. ★⇨授精(人工授精) **人工礁** an artificial [a man-made] reef. **人工水路** a sluiceway. **人工生殖** artificial reproduction. **人工生命** ⇨生命 **人工染色体** an artificial chromosome. **人工巣塔** a nest tower. **人工増殖** artificial breeding. **人工着色料** 《contains》artificial coloring matter; synthetic dye. / ～着色料含

有．〔食品の表示〕Artificially colored. | Contains artificial coloring. **人工調味料** a synthetic flavoring [seasoning]. **人工添加物** (an) artificial additive. **人工島** an artificial [a man-made] island. **人工渚** =人工海岸. **人工日光** artificial daylight. **人工波** machine(-made) waves; mechanically induced waves. **人工培養** artificial culture. **人工ビーチ** =人工海岸. **人工干潟** artificial [man-made] wetlands. **人工雪** artificial [man-made] snow. ★⇨人工降雪機 **人工林** a man-made [an artificial] forest.

人災 a man-made disaster [calamity]; a disaster caused by human error. ▶それは天災ではなく〜だった. It was not a natural disaster but a man-made one. ★⇨天災

新種 〔種〕a new species;〔変種〕a new variety;〔型〕a new type. ★⇨ UPOV(和英篇末尾) ▶〜のチューリップ a new variety of tulip. / ... の〜をつくり出す grow [cultivate] a new variety of....

人獣共通感染症 a zoonosis《複数形 zoonoses》; an anthropozoonosis《複数形 anthropozoonoses》.

浸食 erosion; corrosion. ▶〜する erode; corrode; eat away;〔水が〕wash out;〔海が陸地を〕gain [encroach] on. / 海岸一帯にわたって海が陸地を〜している. The sea is gaining [encroaching] on the land all along the coast. / 雨[風, 水]による〜 rain [wind, water] erosion. / その木は〜防止用に広く使われている. The tree is used widely for erosion control. ¶**浸食作用** erosion; erosive [corrosive] action. / 酸による〜作用で引き起こされた caused by the corrosive action of an acid. **浸食台地** an eroded plateau. **浸食平野** an erosional plain. **浸食輪廻** an erosion cycle.

親水(しん) ¶**親水権** 〔環境権の1つ〕the right of access to water. **親水公園** a water garden.

深層 ⇨深海, 海洋(海洋深層水) ¶**深層海流** a deep ocean current. **深層循環** 〔深海における地球的規模での海水循環〕deep water circulation; abyssal circulation. **深層地下水** deep groundwater. **深層天然ガス** deep earth natural gas. **深層熱水** deep geothermal water. **深層風化** deep weathering. **深層流** a deep (sea) current.

人体 a [the] human body;《口》the system;〔肉体〕flesh. ▶エックス線が〜に与える[及ぼす]

影響 the effect of x-rays on the human body. ¶**人体解剖学** human anatomy.　**人体解剖図** an anatomical chart.　**人体寄生虫** a human parasite; a parasite on the human body.　**人体計測法** anthropometry.　形 anthropometric(al).　**人体効果[容量]** body effect [capacity].　**人体実験** human experimentation; testing on humans; human testing.　**人体測定** anthropometry.　**人体模型** 〔解剖模型〕a model of the human anatomy.

人畜 men and [or] beasts; men and [or] cattle [livestock]; man and beast; living creatures [things].　▶〜に無害．No harm to man or [and] beast. / 〜に損害なし．There was no damage to man or beast. / 〜に死傷なし．There were no deaths or casualties among men or beasts.

浸透 ▶〜する permeate; infiltrate; penetrate; percolate; spread《into [through]…》． / 水が徐々に土壌の中に〜してゆく．Water slowly percolates [filters, spreads] down through the soil.　¶**浸透水** seepage (water).　**浸透ます** 〔浄化槽からの汚水を地面に吸収させるための〕a seepage pit.

振動 (an) oscillation; (a) vibration; (a) swing; (a) shaking; (a) to-and-fro movement; (a) movement from side to side.　▶〜する oscillate; swing; vibrate; librate; move back and forth [to and fro];　¶**振動規制法** the Vibration Regulation Law.　**振動公害** vibration pollution.　★⇒低周波空気振動

塵肺(じんぱい)　pneumoconiosis; pneumokoniosis.　¶**塵肺症** pneumoconiosis; coniosis.　**塵肺訴訟** a pneumoconiosis case [(law) suit].　**塵肺法** the Pneumoconiosis Law.　**トンネル塵肺** pneumoconiosis in tunnel workers.

侵略的外来種　an invasive (introduced) species.　★⇒外来種

森林　a forest; a wood.　¶**森林インストラクター**　a forest instructor.　**森林害虫**　a forest pest.　**森林開発**　forest use [exploitation]; the exploitation of (a) forest; using (the) forest(s); forest development; development of a forest; planting [growing] forests.　**森林学** ⇒林学(りんがく)．　**森林火災**　(a) forest fire.　▶各地で〜火災が発生した．Forest fires broke out in many places.　**森林官**　a forester.　**森林(環境)税**　(a) forest environment tax.　**森林**

森林

監視員 a forester; a (forest) ranger. ★⇨英和 forest ranger　森林管理 forestry (management); management of a forest [the forests].　森林管理協議会 〔世界的な森林認証団体〕the Forest Stewardship Council (略: FSC). ★⇨英和 Forest Stewardship Council　森林管理局 〔林野庁の部局〕Regional Forest Offices.　森林管理者 a forest steward.　森林管理認証 ⇨英和 forest management certification　森林吸収源 〔温室効果ガスの吸収体としての森林〕a forest sink. ★⇨吸収　森林吸収量 〔二酸化炭素の〕the amount of carbon dioxide [CO_2] absorption by forests; the amount of forest CO_2 absorption.　森林行政 forest administration.　森林組合 a forestry (owners') association.　森林計画 forestry planning; 〔その一計画〕a forestry plan.　森林警備員 ⇨森林監視員　森林限界線 a [the] tree line; a timberline.　森林公園 a forest park.　森林資源 forest [timber] resources.　森林整備 forest management.　森林整備法人 a forest management corporation.　森林総合研究所 the Forestry and Forest Products Research Institute (略: FFPRI).　森林帯 a forest zone.　森林地 woodland(s); a forested [wooded] area.　森林地方[地帯] a wooded region; forest land; woodland; a timber region; timberland.　森林鉄道 a forest railway.　森林動物 a forest animal.　森林土壌 forest soil.　森林認証 〔木材の生産地の保証〕forest certification. ★⇨英和 chain of custody　森林認証制度 a forest certification system.　森林破壊 deforestation; forest destruction; destruction of (the) forests. / 世界中で恐ろしい速さで〜破壊が進んでいる. Deforestation [Destruction of the forests] is proceeding at a terrifying pace all over the world.　森林パトロール a forest patrol.　森林被害 forest damage; tree damage; damage to forests [trees]. / 台風[酸性雨]による〜被害 forest damage due to typhoons [acid rain].　森林文化 (a) forest culture.　森林保険 forest insurance.　森林保護 conservation [conservancy, protection, preservation] of (the) forest(s); protecting [preserving] (the) forests; forest protection [conservation, preservation].　森林保養地 a forest area for health and recuperation; a forest resort.　森林

ボランティア a forest [forestry] volunteer. **森林面積** a forest area; (a) forested area. /世界の〜面積は約 40 億ヘクタールで，陸地の 3 割に当たる．The forested area of the earth covers about 4 billion hectares, which is one third of the whole land surface. **森林乱伐** indiscriminate [reckless, unrestricted] logging [felling of the forests]. **森林療法[セラピー]** forest therapy. **森林療法士** a forest therapist. **森林・林業基本法** the Basic Law for Forests and Forestry. **森林・林業白書** a white paper on forests and forestry;〔林野庁の〕the Annual Report on Trends of Forest and Forestry《2005》.

す

巣 a nest; (ワシなどの) an aerie; (蜜蜂の) a beehive;〔クモの〕a web; a cobweb;〔獣の〕a lair; a den. ★⇨営巣(えいそう)，帰巣(きそう)，群生 ▶巣から落ちたひな a baby bird fallen from the nest. / カッコウはほかの鳥の巣に卵を産み付ける．The cuckoo lays its eggs in the nests of other birds. / ツバメがわが家の玄関先に巣を作った．Swallows built a nest in my front entrance. / (クモが)巣をかける[張る] spin [weave] a web.

スーパー雑草〔厳しい環境や除草剤等に耐性をもつと考えられている草〕a superweed.

スーパーファンド法〔米国で1980年制定の総合環境対策補償責任法〕the Superfund Law. ★正式名称は Comprehensive Environmental Response, Compensation and Liability Act (略: CERCLA). ⇨英和 superfund

スーパー林道〔特定森林地開発用道路〕a super forest road.

水界〔地球の表面の〕the hydrosphere.

水害 damage by [from] a flood; flood damage;〔惨禍〕a flood disaster. ★⇨保安林(水害防備保安林) ▶この地方では毎夏多少の〜がある．This district suffers some flood damage every summer. / 〜にあう[見舞われる] be hit by a flood; suffer flood damage; be damaged by a flood. / 〜をもたらす cause flood damage. / 〜を防ぐ prevent [protect against] flood damage. / 水害に備えた堤防 an embankment erected as a flood defense. / 〜の多い地帯 a flood-prone area. ¶**水害対策**〔防止の〕a flood-control measure;〔救済の〕a relief measure for flood victims. **水害地** a flooded [flood-stricken] district. **水害防止** prevention of floods; flood control. **水害保険** flood insurance. **水害罹災民** flood victims.

吸い殻 a cigarette [cigar] butt [stub, end]; a cigarette [dog] end. ▶〜の不始末による火事 a fire caused by not making sure a cigarette butt was out. / 路上の〜を拾う pick up (thrown-away) cigarette butts in the street. / 〜を捨てる toss [throw] away [down] a cigarette end. / 車の窓から〜を捨てる flip a cigarette butt out the car window. ★⇨ポイ捨て, 路上喫煙　¶ **吸い殻入れ**　an ashtray;〔携帯用の〕a pocket ashtray.

水気耕栽培　〔培地が固形でない水耕栽培の一種〕aerohydroponics. ★⇨水耕栽培

水銀　mercury; quicksilver. ▶総〜〔環境基準の1つ〕total mercury.　¶ **水銀汚染**　mercury contamination [pollution]. **水銀温度計**　a mercury thermometer; a mercury-in-glass thermometer.　**水銀柱**　a column of mercury; a mercurial column.　**水銀柱ミリメートル**〔圧力の単位〕a millimeter of mercury（記号: mmHg）．**水銀中毒(症)**　mercurialism; mercurial [mercury] poisoning; hydrargyrism.　**水銀電池**　a mercury battery [cell].　**水銀農薬**　an agricultural chemical containing mercury.　**有機水銀(化合物)**　an organic mercury compound.

水系　a water system. ▶石狩川〜流域に in the Ishikari River system catchment area. / 利根川〜にある8つのダムの貯水量が低下している．The water reserves at the eight dams in the Tone River system are falling.

水源　the head [source]《of a stream》; a riverhead; a fountainhead; a headspring;〔水道の〕the source of water supply;（貯水池）a reservoir. ▶〜が枯れてしまった．The water source [source of water] dried up. / 〜を探る search for the fountainhead. / 川の〜をたどる trace (up) a river to its source. / この川の〜(地)はどこですか．Where does this river rise [take its rise]? / その川の〜は諏訪湖だ．The river issues from Lake Suwa.　¶ **水源開発**　water source development.　**水源(環境保全)税**　a headwater conservation tax.　**水源(涵養)林**　a watershed (protection) forest. ★⇨保安林

水耕　¶ **水耕栽培[法]**　hydroponics（形 hydroponic）; water [hydroponic, soilless] culture; aquaculture.　**水耕栽培者**　a hydroponist.　**水耕農場**　a hydroponic farm.

水産　▶〜物に富む be rich in aquatic [marine] products.　¶ **水産エコラベル**　〔マリンエコラベル(MEL)ジャパンが認定する日本版 MSC ラベル(⇨和英篇末尾)〕the sustainable seafood ecolabel.　**水産学**　fishery [fisheries] science.　**水産学部**　the department of fishery [fisheries] science.　**水産加工業**

the marine product processing industry.　**水産加工品**　processed marine products.　**水産基本法**　the Basic Law on Fisheries.　**水産業**　fisheries.　**水産業協同組合**　a fisheries cooperative association.　**水産業協同組合法**　the Fisheries Cooperative Association Law.　**水産組合**　a fisheries union.　**水産研究所**　a fisheries laboratory.　**水産国**　a fishing country; a nation rich in marine resources.　**水産資源**　(living) aquatic resources; marine resources [products].　**水産資源保護法**　the Fishery Resources Conservation Law.　**水産試験場**　an experimental fishery station; a fisheries experimental station.　**水産食(料)品**　seafood.　**水産大学[学校]**　a fisheries college [school].　**水産大学校**　National Fisheries University.　**水産庁**　〔農林水産省の〕the Fisheries Agency.　**水産白書**　a white paper on fisheries;〔農林水産省の〕the Annual Report on Developments in the Fisheries Industry in 《2005》.

水質　▶～を検査する test [analyze] the water quality《of a well》. ¶**水質汚濁[汚染]**　water contamination [pollution]. / 定期的に実施している～汚染調査で東京湾はきれいになってきていることがわかった. From the regular surveys carried out to establish levels of water pollution, we have learned that Tokyo Bay has become cleaner.　**水質汚濁基準**　a water pollution standard.　**水質汚濁[汚染]防止**　prevention of water contamination [pollution]; water contamination [pollution] prevention.　**水質汚濁防止法**　the Water Pollution Prevention Law. ⇨英和 Clean Water Act　**水質階級**　a water quality class.　**水質環境基準**　environmental standards for [on] water quality.　**水質管理**　water quality management [control].　**水質基準**　water quality criteria [standards].　**水質検査[調査]**　a water quality test; examination of water;〔分析〕(a) water analysis; an analysis of water quality. / ～調査をする carry out a water quality survey.　**水質軟化剤**　〔硬水の〕a water softener.　**水質分析**　water analysis.　**水質保全**　water quality conservation.　**水質保全法**　the Water Quality Conservation Law.

水生, 水棲(すいせい)　▶～の aquatic; living in the water. ★⇨水中, 水底(すいてい) / 半～の semiaquatic. ¶**水生昆虫**　a water bug.　**水生**

シダ an aquatic fern; a hydropterid. 水生植物 an aquatic plant; a water plant; a hydrophyte. 水生生物 〈集合的に〉aquatic life. 水生生物学 hydrobiology. 水生動物 an aquatic (animal).

水洗 ▶トイレを〜式にする convert the lavatory into a flush toilet. /〜化率 the percentage (of households) converted to flush toilets. ¶水洗トイレ[便所] a flush toilet; a water closet.

水素 hydrogen. ▶〜の hydrogenous; hydric. ¶水素イオン a hydrogen ion. 水素イオン検出器 a hydrogen flame ionization detector (略: FID). 水素イオン指数 a hydrogen ion exponent (略: pH). 水素イオン濃度 hydrogen-ion concentration. 水素移動 hydrogen migration [transfer]. 水素エネルギー hydrogen energy. 水素(エネルギー)社会 a hydrogen-based [hydrogen energy] society; a society in which hydrogen is [replaces fossil fuels as] a primary form of energy. 水素炎 a hydrogen flame. 水素エンジン a hydrogen engine. /〜ロータリーエンジン a hydrogen rotary engine. /〜エンジン車 a hydrogen-engine vehicle; a hydrogen car. 水素化 hydrogenation. 水素化触媒 a hydrogenation catalyst. 水素ガス hydrogen gas. 水素(ガス)センサー a hydrogen (gas) sensor. 水素化脱硫 〔石油精製の〕hydrodesulfurization. 水素化物 a hydride. 水素化分解 〔加水素分解〕hydrogenolysis; 〔石油の〕hydrocracking. 水素吸蔵合金 an alloy for hydrogen storage; a hydrogen storage alloy;〔水素を含んだもの；金属水素化物〕a metal hydride. 水素供与体 a hydrogen donor. 水素極 〔燃料電池の〕a hydrogen pole; an anode electrode. 水素結合 a hydrogen bond; hydrogen bonding. 水素自動車 a hydrogen car [vehicle]; a hydrogen-fueled vehicle [car]. 水素受容体 a hydrogen acceptor. 水素ステーション 〔燃料電池車に水素を供給する施設〕a hydrogen (filling) station. 水素脆性[脆化] hydrogen embrittlement. 水素貯蔵 hydrogen storage. 水素貯蔵合金 ＝水素吸蔵合金. 水素電極 a hydrogen electrode. 水素電池 a hydrogen battery [cell]. 水素内燃機関[エンジン] a hydrogen internal combustion engine (略: ICE, HICE). 水素熱イオン検出器 a hydrogen flame thermoionic detector

(略: FTD). **水素燃料電池** a hydrogen fuel cell. **水素ハイブリッド車** a hydrogen hybrid vehicle [car]. **水素プラズマ** (a) hydrogen plasma. **水素メーザー** a hydrogen maser. **水素レーザー** a hydrogen laser. **過酸化水素** hydrogen peroxide. **重水素** heavy hydrogen;〔二重水素〕deuterium;〔三重水素〕tritium. **脱水素** dehydrogenation. **硫化水素** hydrogen sulfide.

水中 ▶～の underwater; subaqueous; aquatic ★⇨水生 / ～の微生物 a water microorganism. ¶**水中根** a water root. **水中植物** a water plant; an aquatic plant. **水中生物** 〈集合的に〉underwater life;〔海中の〕marine life. **水中葉** underwater leaves.

垂直分布 ⇨分布

水底 the bottom of the water [sea, river]. ▶～にすむ魚 a groundfish; a groundling. / ～性の living on the bottom of the water; benthic; benthal; benthonic. / ～生物 benthos.

水田 a paddy field; a rice paddy; an irrigated rice field. ¶**水田地帯** a paddy region; paddy lands. **水田土壌** paddy soils.

水道 〔設備〕waterworks; water service [supply];〔導水路〕an aqueduct; a water conduit;〔用水〕city [tap, running] water;〔水路〕a water course; a waterway;〔海峡〕a channel; a gut. ▶～(の)水 city [running, tap] water. / ～の水をやかんに入れる fill a kettle (with water) at the tap. / ～の出が悪い. The flow of water is poor. | Water doesn't flow [come out] well. / ～を引く have water pipes laid; have water supplied; lay on water. / ～を出す[止める] turn on [off] the tap. / お宅は～ですか井戸ですか. Do you use city or well water? / 豊後～ the Bungo Channel. ¶**水道管** a water [service] pipe;〔本管〕a water main. **水道局** 〔都の〕the Waterworks Bureau; the Bureau of Waterworks. **水道工事** waterworks. **水道事業** a water supply business. **水道栓**(せん) a hydrant; a tap. **水道法** the Waterworks Law. **水道メーター** a water meter. **水道屋** a plumber. **水道料(金)** water rates [charges]. / ～料金がずいぶん高い. The water bill is very high. **上下水道** plumbing; water supply and sewerage. **上下水道事業** water supply and drainage operations. ★⇨上水道, 下水(下水道), 中水道

水平移動 〔生物の〕horizontal

migration.

水平分布 〔生物の〕horizontal distribution. ★⇨分布

水脈 〔地下水の通る道〕a water vein. ▶〜を掘り当てる strike (a vein of) water.

水門 a floodgate; a penstock; a sluice (gate);〔運河の〕a lock gate. ▶〜を開ける［閉じる］open [shut] a floodgate [sluice]. /（船が）〜を通過する pass through a lock. ¶**水門管理人** a lock keeper. **水門通行税** lockage.

水文学(すいもんがく) hydrology. 形 hydrologic. ¶**水文学者** a hydrologist.

水溶 ¶**水溶液** aqueous solution [liquor]; water solution. ▶酸性［アルカリ性，中性］の〜液 an acid(ic) [an alkaline, a neutral] solution. **水溶性食物繊維** water-soluble dietary fiber. **水溶性ビタミン** a water-soluble vitamin. **水溶性肥料** water-soluble manure.

水利 〔水の利用〕utilization of water;〔給水〕water supply;〔灌漑〕irrigation;〔舟運の便〕navigability《of a river》. ¶**水利組合** an irrigation association. **水利権** water rights. **水利施設** a water-use facility. **水利使用** utilization [use] of water《for irrigation》; water use [usage]. **水利使用許可** permission [a permit] to utilize water.

水量 the quantity [volume] of water. ★⇨貯水 ¶**水量計** a water gauge [meter].

水路 a waterway; a water course; a water conduit; an aqueduct;〔港口の〕a fairway; a (water) channel. ¶**水路学** hydrography. **水路学者** a hydrographer. **水路橋** an aqueduct (bridge). **水路誌** a pilot; sailing directions. **水路式発電** 〔ダムを使わず水路で落差をつくる〕conduit type power generation. **水路図** a hydrographic map. **水路測量** a hydrographical survey. **水路測量学[術]** hydrography. **水路標識** a beacon.

図鑑 a picture [pictorial] book; an illustrated reference book; an《insect》identification manual. ▶(野生生物などを)〜で調べる look sth up in a field guide. /「日本植物〜」〔書名〕An Illustrated [A Pictorial, A Field] Guide to the Japanese Flora.

杉 a Japanese (red) cedar; a *sugi*. ★日本の杉はスギ科ないしヒノキ科で日本固有種．英語の cedar はマツ科ヒマラヤスギの仲間で杉とは全くの別種だが便宜的

に使われる．▶杉花粉のアレルギーである be allergic to *sugi* [cedar] pollen. ¶杉花粉症 an allergy to Japanese cedar pollen; Japanese cedar pollinosis. ★⇨アレルギー，花粉　杉並木 an avenue of *sugi* [cedar] trees.　杉林 a *sugi* [cedar] forest.

杉並病〔1996年から東京都杉並区にあったごみ処理施設周辺で発生した健康被害〕Suginami disease; various physical disorders suffered by residents of Suginami Ward, Tokyo, supposedly due to combustion of toxic chemicals.

すぐやる課〔役所の〕the "Do-it-now" Section. ▶市民の苦情をできるだけ早く解決するために「〜」が設置された．A "Do-it-now" section has been set up to deal speedily with citizens' problems.

スコーピング〔予定される開発計画などについて事前に環境影響評価の内容を公表し，それについて関係者の意見を聴取することによって環境影響評価の内容を絞り込む手続き〕scoping.

煤(ｽｽ)　soot. ¶煤粒子 a soot particle. ★⇨英和　black carbon

スタッドレスタイヤ　a studless tire. ★⇨スパイクタイヤ

捨てる〔投げすてる〕throw [fling, toss, cast] away;〔ごみなどを〕dispose of; dump;《口》chuck (out); junk. ★⇨ポイ捨て，吸殻，分別，ごみ，投棄，廃棄 ▶ごみを〜 throw away trash; dump refuse. / ごみを分別して〜 separate garbage before disposal. / (ごみの)捨て場(所) a dumping ground [place]《for refuse》; a dump. / ごみの捨て場(所)がない have no place to throw rubbish. / ごみの捨て場に困っている．I don't know where to dump my refuse. / 不要になった本を〜 discard a no-longer-needed book. / 空缶をポイと〜 toss out [away] an empty can. / 産業廃棄物を不法に〜 dump industrial waste illegally. / 猫を〜 abandon a cat. / 捨て猫 a stray cat; an unwanted [ownerless] cat; an abandoned cat [kitten]. / 捨て犬 a stray dog; an unwanted [abandoned, ownerless] dog. / 捨て犬を拾ってくる rescue a stray dog. / がらくたを捨てないでとっておく keep *sth* around; hold on to odds and ends. / ガムをかんだ後は紙に包んでゴミ箱に捨てましょう．After you chew gum, wrap it in paper and put [place] it in a wastebasket. / 各自治体の指示に従って捨ててくだ

さい．Discard as instructed by your local government. / 父は捨ててあったテレビを拾ってきて修理した．My father brought home a TV set that had been thrown away and fixed it. / 母は物を捨てられない性分の人だ．My mother is the type who can't throw anything away.

ストックホルム会議 ⇨国連(国連人間環境会議)

ストックホルム条約 ⇨英和 Stockholm Convention

砂地 a sandy place; the sands. ▶〜に住む[生える] living [growing] in sandy places; arenaceous. ¶砂地植物 a plant thriving on sandy soil; a psammophyte.

砂ぼこり dust; a dust storm. ★⇨黄砂(こうさ) ▶〜が立つ a cloud of dust blows up. / 〜を立てる raise a cloud of dust. / ひどい〜ですね．What dust!

スパイク・タイヤ 〔鋲つきの滑り止め用タイヤ；路面が削られ粉塵公害が起きるため，鋲のないスタッドレスタイヤが開発された〕a studded tire.

スマートグリッド ⇨英和 smart grid

炭 charcoal; soot《on a kettle》．▶炭を焼く burn [make] charcoal. / 炭には水を浄化する作用がある．Charcoal acts to purify [has a purifying effect on] water. ¶炭石けん charcoal soap. / 竹炭石けん bamboo charcoal soap. 炭焼きがま a charcoal kiln. 炭焼き小屋 a charcoal burner's lodge.

住む，棲む live [dwell, reside]《in [at]...》; inhabit《a place》; people《a district》; occupy《a house》;〔居住する〕have one's home《in [at]...》; take up residence《in [at]...》. ★⇨生活，生息，分布 ▶森に棲む鳥 birds inhabiting a wood. / 川に棲む魚 fish living in a river. / ホタルの棲む里 a village where fireflies dwell. / 住むに適した，住[棲]める inhabitable; good [fit] to live in. / 住むに適さない，住[棲]めない uninhabitable; unfit to live in. / アユも棲めない汚い川 a river too dirty for *ayu* [sweetfish] to live in. /〔環境が悪化して〕動物たちはその土地には住[棲]めなくなった．The animals lost their habitat there. | The area became uninhabitable to those animals.

棲[住]み分け segregation of niche; habitat segregation [isolation]; coexistence. ▶棲み分ける segregate; differentiate. ★⇨住む / 棲み分けというのはそれぞれの縄張りの中で生きるということである．Segregation means different groups living within their own separate niches. / 多様な生物がそれぞれに適した環境ごとにうまく棲み分けている．Diverse life-forms are effectively segmented into suitable environmental niches.

スモッグ smog. ★⇨光化学スモッグ ▶～におおわれた都市 a smog-laden city. / ～の多い[かかった] smoggy.

スラグ 〔鉱滓(こうさい)〕slag. ¶スラグ・セメント slag cement. スラグ粉 crushed slag. スラグ煉瓦 a slag brick. スラグ炉 a slag furnace [hearth].

スラッジ ⇨英和 sludge

3R(スリーアール) 〔循環型社会構築のための3要素〕the three Rs; the 3Rs. ★⇨英和 three R's ¶**3R政策** a "3R" policy; a "reduce, reuse and recycle" policy. **3Rイニシアティブ閣僚会議** the Ministerial Conference on the 3R Initiative. ★2005年東京で開かれた循環型社会の実現をめざす国際会議．

スロー ▶～ライフを送る live a slow life. ¶スロー・フード 〔ファースト・フードに対し〕slow food. ★⇨英和 slow food movement スロー・ツーリズム slow tourism.

せ

生育 ⇨成長, 環境(生育環境) ¶**生育期** a growing [vegetative] period. **生育地** a habitat. **生育障害** a growth [development(al)] disorder. **生育条件** growth conditions. **生育不足** 〔動植物の〕insufficient growth.

静穏権 〔騒音のない静かな生活を送る権利〕the right to a life of peace and quiet.

静音化 noise [sound] reduction; muting; damping; muffling.

静音性 〔モーター音などがしないこと〕quietness; silence.

静音冷却 quiet [silent] cooling.

生化学 biochemistry. 形 biochemical. ★⇨化学 ¶**生化学的酸素要求量** biochemical oxygen demand (略: BOD). ★⇨英和 biochemical oxygen demand

生活 living; (a) life; (an) existence. ★⇨樹上生活, 生態 ▶〜する live. / 森に〜する動物 animals that live in the forest. / アリの生活 the lives of ants. / 集団で生活を営む live in a community. ¶**生活安全条例** a community safety regulation. **生活音** household noises [sounds]; the noises [sounds] of daily domestic life. **生活科** 〔小学校の学科〕socioenvironmental studies. **生活環** a life cycle. **生活環境** *one's* living environment; the environment *one* lives in. ★⇨環境 **生活環境整備法** the Living Environment Maintenance Law. ★正式名称は「防衛施設周辺の生活環境の整備等に関する法律」(the Law Concerning Adjustment, etc. of the Living Environment in the Environs of Defense Facilities). **生活形** 〔生活様式の類型化〕a life form. **生活圏** a zone [sphere] of life; a livelihood zone. **生活権** living rights. **生活ごみ** household garbage [rubbish, trash, refuse]; domestic garbage; everyday garbage. **生活史** (a) life history 《of a species》; a life cycle. **生活騒音** daily (neighbourhood) noise; noise encountered in everyday (domestic) life. **生活道路** a road [street] serving a community; a neighborhood road [street]. **生活廃棄物** daily waste [garbage, refuse]. **生活排水[廃水]** 〔台所・浴室などからの〕domestic waste water [wastewater]; household effluent;〔水洗便所の〕domestic sewage. / 〜雑排水 miscellaneous household waste water. **生活用水** water for daily life; a domestic water supply;〔地域の〕a community water supply.

生気候学 bioclimatology; bioclimatics. ▶〜の bioclimatological, bioclimatic. ¶**生気候学者** bioclimatologist.

制御音 〔騒音を低減するための逆位相の音波〕a (noise) control [cancellation] wave.

生産 production; output. ▶〜する produce《food》; make; turn out《cars》; put out《a product》. ★⇨大量(大量生産) / 工場で〜する make [produce]《goods》in a factory. / 〜を一時ストップする temporarily stop production; halt [cease] production for a while;〔生じた公害問題などが〕hold up production. ¶**生産業**〔鉱・農・漁などの〕a primary industry. **生産者** a producer; a maker; a grower. ★⇨英和 producer, traceability / 〜者の顔が見える[見えない]. You feel [don't feel] you know who made [grew] it. / 道の駅なんかに行くと，作った人の住所氏名，顔写真まで貼ってあったりして，文字通り〜者の顔が見えるジャガイモやニンジンが並んでいる．When you go to one of those local tourist rest stops, you see potatoes and carrots all lined up with the names and addresses of the farmers and even photos of them. You can really see who produced the food. / 指定〜者 a designated [an approved] maker [grower]. **生産者価格** a producer's [maker's, grower's] price. **生産者価格指数** a producer price index (略: PPI). **生産者直送** (a) direct dispatch (from the maker [producer, grower]) to the consumer; sending goods direct to the consumer; (a) drop shipment. ★⇨産地直送 **生産者責任** producer responsibility (略: PR). ★⇨製造 / 拡大〜者責任〔生産者・輸入業者は製品のライフサイクルを通じてそれが環境に及ぼす影響に関して応分の責任を負うという考え方〕extended producer responsibility (略: EPR). **生産者団体** a《rice》producers association [organization, group]. **生産者番号** a manufacturer('s) [producer('s)] identification number. **生産者物価指数**〔米国の〕the Producer Price Index (略: PPI). **生産農家** a《tomato》grower [producer, farmer]; a producer who grows《orchids》. /～農家による茶の販売 sales of green tea direct from the [a] farm [producer]. **生産緑地** 〔市街化区域内の農地〕a productive green zone (to protect agricultural land in urban areas). **生産緑地法** the Productive Green Zone Law.

生殖 reproduction; procreation; generation. ▶有性[無性]～ sexual [asexual] reproduction. ¶ **生殖羽** nuptial plumage [dress]; breeding plumage. **生殖窠**(か), **生殖器巣**〔藻類の〕a conceptacle. **生殖期** a period of reproduction. **生殖器** the organs of generation [reproduction]; the genital [reproductive, generative, sexual] organs; the genitals; the genitalia. **生殖群泳** 〔魚の〕reproductive swarming. **生殖原細胞** a gonium《複数形 -nia》. **生殖細胞** a reproductive [generative, germ] cell. **生殖作用[機能]** a generative [reproductive] function; reproduction. **生殖質** germ plasm. **生殖(生物)学** reproductive [reproduction] biology; the biology of reproduction; genesiology.

生殖腺 a genital [sex, sexual] gland; a gonad. **生殖体** 〔ヒドロ虫類などの〕a gonophore. **生殖的隔離** reproductive isolation. **生殖囊**(のう)[**褶**(ちゅう)] a genital bursa [fold]. **生殖能力** reproductive ability [potential]. **生殖母細胞** a gonocyte; a gametocyte. **生殖力** generative [procreative] power;〔女性の〕fecundity;〔男性の〕virility. /～力のある generative; reproductive; progenitive. **環境生殖学** environmental reproductive biology. **動物生殖学** animal reproductive biology.

清掃 cleaning; a cleanup;〔街路の〕road-cleaning; keeping the roads clean. ★⇨掃除 ▶市では各自治会の協力により年一回市内一斉～を実施しています. The city conducts [carries out] a simultaneous street-cleaning operation once a year with the cooperation of all its self-governing associations. / 道路を～する clean [scavenge] a street. ¶ **清掃工場** a garbage processing plant; a garbage incineration plant. **清掃作業員** 〔ゴミの〕a sanitation worker; a sanitation man; a garbage collector; a dustman;〔街路の〕a street cleaner; a street sweeper. **清掃事業** the waste disposal industry. **清掃車** a garbage wagon [truck]; a refuse cart. ★⇨道路(道路清掃車) **清掃登山** a《Mount Fuji》cleanup climb; a《Mount Everest》cleanup expedition.

製造 ¶**製造者責任** 〔欠陥製品などに対する〕(a) manufacturer's responsibility;〔製品が環境に及ぼす影響に対する〕⇨生産(生産者責任). **製造物** a product;〈集合的に〉manufactures. ★⇨製品 **製造物責任** product liability (略: PL). ▶…を相手取って～物責任訴訟を起こす file [bring] a product liability suit against…. **製造物責任制度** a product liability system. **製造物責任法** the Product Liability Law. ★通称 PL 法.

成層圏 the stratosphere. ¶**成層圏オゾン** stratospheric ozone; ozone in the stratosphere.

生息 ▶～する ⇨住む / そのモグラは関東以東に～する. That (species of) mole inhabits the Kanto (Plain) and parts east of it. / この鳥は主に東京とその西側に～している. For the most part, that bird inhabits Tokyo and points west [further west]. | That bird's territory is chiefly Tokyo and the area west of Tokyo. ¶**生息域** a habitat; a range; a territory. **生息環境** ⇨環境 **生息状況** 〔野生生物などの〕the state of survival《of a species》. **生息数** the number《of tigers》in the wild; (the size of) a population;〔希少動物などの〕the number of individuals surviving. / 野生のトラの楽園といわれるインドでトラの～数が激減している. In India, said to be a paradise for wild tigers, the number of surviving individuals has declined dramatically. **生息地** a habitat; a haunt; a home. / 野生生物の～地破壊 the destruction of wildlife habitat. / 自然の～地にいるライオン lions in their natural habitat. / その島は多くの希少動物の～地となっている. The island is home to many rare species. **生息地保護** habitat protection. **生息調査** 〔野生生物などの〕a population survey. **生息動物** an inhabitant《of the wood》. **生息範囲** (the range [extent, size] of) a habitat; a range. / 地球温暖化のせいで, 熱帯の生物の～範囲が北に広がっている. With global warming, tropical organisms are extending their habitats north(wards). **生息分布** the population distribution《of a species》. **生息密度** (a) population density.

生態 an ecology. ★⇨生活 ▶昆虫の〜 the ecology of an insect. ¶**生態画** an ecology [ecological, ecosystem] picture; an ecopicture. / 動物[植物]〜画 a picture of animal [plant] life. **生態学** ecology; bionomics. 形 ecological; ethological; bionomic(al). / 〜学者 an ecologist; an ethologist; a bionomist. / 各個〜学 autecology. / 群集〜学 synecology. / 古〜学 paleoecology. / 植物〜学 ecological botany. / (品)種〜学 genecology. / 数理〜学 mathematical ecology. **生態型** an ecotype. 形 ecotypic. **生態気候** ecoclimate. **生態機能**〔環境への対応機能〕(an) ecological function. **生態系**〔特定地域における〕an ecosystem;〔生態環境一般〕ecology. ★⇨自然, 環境, 化石, 英和 ecosystem. / 〜系を破壊する destroy an ecosystem. / 〜系を乱す disrupt [disturb] an ecosystem. / 〜系の崩壊をまねく invite the breakdown of an ecological system [ecosystem]. / 地球の〜系を守る protect the global ecosystem. / 地域〜系を脅かす帰化生物 an immigrant species that threatens a regional [local] ecosystem. / この工事が進めば, この地域の微妙な〜系のバランスが崩れてしまうだろう. If the construction goes forward, the delicate ecological balance in the area will be destroyed. / わずかな海水温の上昇も〜系に著しい影響を及ぼす. Even a small rise in the ocean temperature has a noticeable effect on the ecology. **生態研究** ecological research;《study》ecology. / 日本猿の〜研究を行う do research on the ecology of the Japanese monkey; carry out an ecological study on the Japanese macaque. **生態種** an ecospecies《複数形 ecospecies》. **生態人類学** ecological anthropology. **生態生理学** ecophysiology. **生態的地位** an ecological niche; an econiche; a niche. **生態的ピラミッド** an ecological pyramid. **生態分布** ecological distribution. **生態(的)展示** ⇨展示.

生体 a living body; an organism. ¶**生体解剖** vivisection. ▶〜解剖をする[に付する] vivisect. **生体解剖反対論者** an antivivisectionist. **生体肝[腎, 肺(など)]移植** (a) liver [kidney, lung, etc.] transplant from a living donor. / 〜部分肝移植 (a) partial liver transplant from a living donor. **生体工学** bionics; bioengineering. **生体高分子** a biopolymer; a biomacromolecule. **生体材料** ⇨ 生体適合材料. / ～材料学 biomaterials science. **生体色素** a biochrome. **生体自己制御** biofeedback. **生体実験** an [a medical] experiment on a living person [subject]. **生体触媒** a biocatalyst. **生体(用)シリコン** biocompatible silicone. **生体資料** 〔鑑定・検査の目的で人体から採取された体液・組織等〕a biosample. **生体信号** a biological signal. **生体染色(法)** vital staining. **生体組織検査** 《do》a biopsy. **生体適合材料** a biocompatible material; a biomaterial. **生体電位** bioelectric potential. / 植物～電位 the bioelectrical potential of plants. **生体時計** ⇨体内(体内時計) **生体反応** a vital reaction. **生体分子** 【生化】a biomolecule. **生体弁** 〔人工弁の一種; ブタやウシの組織から作る〕a tissue valve. **生体防御** biodefense; biophylaxis. **生体防御機能** a biodefense function [mechanism]. **生体膜** a biomembrane; a biological membrane. **生体模倣化学** biomimetic chemistry. **生体模倣技術** biomimetics. 〚形〛 biomimetic. **生体用金属材料** a biocompatible metal; metallic biomaterials. **生体リズム** biorhythm. **生体流体力学** biofluid mechanics.

製品 manufactured goods [articles]; a product; 〔完成品〕finished goods. ¶**製品アセスメント** product assessment. **製品安全性** product safety. **製品安全データシート** 〔指定された化学物質を含む化学製品の毒性や安全な取り扱いに関する説明書〕a material safety data sheet (略: MSDS). ★⇨英和 material safety data sheet **製品安全マーク** ⇨SGマーク(和英篇末尾) **製品回収** 〔不良品・欠陥品などの〕(a) product recall. ★⇨使用, 回収 **製品評価技術基盤機構** the National Institute of Technology and Evaluation (略: NITE).

政府開発援助 ⇨ODA(和英篇末尾)

生物 a living [an animate] thing; an organism; a creature; 〈集合的に〉life. ★⇨生き物, 生命 ▶様々な～ many forms of life. / 森林の～ the life of a forest. ¶**生物遺伝資源部門** 〔製品評価技術基盤機構の〕the NITE Biological Resource Center (略: NBRC). **生物エネルギー** bioenergy. **生物界** the biological world; animate nature; life. **生物海洋学** biological oceanography. **生物海洋学者** a

biological oceanographer. **生物科学** bioscience. / 〜化学者 bioscientist. **生物化学** biological chemistry. **生物化学的酸素要求量** biochemical oxygen demand ★⇨生物学的酸素要求量 **生物化学素子** a biochip **生物化学兵器** biological and chemical weapons. **生物学** biology. 形 biologic(al). **生物学者** a biologist. **生物学主義** biologism. **生物学的応答調節物質** a biological response modifier (略: BRM). **生物学的許容漁獲量** an allowable biological catch (略: ABC). **生物学的検定(法)** biological assay; bioassay. **生物学的酸素要求量** (a) biochemical oxygen demand; (a) biological oxygen demand (略: BOD). ★英和⇨biochemical oxygen demand **生物学的時間** biological time. **生物学的種** a biological species. **生物学的製剤** biological products; biological preparations [drugs]; biologicals. **生物学的多様性** biological diversity. ★⇨生物多様性 **生物学的同等性** bioequivalence; biological equivalence. **生物学的同等製剤** biological equivalents; bioequivalents. **生物学的封じ込め**〔細菌などの〕biological containment. **生物学的要因** a biological factor. **生物(学的)療法** biotherapy. **生物学的利用能** bioavailability. **生物学的半減期** a biological half-life. **生物活性** biological activity. **生物活性物質** a bioactive substance; a biologically active substance. **生物岩** biogenetic rock; biolith; biolite. **生物環境** a [the] bio-environment. **生物気候学** ⇨生気候学 **生物季節学** phenology. **生物季節観測** 《a global》phenological observation. **生物群集** a biotic community. **生物経済学** bioeconomics. **生物圏** the biosphere. **生物圏保護区**〔ユネスコの「人間と生物圏計画」に基づいて指定された国際的な保護区〕a biosphere reserve. **生物検定** bioassay. ★英和⇨bioassay / 〜検定を行う perform a bioassay [biological assay]《on...》; bioassay. **生物工学** bionics; biotechnolgy; ergonomics. **生物剤**〔生物兵器用の〕a biological agent. **生物製剤**〔生物材料を起源とする医薬品〕biologics. **生物災害** a biohazard. ★⇨英和 biohazard **生物指標** a biological indicator; a bioindicator;〔水中生物の〕an aquatic life index. ★⇨指標 **生物社会学** biosociology. **生物情報学[科学]** ⇨生命(生命情報学) **生物静学** biostatics. **生物**

生態学 bioecology.　**生物相** a biota. / 陸生［地中, 底生］〜相 a land [soil, benthic] biota.　**生物測定学** biometry; biometrics.　**生物(体)量** (a) biomass.　★⇨英和 biomass　**生物多様性** biodiversity.　★⇨国際(国際生物多様性の日), 英和 biodiversity

生物多様性国家戦略〔生物多様性条約締結国政府がそれぞれ策定する戦略〕a national biodiversity strategy. / 新・〜多様性国家戦略〔日本政府が2002年に改定したもの〕the New National Biodiversity Strategy (of Japan).

生物多様性条約 the (UN) Convention on Biological Diversity (略: CBD).　★⇨カルタヘナ議定書　**生物蓄積**〔ダイオキシン・農薬などの〕bioaccumulation.

生物地理学 biological geography; biogeography.　**生物的環境** a [the] biotic environment.　**生物テロ** biological terrorism.　**生物電気** bioelectricity.　圏 bioelectric(al).　**生物電気学** electrobiology.　**生物(燃料)電池** a biological battery; a bioelectrochemical cell.　**生物統計学** biostatistics.　**生物毒** a biological toxin.　**生物・毒素兵器[生物兵器]禁止条約** the Biological Weapons Convention (略: BWC).　★正式には「細菌兵器(生物兵器)及び毒素兵器の開発, 生産及び貯蔵の禁止並びに廃棄に関する条約」(the Convention on the Prohibition of the Development, Production and Stockpiling of Bacteriological [Biological] and Toxin Weapons and on Their Destruction) という.　**生物時計** a biological clock; a circadian clock.　★⇨体内(体内時計)　**生物特許** a biotechnological patent; a bio-patent.　**生物濃縮**〔有害物の体内蓄積〕bioaccumulation; bioconcentration; biomagnification.　**生物農薬** a biotic pesticide; a biological pesticide.　**生物発光** bioluminescence.　**生物発生説** (the theory of) biogenesis.　**生物発電** bioelectrogenesis.　**生物フォトン**〔生物微弱発光〕a biophoton.　**生物物理化学** biophysical chemistry.　**生物物理学** biophysics; biological physics.　**生物物理学者** a biophysicist.　**生物分解性** biodegradability.　★⇨生分解性　**生物分布学** chorology.　圏 chorological.　**生物分類** biological classification.　**生物兵器** a biological weapon; a bioweapon.　★⇨生物・毒素兵器禁止条約　**生物兵器テロ** biological terrorism.　**生物平衡** (a) biotic balance [equilibrium].　**生物変移説** (the theory of) transformism.　**生物モニタリン**

グ biological monitoring.　**生物模倣化学**　⇨生体(生体模倣化学).　**生物薬剤学**　biopharmaceutics; biopharmacy.　**生物有機化学**　bioorganic [bio-organic] chemistry.　**生物由来製品**　〔医薬品の〕a biologically derived [bio-derived] product. / 特定～由来製品 a specified bio-derived product.　**生物由来製品感染等被害救済制度**　a system to protect victims of infection arising from biologically derived products.　**生物理学**　biophysics.　**生物リズム**　(a) biological rhythm; a biorhythm.　**生物量**　⇨生物体量

生分解性　biodegradability. ★⇨分解　▶～の biodegradable ¶**生分解性高分子**　a biodegradable polymer.　**生分解性樹脂**　biodegradable resin.　**生分解性素材**　(a) biodegradable material.　**生分解性プラスチック**　(a) biodegradable plastic.　**生分解性ポリマー**　a biodegradable polymer.

生命　life. ★⇨生物　▶海の～ marine life; life in the sea. / ～の起源 the origin of life. / ～の尊厳 the dignity of life. / サケの卵が孵化した，小さな～の誕生である．The salmon spawn hatched. It was the start [birth] of a little life. / きっと宇宙のどこかに～の存在する星がある．There must be some planet in the universe where there is life. / 地球外の知的～体 intelligent extraterrestrial life [forms of life, organisms]. / 人工～体 an artificially created form of life [(living) organism]; a computer which is 'alive'. ¶**生命科学**　life science; the life sciences.　**生命現象**　a life phenomenon; a vital phenomenon.　**生命工学**　biotechnology.　**生命誌**　biohistory.　**生命情報学[科学]**　bioinformatics.　**生命力**　vital force; vitality; animus; the power [ability] to stay alive [survive]. / 雑草の旺盛な～力 the rampant force of weeds. / ～力が強い[弱い] have (a) strong [weak] vital force. / 動物の～力の強さに驚嘆する．be filled with wonder at animals' ability to survive [life force].　**生命倫理(学)**　bioethics; ethics in human life; life ethics. / クローン技術は～倫理の上からまだ大いに疑問の余地がある．Cloning technology is bioethically still open to question.　**生命倫理懇談会**　a council on bioethics.　**生命倫理学者**　a bioethicist.　**生命倫理専門調査会**　〔内閣府，総合科学技術会議の〕the Bioethics Committee of the Council for

世界

Science and Technology Policy.

世界 the world; 〔地球〕the earth; the globe. ▶〜的な, 〜中の worldwide; international; global; universal. / 〜でも有数の地震国 one of the most earthquake-prone countries in the world; one of the world's most earthquake-prone countries. / 〜でもここにしかいないめずらしいカメ a rare turtle that only inhabits this part of the world; a globally rare turtle that is only found around here. / 今デパートで「〜の蝶展」が開かれている. The department store is holding an exhibition of "Butterflies of the World." ¶ **世界遺産** 〔世界遺産条約で指定される〕a World Heritage (site). ★⇨ 危機(危機遺産) / 1987 年にその公園は〜遺産として登録された. In 1987 the park was placed [inscribed] on the World Heritage List. / 〜自然遺産 a World Natural Heritage Site. **世界遺産委員会** the World Heritage Committee. **世界遺産条約** 〔ユネスコにより 1972 年に採択された〕the World Heritage Convention. ★正式名称は,「世界の文化遺産及び自然遺産の保護に関する条約」the International Convention for the Protection of the World Cultural and Natural Heritage. 〔国連の〕**世界遺産リスト** the United Nations World Heritage List. / 〜遺産リストに載っている be on the United Nations World Heritage List.

日本の世界遺産とその登録年は以下のとおり.

1993 法隆寺地域の仏教建造物 the Buddhist Monuments in the Horyu-ji Area

1993 姫路城 Himeji Castle

1993 屋久島 Yakushima Island (★自然遺産)

1993 白神山地 the Shirakami Mountains (★自然遺産)

1994 古都京都の文化財 the Historic Monuments of Ancient Kyoto (Kyoto, Uji and Otsu Cities)

1995 白川郷・五箇山の合掌造り集落 the Historic Villages of Shirakawa-go and Gokayama

1996 原爆ドーム the Hiroshima Peace Memorial (Genbaku Dome)

1996 厳島神社 Itsukushima Shinto Shrine

世界

1998 古都奈良の文化財 the Historic Monuments of Ancient Nara

1999 日光の社寺 the Shrines and Temples of Nikko

2000 琉球王国のグスク及び関連遺産群 Gusuku Sites and Related Properties of the Kingdom of Ryukyu

2004 紀伊山地の霊場と参詣道 the Sacred Sites and Pilgrimage Routes in the Kii Mountain Range

2005 知床 Shiretoko (★自然遺産)

2007 石見銀山遺跡とその文化的景観 Iwami Ginzan Silver Mine and its Cultural Landscape

¶ **世界エネルギー会議** the World Energy Council (略: WEC). **世界エネルギー展望** 〔国際エネルギー機関が毎年発表する報告書〕the World Energy Outlook 《2007》. **世界開発金融** 〔世界銀行が毎年発行する報告書〕Global Development Finance《2006》(略: GDF). **世界開発情報の日** 〔10月24日; 国連制定〕World Development Information Day. **世界開発センター** 〔米国のシンクタンク〕the Center for Global Development (略: CGD). **世界海洋観測システム** the Global Ocean Observing System (略: GOOS). **世界海洋循環実験計画** the World Ocean Circulation Experiment (略: WOCE). **世界環境デー** 〔6月5日; 国連制定〕⇨英和 World Environment Day **世界気候会議** the World Climate Conference. **世界気候計画** the World Climate Programme (略: WCP). **世界気候研究計画** the World Climate Research Programme (略: WCRP). **世界気候データ[資料]・監視計画** the World Climate Data and Monitoring Programme (略: WCDMP). **世界気候変動** global climate change. **世界気象会議** the World Meteorological Convention. **世界気象監視計画** the World Weather Watch (略: WWW). **世界気象機関** the World Meteorological Organization (略: WMO). **世界気象の日[気象デー]** 〔3月23日; 国連制定〕World Meteorological Day. **世界銀行** the World Bank. 国際復興開発銀行 (the International Bank for Reconstruction and Development (略: IBRD)) の通称. **世界原子力発電事業者協会** the World Association of Nuclear Operators (略: WANO). **世界湖沼会議** the International Conference on the Conservation and Management of Lakes.　**世界作物多様(性)財団**　⇨グローバル(グローバル作物多様性トラスト)　**世界資源研究所**　〔米国 ワシントン DC に本拠を置く環境問題についてのシンクタンク〕the World Resources Institute (略: WRI).　**世界自然遺産**　⇨世界遺産　**世界自然保護基金**　〔国際自然保護団体〕the World Wide Fund for Nature (略: WWF; ⇨英和 World Wide Fund for Nature). ★日本支部は世界自然保護基金ジャパン the World Wide Fund for Nature Japan (略: WWF Japan).　**世界食糧計画**　〔国連の〕the World Food Program (略: WFP).　**世界食糧サミット**　the World Food Summit.　**世界食糧デー**　〔10月16日〕World Food Day.　**世界人口会議**　〔国連の〕the 《Montreal》World Population Conference (略: WPC).　**世界人口行動計画**　〔国連の〕the World Population Plan of Action (略: WPPA).　**世界人口・住宅センサス**　〔国連の〕the 《2000》World Population and Housing Census Programme.　**世界人口デー**　〔7月11日; 国連制定〕World Population Day.　**世界人口年**　〔1975年; 国連制定〕International Population Year.　**世界人口白書**　〔国連の〕the State of World Population Report.　**世界農産物生産高予想**　〔米国農務省が毎月刊行する〕the World Agricultural Supply and Demand Estimates (略: WASDE).　**世界農林業センサス**　〔国連の〕the World Census of Agriculture and Forestry.　**世界水会議[フォーラム]**　★⇨英和 World Water Council [Forum]　**世界水の日**　〔3月22日〕World Water Day.

堰(せき)　a dam; a barrage; 〔流水量測定用の〕a weir.　★⇨河口(河口堰)　▶川に堰を作る construct [build] a dam across a river; dam (up) a stream. / 堰を切って落とす break [burst] a dam. ¶ **堰板**　〔土止め用の〕a sheeting (board); 〔水位を高くするための〕a flashboard.　**洗い堰**　a fixed dam.　**可動堰**　a movable dam [weir].　**取水堰**　a diversion weir; an intake dam.　**本[翼]堰**　a main [wing] dam.

石炭　coal.　▶〜を掘る mine [dig out] coal. / 〜を燃やす[たく] burn coal. / 〜を燃料にする use coal (for fuel); be fuelled by coal.　¶ **石炭液化油**　coal liquefaction oil; liquefied coal.　**石炭ガス**　coal gas.　**石炭ガス化**　coal gasification; gasification of coal.　**石炭ガス化燃料電池複合発**

電　an integrated coal gasification fuel cell combined cycle（略：IGFC）．　**石炭ガス化複合発電**　an integrated coal gasification combined cycle（略：IGCC）．　**石炭ガス化炉**　a coal gasification furnace.　**石炭ガス(製造)工場**　a coal gasification plant.　**石炭火力発電所**　a coal-fired [coal-burning] power plant.

脊椎(せきつい)動物　a vertebrate (animal).

責任　(a) responsibility; (a) liability;〔責務〕(an) obligation; (a) duty.　★⇨管理(管理責任)，企業(企業責任)，製造(製造物[者]責任)，生産(生産者責任)，排出(排出者責任)，発生(発生責任) ▶営造物〜 government [local government] liability for damage arising from faults in public works; public liability. / メーターから屋内の配管は各ご家庭の〜で管理してください．Each house [The householder] is responsible for the pipes between the meter and the house. | We cannot take responsibility for [look after] the plumbing beyond the meter. / 事故に対する〜 liability for an accident / 狂牛病対策の遅れに関しては農水省の〜が重い．The Ministry of Agriculture, Forestry and Fisheries carries grave responsibility for not having developed a BSE policy earlier.

石綿(せきめん)(いしわた)　asbestos;〔上質の〕amiantus.　★⇨アスベスト ▶〜を使わない[用いない]ブレーキライニング[摩擦材] asbestos-free [non-asbestos] brake linings. ¶ **石綿関連疾病**　(an) asbestos-related disease.　**石綿胸水**　asbestos pleural effusion.　**良性石綿胸水**　benign asbestos pleural effusion（略：BAPE）．　**石綿胸膜炎**　asbestos pleurisy.　**石綿健康被害救済法**　the Asbestos Health Damage Compensation Law; the Law for Compensation of Patients with Asbestos-Related Diseases.　★正式名称は「石綿による健康被害の救済に関する法律」．　**石綿障害予防規則**　the Ordinance on Prevention of Hazards due to Asbestos.　**石綿小体**　an asbestos body.　**石綿条約**　the Asbestos Convention.　★正式名称は「石綿の使用における安全に関する条約」(the Convention Concerning Safety in the Use of Asbestos)．　**石綿紙**　asbestos paper.　**石綿(沈着)症**　(pulmonary) asbestosis.　**石綿スレート**　asbestos slate.　**石綿セメント**　asbestos cement.　**石綿繊維**　asbestos fiber.　**石綿対策全国連絡会議**　Ban Asbestos

Network Japan〔略：BANJAN〕． **石綿テープ** an asbestos tape． **石綿肺** asbestos lung; pulmonary [lung] asbestosis． **石綿パイプ** (an) asbestos cement pipe． **石綿板** an asbestos board． **石綿布** an asbestos blanket． **石綿フェルト** asbestos felt． **石綿プラスター**〔防火断熱材〕asbestos plaster． **石綿粉末** asbestos powder． **石綿保温被覆** (an) asbestos covering． **石綿ロープ** an asbestos rope． **青石綿** blue asbestos; crocidolite． **木状石綿** rockwood． **白石綿** ＝温石綿（おんじゃくめん）chrysotile． **茶石綿** brown [amosite, gray] asbestos; amosite． **飛散性石綿** friable asbestos． **非飛散性石綿** non-friable asbestos． **吹き付け石綿** sprayed asbestos.

石油 petroleum（★他種の油と混同の恐れがなければ単に oil でもよい）；〔灯火用〕kerosene; paraffin．★⇨石油代替エネルギー，脱石油，代替，埋蔵 ▶～の枯渇 depletion of oil (resources)．/ いずれ世界中の～が掘り尽くされてしまう時が来る．The world's oil supply will be exhausted sooner or later [someday,《口》sometime down the track]．★⇨英和 peak oil，石油資源 / ～の豊かな oil-rich．/ 海底～の採掘 offshore oil drilling．/ ～を掘り当てる strike oil．/ ～を掘る drill for oil．/ ～を燃料にする use oil as [for] fuel; fuel《an engine》with oil．/ ～を豊富に産する petroliferous．/ 国際～カルテル〔資本〕an international oil cartel [oil major]．¶**石油アスファルト** petroleum asphalt． **石油安定化基金** a petroleum stabilization fund． **石油依存社会** an oil-dependent society; a society (wholly) dependent on oil． **石油依存度** (the degree of) dependence on oil． **石油井戸**〔油井〕an oil well． **石油インフレ** inflation caused by an oil shortage; oil-induced inflation． **石油エーテル**〔石油低沸点留分〕petroleum ether． **石油エンジン[発動機]** an oil [a petroleum, a kerosene] engine． **石油王** an oil baron [king]． **石油卸大手** a major oil wholesaler． **石油温風機[器]** an oil fan heater; an oil (space) heater [stove]． **石油会社** an oil company．/ 国際～会社 a major international oil company． **石油価格** the price of oil;《a huge drop in》oil prices．/ 国際～価格 international oil prices． **石油化学** petrochemistry． **石油化学工業** the petrochemical industry． **石油化学工場[プラン**

ト〕 a petrochemical factory [plant]. **石油化学コンビナート** a petrochemical complex. **石油化学製品** petrochemicals. 形 petrochemical. **石油化学メーカー** a petrochemical producer. **石油化学工業協会** the Japan Petrochemical Industry Association (略: JPCA). **石油ガス** petroleum gas. / 液化〜ガス liquefied petroleum gas (略: LPG); LP-gas; 〔携帯用ボンベ入りガス〕 bottled gas. **石油ガス税** the liquefied petroleum gas tax. **石油株** oils; oil shares. **石油缶** an oilcan; a kerosene can; a paraffin tin. **石油関連(諸)税** oil-related taxes. **石油危機** an oil [a petroleum] crisis. **石油業** the oil [petroleum] industry. **石油業者** an oilman. **石油協定** a petroleum convention. **石油業法** the Petroleum Industry Law. **石油掘削機** an oil (drilling) rig; a rig. **石油系鉱物油** petroleum mineral oil. **石油坑** an oil [a petroleum] well. **石油工業** the petroleum [oil] industry. **石油鉱区** an oil field; an oil drilling area; an oil exploration block. **石油公示価格** a posted oil price. **石油鉱床** an oil deposit; oil deposits. **石油工場** a petroleum plant. **石油合成菌** ⇨石油分解菌. **石油公団** the Japan National Oil Corporation (略: JNOC). **石油コークス** petroleum coke. **石油国家** 〔経済力を石油生産に依存する国家〕 a petro-state. **石油コンビナート** ⇨石油化学コンビナート. **石油コンロ** an oilstove (for cooking); a kerosene cooking stove; a paraffin cooking stove. **石油採掘** oil(-well) drilling; drilling for oil. **石油採掘権** 《hold》 an oil concession; oil-drilling rights. / 〜採掘権を更新する renew (oil-)drilling rights. **石油先物市場** an oil futures market. **石油酸** 〔石油中の酸性物質〕 petroleum acid. **石油産業** the oil [petroleum] industry. **石油産業活性化センター** the Japan Petroleum Energy Center (略: JPEC, PEC). **石油産出国** an oil-producing country [nation]; an oil producer. **石油資源** petroleum [oil] (resources). / 〜資源が豊かである be rich in oil (resources); have a lot of oil. / 〜資源を開発する develop [exploit] petroleum resources. / このまま使い続ければ、いつか〜資源はなくなる. If we keep on using them this way, petroleum [oil] resources will someday dry up.

石油

石油週報 〔石油連盟 (the Petroleum Association of Japan; PAJ) が発表する〕PAJ Oil Statistics Weekly. ★正式名称は「原油・石油製品供給統計週報」.
石油証券 petroleum securities.
石油消費国 an oil-consuming country [nation]; an oil consumer. **石油情報センター** 〔財団法人；日本エネルギー経済研究所の付属機関〕the Oil Information Center. **石油植物** 〔アオサンゴ・ユーカリなど，低級炭化水素化合物を多く含む植物〕a petroleum plant; a petroplant.
石油ショック an oil crisis. **石油ストーブ** an oilstove. **石油精** petroleum spirits. **石油税** (a) petroleum tax. **石油精製** oil refining. **石油製品** 〔ナフサ・重油・灯油・ガソリンなどの総称〕petroleum products. **石油製品価格** prices of petroleum products; petroleum product prices. **石油製品市況動向調査** 〔財団法人石油情報センターによる〕the survey of market conditions for oil products [oil and gasoline prices] at service stations. **石油石炭税** (a) petroleum and coal tax. **石油・石炭製品** petroleum and coal products. **石油戦争** an oil war; a war for [over] oil. **石油代替エネルギー** an energy alternative [alternative energy] to oil.
石油代替エネルギー法 〔石油代替エネルギーの開発及び導入に関する法律の通称〕the Law Concerning Promotion of the Development and Introduction of Alternative Energy. **石油代替燃料** an alternative fuel to oil.
石油タンク an oil storage tank.
石油探査 oil exploration. **石油たんぱく(質)** (a) petroprotein.
石油地質学 petroleum geology.
石油天然ガス・金属鉱物資源機構 the Japan Oil, Gas and Metals National Corporation (略：JOGMEC). **石油特会** 〔石油及びエネルギー需給構造高度化対策特別会計〕the Special Account for Petroleum; the Special Account for Petroleum and a More Sophisticated Policy for the Structure of Demand and Supply. **石油トレーダー[ブローカー]** an oil trader [broker].
石油乳剤 petroleum emulsion.
石油バーナー an oil burner. **石油販売権** the right to sell oil; 〔国連の対イラク人道支援事業における〕a right to buy oil [petroleum] (under the UN Iraq program Oil-for-Food). **石油備蓄** 〔備え蓄えておくこと〕stockpiling of oil; oil storage; storing of petroleum;〔備え蓄えられた石油〕oil stocks [reserves]; pe-

troleum stocks [reserves]. / 〜備蓄の増強 increasing [an increase in] oil stocks; stocking more oil. / 90日分の石油〜を義務づける make it obligatory to stock sufficient oil for 90 days [90-day oil stocks obligatory]. / 〜備蓄を放出する release 《thirty million barrels of》oil from oil stocks.　**石油備蓄基地** an oil (storage) depot.　**石油備蓄法** the Petroleum Reserve Law.　**石油ピッチ** petroleum pitch.　**石油ピッチコークス** petroleum pitch coke.　**石油ブーム** an oil boom.　**石油分解菌** oil-decomposing bacteria [microorganisms].　**石油ベンジン**〔揮発油〕petroleum benzine.　**石油ポンプ**〔油井から石油を汲み上げる〕an oil pump;〔家庭用の〕a《battery-operated》kerosene [paraffin] (siphon) pump.　**石油埋蔵量** oil deposits.　**石油元売り業者** a refiner-marketer.　**石油輸出国機構** the Organization of Petroleum Exporting Countries (略: OPEC).　**石油輸送管**〔1本の〕an oil pipe;〔輸送管路〕an oil pipeline.　**石油洋上備蓄** offshore oil storage.　**石油ランプ** an oil [a kerosene, a paraffin] lamp.　**石油流出(事故)** an oil spill.　★⇨流出　**石油連盟** the Petroleum Association of Japan (略: PAJ).

施業林(せぎょうりん)　a managed forest; managed woodland.

脊梁(せきりょう)**山脈**〔長く連なる高地〕a mountain ridge.　▶本州の〜山脈 the backbone of Honshu's mountains.

世代〔一つの個体の一生〕a generation;〔同時期に生まれた個体群〕a cohort.　¶**世代促進**〔品種改良などのため，冬場温室を用いたりして育種に要する年月を短縮する手法〕rapid generation advancement (略: RGA).　**無性世代** an asexual generation.　**有性世代** a sexual generation.

節水　▶〜する save [economize on] water; reduce water consumption.　★⇨取水制限，給水制限 / 市民に5%の〜を呼びかける call on residents to reduce water consumption by 5%. / 自主〜を促す urge voluntary water conservation.　¶**節水技術** water-saving technology.　**節水効果** a water-reduction [water-saving] effect. / 〜効果の高い新型トイレ a new kind of toilet which uses much less water [is highly economical with water].　**節水対策** water conservation measures.　**節水率** a water-saving rate《of 10%》.

石鹸(せっけん)　(a) soap.　▶〜1個 a cake [bar, piece] of soap. / 〜で

洗う wash 《one's hands, a baby, a cloth》 with soap and water. / 〜の泡 soap bubbles; soapsuds; suds; (a) lather. ¶ **石鹸圧出機** a soap compressor. **石鹸入れ[箱]** a soap dish [box]. **石鹸型打機** a soap press. **石鹸工場** a soap works. **石鹸水** soapy water; 〔泡立った〕soapsuds; suds. **石鹸製造** soapmaking; soap manufacture [production]; soap boiling. **石鹸製造釜** a soap pan. **石鹸フレーク** soap flakes. **浮き石鹸** floating soap;《a cake of》soap that will float. **海水用石鹸** marine soap. **紙石鹸** soap paper; a thin leaf of soap. **カリ石鹸** soft soap. **逆性石鹸** invert [cationic] soap. **金属石鹸** metallic soap. **化粧石鹸** toilet soap. **粉石鹸** washing powder; soap powder [flakes]; powdered detergent. **洗濯石鹸**《a bar of》washing [laundry] soap. **中性石鹸** neutral soap. **棒石鹸** a stick [bar] of soap. **マルセル石鹸** 〔中性石鹸の一種〕Marseille(s) soap. **水石鹸** liquid soap. **薬用石鹸** medicated soap.

摂取 absorption; adoption; intake; ingestion; 〔食物の〕ingestion. ▶〜する take; absorb; ingest; take in…; swallow; incept; imbibe; 〔同化する〕assimilate. ¶ **摂取量** an intake《of…》. / カロリーの〜量 a caloric intake. / あなたの毎日のカロリー〜量はどれくらいですか. What is your daily intake of calories? / 許容(上限)〜量〔栄養素の〕a [the] maximum acceptable intake. / 許容日常[一日]〜量 an [the] acceptable daily intake (略: ADI). / 耐容日常[一日]〜量 a [the] tolerable daily intake (略: TDI).

接触槽 〔浄化槽の〕a contact tank.

接触濾床 〔排水処理技術の〕a contact bed.

節電 saving electricity; power saving; reducing [a reduction in] power consumption. ▶〜する save [use less] electricity; cut down on [reduce] electricity (consumption). / 〜のヒント a tip for saving electricity. ¶ **節電週間** (a) power-saving week. **節電装置** a power-[an electricity-]saving device.

設備 equipment; facilities. ★⇨安全(安全設備), 衛生(衛生設備), 基準, 公害(公害防止), 浄化設備, 下水

説明会 a briefing; an explanatory meeting. ★⇨住民

絶滅 extinction《of the dinosaurs》; extermination; annihilation; eradication; total destruction《of mankind》. ★⇨全滅 ▶〜する〔絶やす〕exterminate; extirpate; eradicate; wipe [stamp] out; annihilate; totally destroy; extinguish;〔絶える〕become extinct; die out; cease to exist. / 〜の一途をたどる be on the road [way] to extinction; be destined to die out. / 〜の淵に追い込む drive *sth* to the verge of extinction; threaten *sth* with extinction. / 乱獲で〜の危機にある be in danger of extinction from overfishing [indiscriminate hunting]; be hunted (nearly) to the point of extinction. / この辺のホタルは農薬でほとんど〜してしまった. The fireflies around here have been practically wiped out by pesticides [agricultural chemicals]. / 〜のおそれのある野生動植物の種の国際取引に関する条約（ワシントン条約）⇨英和 CITES ¶ **絶滅危惧**(きぐ)**種** an endangered species. / 〜危惧種に指定される be designated as an endangered species. **絶滅種** an extinct species. **絶滅動物** extinct animals. **地域絶滅** (a) local extinction. **野生絶滅** extinction in the wild（略：EW）. ★⇨大量（大量絶滅）

環境省の「レッド・データブック」による絶滅を危惧されている種の分類は以下のとおり．★⇨英和 red data book, Red List

「絶滅」Extinct（略：EX） わが国ではすでに絶滅したと考えられる種

「野生絶滅」Extinct in the Wild（略：EW） 飼育・栽培下でのみ存続している種

「絶滅危惧Ⅰ類」Critically Endangered（略：CR+EN） 絶滅の危機に瀕している種

「絶滅危惧ⅠA類」Critical（略：CR） ごく近い将来における絶滅の危険性が極めて高い種

「絶滅危惧ⅠB類」Endangered（略：EN） ⅠA類ほどではないが，近い将来における絶滅の危険性が高い種

「絶滅危惧Ⅱ類」Vulnerable（略：VU） 絶滅の危険が増大している種

> 「準絶滅危惧」Near Threatened (略: NT) 現時点では絶滅危険度は小さいが，生息条件の変化によっては「絶滅危惧」に移行する可能性のある種
>
> 「情報不足」Data Deficient (略: DD) 評価するだけの資料が不足している種
>
> 「絶滅のおそれのある地域個体群」Threatened Local Population (略: LP) 地域的に孤立している個体群で，絶滅のおそれが高いもの

節約 economy; saving;〔倹約〕thrift; husbandry; frugality. ▶ 〜する economize《on…, in…, with…》; be economical《with…》; save《on electricity》; do not waste《fuel》;〔資金を〕husband《*one's* resources》;〔節減する〕curtail; cut; cut down《on…》. / 電気を〜する economize on (electric) power. ★⇨節電, 省エネ / 資源を〜する economize on [avoid wasting] natural resources. ★⇨資源 / 燃料〜装置 a fuel economizer. / ガソリンの大幅な値上げで消費者の〜意識が高まった. With a major rise in gasoline prices, consumers grew more conscious of the need to economize. ¶**節約志向** economy-mindedness; budget-mindedness; (a) consciousness of the need to economize.

セメント固化〔放射性廃棄物の〕cement solidification. ¶**セメント固化体**〔放射性廃棄物の〕cement-solidified (radioactive) waste.

セリーズ原則〔企業に環境責任を求める〕CERES principles. ★米国の，環境に責任を持つ経済機構のための協議会(Coalition for Environmentally Responsible Economies)の略をローマ女神の名にかけた称. ★⇨英和 Valdez Principles

ゼロエミッション〔産業廃棄物をゼロにすること〕zero-emission. ★⇨ごみ(ごみゼロ) ¶**ゼロ・エミッション・カー[ビークル]**〔無公害車〕a zero-emission vehicle (略: ZEV). **ゼロ・エミッション計画**〔国連大学が提唱している産業廃棄物ゼロを目指した計画〕the Zero Emission Recycle Project. **ゼロ・エミッション工場**〔産業廃棄物をすべてリサイクルに回している工場〕a zero emission factory. **ゼロ・エミッション事業** a zero-emission(s) project [enterprise]. **ゼロ・エミッション都市** a zero-emission city.

ゼロメートル地帯 a sea-level

繊維　a fiber. ¶繊維飲料　a fiber drink [beverage]. ★⇨食物(食物繊維), 水溶性食物繊維　繊維強化プラスチック　【化】fiber-reinforced plastics（略：FRP）. バイオ繊維　biofiber.

遷移(せんい)　succession. ¶偏遷移　deflected succession.

浅海(せんかい)　a shallow sea. ¶浅海魚　a shallow-sea fish.　浅海漁業　shallow-water fishing.

全球　¶全球海洋観測システム　the Global Ocean Observing System（略：GOOS）. 全球気候観測システム　the Global Climate Observing System（略：GCOS）. 全球数値予報モデル　a global numerical weather prediction model.　全球大気監視計画　the Global Atmosphere Watch（略：GAW）.

全国　¶全国危険物安全協会　the Japan Association for Safety of Hazardous Materials.　全国地球温暖化防止活動推進センター　the Japan Center for Climate Change Action（略：JCCCA）. 全国治水砂防協会　the Japan Sabo Association（略：JSA）. 全国都市清掃会議　the Japan Waste Management Association（略：JWMA）.

洗剤　a cleaner; cleaning material; a cleanser; a detergent. ★⇨残留(残留洗剤), 合成(合成洗剤), 中性洗剤, 無燐洗剤, 油汚れ　¶洗剤不要[無洗剤]洗濯機　a detergent-free washing machine.　台所用洗剤　(a) kitchen detergent.

染色体　a chromosome. 形 chromosomal. ▶異常〜　a heterochromosome; a heterosome. ¶染色体異常　a chromosomal [chromosome] abnormality [aberration].　染色体異常誘発　clastogenesis.　染色体欠失　chromosomal deletion.　染色体検査　chromosome testing.　染色体工学　chromosome engineering.　染色体操作　chromosome manipulation.　染色体地図　a chromosome map.　染色体突然変異　chromosome mutation.

喘息(ぜんそく)　asthma;〔動物の〕whistling;〔馬の〕broken wind. ★⇨アレルギー ▶〜にかかった馬　a whistler. /〜(性)の　asthmatic. /〜の発作　an attack of asthma; an asthma attack. /〜を患っている　have [get, suffer from] asthma. / 工場煤煙による四日市〜　Yokkaichi asthma《from factory smoke》. ¶喘息患者　an asthmatic; an asthmatic patient.　アトピー性喘息　atopic asthma.　気管支喘息　bronchial asthma. 小児喘息 infantile asthma. 塵埃喘息 dust asthma. 神経性喘息 nervous asthma. 心臓喘息 cardiac asthma. 成人発症喘息 adult-onset asthma.

選択 (a) selection; (a) choice; (an) option. ★⇨自然(自然選択), 人為選択, 雌雄(しゆう)(雌雄選択), 触媒 ▶〜的に作用する work selectively; have a selective effect. ¶**選択透過性** selective permeability; permselectivity. **選択毒性** selective toxicity. **選択的農薬** a selective insecticide. **社会選択** social selection.

全窒素(ぜんちっそ) ⇨窒素

剪定(せんてい) ▶〜する trim; prune; cut (off) 《a branch》; cut back 《new growth》. / 〜ゴミを処分する dispose of prunings.

船底塗料 〔海洋汚染の一因となる〕 ship bottom paint.

先天 ▶〜的な inborn; innate; inherent; native; congenital; 〔遺伝的〕hereditary; inherited. ★⇨遺伝 / 〜的な病気 a congenital disease. / 〜的に by nature; naturally; innately; inherently; congenitally. ¶**先天性障害** a congenital handicap. **先天性水俣病** ⇨水俣病 **先天的異常** congenital malformation [anomaly]. **先天免疫** congenital [inborn, innate, native] immunity.

選別処理 〔廃棄物の〕disposal of classified waste. ★⇨分別

全滅 ⇨絶滅 ▶イネが洪水で〜に近い被害を受けた. The rice crop was as good as wiped out by floods. | Flood damage resulted in nearly total destruction of the rice crop.

戦略的環境アセスメント Strategic Environmental Assessment (略：SEA).

線量 〔放射線の〕a dose (of radioactivity). ¶**線量計** a dosimeter. **線量限度** a dose limit. **線量測定** dosimetry. **線量当量** a dose equivalent. **線量当量限度** a dose equivalent limit. **吸収線量** an absorbed dose. **高線量** a high dose. **実効線量** an effective dose. **照射線量** an exposure dose. **低線量** a low dose. **等価線量** an equivalent dose. **ポケット線量計** a pocket dosimeter.

そ

騒音 (a) noise; (a) din; (a) discord; noise pollution. ★⇨うるささ指数, 運転音, 近隣騒音, 交通(交通騒音), 自動車(自動車騒音), 生活(生活騒音), 低周波, 低騒音, 暴騒音, 公害, 消音, 遮音(しゃおん), 防音 ▶町の～ the din and bustle of a town; city sounds; street noises; urban noise pollution. / 耳を聾するような～ a deafening [an ear-splitting] noise. / この近所は夜間は～があまりひどくない. This neighborhood is not so noisy at night. / 45デシベルを越える～レベル a noise level greater than 45 dB. / ジェット機の～に悩まされる suffer from the noise of jet planes. / ～をへらす reduce noise. / 交通～対策 traffic noise control measures. / (機械などが)すさまじい～を発する produce [generate] a tremendous [hideous] noise. / アパートの隣人の～を管理人に訴えた. We complained to the superintendent about the noise from the neighboring apartment. ¶**騒音外傷** noise trauma. **騒音規制** noise control [regulation]. **騒音規制法** the Noise Control Law. **騒音計[測定器]** a noise [sound-level] meter. **騒音公害** noise [sound] pollution. **騒音性難聴** noise (induced) deafness. **騒音被害** noise damage; damage due to noise. / ～被害が深刻化している. The noise damage is becoming serious. **騒音防止** noise prevention. **騒音防止運動** an anti-noise movement [campaign]. **騒音防止条例** a noise-control regulation.

相観 physiognomy.

雑木(ぞうき) miscellaneous trees; a bush; scrub. ¶**雑木林** a thickly wooded area; a thicket; a copse; a coppice.

操業停止 ▶～を命ずる order a stop to operations. ¶**操業停止命令** an order to stop [cease] operations [operating].

象牙(ぞうげ) ivory. ▶～狩りをする hunt ivory.

草原 a grassy [grass-covered] plain; grassland(s); 〔熱帯・亜熱帯の〕a savanna(h); 〔南アフリカの〕a veld(t); 〔北米の〕a prairie; 〔南米アマゾン川以南の〕

the pampas;〔南米アマゾン川以北の〕a llano;〔ロシア・中央アジアの〕a steppe. ★⇨草地(くさち)
¶草原動物 a grassland animal.

総合 ¶総合資源エネルギー調査会 the Advisory Committee for Natural Resources and Energy. **総合食料局** 〔農水省の内局;旧食糧庁〕General Food Policy Bureau. **総合説** 〔進化論の〕the modern (evolutionary) synthesis; the new [evolutionary] synthesis.

掃除(そうじ) cleaning;〔はき掃除〕sweeping (up);〔ふき掃除〕(机のほこりなどの) dusting; wiping; (こすり洗い) scrubbing;〔街路の〕street cleaning [cleansing]. ★⇨清掃 ▶～する clean; sweep; dust; scrub; cleanse. / 室内を～する clean a room (up, out); do a room; sweep a room clean. / 食物繊維はおなかの～をしてくれる. Dietary fibers clean up our digestive system. / 大～をする do a general house cleaning. / どぶ～をする clean (out) the gutter. / 今日の～当番は誰？ Who is on cleaning duty today? | Whose turn is it to clean up today? | Who is the duty cleaner today? / ～当番をさぼる shirk *one's* cleaning duty. ¶**掃除機** a (vacuum) cleaner. / 部屋に～機をかける vacuum a room. / 居間のカーペットに～機をかける run a vacuum cleaner over [vacuum clean] the carpet in the sitting room. **掃除道具[用具]** cleaning equipment; dusting [scrubbing] things. **掃除人** a cleaner; a sweeper; a cleaning man [woman]. / 道路～人 a street cleaner [sweeper].

草食 ▶～の plant-[grass-]eating; graminivorous; herbivorous. ¶**草食動物** a plant-[grass-]eating animal; a grazer; a herbivorous animal; a herbivore.

増殖 ▶～する increase; breed; multiply; propagate; proliferate; replicate; reproduce. ¶**増殖礁** a propagation reef. **増殖炉** 〔原子炉の一形〕a breeder (reactor); (実験用の) an experimental breeder reactor; (発電用の) a power breeder reactor.

総水銀 ⇨水銀

造成 creation; preparation《of a housing site》;〔埋め立て〕reclamation. ★⇨埋め立て ▶～する create《land》; clear; prepare《the ground for housing》; reclaim《land from the sea》;〔整地する〕put《land》in order; stabilize《man-made land》. /

宅地を〜する prepare a housing site; turn the land into housing lots. / 山林を〜する make a woodland; afforest [plant trees on] a mountain. / その沿岸地方に臨海工業地帯が〜された. A coastal industrial zone was created along the shoreline. / 海を埋め立てて道路や公園地帯が〜されることがよくある. Roadways and park areas are often reclaimed from the sea. ¶**造成工事**〔土地の〕creation [clearing] of land. / 宅地〜工事 creation [clearing] of land for residential purposes. **造成地** housing land; a housing site; land prepared for housing development. **造成地域** a development area.

草地(そうち)〔牧草地〕a grassland; a meadow. ¶**草地改良** grassland improvement. **草地学** grassland science. **草地学者** a grassland scientist. **草地造成** grassland [meadow] development.

送電 transmission of electricity; power [electric] transmission; 〔電気の供給〕(electric) power supply. ★⇨供給 ▶〜する transmit electricity《from... to...》; supply (electric) power《to...》. / 〜を断つ cut [shut] off the power supply《to...》. ¶**送電施設** a power transmission facility. **送電線** a power transmission line [cable, wire];〔高圧線〕a power cable. **送電線使用料** a power line rental charge. **送電塔[所]** a (power) transmission tower [site]. **送電力[容量]** (power-)carrying [transmission] capacity. **送電ロス** power transmission loss.

草木(そうもく) trees and plants; plant life; vegetation. ▶山頂には〜がまったくない. The summit of the mountain is bare [denuded] of all vegetation. ¶**草木灰** ashes made from burned vegetation.

総量規制〔汚染物質排出総量の規制〕total volume control; regulations on total allowable volumes of pollutants. ¶**総量規制地域** an area where total volume control on pollutants is enforced; a total emission-regulated area.

造林 afforestation; forestation; timber-planting;〔森林再生〕reforestation. ★⇨砂防造林, 植林 ▶〜する afforest《a mountain, a valley》; plant trees. ¶**造林学** forestry. **造林公社** an afforestation corporation. **造林署** an afforestation office. **造林奨励** encouragement of afforestation. **造林地** afforested land; an afforested area; 〔森林〕a plantation. **造林法** forest management; forestation; forestry; silviculture; sylviculture. **造林面積** an afforested area.

藻類 algae;〔海藻〕seaweeds. ▶〜の algal. ¶**藻類学** algology; phycology. **藻類学者** an algologist; a phycologist.

ソーシャル ¶**ソーシャル・アカウンティング** 〔GNP などの計算を行う社会会計〕social accounting. **ソーシャル・エンタープライズ** 〔社会的企業；ビジネス手法を使って社会貢献をする事業〕a social enterprise. **ソーシャル・マーケティング** 〔社会・環境との関係を考慮したマーケティング〕social [societal] marketing.

ソーラー 〔太陽の；太陽熱・太陽光を利用した〕solar. ★⇨太陽 ¶**ソーラー・エアコン** a solar air conditioner; solar air-conditioning. **ソーラー・エンジン** a solar engine. **ソーラー・カー** a solar(-powered) car. **ソーラーカー・レース** solar car racing;〔1 回の〕a solar car race. **ソーラー・クッカー** a solar cooker. **ソーラー・クッキング** 〔太陽光を利用した調理法〕solar cooking. **ソーラー・コレクター** 〔太陽熱集熱器〕a solar collector. **ソーラー・システム** 〔太陽系〕the solar system;〔太陽熱利用システム〕a solar (heating) system. **ソーラー照明灯** solar lighting. **ソーラー・セル** 〔太陽電池〕a solar cell. **ソーラー電卓** a solar calculator. **ソーラー・ハウス** 〔ソーラー・システムを利用した住宅〕a solar house. **ソーラー・パネル** 〔太陽光発電用〕a solar panel. **ソーラー・パワー** 〔太陽エネルギー〕solar power. **ソーラー飛行船** a solar airship. **ソーラー・プレーン** a solar plane. **ソーラーボート** a solar-powered boat. **ソーラー・ポンド** 〔太陽熱発電用の池〕a solar pond.

属 a genus. 形 generic. ★⇨種(しゅ), 学名 ▶属特有の性質〔属性〕a generic character. / 属の特性記載 a generic description. ¶**属間雑種** an intergeneric hybrid. **属差** a generic difference. **属名** a generic name. / 属名種名同一の学名 a tautonym.

促進 ▶植物の成長を促進する hasten [accelerate, force] the growth of a plant. ★⇨世代促進

促成栽培 ▶forcing culture. / イチゴを〜する force strawberries. / 〜用の温室〔温床〕a forcing house [bed]. / 〜の野菜

forced vegetables.

促成飼育 ⇨飼育

粗大ごみ 《an item of》bulky [large-size(d), oversized] rubbish [trash]. ★⇨ごみ ▶〜のリサイクルは解体や保管にかかるコストがネックとなってなかなか進まない. The cost of dismantling and storing bulky waste [junk] is the bottleneck that holds up progress in recycling such waste. ¶**粗大ごみ収集日** the collection day for [day for collecting] bulky (items of) rubbish.

ソフト ¶**ソフト・エネルギー** 〔太陽熱・風力などによる自然エネルギー〕soft energy. **ソフトエネルギー・パス** 〔ソフトエネルギー戦略〕soft energy paths. **ソフト・テクノロジー** 〔太陽熱・風力などの自然エネルギー利用技術〕soft technology.

た

田 ⇨水田, 棚田(たな), 湛水(たん)

代エネ法 ⇨石油(石油代替エネルギー法)

ダイオキシン dioxin. ★⇨英和 dioxin, brominated dioxin, TCDD ▶周辺の土壌から高濃度の〜が検出された. High concentrations of dioxin were detected in the surrounding soil. /〜の耐容一日摂取量 the tolerable daily intake of dioxin. /〜の放出量を削減する reduce dioxin emissions. /〜汚染 dioxin contamination. /大気中の〜濃度を測定する measure the atmospheric dioxin concentration [concentration of dioxin in the air]. ★⇨母乳
¶**ダイオキシン類対策特別措置法** the Law Concerning Special Measures against Dioxins.

タイガ 〔亜寒帯の針葉樹森林〕a taiga.

体外 ▶〜に outside the body. /〜に(老廃物を)排出する excrete [eliminate] (waste matter) from the body. /〜の external; ectosomatic; extracorporeal.
¶**体外受精** 〔ヒトの〕in vitro fertilization (略: IVF); external fertilization; 〔水生動物の〕external fertilization. /第三者から卵子の提供を受けて〜受精を行う carry out IVF with donor eggs. **体外受精児** a test-tube baby; a baby born by in vitro fertilization. **体外受精専門医** an IVF specialist. **体外受精治療** IVF treatment. **体外受精卵** an in-vitro fertilized egg; a test-tube egg.

体感 ▶気温が高くても湿度が低いと〜的にはさほど暑さを感じない. Even when air temperature is high, if humidity is low it doesn't feel that hot to the [one's] body. ¶**体感温度** sensible temperature. /風速が1メートル増すごとに〜温度は1度下がる. The sensible temperature falls by one degree with every meter-per-second increase in wind-speed. /日本の夏, ネクタイをゆるめるだけで〜温度はかなり低くなる. In summer in Japan you can feel quite a bit cooler just by loosening your tie.

大気 〔地球を取り巻く〕the atmosphere; 〔空気〕the air. ¶**大気圧**

atmospheric pressure. **大気汚染** air pollution. **大気汚染常時監視測定局** a continuous air pollution monitoring station. **大気汚染物質** air [airborne] pollutants. **大気汚染防止法** the Air Pollution Control Law; the Air Pollution Prevention Law. ★米国の「大気浄化[清浄]法」は⇨英和 Clean Air Act **大気汚染予報** an air pollution (weather) forecast. **大気圏** the atmosphere. ▶〜圏内の核実験 an atmospheric nuclear test. **大気質** air quality. **大気大循環** 〔全地球規模の大気運動〕atmospheric general circulation.

待機電力[電流] 〔電流〕standby current;〔電力消費〕standby power consumption.

耐空証明 a certificate of airworthiness; an airworthiness certificate.

堆砂(たいさ) 〔ダムの貯水池の底に土砂が溜まること〕siltation; sedimentation;〔その土砂〕silt; sediment. ▶〜する sediment; accumulate.

第三者認証ラベル 〔日本環境協会が発行する〕a third-party certification label.

代償植生 〔自然植生に対して〕a substitutional vegetation [community].

耐震指針 〔原子力発電所の〕the Seismic Safety Guidelines on the Design of Nuclear Power Plants.

帯水層 〔地下水を含む多孔質浸透性の地層〕an aquifer.

大西洋まぐろ類保存国際委員会 ⇨英和 ICCAT.

代替(だいたい) ¶**代替エネルギー** alternative (forms of) energy. ★⇨石油(石油代替エネルギー) **代替エネルギー源** an alternative source of energy [energy source]. **代替エネルギー資源** alternative energy resources. **代替燃料** alternative fuel. **代替フロン** an alternative (chloro)fluorocarbon; an alternative CFC;〔フロン以外のものも含めて〕a CFC substitute.

大腸菌 a colon bacillus《複数形 colon bacilli》; a coli (bacillus);〔その代表種〕*Escherichia coli*; *E. coli*. ¶**大腸菌群** coliform bacteria [bacilli]. ▶糞便性〜群 fecal coliform bacteria. **大腸菌(群)数** a coliform (bacteria) count. **大腸菌指数** 〔水などの〕a coliform index. **大腸菌症** colibacillosis. **大腸菌腸毒素** *Escherichia coli* enterotoxin.

体内 the interior of the body. ▶〜に in the body [system]. /〜の in the body [system]; intestine; internal. / 〜の糖分 body [tissue] sugar. ¶ **体内汚染** internal pollution. **体内温度** (an) internal body temperature. **体内環境** the internal environment of the human [*sb's*] body; the body's [*sb's*] internal environment. **体内受精** internal [entosomatic] fertilization. **体内時計** a biological [physiological] clock; an internal clock; a body clock. **体内年齢** (a [*one's*]) body [biological] age. **体内被曝** internal exposure. **体内リズム** an [*one's*] internal (body [biological]) rhythm.

堆肥(たいひ) compost; manure; (organic) fertilizer; muck. ▶ 〜を施す compost 《the ground》; manure; fertilize; spread muck. ¶ **堆肥積み** a compost heap; a pile of manure; a dunghill.

太陽 the sun. ¶ **太陽エネルギー** solar energy. **太陽活動** solar activity. **太陽(活動)周期** a solar cycle. **太陽観測** solar observation. **太陽光線** a ray of sun(light); a sunray; a sunbeam; solar rays. ▶人工〜光線 artificial sunlight. **太陽光発電** 〔太陽電池による〕photovoltaic (power) generation; solar power generation (with a photovoltaic cell). / 住宅用〜光発電システム a residential solar cell system. / 〜光発電パネル a solar [photovoltaic] power (generation) panel. / 〜光発電の普及には，売電価格の引き上げや自治体の補助制度など，ソフト面を充実させることも大切だ．For solar power to become more popular, enhancements that go beyond the technology itself, such as higher prices for electricity sold and subsidies from local governments, are also important. **太陽光(線)吸収率** solar absorptance. **太陽光(線)透過率** solar transmittance. **太陽光(線)反射率** an albedo value; a solar reflectance. **太陽(光)発電衛星** a solar power satellite (略：SPS). **太陽光発電協会** the Japan Photovoltaic Energy Association (略：JPEA). **太陽光発電所** a solar power plant [station]; a photovoltaic [PV] power plant [station]. / 大規模〜光発電所 a large-scale [mega] solar power plant. **太陽黒点** a sunspot; a solar spot. **太陽黒点周期** a sunspot cycle. **太陽黒点説** 〔気象などの変化の〕 the sunspot theory. **太陽コンパス** a solar [sun] compass. **太陽コンパス定位** sun compass orientation. **太陽蒸溜器** a solar still.　**太陽電池**　a solar cell [battery].　**太陽電池パネル**　a solar (cell) panel. / 両面受光〜電池（パネル）a double-sided solar cell panel.　**太陽電波**〔通信に混入する〕solar noise;〔一般に〕solar radio waves; solar radio-frequency radiation.　**太陽熱**　solar heat [energy]; the heat of the sun's rays. / 〜熱利用のプール a solar-heated swimming pool. / 〜熱で温められた水 sun-warmed water. / 〜熱を利用する utilize [make use of] solar heat [energy].　**太陽熱温水器**　a《rooftop》solar unit [collector]; a solar water heater.　**太陽熱暖房**　solar heating.　**太陽熱発電**〔太陽光を集束させる〕solar thermoelectric power generation.　**太陽発電**〔太陽光発電・太陽熱発電〕solar power generation.　**太陽発電装置**　a solar-power generator.　**太陽風**　the solar wind.　**太陽放射**[輻射]　solar radiation.　**太陽炉**　a solar furnace.

大洋　the ocean.　¶ **大洋水深総図**〔世界の海底地形図を作成する国際プロジェクト〕the General Bathymetric Chart of the Oceans（略: GEBCO）;〔その図〕a chart produced by GEBCO.　**大洋性気候**　an oceanic climate.　**大洋中央海嶺**　the mid-oceanic ridge.　**大洋底**　the ocean bottom [floor].　**大洋島**　an oceanic island.

耐容摂取量　⇨摂取

大陸　a continent.　¶ **大陸棚**　a continental shelf.　**大陸棚延長**　extension to the edge of the continental shelf.　**大陸棚開発**　continental shelf development.　**大陸棚資源**　continental shelf resources.　**大陸棚調査**　a continental shelf survey.　**大陸島**　a continental island.　**大陸氷河**　a continental glacier.　**大陸プレート**　a continental plate.

対流　¶ **対流雲**　convection cloud.　**対流圏**　the troposphere.　**対流圏オゾン**　tropospheric ozone.　**対流圏界面**〔圏界面〕the tropopause.　**対流式暖房機**〔輻射式に対して〕a convection heater; a convector.　**対流層**　a convective zone.

大量　a large [an enormous, a huge, a massive] quantity [amount, volume]《of…》; a lot《of…》;《口》masses [tons, piles, stacks, loads, heaps]《of…》.　▶〜の extensive; wholesale; mass; massive; large-scale; a large [huge, massive] quantity [amount] of…; a lot of…;《口》masses [loads, tons, piles, stacks, heaps, mountains] of… / 〜のゴミ masses of trash. ⇨ごみ，放棄 / 〜の火山ガス an enormous volume of volcanic gas ¶**大量供給** a huge [large-scale] supply; massive supplies 《of crude oil》. **大量死** massive deaths [fatalities]. ★⇨流出 **大量使用** mass [massive, large-scale, extensive] use. / アスベストの〜使用 use of asbestos on a large scale. **大量消費** mass consumption. **大量消費(型)社会** a [the] mass consumption [consumer] society. **大量処分** mass [large-scale, major, sweeping] disposal 《of dead animals [infected birds, equipment, ammunition]》. **大量処理** mass [large-scale] processing 《of garbage》. **大量生産** mass [quantity, large-scale, bulk] production. / 〜生産する mass-produce; produce *sth* in commercial quantities. / 安い労働力で[機械を使って]製品を〜生産する mass-produce goods with cheap labor [by using machinery]. / その製品の〜生産に入る go into full commercial production of the product. / 〜生産の mass-produced. / 〜生産者 a mass-producer. **大量摂取** (a) massive intake 《of alcohol》. **大量絶滅** a mass extinction; an extinction event. **大量増殖** mass propagation; [植物の微細繁殖] micropropagation. **大量廃棄** mass disposal. **大量破壊兵器** a weapon of mass destruction (略: WMD); a mass destruction weapon. **大量発生** [害虫の] a plague of harmful insects; [プランクトンの] a massive proliferation; [食中毒の] a mass outbreak of food-poisoning. ★⇨プランクトン, 発生 **大量販売** mass marketing; volume sales; bulk [block] sale; cash and carry. / 〜販売する mass-market 《a product》; sell in volume [bulk]. **大量放出** (a) mass [massive] discharge 《of waste water》; (a) massive release 《of radiation into the atmosphere》; mass emission 《of hydrocarbons》. / 大気中に二酸化炭素を〜放出する release massive amounts of CO_2 into the atmosphere. ★⇨垂れ流す **大量輸送** mass transit; mass transportation. **大量輸送機関[手段]** a mass transit [transportation] system.

宅地 [家を建てるための] land for housing; housing [building] land; a house [home, building] lot [plot]; a housing [building] site; [すでに家のある] residential land; a residential site; grounds. ▶山林の〜化が進んでいる．Rural land is being steadily built up. / 農地を〜化する develop agricultural land for housing; turn agricultural land into housing lots. / 戦後このあたりの〜開発が進んだ．It was after the war that this area began to be built over in housing lots. ★⇨開発，造成

濁度(だく) turbidity; cloudiness; murkiness. ¶**濁度計** a turbidimeter.

択伐(たくばつ) 〔目的に合った木だけを選んで伐採すること〕selective logging [cutting, felling] (of trees). ▶〜する do [practice] selective logging [cutting, felling].

多自然型川づくり neo-natural river reconstruction; a method of river conservation work that avoids, or keeps to a minimum, damage to the natural environment.

立ち枯れ ▶〜する be blighted [wilt, wither, die] while remaining upright. / 農作物が〜している．The crops are wilting on stalk and vine. / 公園の木がほとんど〜になっている．Almost all the trees in the park have suffered blight. / 〜の木 a blighted [withered] tree. ¶立ち枯れ病 damping-off.

立ち木 〔樹木〕a living [standing, growing] tree; 〔木材〕timber growth; (standing) timber. ▶〜が多い．Timber growth is heavy. | There is a lot of standing timber (on the property). / 〜のまま買う buy on the stump. ¶**立ち木トラスト** a timber trust. / 〜トラスト運動 a movement to form timber trusts.

脱(だつ)… ▶脱工業化社会 (a) postindustrial society; a society no longer based on industry. / 脱原発の風潮 an increasing tendency to escape from dependence on nuclear power. ★⇨原発，温暖化，脱ダム宣言，脱石油 / 脱車社会を目指す strive for a post-automobile society.

脱塩 desalinization; desalination; desalting; depickling; demineralization. ▶〜する desalinate; desalt. ¶**脱塩工場** 〔海水を真水に変える〕a desalination plant. **脱塩水** desalinated [desalted, demineralized] water. **脱塩率** a demineralization rate.

脱石油 ¶**脱石油(依存)社会** a post-petroleum society; a society no longer dependent on oil. **脱石油時代** a post-petroleum age.

脱ダム宣言 a "no more dams"

脱窒(だっ) denitrification. ▶～する denitrify.

脱硫(だつりゅう) desulfurization; desulphurization; desulfuration;〔石炭乾留ガスからの〕purification;〔加硫ゴムを可塑化する〕devulcanization. ▶～する desulfurize; desulfur; purify; devulcanize. ¶**脱硫器** a desulfurizer; a purifier. **脱硫剤** a desulfurizing agent; a devulcanizing agent. **脱硫設備[装置]** a desulfurization system; desulfurization equipment. **排煙脱硫** desulfurization of exhaust (gas). **重油脱硫** desulfurization of fuel oil.

脱燐(だつりん) dephosphorization. ▶～する dephosphorize.

棚田(たなだ) terraced paddy [rice] fields.

たばこ，煙草 〔紙巻きたばこ〕a cigarette;〔葉巻き〕a cigar;〔刻みたばこ〕tobacco. ★⇨喫煙，禁煙，ポイ捨て，煙害，無煙 ▶歩き～禁止．〔掲示〕Smoking while walking not allowed [permitted]. / 寝～はご遠慮ください．Please don't smoke in bed. / お～はご遠慮ください．〔掲示〕Kindly refrain from smoking. | Thank you for not smoking. / 人ごみでのくわえ～は危ない．Walking through a crowd of people with a cigarette in your mouth is dangerous. / ～を吸う smoke (a cigarette); have [take] a smoke. / ～を吸ってもよろしいでしょうか．Would you mind if I smoke? / ～を控える〔吸わないでいる〕refrain from smoking;〔ほどほどにする〕smoke in moderation; smoke less; cut down on smoking. / ～をやめる give up [quit] smoking; cut out smoking altogether. / ～は健康によくない．Smoking is bad for [injurious to] your health. / この部屋は～くさい．This room smells of tobacco smoke. / 一日その会議に出ると，～のにおいがすっかり服に移ってしまう．If you spend a day at those meetings, the smell of tobacco permeates your clothes [your clothes will smell of tobacco]. ¶**たばこ依存症** tobacco dependence. **たばこ公害** tobacco smoke pollution. **たばこ税** a cigarette [tobacco] tax. **たばこ病** (a) tobacco disease. **たばこ病訴訟** a tobacco lawsuit.

食べ物 food; a thing to eat. ★⇨食生活，食料，食品，食物 ▶旅行中は飲み物や～に気をつけなさい．While traveling be careful what you eat and drink. / (動物が残飯などの)〜をあさる scavenge around [rummage, poke around] for food. / 〜を粗末にしてはいけない．Don't waste (your) food.

ダム a dam; a barrage. ★⇨脱ダム宣言, へどろ, 堆砂(_{たいさ}), 緑(緑のダム) ▶川に〜を造る build [construct] a dam across a river; dam a river. / 〜は河川の流量を調節する．A dam controls the flow of a river. ★⇨治水(治水ダム) / 日照り続きで〜の水位がだいぶ下がっている．The fine weather has brought the water level in the dam pretty low. | The water level in the dam has been significantly lowered by the drought. / 〜が決壊した．The dam gave way [collapsed]. / その〜は公益性が低いと判断され，建設が見送られた．The dam was judged to have little public benefit, and construction was put off. / 〜建設の是非 the pros and cons of building the dam. / 〜に水没した村 a village submerged by a dam. / 上流に〜が建設された．そのために魚の環境が一変した．A dam was constructed upriver, as a result of which the environment for fish was transformed. ¶**ダム建設用地, ダムサイト** a dam site.　**ダム湖** a dammed (up) lake.　**アーチ式ダム** an arch dam.　**穴あきダム** a flood control dam.　**水力発電用ダム** a hydroelectric dam.　**多目的ダム** a multipurpose dam.

多様性 ⇨生物(生物多様性)

垂れ流す 〔汚染物質などを〕spill; carelessly [inadvertently] discharge [release]. ▶工場廃水を川に〜 discharge untreated industrial waste water into a river. / 長年にわたって有機水銀が水俣湾に垂れ流されていた．For many years, organic mercury had been dumped untreated into Minamata Bay. ¶**垂れ流し** 〔汚染[有害]物質などの〕a careless discharge 《of waste》; an unchecked effluence. ★⇨公害 / 化学工場から出る有害廃棄物の垂れ流し effluence of poisonous waste from a chemical factory. / その工場はカドミウムを含んだ廃水を川に垂れ流しにしていた．The factory had been discharging waste water containing cadmium into the river.

単一耕作[栽培] monoculture.

タンカー a tanker; a tankship; 〔石油の〕an oiler; a [an oil] tanker. ▶相次ぐ〜の座礁事故 frequent tanker wrecks; frequent incidents in which tankers have been grounded. / 〜の

座礁事故で大量の原油が湾内に流出した．When the tanker ran aground, a huge volume [quantity] of crude oil spilled out into the bay. ¶マンモス[大型]タンカー a mammoth tanker; a supertanker.

炭酸 carbonic acid. ¶炭酸ガス〔二酸化炭素〕carbonic acid gas; carbon dioxide. ★⇨二酸化炭素, 英和 carbon dioxide 炭酸(ガス)固定 〔光合成などによる〕carbon dioxide fixation. 炭酸同化作用 carbon assimilation.

淡水 fresh water. ▶塩水を〜にする turn salt water into fresh water. / 〜(産)の freshwater; limnetic. ¶淡水化 desalination; desalinization; desalting. / 〜化する desalinate 《seawater》; desalt; desalinize. / その装置は1日に10万トンの塩水を〜化する能力がある．The facility has a desalination [desalting] capacity of 100,000 tons a day. | The facility can purify [desalinate] 100,000 tons of salt water per day. 淡水化プラント a desalination plant. 淡水魚 a freshwater fish. 淡水湖 a freshwater lake.

湛水(たん) flooding; filling; inundation; submerging. ▶〜する flood 《a rice paddy》; fill 《a dam》; inundate; submerge. ¶湛水灌漑 flood irrigation. 湛水期間 a flooding [filling] period. 湛水直播 wet (direct) seeding [sowing]; direct planting of rice in a flooded paddy. 湛水土壌 flooded [waterlogged, puddled] soil. 冬期湛水 〔田に冬場も水を張っておくこと〕winter flooding (of rice fields). / 冬期〜水田 a winter-flooded rice field.

断水 a water cutoff; (a) suspension of the water supply;〔水枯れ〕a water shortage. ★⇨給水, 水道 ▶(水道局が)〜にする cut [shut, turn] off the water (supply); stop the supply of water. / 現在〜中である．The water (supply) is cut off right now. / 昔はよく〜したものだ．The water used to get cut off a lot. / 全市にわたる〜 a failure of the water supply over the whole city; a city-wide water cutoff. / 明日都内の一部で〜がある．The water supply will be shut [cut] off tomorrow in parts of Tokyo. / 〜に備えて汲み置きする store water in case the water supply is cut off. ¶断水区域 an area affected by a water cutoff; a section for the water supply to be cut off.

炭素 ⇨英和 carbon, 低炭素社会 ¶炭素隔離リーダーシップ・フォー

ラム〔二酸化炭素を地下に貯留する技術に関する国際会議〕the Carbon Sequestration Leadership Forum（略：CSLF）. ★⇨英和 carbon sequestration **炭素基金** ⇨英和 Prototype Carbon Fund **炭素クレジット**〔取引可能な温室効果ガスの排出削減権〕a carbon credit. **炭素循環[サイクル]**〔地球環境中での〕the carbon cycle. **炭素税**〔環境税の一種〕a carbon tax. ★⇨英和 carbon tax **炭素同化** ⇨炭酸（炭酸同化作用） **炭素排出量** carbon emissions.

暖冬 a warm [mild, green, clement] winter. ▶〜異変 an abnormally warm [mild] winter./ 世界的な〜傾向がここ数年続いている. A worldwide trend toward warmer winters has continued for some years. | For some time now winters have tended to be warmer./ まだ冬に入ったばかりだが，今年も〜傾向が顕著である. Though the winter has just started, it promises to be another warm one.

弾頭 a warhead. ▶化学〜搭載ミサイル a missile equipped with a chemical warhead./ 通常〜〔核弾頭でない〕a conventional warhead. ★⇨核（核弾頭）

断熱 (thermal [heat]) insulation. ▶〜する insulate (from heat)./ 不完全な〜 inadequate insulation./ 〜が悪いといくら暖房してもちっとも暖かくならない. If the insulation is bad, the room won't warm up at all no matter how much heating you use. ¶**断熱減率** an adiabatic lapse rate./ 乾燥[湿潤]〜減率 a dry [moist] adiabatic lapse rate. **断熱効果** insulation (effect). **断熱材** (heat-)insulating material; (thermal [heat]) insulation; a thermal [heat] insulator; an insulant. **断熱サッシ** an insulated window sash. **断熱性** heat-insulating properties; nonconductiveness. **断熱れんが** a heat-insulating brick; an insulating brick.

田んぼ ⇨田

暖房 (indoor) heating. ★⇨冷暖房，温度 ▶〜が入っている[いない] the heating is on [off]./ この部屋は〜がよくきいている[あまりきいていない]. The heating in this room works well [doesn't work very well]. ★⇨断熱 / この部屋は〜がきき過ぎている. This room is overheated. | The heat is on too high in here./ この会社では11月にはもう〜が入る. In this company the heating comes on as early as November./ 部屋を〜する

暖房

heat a room. / この壁材だと熱が逃げないので〜費の節約になる. With this wall material, the heat will not leak out and you can save on your heating bills. / 〜を低め[弱]にする turn the heating (down) to low; set the thermostat on the low side. / 〜中は窓を開け閉めして換気をよくしてください. When you have the heating on frequently open the window(s) for a while and let in some fresh air. / この冬はずっと〜はこたつだけで過ごしてみた. This winter I tried to go through the whole winter with just a *kotatsu* for heating. ¶暖房効率 heating efficiency. 暖房設備 heating equipment. / 〜設備のある[ない]建物 heated [unheated] building / この建物には〜設備がない. There is [We have] no heating system in this building. 暖房装置[器具] a heating apparatus [device, system]; a heater. ガス[電気]暖房 gas [electric] heating. 集中暖房 concentrated heating. スチーム[温風, 温水]暖房 steam [warm-air, hot-water] heating. 地域暖房 local heating. 輻射[放射](式)暖房 panel [radiant] heating. 床暖房 floor heating.

ち

地域 〔地理的区域〕a region; an area; a zone; a district;〔地域社会〕a (local) community. ★⇨生態(生態系) ▶～社会を再生する reactivate the local community. / 観光によって～活性化を図る try to revitalize [bring life to] an area through tourism. ¶ **地域エゴ** regional self-interest. ★⇨ニンビー **地域温室効果ガス・イニシアティブ** 〔米国北東部の諸州が連合して行う，火力発電所の二酸化炭素排出についての自主規制〕the Regional Greenhouse Gas Initiative (略: RGGI). **地域開発** community [regional] development. **地域貢献活動** an activity contributing to a community. **地域情報システム** a geographic information system (略: GIS). **地域(地区)制** 〔都市計画の〕zoning. **地域相関研究** 〔生態学的研究〕an ecological study. **地域通貨** 〔同じ地域の複数国で使う〕a regional currency;〔ある地域やグループの中で，物やサービスの交換に使われる独自の通貨〕a local [community] currency. **地域猫** a (stray [homeless]) neighborhood cat. ★⇨糞尿被害 **地域熱供給(システム)** district heating and cooling (略: DHC); a district heating and cooling system. **地域ボランティア** 〔活動〕local [neighborhood] volunteering;〔人〕a local [neighborhood] volunteer. **地域密着型の** community-based. / ～密着型の広報活動 public-relations activity oriented to the local community. **地域冷暖房(システム)** a regional air-conditioning system.

地衣類 lichens.

地下 〔地面の下〕the underground; the region below the ground. ★⇨地中化 ¶ **地下核実験** an underground nuclear test. **地下核爆発** an underground nuclear explosion. **地下溝** a culvert; a drain;〔下水〕a sewer. **地下資源** underground resources. **地下室マンション** a condominium on a slope with residential floors designated as basements in order to get around zoning limits on height; a multibasement condominium. **地下水**

groundwater; underground water; a subsurface flow of water. ★⇨地盤沈下　▶〜水の汚染 groundwater contamination [pollution]; the contamination of groundwater 《by industrial waste》.　**地下水位**　the groundwater level; the water table. / 〜水位の低下 a lowering of the water table. / 〜水位面 the water table; the groundwater level.　**地下生物圏**　the deep biosphere.　**地下ダム**　a groundwater reservoir.　**地下貯留**　〔雨水などの〕underwater storage.　**地下貯留槽**　an underground storage tank.　**地下熱**　subterranean heat; geothermy.

地球　the earth; the planet Earth; the globe.　▶〜以外の場所に生物が存在する可能性 the possibility of extraterrestrial life. / 〜上の［で］on (the face of) the earth [globe]. / 〜の terrestrial; global; earthly. / 〜規模の企業 a global-scale enterprise. ★⇨グローバル / 〜規模の環境保護運動 a worldwide environmental protection movement. / 〜にやさしい environment-friendly; earth-friendly. ★⇨環境 / 〜に優しい建築材料 environmentally friendly [environment-friendly] building materials. / 〜に優しいハイブリッド・カー an ecologically friendly hybrid vehicle. / 私たちは「〜に優しい」をキーワードにさまざまな〜環境問題に取り組んでいます． We use the term "environmentally friendly" as a watchword [key word] when we wrestle with all kinds of global environmental problems. / 〜環境を守る protect the global environment. / 〜を救おうなんて大それたことを言うつもりはないが，日々の生活で資源を無駄にしない努力くらいはしたいものだ． I'm not saying anything so preposterous as that we could save the planet. But I do think that we could try at least not to waste natural resources in our day-to-day lives. / 私たちは未来への遺産としてこの美しい〜を残さなければならない． We must leave this beautiful globe as a heritage [legacy] to future generations. ¶ **地球温暖化**　global warming; the warming of the earth. ★⇨温暖化 / 〜温暖化が進む warming of the globe progresses. / 〜温暖化による気候変動 climatic changes resulting from the warming of the earth [globe]. ★⇨気候 / 〜温暖化のメカニズム the global warming mechanism. / ～温暖化の影響 an effect of the warming of the earth [globe]. / 二酸化炭素が～温暖化の原因となっている．Carbon dioxide is the cause of the warming of the earth [globe]. ★⇨温室(温室効果ガス) / ～温暖化を防止する prevent the warming of the earth [globe]. **地球温暖化係数** a global warming potential (略: GWP). **地球温暖化対策(推進)大綱** 〔環境省の〕(the) Guidelines for Measures to Prevent Global Warming. **地球温暖化対策推進法** the Law for the Promotion of Measures to Deal with Global Warming. **地球温暖化対策・ヒートアイランド対策モデル地域** a model district for measures against global warming and heat islands. **地球温暖化防止会議** the Conference of Parties of the UN Framework Convention on Climate Change (略: UNFCCC). ★⇨英和 United Nations Framework Convention on Climate Change **地球温暖化防止京都会議** the Third Conference of the Parties [COP3] to the UN Framework Convention on Climate Change in Kyoto in 1997; COP3 in Kyoto. ★⇨英和 Kyoto Protocol, COP **地球環境概況** 〔国連環境計画が発行する報告〕the Global Environment Outlook (略: GEO). ★⇨国連 **地球環境学** global environment studies [science]; whole earth science(s). **地球環境基金** 〔環境保護活動に取り組む NGO を支援する〕the Japan Fund for Global Environment. **地球環境局** Global Environment Bureau. ★⇨環境(環境省) **地球環境行動会議** Global Environmental Action (略: GEA). **地球環境国際議員会議** Global Legistlators Organization for a Balanced Environment(略: GLOBE). **地球環境国際議員連盟** the Global Legislators Organization for a Balanced Environment (略: GLOBE). **地球環境サミット** ⇨地球サミット **地球環境産業技術研究機構** 〔日本の財団法人〕the Research Institute of Innovative Technology for the Earth (略: RITE). **地球環境戦略研究機関** 〔財団法人〕the Institute for Global Environmental Strategies (略: IGES). **地球環境パートナーシッププラザ** the Global Environment Information Centre (略: GEIC). **地球環境ファシリティ** the Global Environment Facility (略: GEF). ★1991 年に設けられた発展途上

国の地球環境保全のための資金援助機関. **地球環境変化の人間社会側面に関する国際研究計画** International Human Dimensions Programme on Global Environmental Change (略: IHDP). **地球環境保護(主義)** planetism. **地球環境保護基金** ⇨地球環境ファシリティ **地球環境モニタリング・システム** 〔国連環境計画の〕the Global Environment Monitoring System (略: GEMS). **地球観測衛星** an earth observation [survey] satellite. **地球儀** a (terrestrial) globe; a globe of the world; a world globe. **地球規模生物多様性概況** the Global Biodiversity Outlook (略: GBO). **地球圏・生物圏国際協同研究計画** the International Geosphere and Biosphere Program (略: IGBP). **地球サミット** (the) Earth Summit. ★⇨英和 United Nations Conference on Environment and Development **地球シミュレーター** 〔地球環境の変動などの予測に使うスーパーコンピューター〕the Earth Simulator. ¶**地球市民** a global citizen; a citizen of the earth. **地球人** an earthling; an earthman; an earthwoman. **地球深層ガス** 〔石油生成に関する仮説上の〕(a) deep-earth gas.

地球深部 earth's depths; the depths of the earth; areas [places, regions] deep in the earth. **地球深部探査船** a deep-sea drilling vessel. **地球大気開発計画** Global Atmospheric Research Program (略: GARP). **地球地図** 〔日本の提唱による地球環境問題解明のためのプロジェクト〕Global Mapping; the Global Mapping Project;〔作成された地図〕a global map. **地球地図国際運営委員会** the International Steering Committee for Global Mapping (略: ISCGM). **地球の友** ⇨英和 Friends of the Earth **地球の日** 〔4月22日〕⇨英和 Earth Day **地球平和監視時計** 〔直近の核実験からの日数を表示する時計; 広島市の原爆資料館にある〕the Peace Watch Tower. **地球村** the global village. ★⇨宇宙船地球号 **地球惑星科学** earth and planetary sciences.

畜産 stock raising [farming]; stockbreeding; animal husbandry; the livestock industry. ¶**畜産公害** animal waste pollution. **畜産食品** livestock products [foodstuffs]. **畜産農家** a livestock farmer.

蓄積 accumulation; stockpiling;〔薬物の〕accumulation. ▶長期にわたる農薬の体内〜 the long-term accumulation of agricultural chemicals inside the body. / 〜する accumulate; amass; store (up); stockpile; hoard up. / 〜進行性の疾患〔加害行為あるいは有害物質の摂取の終了後かなりの期間が経過してから発症する病気〕a progressive, late-onset disorder (resulting from accumulation of a substance).

治山(ちさん) forestry conservancy [conservation, protection]; antiflood [flood control] afforestation. ★⇨治水

地産全消 (the) nationwide consumption of locally produced goods [local products].

地産地消 the local consumption of locally produced goods [products]. ★⇨英和 locavore

致死 ▶〜の fatal; mortal; deadly; lethal. ¶**致死遺伝子** a lethal gene. **致死線量**〔放射線の〕a lethal dose. **致死量**〔薬物の〕a lethal dose (略: LD).

治水 river [riparian] improvement; river engineering; river training;〔洪水防止〕flood control. ★⇨治山 ▶利根川の〜 flood control of the river Tone. ¶**治水委員会** the River Conservation Committee. **治水計画** a water-control [flood-control, river engineering] project. **治水工学** hydraulic engineering; river engineering. **治水工事** embankment [river engineering, flood prevention] works. **治水対策** flood control [prevention] measures. **治水ダム** a flood-prevention dam. **治水問題** the flood control problem [issue]. **治水容量**〔ダムの洪水調節容量〕a flood control volume. **治水緑地** a greenbelt flood buffer zone [area].

地層 a stratum《複数形 strata》; a layer; a geologic(al) formation; a bed;《coal》measures. ¶**地層処分**〔放射性廃棄物の〕geological [formation] disposal (of radioactive waste). **深地層処分** deep geological disposal (of radioactive waste).

地中化〔電線類の〕burying electric cables (underground). ▶電線を〜する bury a power line; run an electric cable underground.

窒素(ちっそ) nitrogen. ▶全〜〔水中に含まれる窒素化合物の総量；環境基準の１つ〕total nitrogen (略: TN). ¶**窒素化合物** a nitrogenous compound. **窒素ガス** nitrogen gas. **窒素酸化物** (a) nitrogen oxide (略: NOX). **窒素循環** the nitrogen cycle. **窒素同化** nitrogen assimilation. **窒素肥料** (a) nitrogen [nitrogenous] fertilizer.

地熱 terrestrial heat; the heat of the earth; geothermy. ▶〜の geothermal; geothermic. ¶地熱エネルギー geothermal energy. 地熱井(せい) a geothermal well. 地熱探査[探鉱] geothermal exploration; geothermal prospecting. 地熱発電 geothermal (electric) power generation. 地熱発電所 a geothermal power plant [station].

地被植物 a ground cover.

着色料 ⇨人工(人工着色料), 合成(合成着色料)

中央環境審議会 〔環境省の諮問機関〕the Central Council on the Environment; the Central Environment Council.

中間処理 〔産業廃棄物の〕intermediate (waste) processing. ▶廃棄物を〜する subject waste to intermediate processing. ¶中間処理業者 an intermediate (waste) processing company. 中間処理施設, 中間処理場 an intermediate (waste) processing plant [disposal facility].

中間貯蔵 ▶使用済み核燃料の〜施設 an intermediate storage facility for spent nuclear fuel.

中継施設 〔ごみ・廃棄物などの〕a refuse transfer station (略: RTS); a garbage transfer station.

抽水植物 an emergent plant.

中水道 an industrial [a nonpotable, a grey] water pipe. ★ ⇨英和 gray water.

中性 ¶中性花 a neutral [sterile] flower. 中性紙 neutral paper; acid-free paper; archival paper; alkaline paper. 中性洗剤 a neutral detergent. 中性肥料 a neutral fertilizer.

沖積(ちゅうせき)**層** an alluvium; an alluvial bed. ¶沖積平野 an alluvial plain; a floodplain.

中毒 poisoning; toxication; intoxication;〔麻薬などの〕addiction. ★⇨一酸化炭素中毒, 鉛毒(えんどく), 水銀中毒, 毒 ▶キノコを食べて〜する be [get] poisoned by mushrooms. / 魚で〜する suffer from fish poisoning; get poisoned by fish. / 食べた魚で〜した. The fish I ate disagreed with me [upset my stomach]. / 〜症状を呈する develop [present, show] toxic symptoms [symptoms of poisoning]. / 食〜になる get food poisoning. / 夏は食〜の季節である. Food poisoning is most common in summer. / 病院に収容された食〜患者 a hospitalized victim of food poisoning. / 食〜症状を訴える complain of (having) symptoms of food poisoning.

中皮腫(ちゅうひしゅ) mesothelioma. ¶ **悪性胸膜[心膜, 腹膜]中皮腫** (a) malignant pleural [pericardial, peritoneal] mesothelioma.

潮間帯 the intertidal zone [region]. ▶〜の岩 intertidal rock.

調査 (an) investigation; (an) examination; (an) inquiry; (a) survey; research; stocktaking; a checkup. ★⇨捕鯨（調査捕鯨）▶〜する investigate; examine; inquire [look, search, go] into…; survey; make an investigation [inquiry]《into…》; probe. / ツシマヤマネコの生態を〜する investigate the ecology of the Tsushima cat. / 事前環境[環境事前]〜を行う conduct a preliminary environmental survey. ¶ **調査船** 〔海洋学の〕 a research vessel. **学術調査** a scientific [an academic] investigation. **現地調査** a field survey [study, investigation]; fieldwork. ★⇨環境（環境対策）/ 現地〜をする conduct a field survey [an on-site study]; do [perform] fieldwork. / 現地〜班 a field party. **全国調査** a nationwide survey [investigation]; a survey covering the whole country.

鳥獣 birds and animals [beasts]; wildlife. ¶ **鳥獣保護** wildlife conservation [preservation]. **鳥獣保護区** a wildlife sanctuary. **鳥獣保護法** the Wildlife Protection and Hunting Law. ★正式名称は「鳥獣の保護及び狩猟の適正化に関する法律」. **鳥獣輸入証明書** a bird and animal [veterinary] import certificate.

潮汐(ちょうせき) 〔干満〕 ebb and flow; 〔潮〕 a tide. ★⇨潮(しお) ¶ **潮汐エネルギー** tidal energy **潮汐波** a tidal wave. **潮汐発電** tidal power generation. **潮汐発電所** a tidal power plant. **潮汐表** a tide table.

超低周波騒音 ultralow-frequency noise.

眺望(ちょうぼう) a view; a prospect; a lookout. ★⇨景観 ¶ **眺望権** a [the] right to a view. **眺望保全** preservation of a view ▶ホテルから見える富士山の〜保全 preserving the hotel's view of Mount Fuji.

潮力 tidal energy [power]. ¶ **潮力発電** tidal power generation. **潮力発電所** a tidal-powered electric plant [station].

鳥類 birds; fowls. ★⇨鳥獣, 野鳥, 鳥 ¶ **鳥類学** ornithology; birdlore. **鳥類学者** an ornithologist. **鳥類相**(そう) the avifauna [bird fauna] 《of a region》.　**鳥類標識法**　bird banding; bird ringing. ★⇨足輪(あしわ)　**鳥類保護**　bird protection.　**鳥類保護区**　a bird sanctuary; bird reserve.

直接処分場　〔使用済み核燃料の〕a direct disposal site [facility].

貯水　water storage; storage of water. ★⇨ダム　¶**貯水槽**　a (water [water storage]) tank; a cistern;〔トイレなどの〕a flush tank.　**貯水組織**　〔植物の〕(water) storage tissue.　**貯水池**　a (water) reservoir. ▶～池のダム　a storage [reservoir] dam. /～池の水量がかなり減った．The reservoir has sunk considerably below its usual level. / 補充～池　a compensation [supplementary] reservoir.　**貯水塔**　a water tower.　**貯水率**　a [the] water storage rate; the ratio of reservoir water to demand; water availability.　**貯水量**　water reserves [supplies]; the volume [amount] of water available (for use).

貯蔵プール　〔使用済み核燃料の〕a spent fuel storage pool.

地理的表示　〔原産地名がついた商品表示〕a geographical indication.

地理的品種　a geographic race.

沈水植物　a submerged plant.

沈黙の春　⇨英和 silent spring

つ

追加性 ⇨英和 additionality

通気 ventilation;〔空気にさらすこと〕airing; aeration. ¶通気管 an air pipe. 通気孔 a vent (hole). 通気坑 a ventilation shaft;〔鉱山の〕an air pit. 通気性 (air-)permeability. ▶〜性がよい easily permeated. / 〜性のよい建物 an air-permeable building. / このシャツの素材は〜性がいい[悪い]. This shirt is made of material that breathes [doesn't breathe]. / 〜性のある breathable 《cloth, fabric, leather, material》; poromeric; permeable. 通気性防水 porous waterproofing. 通気組織〔植物の〕an aerenchyma; a pneumatic system.

通風 ventilation; airing; draft. ▶〜のよい[悪い]部屋 well-[ill-]ventilated room. ¶通風管 an air pipe [line]; a vent pipe; a ventiduct. 通風乾燥機 a grain drier. 通風器 a ventilator; an aerator. 通風権 the right to ventilation. 通風坑〔鉱山の〕an air pit. 通風口〔旅客機の座席上の〕a fresh air inlet. 通風孔 a draft [ventilating] hole; a ventilation opening [hole, (四角の) square]; a funnel; a vent. 通風湿度計 aspiratory psychrometer. 通風設備 a ventilation system. 通風装置 a ventilation arrangement [device, apparatus];〔ストーブ・炉などの〕a register. 通風塔 a ventilation tower. 通風筒 an air duct; a ventilation duct; a vent.

使い捨て ▶〜の disposable; throwaway; single-use; one-time-use. / 〜の紙おむつ a disposable diaper [nappy]. ★⇨紙 / 〜の時代 the "throwaway age"; an age in which people discard things that are still useable. ¶使い捨て型ロケット an expendable launch vehicle (略: ELV). 使い捨てかみそり a disposable razor. 使い捨てカメラ a single-use [disposable, throwaway] camera. 使い捨てコンタクト・レンズ disposable contact lenses. 使い捨て注射器 a disposable syringe. 使い捨て文化 the "throwaway culture." 使い捨てライター a disposable [throwaway] (cigarette) lighter.

つけっぱなし ▶テレビを〜にする leave a TV on. / 帰宅したらテレビが〜だった．When I got home the TV was still on. / エアコンが〜になっている．The air conditioner is still on. ★⇨消費(消費量) / 外灯が〜だ．The outside light has been left on [hasn't been turned off]. / 電気を〜で眠る sleep with the light on.

ツボカビ症 〔両生類の新興感染症〕chytridiomycosis; amphibian chytrid fungus disease.

詰め替える refill. ▶詰め替え可能な容器 a refillable container. / 詰め替え用の洗剤 a detergent refill.

釣り糸 a fishing line; a fishline. ▶〜が足に絡んだ鳥 a bird whose legs are tangled in a fishing line.

ツンドラ 〔凍土帯〕the tundra. ¶**ツンドラ気候** 〔寒帯気候の一〕tundra climate.　**ツンドラ地帯** a [the] tundra area.

て

低硫黄(ていいおう) ¶**低硫黄ガソリン** (a) low-sulfur gasoline（略：LSG）. ▶**超〜ガソリン** (an) ultra-low-sulfur gasoline（略：ULSG）. **低硫黄軽油** low-sulfur diesel (fuel). / **超〜軽油** ultra-low-sulfur diesel (fuel)（略：ULSD）.

ディーゼル ¶**ディーゼル・エンジン** a diesel engine [motor]; a diesel. **ディーゼル・カー[機関車]** a diesel(-electric) locomotive; a diesel(-electric). **ディーゼル車** a diesel (car); a diesel(-engine) car. **ディーゼル(車)規制** diesel (vehicle) restrictions [regulations]. ▶**首都圏〜車規制条例** the Regulations for Diesel-Powered Vehicles in the Environmental Security Ordinance. **ディーゼル排気微粒子** diesel exhaust particles（略：DEP）. **ディーゼル(排気)微粒子除去装置** a diesel particulate filter（略：DPF）. **ディーゼル・ハイブリッド車** a diesel hybrid (car [vehicle]). **ディーゼル粉塵[煤塵, 微粒子]** diesel particles [soot]. **ディーゼル油** diesel oil [fuel].

ディープ・エコロジー 〔長期的な環境保護〕deep ecology. ¶**ディープ・エコロジスト** deep ecologist.

低環境負荷 a low environmental load [impact, burden]. ★⇨環境(環境負荷) ¶**低環境負荷エネルギー** low-environmental-impact energy. **低環境負荷エンジン** a low-environmental-impact engine. **低環境負荷(型)社会** a low-environmental-impact society; a society with a low impact on the environment. **低環境負荷技術** low-environmental-impact technology. **低環境負荷製造** low-environmental-impact manufacturing.

定期安全レビュー 〔電力会社が, 原発設備の信頼性などについて約10年ごとに総合評価する〕《conduct》a periodic safety review（略：PSR）.

定期事業者検査 〔原発設備に対して電力会社が行う〕(a) regular inspection《of a nuclear power station》by a power company.

締結 ▶**…と条約を〜する** conclude [enter into] a treaty

with … ★⇨条約　¶締結国〔条約などの〕a contracting state [power, party, country]. / 国連海洋法条約の〜国 a contracting party to the United Nations Convention on the Law of the Sea.

低公害　¶低公害エンジン〔有毒物質排出量の少ない〕a clean [low-emission, low-pollution] engine.　低公害車　a low-emission vehicle（略：LEV）; a low-pollution car.　超低公害車　an ultra-low-emission vehicle（略：ULEV）. ★⇨公害, 無公害

底質〔水底の表層にある物質〕bottom material.

低周波　low frequency（略：L.F., LF, l.f.）. ¶低周波空気振動[騒音] low-frequency air vibration [noise].　低周波公害　low-frequency pollution.

定常走行騒音　the noise of a car running at a constant speed.

底生生物　a benthic [benthonic] organism;〈集合的に〉benthos. ¶底生生物相　(a) benthic [benthonic] biota.　底生動物　a benthic [benthonic] animal;〈集合的に〉zoobenthos.　底生動物相　(a) benthic [benthonic] fauna; (a) bottom fauna.

低騒音　low noise. ¶低騒音化　noise reduction.　低騒音舗装　low-noise pavement.

低層湿原　(a) low moor.

低炭素社会　a low-carbon society. ★⇨英和 low-carbon society

定点　¶定点観測　weather observations at a specific point at sea; fixed point observation.　定点観測船　a ship weather station; an ocean station vessel; an ocean weather ship.　定点調査　(a) fixed-point observation [survey, research].

泥土(でい)　¶建設泥土　construction-site mud; earth removed from a construction site.　浚渫(しゅんせつ)泥土　dredged(-up) mud.　廃泥土　waste mud. ★⇨汚泥

低燃費　low fuel consumption. ★⇨燃費 ▶自動車の〜化 increasing fuel efficiency in cars; making cars more fuel-efficient. ¶低燃費車　a fuel-efficient car.　低燃費走行　fuel-efficient driving.

低濃縮ウラン　slightly enriched uranium.

低農薬　¶低農薬栽培　cultivation using small amounts of agricultural chemicals.　低農薬米[野菜]　rice [vegetables] grown with small amounts of agricultural chemicals. ★⇨無農薬

低排(出)ガス車　a low-emission vehicle（略：LEV）. ¶低排出ガス車認定制度　a certification system for low-emission vehicles; an LEV certification system.

ディルドリン 〔有機塩素系の農業用殺虫剤〕dieldrin.

デオキシリボ核酸 ⇨ DNA(和英篇末尾)

適応 (an) adaptation. ★⇨順応 ▶～する adapt; adjust; adapt [adjust] *oneself*《to...》; acclimatize; acclimate. / 環境に～する adapt to the [an] environment. / 樹上の生活に～した猿の手足 monkey paws that are adapted to life in the trees; the paws of the monkey, which are adapted to arboreal existence. / 寒さに～して彼らの毛は長くなった．Their hair lengthened in adaptation to the cold. | They developed long hair as an adaptation to the cold. / この突然の環境の変化に～しえたものだけが生き残った．Only those who could adapt to this sudden change in the environment survived. | The only forms of life that survived these sudden environmental changes were those able to adapt [capable of adaptation]. ¶**適応機制** an adaptive [adjustment] mechanism. **適応形質** an adaptative characteristic; an adaptation. **適応策** 〔地球温暖化による影響への対策の１つ〕(climate change) adaptation strategies. **適応進化** adaptive evolution. **適応戦略** an adaptive strategy. **適応放散** adaptive radiation.

適合 〔適応〕adaptation; accommodation. ▶基準に～する ⇨ 環境(環境基準) / 生息環境に～した体型 a (physical) shape suited to *one's* habitat. ¶**(群落)適合度** fidelity.

デシベル a decibel (略: dB, db). ¶**デシベル数** a decibel count. **デシベル測定器, デシベル計** decibeloscope

撤去 removal. ▶～する remove; take away. / 幾年も野ざらしになっていた古タイヤがやっと～されることになった．At last they are going to dispose of the old tires that have been lying out in the open for years and years. / 駐輪禁止区域にとめてある自転車は～します．〔掲示〕We will remove any bicycles parked in the No Bicycle Zone. ★⇨放置(放置自転車) / 違法な建物を～してそこに小さな公園を造った．They removed the illegal building and turned the area into a small park. ¶**撤去費(用)** the removal [evacuation] cost. ★⇨解体(解体撤去) / 座礁した船の～費用はどこから出

るのか. Where are the funds to dismantle the grounded ship going to come from?

鉄道騒音 train [railway, railroad] noise.

テトラクロロエチレン ⇨英和 tetrachloroethylene

デポジット制度 〔空缶・空瓶のリサイクルシステム〕a deposit system.

添加 addition. ▶~する add. ¶**添加物** an additive. / 当店の製品には~物は一切用いていません. Our food products contain no additives. ★⇨無添加 **添加物表示** additive labeling. **既存添加物** 〔広く使われてきたため例外規定で使用を認められる,天然の食品添加物;食品衛生法に基づく分類〕an existing food additive. **指定添加物** 〔国により使用が認められた食品添加物;食品衛生法に基づく分類〕a designated food additive. **食品添加物** a food additive. **飼料添加物** feed additives.

転換 conversion; a switch; a turnabout; a flip-flop. ★⇨ウラン(ウラン転換) ▶~する convert; switch. / 産業構造を~する switch [convert] to a new industrial structure. / 火力発電から原子力発電への~ a conversion [switch] from thermal power generation to atomic power generation. / 米作から果樹栽培への~ shifting from rice growing to fruit cultivation. / 軍需産業の平和産業への~ the conversion of the munitions industry to peacetime industry; the switching from war industries to peace industries. ¶**転換期間中有機農産物** transitional organic farm products. **転換作物** a replacement crop. **転換比[率]** 〔核分裂の〕a conversion ratio [factor]. **転換炉** a converter [convertor] (reactor).

電気 electricity;〔電灯〕a [an electric] light. ★⇨電力,電源,発電,充電,送電 ▶大量の~を消費する consume a large amount of electricity. ★⇨消費(消費電力) / あまり~を食わない調理器具 cooking utensils that do not eat up a lot of electricity. ★⇨省エネ,節電 / ~を節約する economize on [save] electricity. / ~は小まめに消すようにしなさい. Be (more) careful to (always) turn off the lights. ★⇨つけっぱなし ¶**電気エネルギー** electric energy ★⇨エネルギー(エネルギー消費) **電気事業** the electricity business. / ~事業者 a [an electric] power company. / 卸(おろ)~事業 (the business of) supplying electricity to electric utility companies; wholesale electricity [power] provision. / 卸～事業者 an electricity wholesaler; an electricity producer which supplies electricity to electric utility companies. / 特定規模～事業者 a power producer and supplier. **電気自動車** an electric car [vehicle (略: EV)]; a battery car. / ～自動車対応型の EV-compatible; fitted for [suitable for] EV [electric vehicles]. **電気自動車対応住宅** housing with (domestic [residential]) electric vehicle [EV] charging equipment. **電気集塵** electric [electrostatic] precipitation. **電気集塵機[器]** an electrostatic (dust) precipitator; 〔採取用の〕an electrostatic sampler. **電気保安協会** an electrical safety inspection association. / 関東[関西(など)]～保安協会 the Kanto [Kansai, etc.] Electrical Safety Inspection Association. **電気用品安全基準** ⇨ PSE (和英篇末尾) **電気用品安全法** 〔電気用品安全基準を定める〕the Electrical Appliance(s) and Material(s) Safety Law.

電源 〔機器への電力供給源〕a power source; a source of electricity [electrical power]; 〔水力発電の資源〕hydroelectric resources. ▶地震の時には～が切れる. When an earthquake strikes the power (supply) cuts off. ¶**電源開発** the development of electrical power. **電源開発株式会社** Electric Power Development Co., Ltd.; 〔略称〕J-POWER. **電源開発促進税** an electric power development promotion tax. **電源開発促進税法** the Electric Power Development Promotion Tax Law. **電源開発促進対策特別会計** the Special Account for Electric Power Development Promotion. **電源開発促進対策特別会計法** the Special Account Law for Electric Power Development Promotion. **電源三法** the three laws relating to electric power development. ★電源開発促進税法，電源開発促進対策特別会計法，発電用施設周辺地域整備法の三法. **電源三法交付金** grants-in-aid and subsidies under the three electric power development laws. **電源地域** a power supply region. **電源立地地域対策交付金** a government grant for the area where a power plant is located. **3 電源方式** 〔AC 電源・電池・カーバッテリーの３つなどを利用できる〕three-way power supply. **分散型電源** 〔ビルや工場の自家用

発電設備〕a dispersed [decentralized] power source.

天災 a natural calamity [disaster];（不可抗力）an act of God. ★⇨人災 ▶〜による損害 damage [losses] caused by a natural disaster. /〜は忘れたころにやってくる．A natural disaster visits us when we're least thinking about it. | Disasters happen when they're least expected.

展示 ¶**形態展示**〔動物園の〕exhibiting [displaying]《animals》so that their form can be easily seen. **行動展示** ⇨行動 **生態展示**〔動物園の〕(an) ecologically realistic display; exhibiting《a species》in a realistic ecological environment.

電子ごみ〔廃棄されたハイテク電気製品〕electronic waste; e-waste; hi-tech waste. ★⇨英和 WEEE

電磁波 electromagnetic waves;〔通信〕a radio wave. ¶**電磁波過敏症** electromagnetic hypersensitivity. **電磁波シールド材** an electromagnetic shielding material.

伝統的建造物群保存地区〔文化財保護法上の〕a historic buildings preservation district.

天然 ⇨自然 ▶〜系化粧品 cosmetics made of natural ingredients. / 原料は100%〜系です．The ingredients are 100% natural. ¶**天然アスファルト** natural asphalt. **天然鮎** a wild *ayu*. **天然ウラン** natural uranium; naturally occurring uranium. **天然栄養素** a natural nutrient. **天然塩** natural salt; native salt. **天然温泉** ⇨温泉 **天然化合物** a natural compound. **天然ガス** natural gas. /〜ガス自動車 a natural gas-powered[-engine] car; a car [a motor vehicle, an auto] powered by natural gas. /〜ガスの埋蔵量 natural gas reserves [deposits]. / 相当な埋蔵量の〜ガスが発見された．Substantial natural gas deposits were discovered. / 圧縮〜ガス compressed natural gas（略：CNG）. / 圧縮〜ガス車〔CNG 車〕a CNG car [vehicle]. / 液化〜ガス ⇨LNG（和英篇末尾）/〜ガス化学工業 the natural gas chemical industry. /〜ガス田［埋蔵地］a natural gas field. /〜ガス・コジェネレーション natural gas cogeneration. **天然ガス対策法** Natural Gas Processors Association;〔米国の〕Natural Gas Policy Act（略：NGPA）. **天然ガソリン** natural gasoline; natural gas liquids（略: NGL）． **天然記念物** a natural monument;〔動〔植〕物〕a protected species of animal [plant];〔動〔植〕物の生息地〕a protected habitat;〔地質〕a protected geologic formation. / 国はその桜を〜記念物に指定した．The Government has designated that particular cherry tree as a natural monument. / 特別〜記念物 a special natural monument;〔動物〕a species of animal [habitat] designated for special protection. / 指定〜記念物〔国または地方自治体指定の〕an animal [a bird] designated as a protected species. / 特別史跡名勝〜記念物〔文化財保護法で指定する〕a special historic landmark, battleground or natural monument. / 国〔県，市〕指定〜記念物 an animal [a bird, a plant] designated as a nationally [prefecturally, municipally] protected species. / 〜記念物指定区域 a (designated) protected area [habitat]． **天然魚** a wild fish. **天然現象** a natural phenomenon. **天然原子炉** a natural reactor. **天然港** a natural harbor. **天然更新** natural regeneration. **天然酵母** natural yeast. / 〜酵母パン natural yeast bread. **天然香料** a natural scent [perfume];〔食品衛生法に基づく分類〕a natural flavoring agent. **天然ゴム** natural rubber. **天然材** natural [native] wood. **天然色素** natural color [dye, pigment]． **天然資源** natural resources. ★⇨資源 **天然資源保護評議会[協議会]**〔米国の〕the Natural Resources Defense Council（略: NRDC）． **天然磁石** a natural magnet. **天然樹脂** (a) natural resin. **天然植物油** (a) natural vegetable oil. **天然真珠** a natural pearl. **天然水** natural water. **天然成分** a natural ingredient. **天然石** natural [native] rock. **天然繊維** a natural fiber [fabric]． **天然染料** a natural dye. **天然素材**〔衣類・家具などの〕(a) natural material;〔化粧品・食品などの〕a natural ingredient. **天然ダム** a natural dam. **天然着色料** a natural colorant [coloring agent]． ★⇨人工(人工着色料), 合成(合成着色料) **天然調味料** a natural flavoring [seasoning]． **天然添加物** a natural additive. **天然バリア**〔放射性廃棄物埋設地の岩盤などが形成する障壁〕a natural barrier. **天然肥料** a natural fertilizer. **天然木**〔樹木〕a native tree;〔材木〕natural [native] wood. 天

然物 〔魚〕a wild fish;〔野菜〕a wild vegetable. / 養殖ものの魚もいいが，やはり旬の〜ものが一番だ. Farmed fish are okay, but nothing is better than wild fish in season. / このワラビは〜ものだ. This is wild bracken. | This bracken was harvested in the wild. ★⇨栽培(栽培物)，養殖(養殖物)　**天然林**　a natural forest.

10-15(テンフィフティーン)**モード**　〔燃費・排出ガスなどを算定するための走行パターン〕the 10-15 mode cycle. ¶**10-15モード燃費**　10-15 mode fuel efficiency.

電力　electric(al) power [energy]; electricity; power. ★⇨電気，消費(消費電力)，待機電力　▶〜の浪費[節約]　waste [economy] of electric power. ¶**電力卸**(おろ)**売り**　⇨電気(卸電気事業)　**電力会社**　a [an electric] power company; a [an electricity] generating company.　**電力開発**　electric power development. ★⇨電源(電源開発)　**電力管理**　control of electric power; power control.　**電力危機**　electric power crisis.　**電力供給**(**量**)　(electric) power [electricity] supply.　**電力計**　a wattmeter.　**電力小売り**　electric power retailing; the retail sale of electric power. / 〜小売りの自由化　the deregulation [liberalization] of electric power retailing.　**電力小売り事業者**　a retail (electric) power supplier.　**電力債**　a power company bond.　**電力事情**　(electric) power conditions.　**電力自由化**　electric power deregulation.　**電力需要**　(electric) power [electricity] demand.　**電力消費**[**使用**](**量**)　(electric) power [electricity] consumption; an amount of electricity [power] used [consumed].　**電力貯蔵**　(electric) power [electricity] storage.　**電力低下**　a fall in [a drop in, a lowering of] (electric) power.　**電力取引所**　an electric power exchange.　**電力ピーク**　(the) peak load.　**電力不足**　an (electric) power shortage; a short supply of electricity. / 〜不足の地域　power-short areas.　**電力割当制**　the electric power allocation system.　**工業用電力**　industrial electric power.　**深夜電力**　late-night [nighttime] (electrical) power.

と

トイレ a toilet. ★⇨紙, 環境(環境トイレ), 再生(再生紙), 水洗, ポータブルトイレ, 節水(節水効果) ¶**自己処理型トイレ** 〔汚物を外に流さず内部で処理するトイレ〕a self-treating toilet; a non-discharging toilet. **トイレカー** a (mobile) toilet vehicle.

等価感覚騒音レベル equivalent continuous perceived noise level (略: ECPNL). ★⇨英和 WECPNL

投棄 abandoning; dumping; disposal. ★⇨廃棄, 放置, 海洋(海洋投棄), 不法(不法投棄) ▶〜する discard; dispose of; abandon; give up; throw [cast] away; dump. / 産廃〜による土壌汚染は年々深刻さを増している. The severity of soil pollution brought about by disposal of industrial waste is growing by the year.

冬期湛水(たんすい)水田 ⇨湛水

透水性舗装 porous [water permeable] pavement.

頭数管理 《animal》numbers management [control]; 〔家畜の〕herd size management [control].

淘汰(とう)(た) 〔種の〕selection. ★⇨自然(自然淘汰), 人為(人為淘汰), 雌雄(しゆう)(雌雄淘汰)

同定 〔生物の属・種の決定〕identification. ▶〜する identify.

動物 an animal; a (living) creature; 〔獣〕a beast; a brute; (一地方分布の) fauna;〈総称的に〉animal life. ★⇨鳥獣, 草食, 肉食, ペット, 生物 ▶〜にえさを与えないでください. 〔掲示〕Please don't feed the animals. / 〜の権利 animal rights. ★⇨英和 animal rights / 実験〜⇨動物実験 / 下等[高等]〜 the lower [higher] animals;〔個々の〕a lower [higher] animal. / 二足[四足]〜 a biped [quadruped]. ¶**動物愛好家** an animal lover. **動物愛護** ⇨愛護 **動物愛護協会** the 《Kanagawa》Society for the Prevention of Cruelty to Animals. ★⇨英和 SPCA, HSUS **動物愛護週間** Be Kind to Animals Week. **動物愛護団体** an animal welfare group; a society for the prevention of cruelty to animals. **動物愛護の日** Be Kind to Animals Day. **動物愛護[福祉]論者** animal

動物

welfarist. **動物愛護(管理)法** the Animal Protection Law. ★正式名称「動物の愛護及び管理に関する法律」(the Law Concerning the Protection and Control [Law for the Humane Treatment and Management] of Animals). **動物遺伝学** animal genetics. **動物医薬品検査所** National Veterinary Assay Laboratory. **動物慰霊祭** a spirit-consoling service for animals. **動物衛生研究所** 〔農水省の〕the National Institute of Animal Health (略: NIAH). **動物園** a zoo; zoological gardens. / 恩賜上野〜園 Ueno Zoological Gardens;〔略称〕Ueno Zoo. / 〜園で生まれたヒョウ a zoo-born leopard. / 〜園の園長 the curator of a zoo. / 〜園のライオン a (caged) lion at the zoo; a zoo-kept lion. / 〜園の動物 a zoo animal. **動物介在活動** 〔介護施設などでの〕(an) animal-assisted activity (略: AAA). ★⇨動物療法 **動物学** zoology. / 〜学的な[上の] zoological. / 〜学的に見て from the zoological point of view. **動物学者** a zoologist. **動物看護師** an animal health technician (略: AHT). **動物感染症** 〔動物間の感染症〕an animal infectious disease;〔動物がヒトに媒介する感染症〕=動物由来感染症. **動物季節観測** phenological observation of animals. **動物虐待** animal cruelty; cruelty to animals. **動物嫌い** a dislike [hatred] of animals;〔病的な〕zoophobia;〔人〕a person who dislikes animals; an animal hater. **動物形態学** animal morphology. **動物検疫** animal quarantine. **動物検疫犬** an animal-products sniffer dog; a customs dog trained to sniff out animal products. **動物原性感染症** ⇨人獣共通感染症. **動物公園** a zoo; a zoological park; a game park; a safari park. / 自然〜公園 a wildlife park. **動物考古学** arch(a)eozoology. **動物行動学** ethology. **動物行動学者** an ethologist. **動物小屋** 〔研究所などの〕an animal house. **動物細胞** an animal cell; a zooblast. **動物誌** a fauna; zoography. **動物磁気** animal magnetism; zoomagnetism. **動物試験** a biological test. / 医薬品の〜試験を行う subject a medicine to a biological test. **動物質** animal matter. **動物実験** an animal experiment [test]; an experiment on [using] an animal [animals];〈総称的に〉animal testing [experimentation]. / 〜実験をやめさせる会 an association [a society, a group] for stopping animal experiments; an association to halt experiments on animals. ★⇨英和 animal rights / 〜実験を行う experiment [conduct experiments, do tests] on animals. / その新薬の効果は〜実験では確かめられているが，実際の応用は未知数である．The efficacy of the new drug has been confirmed by experiments on animals, but its practical application remains an unknown quantity. / 実験〜 a laboratory animal; an experimental [a test] animal. ★⇨モルモット / 実験〜慰霊祭 a service for the (repose of the) souls of laboratory animals. / 〜実験代替法 alternatives to animal experimentation. **動物社会** animal society. **動物社会学** animal sociology. **動物写真家** an animal photographer. **動物受粉植物** a zoophilic [zoophilous] plant; a zoophile. **動物商** an animal dealer;〔愛玩動物の〕a pet dealer. **動物飼養場** 〔自然の棲息状態に近くした〕a vivarium. **動物心理学** animal psychology; zoopsychology. **動物生態学** zoo-ecology. **動物生理学** animal physiology. **動物生理学者** an animal physiologist. **動物占**有者 the possessor [owner] of an animal. **動物相** fauna. / 沿岸[海産]〜相 littoral [marine] fauna. **動物測定学** zoometry. **動物組織** animal tissue. / 〜組織の人体移植 zoografting. **動物地理(学)** zoogeography; zoography; animal geography; geographical zoology. **動物地理区** a faunal region [area]; a zoogeographic(al) region. **動物塚** an animal burial mound. **動物電気** animal electricity. **動物取扱業者** a business dealing with animals; an animal-related business; an animal dealer. **動物取扱責任者** a person responsible for (dealing with) animals. **動物病院** a veterinary [an animal] hospital. **動物プランクトン** zooplankton. **動物分類学** zootaxy; zoological taxonomy. **動物保険** animal insurance;〔家畜の〕livestock insurance;〔ペットの〕pet insurance. **動物保護センター** an animal shelter; a pound. **動物油脂** animal oils and fats. **動物由来感染症** ⇨人獣共通感染症 **動物用医薬品** a drug for animals; an animal drug; an animal medicine; a veterinary product; a veterinary medicine. **動物力学** zoodynamics. **動物(介在)療法**

透明度 transparency; the degree of transparency [clarity, clearness]. ▶〜度が高い be highly transparent; have a high degree of clarity [transparency]. / この湖は最近〜度が増してきた. The transparency of this lake has improved lately. / 摩周湖は〜度が日本一だ. Lake Mashu is the best in Japan as far as water clarity is concerned.

道路 a road; a thoroughfare;〔街路〕a street;〔街道〕a highway. ★⇨高速(高速道路) ▶〜の妨害 obstruction of traffic [the public road]. / 〜の補修 road repairs [maintenance]. / 〜を開く[ふさぐ] open [block] a road. / 〜を作る construct [build, make] a road. / 〜を掘り返す dig [tear] up a road. ¶**道路安全システム** the Advanced Highway Safety System(略: AHSS). **道路位置指定** designation [adoption] of a private road as a public road. **道路運送業** the road transportation [transport] industry. **道路運送法** the Road Transportation Law. **道路延長** extension of a road; (a) road extension. **道路改正** street improvement. **道路陥没** road subsidence. **道路管理者** a road authority [administrator]. **道路橋** a road bridge. **道路行政** road administration. **道路行政マネジメント** road administration management. **道路局**〔国土交通省の〕Road Bureau. **道路計画** a road plan. **道路建設** road building; street [road] construction. **道路建設機械** road-building equipment. **道路建設技師** a road engineer. **道路建設計画** a road-building program. **道路元標** a milestone; a marker [milestone] at the start and end of a route, and at important points on it. **道路公害** road (traffic) pollution. **道路公債** a public bond [loan] for the improvement of roads. **道路工事**〔建設〕road construction [building]; roadmaking;〔修理〕road repairs; road repair (work); road-mending. **道路工事人** a road construction worker; a roadman;〔補修時の〕a road-mender;〔組〕a road (repair) crew; a highway gang. **道路交通** road traffic. **道路交通渋滞対策** policies [a program] to alleviate traffic congestion. **道路交通情報** traffic information. **道路交通情報通信システム** the Vehicle Information and

Communication System (略: VICS). **道路交通センサス** 〔3-5年間隔で実施される全国道路・街路交通情勢調査〕the Road Traffic Census. ★正式名称は「全国道路・街路交通情勢調査」. **道路交通取締法** the Road Traffic Control Law. **道路交通法** the Road Traffic Law. **道路作業員** a road construction worker; a roadman. **道路事情[状況]** road [traffic, driving] conditions. **道路情報** road [traffic] information; (information on) traffic [road] conditions. **道路照明** road [roadway, highway] lighting. **道路清掃車** a street sweeper [cleaner]. **道路整備** road maintenance and improvement. **道路整備緊急措置法** the Law on Emergency Measures for Road Improvement. **道路整備計画** a road maintenance and improvement project. **道路整備費** expenditure for road maintenance and improvement; road maintenance costs. **道路整備事業** a road maintenance and improvement project. **道路整備事業費** (the) costs of [expenditure on] a road maintenance and improvement project. **道路整備特別会計** a road improvement special account. **道路掃除** street cleaning. **道路法** the Road Law. **道路舗装** (a) road pavement;〔舗装すること〕road paving; paving roads. **道路舗装会社** a road paving company. **道路率** 〔道路面積の割合〕a road to area ratio; a ratio of paved to total area.

ドギー・バッグ 〔レストランなどの食べ残し持ち帰り袋〕a doggy bag.

毒 〔毒物〕(a) poison; a poisonous substance;〔毒薬〕poison; a toxicant;〔毒液〕venom;〔毒素〕a toxin. ★⇨中毒, 有毒 ▶ふぐの毒にあたる be poisoned by eating globefish. / 毒のある poisonous 《mushrooms》; toxic 《substance》; virulent 《toxins》; noxious 《fumes》 ¶ **毒性** toxicity. ★⇨急性, 慢性, 遺伝(遺伝毒性)/ 複合毒性 mixture toxicity; combined toxicity. **毒性学** toxicology. / 環境毒性学 environmental toxicology / 食品化学毒性学 food and chemical toxicology. **毒性緩和** safening; detoxification. **毒性緩和剤** a safener. **毒性基準** a toxicity level; (a level of) toxicity. **毒性試験** toxicity test. **毒素型食中毒** toxic type [toxin-caused] food poisoning. **毒素中和反応** toxin-neutralization reaction. **毒物劇薬取扱者** an authorized handler of toxic substances. **毒物及び劇物取締法** the Poisonous and Deleterious Substances Control Law ★⇨英和 ATSDR

特殊栄養食品 special nutritive foods.

特殊環境生物 organisms in extreme environment.

特定 ⇨指定, 飲用 ¶**特定汚染源** ⇨英和 point source pollution

特定外来生物(被害防止法) ⇨外来

特定化学物質 a specified [designated] chemical substance.

特定化学物質障害予防規則 the Ordinance on Prevention of Hazards due to Specified Chemical Substances. **特定感染症指定医療機関** 〔厚生労働大臣が指定する〕a medical institution (designated by the Ministry of Health and Welfare) for specified infectious diseases.

特定危険部位 〔狂牛病の危険性の高い, 牛の部位〕specified risk material (略: SRM). **特定原材料** 〔食品衛生法で, 使用表示を義務づけられている〕legally specified ingredients (which may be allergenic). **特定市街化区域** a designated urbanization-promotion area [zone]. **特定農薬** 〔農薬取締法で人畜・水産動植物に無害と指定された農薬〕a minimum-risk [low-risk] pesticide. **特定胚** a specific [specified] embryo. **特定物質** 〔使用が規制される〕a controlled substance. ★⇨オゾン(オゾン層保護条約) **特定フロン** ⇨フロン **特定防除資材** ⇨特定農薬. **特定保健指導** 〔メタボリック症候群についての〕a metabolic syndrome-specific health guidance. **特定保健用食品** (a) designated health food. **特定保守製品** 〔経年劣化で重大事故を起こす可能性が高い家庭用器具など〕a product with a specified [statutory] maintenance requirement. **特定有害物質使用制限指令** ⇨英和 RoHS Directive **特定用途制限地域** 〔住環境を損なう可能性のある風俗店・大型店舗などの建設が制限される区域〕a prohibited use (zoning) district [area].

特別 ¶**特別管理産業廃棄物** industrial waste requiring special management [treatment]; specially managed industrial waste. **特別栽培** 〔化学肥料や農薬を通常の50％以上削減して農産物を栽培すること〕agriculture employing less than 50% of normal chemical fertilizers or pesticides. **特別栽培農産物** an agricultural product grown with less than 50% of the normal quantity of chemical fertilizers or pesticides.　**特別(史跡名勝)天然記念物**　⇨天然(天然記念物)　**特別措置法**　a special measures law. ★⇨湖沼(こしょう)(湖沼水質保全特別措置法)　**特別地域**　〔国立公園内などの保護区分の1つ〕a special zone.　**特別保護建造物**　specially preserved buildings.　**特別保護地区**　〔国立公園内などの保護区分の1つ〕a special protection area.　**特別用途食品**　special purpose food.

都市　a city; a town. ▶~の municipal; urban; city. / 人口の~集中傾向 the tendency of populations to move toward urban areas. ★⇨人口 / 巨大~ a megalopolis; a megapolis. / その派手な色のビルは~景観を損ねている．That garishly-colored building mars the cityscape [spoils the view of the city]. ★⇨景観　¶**都市衛生**　urban sanitation.　**都市化**　urbanization. / ~化する urbanize.　**都市改造**　urban renewal.　**都市開発**　urban development.　**都市開発区域**　an urban development zone [area].　**都市改良**　civic improvement.　**都市ガス**　city [town, municipal] gas.　**都市型ケーブルテレビ**　〔農村型に対して〕urban cable television [TV].　**都市(型)災害**　an urban natural disaster.　**都市型社会**　an urbanized society.　**都市型住宅**　〈集合的に〉urban-type housing.　**都市型水害**　urban flood damage.　**都市型犯罪**　(an) urban crime.　**都市環境学**　urban environment(al) studies [science(s)].　**都市気候**　an urban climate.　**都市基盤**　the urban infrastructure.　**都市居住**　living in a city.　**都市居住者**　an urban resident.　**都市行政**　municipal administration.　**都市計画**　urban planning; city planning; town planning. / この~計画には住民の考えが織り込まれていない．Residents' views have not been considered [taken into account, factored in] in this city-planning project. / ~計画の専門家 an urbanist. / ~計画の変更 alteration of the city plan. / ~計画を施行する carry out a city plan.　**都市計画委員会**　the City Planning Committee.　**都市計画課**　the City Planning Section.　**都市計画区域**　a city planning area.　**都市計画事業**　a city planning project.　**都市計画施設**　〔道路・公園など〕city planning facilities.　**都市計画審議会**　the City Planning Council.　**都市計画税**　city planning tax.　**都市計画制限**　〔都市計画のための私権制限〕city planning restrictions.　**都市計**

都市

画法 the City Planning Law. **都市圏** an urban area. **都市公園** a municipal [an urban] park. ★⇨公園 **都市公害** urban pollution. **都市工学** urban engineering. **都市鉱山**〔電子機器廃棄物などに含まれる金属類〕an urban mine. **都市コード**〔世界の空港所在都市を示す3文字の略号〕a city code; a three-letter code;〔日本全国の都市を示す5桁の数字〕a city code number. **都市ゴミ** municipal [city] garbage [waste, refuse]. /～ごみを燃やした熱で発電する generate electricity with the heat obtained from burning the city's waste. ★⇨ごみ(ごみ処理) **都市再開発** urban renewal [redevelopment]. **都市再生** urban renewal. **都市再生機構** the Urban Renaissance Agency (略：UR). **都市再生緊急整備地域**〔都市再生特別措置法に基づく〕an urgent urban-renewal area; an area needing urgent urban renewal. **都市再生特別措置法** the Special Measures for Urban Renewal Law. **都市再生特別地区**〔都市再生特別措置法に基づく〕a special urban renewal district; an urban renewal "special district"; an urban district designated for renewal under the provisions of the Law on Special Measures for Urban Renewal. **都市再生本部**〔内閣に設置した〕the Urban Renaissance Headquarters. **都市社会学** urban sociology. **都市住宅** urban housing. **都市人口** urban population. **都市震災軽減工学** urban earthquake disaster mitigation engineering. **都市人類学** urban anthropology. **都市スプロール(現象)** urban sprawl. **都市生活** city [civic] life; urban life. **都市生活者** a city dweller; an urbanite;〈集合的に〉city people. **都市転入** inflow into urban areas. **都市農山漁村交流活性化機構** the Organization for Urban-Rural Interchange Revitalization. **都市廃棄物** municipal waste. **都市廃熱** urban waste heat; waste heat from city buildings. **都市爆発** urban explosion; a rapid increase in urban population. **都市美化運動** a campaign for a cleaner and more beautiful city [to make a city cleaner and more beautiful]. **都市部** urban districts [areas]. **都市問題** an urban problem. **都市緑化** urban greening. **都市緑化基金** an urban greening fund. **都市緑地**

保全法　the Law for the Preservation of Urban Greenery.

土壌　soil. ¶**土壌安定剤**　a soil stabilizer. **土壌雨量指数**　〔土中に溜まっている雨水の量を示す〕a soil water index（略：SWI）. **土壌汚染**　soil pollution; soil contamination. **土壌汚染対策法**　the Soil Contamination Countermeasures Law. **土壌汚染防止法**　the Agricultural Land Soil Pollution Prevention Law. **土壌改良**　soil improvement. **土壌改良剤[薬]**　a soil conditioner. **土壌学**　pedology; soil science. **土壌学者**　a pedologist; a soil scientist. **土壌型**　a soil type. **土壌環境基準**　environmental quality standards for soil pollution [contamination]. **土壌細菌**　soil bacteria. **土壌殺菌剤**　(a) soil germicide. **土壌消毒**　soil disinfection. **土壌浸食[流出]**　(soil) erosion. **土壌水分**　soil moisture. **土壌生物**　edaphon. **土壌調査**　an agronomical survey. **土壌作り**　soil preparation; preparing the soil《for...》 **土壌微生物**　soil microbes; soil microorganisms. **土壌病害**　(a) soil-borne disease. **土壌分析**　(a) soil analysis. **土壌分析装置**　soil analysis equipment.

土地　〔地面〕land;〔所有地〕property;（狭い）a lot [plot]; an estate. ¶**土地改革**　land reform. **土地会社**　a real-estate agency; an estate agency. **土地開発**　property [land] development. ★⇨開発 **土地開発会社[開発業者]**　a property [land] developer. **土地開発公社**　a land development corporation. **土地改良**　land improvement. ▶～改良を行う improve land. **土地改良区**　a land improvement district. **土地改良事業**　land improvement enterprises. **土地改良法**　the Land Improvement Law. **土地鑑定士**　a property appraiser. **土地区画整理**　land readjustment; land reallocation [replotting]. **土地区画整理組合[事業]**　a land readjustment association [project]. **土地転がし**　hiking the price of a piece of land through repeated sales among insiders; land flipping. **土地資産額**　an assessed land value. **土地収用**　expropriation of land; condemnation [compulsory purchase] of land. **土地収用委員会**　a land expropriation committee. **土地収用権**　(the right of) eminent domain. **土地収用法**　the Land Expropriation Law. **土地条件図**　a map showing the physical characteristics of a piece of land.　**土地使用税**　a land-use tax.　**土地譲渡**　transfer of land; land transfer.　**土地譲渡所得税**　an income tax on land transfer.　**土地譲渡益課税**　taxation on the profit of a land transfer.　**土地所有**　landholding.　**土地所有権**　landownership; a (possessory) title to land.　**土地所有権移転登録**　registration of the transfer of ownership of land.　**土地所有者**　a landowner; a landholder; a landed proprietor; a territorial owner;〈集合的に〉the landed interest.　**土地信託**　land trust.　**土地政策**　a land policy.　**土地税制**　a land taxation system.　**土地税制改正**　reform of the land taxation system.　**土地制度**　the land system.　**土地造成**　land reclamation.　**土地相続税**　a land inheritance tax.　**土地台帳**　⇨ 土地登記簿.　**土地建物**　premises.　**土地投機**　land speculation; speculative investment in land.　**土地登記簿**　a land register; a terrier; a cadastre.　**土地投資動向調査**　〔国土交通省の発表する〕the Land Investment Trends Survey.　**土地成金**　a new-rich land speculator.　**土地売却**　sale of land.　**土地売買**　purchase and sale of land. / 〜売買の自由化 liberalization of the purchase and sale of land.　**土地白書**　a white paper on land;〔国土交通省の〕the《2005》White Paper on Land.　**土地評価額**　the estimated [appraised] value of a piece of land; a real estate [property] assessment.　**土地物件調書**　〔土地収用法に基づく〕a land and property report.　**土地ブローカー**　a real estate agent [agency]; a land agent; an estate agent.　**土地法案**　a land bill.　**土地没収**　escheat.　**土地保有税**　a landholding tax.　**土地本位制**　the "land standard system"; the principle, based on belief in a continued rise in land values, that loans should be secured primarily on land.　**土地面積**　land area.　**土地利用**　land use. ★⇨英和 LULUCF　**土地利用規制**　land-use control; regulation of land use.　**土地利用計画**　land-use planning.　**土地利用図**　a land-use map.　**土地割り当て**　acreage allotment.

突然変異　(a) mutation. ▶〜する mutate; sport. ¶**突然変異育種**　mutation breeding.　**突然変異原**　a mutagen.　**突然変異種**　a mutant species. ¶**突然変異説**　〔進化に関する〕a theory of mutation; mutationism. **突然変異体** a mutant; a sport. **突然変異体遺伝子** a mutant gene. **突然変異率** mutation rate. **前進突然変異** forward mutation. **復帰突然変異** reverse [back] mutation.

トップランナー方式 〔省エネルギー性能が最も優れた現行製品の性能に基づいて省エネ基準を策定する方式〕the top-runner approach [method, system].

度日(ﾄﾞﾓ) ⇨英和 degree-day

渡来地 ⇨飛来

トラフィック 〔野生動植物国際取引調査記録特別委員会〕TRAFFIC; Trade Records Analysis of Flora and Fauna in Commerce.

トラフィック・ゾーン・システム 〔自動車の交通システム〕a traffic zone system.

トランジット・モール 〔バスなど公共交通機関だけが通行できるようにした商店街の通り〕a transit mall.

トラブル trouble(s); a problem; a difficulty. ¶**トラブル隠し** concealing [hiding, covering up] problems; a cover-up. ▶電力会社による原発の〜隠しの実態が次々に明らかになった. A stream of facts came out regarding the power companies' cover-ups of nuclear power accidents.

鳥 ⇨鳥類 ¶**鳥インフルエンザ** avian influenza（略: AI）; bird [avian] flu; chicken flu. **高病原性鳥インフルエンザ** highly pathogenic avian influenza. **鳥インフルエンザ・ウイルス** an avian influenza virus; a bird [an avian] flu virus.

トリクレン ⇨トリクロロエチレン

トリクロロアセトアルデヒド trichloroacetaldehyde.

トリクロロエタン ⇨英和 trichloroethane

トリクロロエチレン trichloroethylene.

取り壊し ⇨解体, 撤去 ▶〜命令 〔違法建築などに対する〕a demolition order.

トリジェネレーション 〔熱電併給（⇨英和 cogeneration）で排気中の二酸化炭素を取り出して温室栽培などに利用するもの〕trigeneration.

トリハロメタン a trihalomethane（略: THM）.

トリフェニルスズ化合物 a triphenyltin compound.

トリブチルスズ tributyltin. ★⇨英和 TBT ¶**トリブチルスズオキシド** 〔船底・漁網防汚剤〕tributyltin oxide（略: TBTO）.

トリブチルスタンナン tributylstannane; tributyltin hydride.

トリメチルアミン trimethylamine.

ドリン剤 〔殺虫剤〕drin-[Drin-] insecticides. ★アルドリン (aldrin), エンドリン (endrin) など. ⇨ディルドリン

な

内海 an inland sea; an arm of the sea. ▶瀬戸〜 the Seto Naikai; the Inland Sea (of Japan).

内部 ▶…を〜告発する inform on … from within (an organization). /〜告発者 an internal informer; a whistle-blower. ¶**内部環境** internal environment. **内部リサイクル制度**〔社内などでの〕in-house recycling system.

内分泌 internal secretion; incretion. ▶〜の endocrine; endocrinous; endocrinal. ¶**内分泌液** an internal [endocrine] secretion; a hormone. **内分泌攪乱(化学)物質** a hormone-disruptive chemical [substance]; a hormone disrupter; an endocrine disrupter. ★⇨環境(環境ホルモン) **内分泌攪乱作用** endocrine disturbance. **内分泌疾患** an endocrine disease. **内分泌障害** endocrinopathy. **内分泌腺** an endocrine gland; a ductless gland.

流し網 a drift net. ★⇨混獲 ▶〜漁師 a drift netter; a drifter. ¶**流し網漁業** drift-net fishing.

流し網漁船 a drifter; a drift boat.

鳴き砂 a quartz sand that produces a squeaking sound when stepped on; singing [whistling, squeaking] sand.

ナショナル・トラスト ⇨英和 National Trust, 日本(日本ナショナルトラスト)

ナチュラル ⇨自然, 天然, 英和 Natural England ¶**ナチュラル・ウォーター**〔天然水〕natural water. **ナチュラル・キラー細胞** a natural killer cell; an NK cell. **ナチュラル・ヒストリー**〔自然誌[史]・博物学〕natural history. **ナチュラル・フード[フーズ]**〔自然食品〕natural foods.

夏日(なつ)〔最高気温25°C以上の日〕a day on which the temperature is between 25°C and 29°C. ★⇨真夏日 ▶今日今年最初の〜を記録した. Today was the first official "summer day" of the year.

生ごみ kitchen waste [garbage]; food scraps; raw garbage. ★⇨ごみ ▶〜を出す日[収集日] a kitchen waste [raw garbage] collection day. / 〜の処理 disposal of kitchen [food] waste. ★⇨自家(自家処理) ¶生ごみ処理機 a garbage disposal (unit). 生ごみ堆肥化装置 ⇨コンポスター. 生ごみ発電 generating electricity by burning garbage [(raw) waste]; (kitchen) waste incineration power generation. ★⇨ごみ(ごみ発電) 生ごみリサイクル kitchen waste recycling.

鉛 lead; plumbum. ¶鉛公害 lead pollution. 鉛中毒 lead poisoning. ★⇨鉛毒(えん)

南極 〔南極大陸〕the Antarctic Continent. ★⇨英和 Antarctica ¶南極海 the Antarctic Ocean. 南極環境保護議定書 the Protocol on Environmental Protection to the Antarctic Treaty. 南極観測 Antarctic research. 南極観測基地 an Antarctic research base. 南極観測船 an Antarctic research ship [vessel]. 南極観測隊 an Antarctic research expedition. 南極環流 ⇨英和 Antarctic Circumpolar Current. 南極区 〔動物地理学上の〕the Antarctic region. 南極圏[帯] the antarctic circle [zone]. 南極条約 ⇨英和 Antarctic Treaty. 南極星 the south pole star. 南極探検 an antarctic expedition [exploration]; a south-polar exploration. 南極地方 the south pole region. 南極点 the (geographical) south pole. 南極の海洋生物資源の保存に関する委員会 the Commission for the Conservation of Antarctic Marine Living Resources (略: CCAMLR). 南極氷床 the Antarctic ice sheet. 南極氷床コア the Antarctic ice core. 南極プレート the Antarctic Plate. 南極捕鯨 ⇨捕鯨

南限 the southern limit [border]. ★⇨北限 ▶リンゴ栽培の〜 the southern limit of cultivation [the range] for apples. / 繁殖[生息]地の〜 the southern border of a breeding area [habitat].

軟水 soft water. ▶(硬水を)〜にする soften water. ★⇨硬水(こう)

南南協力 〔開発途上国同士の協力〕South-South cooperation.

難分解 ⇨分解, 残留 ¶難分解性洗剤 hard detergents.

南洋 the South Seas. ¶南洋漁業 South Seas fisheries. 南洋材 tropical wood; timber from Southeast Asia.

に

匂[臭]い ⇨悪臭, 異臭, 臭気 ¶**におい消し** deodorant. **におい分子** an odor molecule.

肉食 ▶〔動物が〕～性の carnivorous. /〔人が〕～をやめる give up [abstain from] (eating) meat. ★⇨菜食, 草食 ¶**肉食獣** a carnivore; a carnivorous [flesh-eating, predatory] animal; a beast of prey. **肉食鳥** a bird of prey; a predatory bird.

二国間 ▶～の bilateral; between (the) two nations. ¶**二国間援助** bilateral aid [assistance]. **二国間協議** bilateral talks. **二国間協定** a bilateral agreement 《between…》. **二国間協力** bilateral cooperation. **二国間交渉** bilateral negotiations. **二国間条約** a bilateral treaty. ★⇨条約

二酸化硫黄(いおう) sulfur dioxide.

二酸化炭素 carbon dioxide. ★⇨CO_2(和英篇末尾), 英和 carbon dioxide ▶～の増加は環境に深刻な影響をもたらすに違いない. The increase in carbon dioxide will undoubtedly have a severe effect on the environment. /～吸収源としての森林の価値 the value of forests as carbon sinks. ★⇨吸収(吸収源) ¶**二酸化炭素回収・貯留[隔離]** carbon (dioxide) capture and storage [sequestration] (略: CCS). ★⇨英和 carbon sequestration **二酸化炭素換算** 〔温室効果ガス排出量の〕a carbon dioxide equivalent; a measure of the global-warming potential of a greenhouse gas expressed in terms of the amount of carbon dioxide that would have the same global-warming potential. **二酸化炭素固定** carbon dioxide [CO_2] fixation. **二酸化炭素削減目標** 《achieve》a CO_2-reduction goal. **二酸化炭素泉** a carbonated spring. **二酸化炭素排出権** the [a] right to emit carbon dioxide. ★⇨排出 **二酸化炭素排出原単位** a (basic) unit of CO_2 emission(s). **二酸化炭素排出抑制** carbon dioxide [CO_2] emission control. **二酸化炭素排出抑制対策** measures to control carbon dioxide [CO_2] emissions. **二酸化炭素排出量** 《reduce》CO_2 emissions; emissions of CO₂. ★⇨英和 carbon footprint

二酸化窒素 nitrogen dioxide. ¶二酸化窒素中毒 nitrogen dioxide intoxication.

二次 ¶二次エネルギー secondary energy. 二次汚染 secondary contamination. 二次感染 (a) secondary infection. 二次公害 secondary pollution. 二次災害 a secondary disaster. 二次消費者 a secondary consumer. ★⇨英和 secondary consumer 二次処理 〔汚水などの〕secondary treatment. 二次遷移 secondary succession. 二次遷移系列 a subsere. ★⇨英和 subsere 二次体腔 a secondary body cavity; deuterocoel. 二次被害 secondary damage. 二次被曝 exposure to residual radiation; residual-radiation exposure. 二次林 a secondary [second-growth, substitution] forest. ★⇨英和 second growth 二次冷却水 〔原子炉などの〕secondary cooling water. 二次冷却装置 〔原子炉などの〕a secondary cooling system.

日米原子力協定 the Japan-US Atomic Energy Agreement.

日照 sunshine; insolation. ★⇨日当たり ¶日照計 a heliograph; a sunshine recorder. 日照権 the right to (sun)light; the sunshine right; access to sunlight; the right to enjoy sunshine. 日照時(間) hours of sunlight [sunshine]. ▶当地の〜時間は年平均 2,500 時間である. This area has an annual average of 2,500 hours of sunlight [sunshine]. 日照調整 〔建物の〕sun control. 日照不足 a lack of sunshine. 日照率 a sunshine rate; a percentage of sunshine. 日照量 the amount of sunlight [solar radiation, insolation].

ニッチ 〔生態的地位〕(ecological) niche.

日本 ¶日本愛玩動物協会 the Japan Pet Care Association. 日本遺伝子診療学会 the Japanese Society for Gene Diagnosis and Therapy (略: JSGDT). 日本エネルギー経済研究所 〔財団法人〕the Institute of Energy Economics, Japan (略: IEEJ). 日本園芸協会 the Japan Gardening Society (略: JGS). 日本温暖化ガス削減基金 the Japan GHG Reduction Fund (略: JGRF). ★GHG は greenhouse gas(es) (温室効果ガス) の略. 日本海溝 the Japan Trench. 日本海洋学会 the Oceanographic Society of Japan (略: JOS). 日本海洋少年団 the Junior Sea Friend's Federation of Japan（略：JSF）． **日本海流** the Japan Current [Stream]． ★⇨黒潮(くろしお) **日本環境化学会** the Japan Society for Environmental Chemistry（略：JEC）． **日本環境協会** the Japan Environment Association（略：JEA）． **日本環境災害情報センター** the Japan Environmental Disaster Information Center（略：JEDIC）． **日本環境変異原学会** the Japanese Environmental Mutagen Society（略：JEMS）． **日本気象学会** the Meteorological Society of Japan（略：MSJ）． **日本気象協会** the Japan Weather Association（略：JWA）． **日本原子力委員会** ⇨原子(原子力委員会) **日本原子力技術協会** the Japan Nuclear Technology Institute（略：JANTI）． **日本原子力研究開発機構** the Japan Atomic Energy Agency（略：JAEA）． **日本原子力研究所** the Japan Atomic Energy Research Institute（略：JAERI）． **日本原子力産業会議** the Japan Atomic Industrial Forum（略：JAIF）． **日本原子力文化振興財団** the Japan Atomic Energy Relations Organization（略：JAERO）． **日本原水爆被害者団体協議会** the Japan Confederation of A- and H-Bomb Sufferers Organizations（略：HIDANKYO）． **日本公園緑地協会** the Parks and Open Space Association of Japan（略：POSA）． **日本公衆衛生学会** the Japanese Society of Public Health（略：JSPH）． **日本産業廃棄物処理振興センター** the Japan Industrial Waste Technology Center． **日本自然保護協会** the Nature Conservation Society of Japan（略：NACS-J）． **日本食品添加物協会** the Japan Food Additives Association（略：JAFA）． **日本植物学会** the Botanical Society of Japan（略：BSJ）． **日本植物生理学会** the Japanese Society of Plant Physiologists（略：JSPP）． **日本生態学会** the Ecological Society of Japan（略：ESJ）． **日本生態系協会** the Ecosystem Conservation Society-Japan． **日本生物工学会** the Society for Biotechnology, Japan（略：SBJ）． **日本鳥類保護連盟** the Japanese Society for Preservation of Birds（略：JSPB）． **日本動物愛護協会** the Japan Society for the Prevention of Cruelty to Animals（略：JSPCA）． **日本動物園水族館協会** the Japanese Association of Zoos and Aquariums（略：JAZA）． **日本動物行動学会** the Japan Ethological Society〔略：JES〕. **日本動物福祉協会** the Japan Animal Welfare Society〔略：JAWS〕. **日本道路公団** the Japan Highway Public Corporation〔略：JHPC〕. **日本道路交通情報センター** the Japan Road Traffic Information Center〔略：JARTIC〕. **日本ナショナルトラスト** the Japan National Trust. ★ National Trust (⇨英和篇) にならって1968年に創設された財団法人．これと別に全国の自然保護団体の連合組織として社団法人の「日本ナショナル・トラスト協会」(the Association of National Trusts in Japan) も 1992 年に結成された．**日本熱帯生態学会** the Japan Society of Tropical Ecology〔略：JASTE〕. **日本農林規格** ⇨農林 **日本爬虫両棲類学会** the Herpetological Society of Japan. **日本繁殖生物学会** the Japanese Society of Animal Reproduction〔略：JSAR〕. **日本ビオトープ管理士会** ⇨ビオトープ **日本哺乳動物卵子学会** the Japanese Society of Mammalian Ova Research〔略：JSMOR〕. **日本水大賞**〔水環境の保護などに取り組む団体や個人に贈られる賞〕the 《10th》Japan Water Award. **日本容器包装リサイクル協会**〔財団法人〕the Japan Containers and Packaging Recycling Association〔略：JCPRA〕.

尿素 urea. ¶ **尿素(系)肥料** a urea (compound) fertilizer. **尿素 SCR システム**〔ディーゼル車の窒素酸化物低減技術の１つ〕a urea-SCR system. ★ SCR は selective catalytic reduction の略. **尿素水**〔尿素 SCR システムで用いる〕(a) urea solution.

人間 a human being; a person; a mortal;〔ヒト科の動物・人類〕man; mankind; humanity. ★⇨自然 ▶～に似た manlike; humanlike; anthropoid. / 最も～人間に近い霊長類の動物 the primates, the closest animals to human beings. / ～の鎖で市役所を包囲する surround the city hall with a human chain; form a human chain around the city hall. ¶ **人間遺伝学** human genetics. **人間開発報告書**〔国連開発計画(UNDP)が発行する〕a Human Development Report〔略：HDR〕. **人間環境会議** ⇨国連(国連人間環境会議) **人間環境宣言**〔1972年，国連人間環境会議で採択された宣言〕the Statement for Human Environmental Quality. **人間工学** ergonomics; human engineering. **人間行動生態学** human behavioral ecology〔略：HBE〕.

人間主義 humanism.　**人間生活工学研究センター** the Research Institute of Human Engineering for Quality Life（略：HQL）.　**人間生態学** human ecology.　**人間中心主義** anthropocentrism; anthropocentricism.　**人間と生物圏計画**〔国連、ユネスコの〕the Man and Biosphere Programme（略：MAB）.　★⇨生物（生物圏保護区）　**人間貧困指数**〔国連開発計画（UNDP）が発表する〕a human poverty index（略：HPI）.

認証　⇨認定.　★⇨ GRAS, MSC（和英篇末尾）　▶ISO 〜取得 certified by the ISO; ISO-certified.　★⇨ ISO（和英篇末尾）　¶**認証排出削減量**〔温室効果ガスの〕(a) Certified Emission Reduction.　★⇨英和 CER　**森林認証**　⇨森林

認定　recognition; acknowledgment;〔承認〕approval;〔認可〕sanction; authorization;〔許可〕permission.　★⇨許可, NPO（和英篇末尾）, 低排ガス車　▶〜する admit; recognize; find; deem; acknowledge; conclude;〔認可する〕authorize; certify; sanction;〔承認する〕approve《of...》. / ...病患者であると認定する recognize as a victim of [suffering from] ... disease. / ... 病患者の司法[行政]〜 court [official, government] recognition [certification] as a ... disease patient. /〔公害病などの〕〜患者 a designated victim of [sufferer from] a disease; a patient registered [designated] as suffering from a disease.　★⇨公害（公害病） / ...病の未〜患者 a patient not registered [designated] as suffering from ... disease.　¶**認定完成検査**　(a) certified completion inspection.　**認定完成検査実施者**　a certified completion inspector.　**認定基準**　certification standards; criteria [standards] for accreditation; accreditation criteria.　★⇨基準　**認定申請**　(an) application for certification. / 原爆症の〜申請をする apply for certification of an A-bomb-related illness [to have an A-bomb-related illness certified].　**認定農業者[農家]**〔農業経営基盤強化促進法に基づいて市町村が認定した農業者〕a designated farmer.　**認定被爆者**　a designated [an officially recognized] *hibakusha*.　**認定保安検査**〔一定基準を満たした高圧ガス使用企業などに認められる自主検査〕(a) certified safety inspection.　**認定保安検査実施者**　a certified safety inspector.　**認定薬**　an approved drug.

ニンビー〔原発・刑務所・ごみ処

分場など地域環境にとって好ましくないものが(よそにならともかく)近所に設置されることに反対する運動〕NIMBY; NIMBYism; 〔人〕a NIMBYist. ★⇨英和 NIMBY, 地域(地域エゴ), 迷惑

ね

ネイチャー ⇨自然　¶**ネイチャー・ゲーム**〔自然との触れ合いに主眼をおいた米国発祥の教育活動〕a nature game.　**ネイチャー・スキー**〔散歩感覚を楽しむスキー〕nature skiing.　**ネイチャー・トレール**〔自然遊歩道〕a nature trail.　**ネイチャー・フォト**〔ジャンル〕nature photography;〔写真〕a nature photo(graph).　**ネイチャー・ライティング**〔自然を題材にしたエッセイなど〕writing on the theme of nature;〔特に1970年代以降の環境保護意識に立脚したノンフィクション文学のジャンル〕Nature Writing.

ネクトン〔海洋の遊泳生物〕a nekton.　★⇨英和 nekton

熱　heat; warmth; 形 thermal. ★⇨太陽(太陽熱), 地熱, 断熱, 熱波, 英和 heat island, heat-trapping, heat index　▶青草を積んでおくと熱が発生する. Green hay heats in a mow. / 工場内に熱がこもらないよう大型換気扇をつけた. We put in an industrial ventilator to keep heat from building up inside the factory. / 熱を出す［発する］generate [produce] heat. /（…に）熱を加える, 熱する apply heat (to…); heat. / 熱を発散［輻射, 吸収］する give off [radiate, absorb] heat. / 熱を貯める［逃がす］store [release] heat. / アルコールは蒸発するとき周囲から熱を奪う. Alcohol takes heat from its surroundings when it evaporates. / 熱を伝える conduct heat. / 熱を通さない建材 a building material impervious to heat. / 気化熱 heat of vaporization.　¶**熱エネルギー**　thermal energy; heat energy.　**熱汚染［公害］**　thermal pollution; heat pollution.　★⇨英和 thermal pollution　**熱回収**〔エネルギー循環のための〕(waste) heat recovery.　★⇨排［廃］熱　**熱回収効率**　heat recovery efficiency.　**熱管理**　thermal management; heat control.　**熱起電力**　thermoelectromotive force.　**熱交換**　heat exchange.　**熱交換器**　a heat exchanger.　**熱交換パイプ**　a heat exchanger tube.　**熱交換量**　an amount of heat exchanged.　**熱効率**　thermal [heat] efficiency. / 正味熱効率

89% net thermal efficiency of 89%. / 総合[理論]熱効率 total [theoretical] thermal efficiency. **熱閃光** 〔核兵器の〕a thermal flash. **熱電池** a thermal battery; a heat-activated battery. **熱排水** ⇨温排水 **熱放散** heat radiation [emission];（放散による熱の消失）heat dissipation [loss]. **熱放射[輻射]** thermal radiation. **熱膨張** thermal expansion. **熱膨張率[係数]** the coefficient of thermal expansion; thermal expansivity. **熱ポンプ** a heat pump. **熱容量** thermal [heat] capacity.

熱帯 the tropics; the [a] torrid [tropical] zone. ▶〜(産)の tropic(al). ¶ **熱帯医学[衛生学]** tropical medicine [hygiene]. **熱帯(降)雨林** a tropical rain forest. **熱帯雨林気候** a tropical rain-forest climate. **熱帯果実** a tropical fruit. **熱帯気候** a tropical climate. **熱帯気団** a tropical air mass. **熱帯降雨観測衛星** the Tropical Rainfall Measuring Mission（略：TRMM）. **熱帯黒色土壌** a vertisol. **熱帯収束帯** the intertropical convergence zone（略：ITCZ）. **熱帯(性)高気圧** tropical high pressure《belt》. **熱帯(性)植物** a tropical plant;〔総称〕tropical flora. **熱帯(性)低気圧** a tropical cyclone. / 弱い〜(性)低気圧 a weak tropical depression; a tropical disturbance. **熱帯前線** a tropical [an intertropical] front. **熱帯草原** ⇨サバンナ. **熱帯多雨気候** a tropical-rainy climate. **熱帯多雨帯** the tropical rain belt. **熱帯多雨林** ⇨熱帯雨林. **熱帯病** tropical diseases. **熱帯暴風雨** a tropical (rain)storm. **熱帯モンスーン気候** a tropical monsoon climate. **熱帯夜** 〔温度が25°C以下にならない夜〕a "tropical night"; a sweltering night when the temperature does not fall below 25°C (77°F). / 23日連続〜夜 23 consecutive nights with temperatures of 25°C or above; 23 consecutive "tropical nights." **熱帯林** a tropical forest. **熱帯林行動計画** ⇨英和 TFAP **熱帯林地** tropical forestland. **熱帯林破壊** tropical deforestation; rain forest destruction.

熱波 a heat [thermal] wave.
熱電(気) thermoelectricity. ¶ **熱電(気)発電** thermoelectric (power) generation. **熱電半導体** a thermoelectric semiconductor. **熱電併給** ⇨英和 cogeneration, トリジェネレーション **熱電併給システム** a

cogeneration system.

燃焼 burning; combustion; ignition. ▶~する burn; ignite. ¶**燃焼温度** burning temperature. **燃焼ガス** combustion gas. **燃焼装置** a burner; a combustion apparatus; combustion equipment. **燃焼残渣[残滓]** 〔燃料の〕combustion residue;〔廃棄物の〕incineration residue. **燃焼式給湯器** a combustion(-type) water heater. **燃焼度** 〔核燃料の〕burnup. **完全燃焼** perfect [complete] combustion. **自然燃焼** spontaneous combustion;〔発火〕spontaneous ignition. **熱核燃焼** thermonuclear burning. **不完全燃焼** imperfect [incomplete] combustion.

燃費 fuel consumption; fuel economy [efficiency]; (fuel) mileage. ▶この車は~がいい[悪い]．This car gets good [poor] mileage. / ~(効率)のよい[低~の](小型)車 a fuel-efficient (compact) car; a (compact) car that gets [gives] good mileage. ★⇨低燃費，エコカー / ~の悪い車 a fuel-inefficient car; a car that gets low mileage;《口》a gas guzzler. / ~を改善する improve [reduce] fuel consumption; improve fuel economy [fuel efficiency, mileage]. / 〔二酸化炭素排出量抑制のための〕~基準を定める set [introduce] a fuel-efficiency standard [target]. / 2010年~基準＋20%達成車 a vehicle exceeding the 2010 fuel-efficiency standard by 20% or more. ¶**燃費規制** fuel efficiency requirements. **燃費計** a fuel economy meter. **燃費性能** fuel efficiency performance. **カタログ燃費** the manufacturer's estimated mileage [fuel consumption]. **公称[公表]燃費** the official [published] fuel consumption [mileage]. **実(用)燃費** the actual mileage [fuel consumption]. **平均燃費** average fuel consumption [mileage]. ★⇨企業(企業平均燃費)，10-15(テン・フィフティーン)モード

燃料 fuel. ★⇨燃料電池 ▶石油[石炭]を~とする use oil [coal] as [for] fuel. / 山の木をみな~に使ってしまった．The trees on the mountain were all used for fuel. / ~を節約する save [economize] (on) fuel. / 飛行機に~を補給する refuel an airplane. ★⇨燃料補給 / このエンジンは多くの~を消費する．This engine consumes a lot of fuel. / 液体[気体，固体]~ liquid [gaseous, solid] fuel. ★⇨核(核燃料)，化石(化石燃料)，石油(石油代替燃

料）¶燃料加工 《MOX [nuclear]》fuel fabrication. 燃料極 〔燃料電池の〕a fuel pole; an anode pole. 燃料計 a fuel gauge; a fuel-level gauge. 燃料警告灯 〔自動車などの〕a (low-) fuel warning light. 燃料経済性[効率] ⇨燃費 燃料庫 〔船の〕a fuel [an oil] bunker. 燃料鉱床 a fuel deposit. 燃料蒸発ガス回収[抑止]装置 onboard refueling vapor recovery（略：ORVR）. 燃料消費率 specific fuel consumption（略：SFC）; fuel consumption rate. ★⇨燃費 燃料消費量 the amount of fuel consumed; fuel consumption. 燃料代 fuel cost(s). 燃料タンク a fuel tank. 燃料チャージ ⇨燃料油価格変動調整金. 燃料チャンネル 〔原子炉の〕a fuel channel. 燃料注入 〔ロケットなどへの〕fueling; fuelling. 燃料調整金 a fuel adjustment charge（略：FAC）. 燃料添加物 a fuel additive. 燃料転換 〔汚染物質発生のより少ない燃料への転換〕fuel conversion. 燃料パイプ(ライン) a fuel pipeline. 燃料板 〔原子炉の〕a fuel plate. 燃料費 the cost of fuel; fuel expenses; a charge for fuel. 燃料費調整制度 〔電気料金の〕a fuel-cost adjustment system. 燃料費比率 〔総経費に占める割合〕the ratio of fuel costs to total costs. 燃料不足 (a) lack of fuel; fuel shortage. 燃料噴射 fuel injection. 燃料噴射装置 a fuel-injection device. 燃料噴射ノズル 〔内燃機関などの〕a fuel injection nozzle. 燃料噴射ポンプ a fuel-injection pump. 燃料ペレット 〔原子炉の〕a fuel pellet. 燃料棒 〔原子炉の〕a fuel rod. 燃料放出装置 〔航空機の〕a fuel dump(ing) [jettison] system. 燃料補給 refueling. /～補給のため入港した. The ship put into port to refuel [take on fuel]. 燃料補給機 a refueling plane. 燃料補給基地 a fueling base. 燃料補給所 a fueling station [depot]. 燃料補給船[艦] a fuel supply vessel [ship, boat]. 燃料ポンプ a fuel pump. 燃料漏れ fuel leakage. 燃料油[ガス, 炭] fuel oil [gas, coal]. 燃料油価格変動調整金 〔航空会社・フェリー会社などが, 燃料油価格の高騰に対処するために利用者に求める上乗せ負担金・サーチャージ〕a fuel surcharge. 燃料要素 〔原子炉の〕a fuel element.

燃料電池 a fuel cell. ¶燃料電池コ(ー)ジェネレーションシステム a fuel-cell cogeneration [co-generation] system. 燃料電池

燃料電池

(電気自動)車　a fuel cell(-powered) vehicle(略: FCV); a fuel cell electric vehicle(略: FCEV). **燃料電池実用化推進協議会**　the Fuel Cell Commercialization Conference of Japan (略: FCCJ). **燃料電池ハイブリッド車**〔燃料電池と二次電池を備えたハイブリッド車〕a fuel cell hybrid vehicle (略: FCHV). **固体高分子型燃料電池**　a solid polymer (electrolyte) fuel cell (略: SPFC); a polymer electrolyte fuel cell (略: PEFC). **固体酸化物型燃料電池**　a solid oxide fuel cell (略: SOFC). **水素燃料電池**　a hydrogen fuel cell. **ダイレクト[直接]メタノール(型)燃料電池**　a direct methanol fuel cell (略: DMFC). **定置用燃料電池**　a residential [stationary] fuel cell. **微生物燃料電池**　a microbial fuel cell. **溶融炭酸塩型燃料電池**　a (molten) carbonate fuel cell (略: (M)CFC). **燐酸型燃料電池**　a phosphoric acid fuel cell (略: PAFC).

の

野犬(のいぬ) ⇨野良犬(のらいぬ), 野犬(やけん), 野生化

農家 〔家〕a farmhouse;〔家庭〕a farm(ing) family [household]; an agrarian home;〔農民・農場主〕a farmer;〔零細農家〕a small-scale farmer; a peasant (farmer). ★⇨園芸(園芸農家), 生産(生産農家), 栽培, 供給 ▶彼は〜に育った. He was brought up [reared] on a farm. / 今は一年中で〜の一番忙しい時です. This is the busiest time of year [season (of the year)] for farmers. ¶**農家経済調査** an annual farm household economic survey. **農家人口** the farming [agrarian, agricultural] population; the number of people involved in agriculture. **農家民宿** a farm inn; a small family-run hotel on a farm. ★⇨農林, ファームステイ **兼業農家** a part-time farming household; a part-time farmer;〔農業を主とするもの〕a farmer with a side job;〔農業を従とするもの〕a person doing farming on the side. **専業農家** a full-time farming household; a household in which everybody works exclusively in farming.

農業 agriculture; farming; the agricultural [farming] industry. ★⇨農家, 農作[産]物, 農地, 農薬, 農用, 農林, 離農 ▶〜が盛んだ have (a) flourishing agriculture. / 〜に従事する[を営む] engage [be] in farming [agriculture]; be a farmer; 《文》follow the plow. / 〜の agricultural; farming; farm; agrarian; rural. / 〜の手伝い assistance in farm work. / 〜の機械化 (the) mechanization of agriculture. / 高度機械化〜 highly mechanized agriculture. / 機械化〜従事者 a worker on a mechanized farm. / 〜の近代化 (the) modernization of agriculture; agricultural modernization. / 〜を継ぐ succeed to a farm; become a farmer (like *one's* ancestors). / 私の実家は〜をしています. My parents are farmers. | I'm from a farming family. ¶**農業委員** an agricultural commissioner. **農業委員会** an agricultural commission [committee]. **農業委員会法** the Agricultural Commission Law. **農業科** an agricultural course. **農業改良助長法** the Agricultural Improvement Promotion Law. **農業改良普及員** an agricultural extension worker. **農業改良普及事業** agricultural extension work. **農業科学** agricultural science(s). **農業化学品** an agrochemical; an agricultural chemical;〔特に殺虫剤, 殺菌剤など〕an agrichemical. **農業革命** an agricultural revolution. **農業学校** an agricultural school. **農業観測** an agricultural outlook; the outlook for agriculture. **農業機械** an agricultural [a farm] machine;〈集合的に〉agricultural [farm] machinery. **農業機械化促進法** the Agriculture Mechanization Promotion Law. **農業機械工業** the agricultural [farm] machinery industry. **農業気候学** agroclimatology. **農業技術** agricultural techniques. **農業技術者** an agricultural engineer. **農業気象学** agricultural meteorology; agrometeorology. **農業基本法** the Agricultural Basic Law. **農業恐慌** (an) agricultural depression; an agrarian crisis; a depression on the farms [in farming, on the land]. **農業共済組合** an agricultural mutual aid association. **農業共済再保険** agricultural mutual aid reinsurance. **農業共済再保険特別会計** an agricultural mutual aid reinsurance special account. **農業共済制度** an agricultural mutual aid system. **農業協同組合(農協)** (Japan) Agricultural Cooperatives (愛称: JA). **農業協同組合中央会** a Central (Prefectural) Union of Agricultural Cooperatives. **農業協同組合法** the Agricultural Cooperative Law. **農業協同組合連合会** the Federation of Agricultural Cooperative Associations. **農業共同経営** cooperative management of agriculture. **農業近代化資金助成法** the Agriculture Modernization Fund Subsidization Law. **農業金融** agricultural financing [credit]; credit for [to] farmers; a farm loan. **農業組合** an agricultural association. **農業経営** agricultural [farm] management. **農業経営基盤強化促進法** the Law for the Promotion of Improved Farm Management. **農業経済(学)** agricultural economics; agronomics; agronomy. **農業工学** agricultural [farm] engi-

農業

neering.　農業高校　an agricultural high school.　農業公社　an [a public] agricultural corporation.　農業構造改善事業　an agriculture structural improvement project.　農業国　an agricultural [a farming] country.　農業災害　(an) agricultural disaster.　農業災害補償　compensation [an indemnity] for agricultural disaster(s).　農業災害補償制度　an agricultural disaster compensation system.　農業災害補償法　the Compensation against Agricultural Loss Law.　農業雑誌　an agricultural [a farm] journal.　農業資金貸付　a rural [an agricultural] loan; loans to farmers.　農業試験場　an agricultural experiment station;〔農林水産省管轄の〕National Agricultural Experiment Stations.　農業施設　agricultural facilities; farm buildings.　農業指導者　an agricultural leader; a leading agriculturalist.　農業者　a farmer; a farm worker; an agricultural worker [producer]. ★ ⇨認定(認定農業者)　農業者年金　a pension [an annuity] for farmers [farm workers]; an agricultural pension; a farmer's annuity.　農業者年金基金　the Farmers' Pension Fund.　農業自由化　liberalization of trade in agricultural products; agricultural (trade) liberalization.　農業従事者　a person engaged in farming [agriculture].　農業集団化　(the) collectivization of agriculture.　農業収入　(an) income from farming [agriculture]; (a) farming [(an) agricultural] income.　農業準専従者　a non-full-time farmer; a person employed in agriculture for more than 60 and not more than 149 days in the year.　農業・食品産業技術総合研究機構　the National Agriculture and Food Research Organization（略: NARO).　農業神　a god of farming.　農業(就業)人口　an agricultural [a farming] population; the population [number of people] engaged in agriculture.　農業振興地域　an agricultural promotion [development] area; a rural development district.　農業振興地域整備法　the Law for Improvement of Agricultural Promotion Areas.　農業政策　(an) agricultural [agrarian] policy; (a) policy for farming.　農業生産　agricultural [farm] production [output]. /～生産性を上げる increase agricultural productivity. **農業生産指数** an agricultural production index. **農業生産法人** an agricultural production corporation. **農業生物学** agrobiology. **農業・生物系特定産業技術研究機構** the National Agriculture and Bio-oriented Research Organization (略: NARO). ★2003年に農業技術研究機構と生物系特定産業技術研究推進機構が統合して発足. **農業生物資源研究所** the National Institute of Agrobiological Sciences (略: NIAS). **農業センサス** an agricultural census;〔国連食糧農業機関の実施する世界農業センサス〕the World Census of Agriculture. **農業専従者** a full-time farmer; a person engaged in farming for at least 150 days in the previous year. **農業専門学校** a vocational school for agriculture. **農業倉庫** an agricultural storehouse [warehouse]. **農業大学** an agricultural college. **農業大国** a major [leading] agricultural country [power]. **農業団体** an agricultural [a farmers'] organization. **農業地域** a farming [an agricultural, an agrarian] district [area, region]. /〜地域類型 an agricultural area type [category]. / 平地[中間, 山間]〜地域 a flat [semimountainous, mountainous] agricultural area. / 中山間〜地域 a mountainous or semimountainous agricultural area. **農業地質学** agricultural geology; agrogeology. **農業地理学** agricultural geography. **農業手形** a promissory note [bill] issued by an agricultural co-operative. **農業統計** agricultural statistics. **農業特区** a special agricultural district; a special structural-reform zone for deregulated agriculture. **農業土木** agricultural civil engineering; irrigation, drainage and reclamation work [engineering]. **農業廃水** agricultural [farm] waste. **農業白書** a white paper on agriculture;〔農林水産省が1998年まで発表した〕the Annual Report on Japanese Agriculture. ★現在名は「食料・農業・農村白書 (⇨食料)」. **農業法人** an agricultural corporation. **農業保険** agricultural insurance. **農業補助金** farm subsidies. **農業輸出国** an agricultural exporting country. **農業輸出補助金** an agricultural export subsidy. **農業用水** agricultural water (supply);〔灌漑用水〕(agricultural) irrigation water. **農業労働者** a farm worker [laborer, hand]; an agricultural laborer [worker].　**集約[粗放]農業**　intensive [extensive] agriculture.　**主穀式農業**　grain-centered agriculture; farming based on grain production.　**大規模農業**　large-scale farming [agriculture].　**多角農業**　diversified farming [agriculture].　**地域支援(型)農業**　community supported agriculture (略: CSA).　**有機農業**　⇨有機　**零細農業**　small-scale[-time] farming.

農作物　a crop; (the) crops; farm produce [products].　★⇨作物　▶〜に害を与える動物[植物]　an animal [a plant] which damages agriculture. / この日照りが長く続くと〜に大きな被害が出るだろう．If this drought lasts long the crops will suffer greatly. / 〜の生産　crop growing [production].

農産物　agricultural [farm] products [produce].　★⇨作物,余剰(余剰農産物)　▶〜が不足している．There is a shortage of farm produce. / 〜に富む　produce a lot of agricultural produce; be rich in farm products. / その国は〜をほとんど輸出していない．The country exports little agricultural produce [few agricultural products].　¶**農産物価格**　agricultural [farm] prices; the price(s) of farm produce.　**農産物価格安定法**　the Agricultural Price Stabilization Law.　**農産物価格支持制度**　an agricultural price support system.　**農産物検査法**　the Agricultural Products Inspection Law.　**農産物知的財産権**　intellectual property rights in agriculture.　**農産物直売所**　a direct sales store [outlet] for agricultural products.　**農産物品評[共進]会**　an agricultural fair [show].　**農産物輸出国**　an agricultural exporter.　**農産物輸入自由化**　the liberalization of the import of agricultural products.

濃縮　concentration;〔ウラニウムなどの〕enrichment; graduation; incrassation.　★⇨牛乳(濃縮牛乳),生物(生物濃縮),濃度　▶〜する　concentrate; enrich. / 高〜ウラン　highly enriched uranium (略: HEU).　★⇨ウラン(濃縮ウラン),低濃縮ウラン / 〜オレンジジュース　orange juice concentrate; concentrated orange juice. / 〜還元フルーツジュース　fruit juice reconstituted from concentrate.　¶**濃縮汚泥**　concentrated sludge.　**濃縮液**　(a) concentrate; (a) concentrated liquid. 濃縮器［装置］ a graduator; a concentrator; a thickener.

農地 farmland; agricultural [farming] land; cropland. ★⇨宅地, 市街, 農用, 廃農地 ▶先祖代々の〜 farmland [a farm] handed down from *one's* ancestors; ancestral farmland. ／〜の宅地並み課税 treating farmland as equivalent to housing land for taxation purposes; a tax on farmland which treats it as equivalent to residential land. ¶**農地委員会** an agricultural land commission [committee]. **農地改革** land [agrarian, farmland] reform(s) [redistribution]. **農地開発** development [creation] of farmland [agricultural land]. **農地解放** 〔日本の〕(the postwar) agricultural land reform; (the postwar) removal of concentrations in land ownership; the transfer of land ownership from landlords to tenant farmers after the Second World War. **農地価格** the price [value] of agricultural land; agricultural land prices [values]. **農地集約** concentration of farmland [(farming) plots]. **農地信託** a farmland trust; a trust for agricultural land. **農地生態系** a cultivated ecosystem. **農地調整法** the Farmland Adjustment Law. **農地転用** conversion of farmland to non-farming uses; diversion of arable land to other uses. **農地法** the Agricultural Land Law.

濃度 density; concentration. ★⇨濃縮, 残留(残留濃度) ▶〜規制〔汚染物質の排出濃度を規制する〕density restriction. ★⇨総量規制／体内に高〜で蓄積されている有機水銀 high concentrations of organic mercury accumulated in the body. ★⇨蓄積／塩化ビフェニールを高〜に含む contain(ing) high-concentration PCB. ／高〜のダイオキシン dioxin in high concentrations; highly-concentrated dioxin. ／従業員の血中から高〜のダイオキシンが検出された. A high level of dioxin was found [detected] in the blood of employees. ／通常の300倍の〜のダイオキシンに汚染された土壌 soil contaminated by a concentration of dioxin 300 times the normal value. ★⇨ダイオキシン ¶**濃度計** a densitometer. ★⇨煤煙(ばいえん)濃度計 **高濃度汚染** highly concentrated [high-concentration] pollution. **高濃度汚染地区** an area of high-concentration pollution. **高濃度放射能汚染物質**

highly radioactive waste (substances).

農民 a peasant; a farmer;〈集合的に〉a [the] farming population; a [the] peasantry. ★⇨農家

農薬 agricultural chemicals; agrochemicals. ★⇨殺菌(殺菌剤), 殺虫(殺虫剤), 除草(除草剤) ▶私の畑では〜は一切使っていない. I don't use any agricultural chemicals on my fields. ★⇨減農薬, 無農薬, 低農薬, 特定(特定農薬), 残留(残留濃度), ゴルフ場農薬 / 〜をまく[散布する] use agrochemicals《on the fields》; spread [disperse, spray] agricultural chemicals. ★⇨散布 / 無登録〜 an unregistered agricultural chemical. ¶**農薬汚染** pollution due to agricultural chemicals. **農薬化学** pesticide chemistry. **農薬科学** pesticide science. **農薬検査所** an Agricultural Chemicals Inspection Station. **農薬公害** contamination [pollution] by agricultural chemicals; agrochemical contamination. **農薬散布機** a crop duster. **農薬使用基準** a standard for the use of agricultural chemicals [pesticides]. **農薬中毒** poisoning by agricultural chemicals. **農薬取締法** the Agricultural Chemicals Regulation Law.

農用 ¶**農用地** land for agricultural use; farm(ing) land. **農用地土壌汚染防止法** the Agricultural Land Soil Pollution Prevention Law. **農用トラクター** a tractor; an agricultural tractor.

農林 agriculture and forestry. ▶〜水産省(農水省) the Ministry of Agriculture, Forestry and Fisheries (略: MAFF). / 〜水産大臣 the Minister of Agriculture, Forestry and Fisheries. / 日本〜規格 the Japanese Agricultural Standards (略: JAS). / 日本〜規格法 the Japanese Agricultural Standards [JAS] Law. ★正式名称は「農林物資の規格化及び品質表示の適正化に関する法律」(the Law Concerning Standardization and Proper Labelling of Agricultural and Forestry Products). / 〜規格検査所〔農水省管轄の〕Agricultural and Forest Products Inspection Institutes. / 〜事業 agricultural and forestry projects. / 〜漁業体験民宿に参加する stay at privately run lodgings where you can try farming and fishing [where guests can have "real" experience of agriculture and fishing at sea]. ★⇨農家(農家民宿) ¶**農林学校** an agricultural and forestry school. **農林漁業信用基金** the Agriculture, Forestry and Fisheries Credit Foundations. **農林漁業団体職員共済組合** the Mutual Aid Associations of Agriculture, Forestry and Fishery Corporation Personnel. **農林水産技術会議**〔農水省の〕Agriculture, Forestry and Fisheries Research Council. **農林水産研修所**〔農水省管轄の〕Training Institute of Agricultural Administration. **農林水産ジーンバンク** the Ministry of Agriculture, Forestry and Fisheries Genebank; the MAFF Genebank. **農林水産消費技術センター**〔独立行政法人〕the Center for Food Quality, Labeling and Consumer Services. **農林水産政策研究所**〔農水省の〕the Policy Research Institute, Ministry of Agriculture, Forestry and Fisheries. **農林水産先端技術産業振興センター** the Society for Techno-innovation in Agriculture, Forestry and Fisheries（略：STAFF）. **農林水産物等輸出促進全国協議会** the Council for the Promotion of Agricultural, Forestry and Fishery Exports. **農林物資規格法** the Agricultural and Forest Commodities Standards Law.

ノーカー運動〔環境保護運動の一環としての〕a campaign to discourage commuting by car; a non-car commuting campaign.

ノーネクタイ ▶〜で without (wearing) a tie. ¶**ノーネクタイ・ノー上着運動**〔夏の軽装化の〕a "No Necktie, No Jacket" campaign. ★⇨クールビズ

ノールトヴェイク宣言〔1989年オランダで行われた環境問題に対する宣言〕the Noordwijk Declaration.

ノニルフェノール ⇨英和 nonylphenol

野猫(のねこ) ⇨野良猫(のらねこ)

野焼き〔春に枯れ草を焼き払うこと〕burning a field; the burning of a field;〔廃材などの野外焼却〕outdoor incineration《of timber waste》▶草原を〜する burn (dead grass) off a field; burn off dead grass. / ごみを〜する burn garbage in the open. / 廃棄物を〜する incinerate waste outdoors.

野山羊(のやぎ) a wild goat. ★⇨野生

野良犬(のらいぬ)〔飼い主のいない犬〕a stray (dog); a homeless [an ownerless] dog;〔捨て犬〕an unwanted [abandoned] dog;〔捨てられて野性化した〕a house [domestic] dog that has gone wild; a wild [feral] dog. ★⇨野犬(けん), 野生 ▶~を拾ってくる rescue a stray dog.

野良猫 〔飼い主のいない猫〕a stray [an ownerless] cat; an alley cat;〔捨て猫〕an unwanted [ownerless] cat; an abandoned cat [kitten];〔捨てられて野性化した〕a house [domestic] cat that has gone wild; a wild [feral] cat. ★⇨地域(地域猫), 野生

法面(のりめん) a slope face; a slope. ¶**法面崩壊** slope failure. **法面保護** slope protection.

ノルマル・ヘキサン normal hexane. ¶**ノルマル・ヘキサン抽出物質** (a) normal-hexane [n-hexane] extract.

ノンフロン冷蔵庫 ⇨フロン

は

パーク・アンド・ライド〔駅まで自家用車で行き、そこで駐車して電車・バスなどの公共交通機関に乗り換える方式〕a park-and-ride service. ¶パーク・アンド・ライド駐車場 a park-and-ride (parking) lot [car park]. パーク・アンド・バスライド a park-and-ride bus service.

パーク・ボランティア〔自然公園の美化・紹介などにあたる〕a park volunteer; an officially registered volunteer who acts as a guide and helps with the maintenance of a nature park.

パーク・レンジャー〔環境省の自然保護官〕a park ranger.

パークロルエチレン perchloroethylene; tetrachloroethylene.

バーゼル条約〔有害廃棄物の越境移動及びその処分に関する条約〕the Basel Convention (on the Control of Transboundary Movements of Hazardous Wastes and Their Disposal). ★⇨英和 Basel Convention

バーチャルウォーター〔ある商品の生産過程において消費されたと考えられる水の総量、仮想水〕virtual water; embodied water; hidden water. ★⇨英和 virtual water

バードウィーク〔愛鳥週間〕Bird Week. ★⇨愛鳥、鳥類、野鳥

バード・ウォッチング〔野鳥観察〕bird watching; birding. ▶〜をする bird-watch; bird; watch wild birds ¶バード・ウォッチャー a bird-watcher; a birder.

バードコール〔鳥を呼ぶ鳥笛・鳴きまね〕a birdcall.

バード・サンクチュアリー〔野鳥の保護区域〕a bird sanctuary; a sanctuary for wild birds.

バード・ストライク〔鳥と飛行機の衝突〕a bird strike.

バードソン〔見た野鳥の種数を競うイベント〕a birdathon.

バードバス〔小鳥用の水鉢〕a birdbath.

パーフルオロカーボン〔代替フロンの一種〕(a) perfluorocarbon (略: PFC). ★⇨フロン

パーマカルチャー〔資源維持・自足を意図した農業生態系の開発〕permaculture.

廃アルカリ ⇨アルカリ

廃液 waste fluid; effluents. ▶工場〜 factory waste(s). ★⇨パル

排煙

プ(パルプ廃液), 汚水, 汚染, 廃水 ¶**廃液処理施設** a liquid-waste treatment facility [plant].

排煙 flue [stack] gas;〔排出された煙〕smoke;〔出すこと〕smoke extraction. ▶焼却炉の〜 smoke from an incinerator. / 〜する let out smoke. ¶**排煙口** a smoke outlet. **排煙設備[装置]** a smoke-extraction system; smoke-extraction equipment. **排煙脱硫** flue-gas desulfurization. **排煙脱硫装置** a flue-gas desulfurization system; flue-gas desulfurization equipment.

煤煙(ばいえん) soot and smoke; (sooty) smoke. ★⇨英和 black carbon ▶〜で黒い be dark with soot and smoke. / 〜に汚れた sooty; smoke-stained; smokestained. / 〜の多い[無い] smoky [smokeless]. ¶**煤煙公害** soot pollution. **煤煙処理施設** sooty smoke treatment facility. **煤煙濃度計** a smoke indicator.

廃塩酸 spent [waste] hydrochloric acid. ★⇨廃酸

バイオ 〔「生物(工学)の, 生物学的な」を意味する接頭辞〕bio-. ★⇨生物 ¶**バイオアッセイ** ⇨生物(生物検定) **バイオアベイラビリティ** ⇨英和 bioavailability. **バイオ安全議定書** ⇨バイオセーフティー議定書 **バイオETBE** 〔バイオ・エチル・ターシャリー・ブチル・エーテル; 植物等を原料とするバイオエタノールと石油系ガスとの合成物〕bio-ETBE; bioETBE; bio-ethyl tertiary butyl ether. ★⇨英和 ETBE **バイオイメージング** 〔生体の器官, 細胞, 分子などを画像で見るための技術〕bioimaging. **バイオ医薬品** (a) biotech medicine; a biotech drug; (a) biomedicine. **バイオインダストリー** 〔生物技術産業〕(a) bioindustry. **バイオインダストリー協会** 〔財団法人〕the Japan Bioindustry Association (略: JBA). **バイオインフォ(ー)マティクス** 〔生命情報科学〕bioinformatics. ★⇨英和 bioinformatics **バイオエコロジー** ⇨英和 bioecology **バイオエシックス** 〔生命倫理〕bioethics. ★⇨英和 bioethics **バイオエタノール** 〔サトウキビやトウモロコシを原料とする燃料〕bio-ethanol. ★⇨英和 bioethanol **バイオエタノール車** an ethanol [a bioethanol] vehicle [car]. **バイオエレクトロニクス** 〔生物電子工学〕bioelectronics. **バイオオートグラフィー** 〔生物検定の一つ〕bioautography **バイオガス** 〔有機物の分解ガス〕(a) biogas. ★⇨英和 biogas **バイオガス発電**

biogas power generation.　**バイオ企業**　a biotech(nology) company.　**バイオ技術**　⇨バイオテクノロジー．　**バイオクリーン・ルーム**　〔微生物のない〕a bioclean room.　**バイオ化粧品**　biocosmetics.　**バイオサイエンス**　〔生命科学〕bioscience; life science.　**バイオ作物**　biotech crops.　**バイオ産業**　(a) biotech(nology) industry; (a) bio(-)industry.　**バイオ産業情報化コンソーシアム**　the Japan Biological Informatics Consortium（略：JBIC）．　**バイオ食品**　biotech foods; bio foods; bio-foods; biofoods;〔遺伝子組み替え食品〕genetically modified [engineered] foods. ★⇨遺伝子(遺伝子組み換え)　**バイオ人工臓器**　a bioartificial organ.　**バイオスフェア**　〔生物圏〕the biosphere. ★⇨英和 biosphere　**バイオ製品**　〔食品・医薬品〕a biological product; a biotech product; biologics.　**バイオセーフティー**　〔遺伝子組み換え動植物の無害性・有害微生物の取り扱い方法の安全性〕biosafety.　**バイオセーフティー議定書**　〔遺伝子組み換え生物の国際取引の規制に関する議定書〕the Biosafety Protocol. ★⇨英和 Biosafety Protocol　**バイオセーフティー・レベル**　a biosafety level（略：BSL）．　**バイオセーフティー・レベル1〔2, 3, 4〕**　〔病原微生物の危険度〕Biosafety Level 1〔2, 3, 4〕（略：BSL1〔2, 3, 4〕）　**バイオセーフティーレベル4施設**　a Biosafety Level 4〔BSL-4〕facility.　**バイオディーゼル燃料**　〔植物油から作られる軽油代替燃料〕biodiesel fuel（略：BDF）．★⇨英和 biodiesel　**バイオテクノロジー**　〔生物工学〕biotechnology;《口》biotech.　**バイオテクノロジー戦略大綱**　〔2002年，政府が策定〕(the Japanese government's) Biotechnology Strategy Guidelines.　**バイオテロ**　〔生物兵器テロ〕bioterrorism; biological terrorism.　**バイオトイレ**　〔微生物を利用して屎尿(にょう)を処理するトイレ〕a biotoilet.　**バイオ特許**　a biotechnology patent; a biotech patent; a biopatent.　**バイオトロン**　〔環境調節実験室〕a biotron.　**バイオナノサイエンス**　〔細胞生物学〕bionanoscience.　**バイオナノテクノロジー**　bionanotechnology.　**バイオニクス**　〔生物工学〕bionics.　**バイオ燃料**　〔生物体燃料〕biofuel. ★⇨英和 biofuel ▶セルロース系〜燃料 (a) cellulosic biofuel. ／第2世代〜燃料〔非食物系原料から作る〕(a) second-generation biofuel. ／非食物系〜燃料 (a) non-food biofuel; (a) biofuel made from non-food sources. ★⇨英和 second generation biofuel. **バイオ農薬** a biopesticide. **バイオパイラシー** ⇨英和 biopiracy **バイオハザード**〔生物研究・医療に由来する環境への危険；生物災害〕a biohazard. ★⇨英和 biohazard **バイオバンク**〔個人遺伝情報を集めたデータベース〕a biobank. **バイオビジネス**〔生物工学関連事業〕a biotech(nology) business. **バイオ肥料** a biofertilizer. **バイオフィードバック**〔生体自己制御〕biofeedback. **バイオフィルム**〔微生物が自己防衛のために寄り集まって作るバリア〕a (microbial) biofilm. **バイオプシー**〔生体組織検査〕a biopsy. **バイオプラスチック**〔バイオマス由来のプラスチック〕(a) bioplastic; bioplastics. **バイオベンチャー**〔バイオテクノロジー分野のベンチャー企業〕a bioventure company; a biotechnology venture business. **バイオポリティクス**〔ヒトのゲノム情報を用いた先端医療や，臓器売買問題・細胞核技術の規制などの生物技術に関する政策論〕biopolitics. **バイオポリマー**〔生体高分子〕a biopolymer. **バイオマス**〔生物資源・生物体燃料〕biomass. ★⇨英和 biomass / セルロース系バイオマス cellulosic biomass. / 木質(系)バイオマス wood [woody] biomass. / 木質バイオマス発電 wood biomass (electricity) generation. / バイオマス・エネルギー biomass energy. / 薪(まき)ストーブは最も単純なバイオマス・エネルギーの活用法だが，都市部では薪の確保が難しい．Wood stoves are the simplest way to extract energy from biomass, but it's hard to get firewood in cities. **バイオマス・エタノール** ⇨バイオエタノール **バイオマス・ガス化発電** biomass gasification (power) generation. / バイオマス・ニッポン総合戦略〔2002年に閣議決定されたバイオマス利活用のための総合的行動計画〕the Biomass Nippon Strategy; the General Biomass Strategy for Japan. **バイオマス燃料** biomass fuel. **バイオマス発電** biomass (power) generation. **バイオマス発電所** a biomass power plant [station]. **バイオマス・プラスチック** (an) organic plastic; a biomass plastic;〔植物プラスチック〕(a) vegetable(-based) plastic; plant-derived plastics. **バイオ野菜** biovegetables. **バイオリージョン**〔生命地域〕a bioregion. ★⇨英和 bioregion **バイオリズム**〔生体リズム〕a biorhythm. **バイオリソースセンター**〔理化学研究所の〕the RIKEN BioResource Center (略: BRC). **バイオレメディエーション** ⇨英和 bioremediation **バイオロジカル・クリーンルーム** a biological clean room. **バイオロジカル・コントロール** 〔天敵などによる害虫防除〕biological control; biocontrol.

排ガス ⇨排気(排気ガス) ★⇨低排ガス車 ¶**排ガス基準** emission standards [norms]; exhaust(-emission) standards [limits]. ▶~基準に準拠した compliant with emission standards [norms]. / ~基準をクリアする meet emission standards. **排ガス規制** exhaust [emission] control. / ~規制車 an emission-controlled vehicle. / 新しい~規制に従う comply with new exhaust restrictions. ★⇨自動車(自動車排ガス規制) **排ガス浄化触媒** an exhaust gas purification catalyst. **排ガス浄化装置** 〔ディーゼル車の〕a diesel particulate filter (略: DPF).

廃家電(製品) waste [used] household (electric) appliances. ★⇨電子ごみ, 廃電気・電子機器 ▶~の不法投棄 illegal disposal [dumping] of waste household (electric) appliances. / ~(の)リサイクル recycling of used household (electric) appliances. ★⇨家電リサイクル法

排気 exhaust(ion); evacuation; 〔排出されたガスなど〕used steam; exhaust (steam [gas]); 〔通風〕ventilation. ▶~する discharge; emit; vent. ¶**排気音** an exhaust sound; an exhaust note. **排気ガス** exhaust gas [fumes]; (engine) exhaust; (gas) emissions; waste gas. / 自動車の~ガスによる大気汚染が特に市街地において目立ってきた. Air pollution caused by automobile emissions has become especially noticeable in urban areas. **排気[排出]ガス再循環, 排気再循環** 〔自動車の〕exhaust gas recirculation (略: EGR). **排気管** an exhaust [eduction] pipe;〔通風管〕a vent pipe. **排気口** an exhaust port; a vent. **排気循環(方)式** 〔掃除機が吸い込んだ空気を吸い込み口のほうに循環させることで排気を減らす方式〕an exhaust recirculation system. / ~循環(方)式掃除機 a vacuum cleaner with exhaust recirculation. **排気扇** an exhaust fan; a ventilation [ventilating] fan. **排気装置** an air exhauster [escape]. **排気筒** an exhaust pipe;〔原子力発電所・汽船などの〕a funnel; a chimney. **排気バルブ** 〔エンジンの〕an exhaust valve. **排気量**

廃棄

〔エンジンの〕(engine) displacement.

廃棄 disposal; throwing away [out]; scrapping. ★⇨廃棄物, 排出 ▶～する dispose of…; throw away [out]…; get rid of…; scrap; discard. / 古い書類[中古電算機器]を～(処分に)する dispose of old documents [used computer equipment]. / 売れ残った野菜を～する throw out unsold vegetables. / 核兵器を～する scrap nuclear weapons. / これまで単に～されていたペットボトルから様々なものを作ることができる．Many things can be made from the plastic bottles that up to now we have simply been throwing away. ¶**廃棄家電(製品)** ⇨廃家電　**廃棄コスト** waste-disposal costs; the cost of waste disposal.　**廃棄自転車** a scrap bicycle.　**廃棄車両** a scrap vehicle. ★⇨廃車　**廃棄車両引き取り業者** a scrap vehicle dealer.　**廃棄二輪車** a scrap motorcycle.　**廃棄ロス** a loss due to the discarding of unsaleable merchandise.

廃棄物 waste (matter). ★⇨医療廃棄物, 核(核廃棄物), 建設(建設廃棄物), 災害廃棄物, 産業(産業廃棄物), 生活(生活廃棄物), 特別(特別管理産業廃棄物), 放射性(放射性廃棄物), 木質(もくしつ)(木質系廃棄物), 廃材, 廃品, ごみ, 垂れ流す, 処分, 海洋(海洋投棄), ゼロエミッション, 排出, 廃物 ▶気体～ gaseous wastes. / 一般～ normal (household) trash [waste]. ★⇨一般 / 未処理～ untreated waste (matter). ★⇨処理, 中間処理 / 産業～を適正に処理する dispose of industrial waste appropriately [properly]. / 実験(系)～ laboratory waste. / 石綿[アスベスト]～ asbestos waste. ★⇨アスベスト / 飛散性[非飛散性]石綿～ friable [non-friable] asbestos waste. ¶**廃棄物減量(化)** (a) reduction in the amount [volume] of waste; waste reduction.　**廃棄物集積場** a garbage dump.　**廃棄物収集運搬** waste collection and transport(ation).　**廃棄物収集運搬業者** a waste collection and transport(ation) business.　**廃棄物焼却炉** a waste-incineration plant. ★⇨焼却　**廃棄物処理** 〔原子炉の〕disposal of (radioactive) waste matter.　**廃棄物処理工学** waste disposal [management, processing, treatment] technology [engineering].　**廃棄物処理工場[処理場]** a waste-disposal plant; a garbage [refuse] dump.　**廃棄物処理(事)業** a (solid-)waste-management business; a (solid-)waste-disposal project. **廃棄物処理施設** a waste processing facility. **廃棄物処理法** the Waste Disposal Law. ★正式名称は「廃棄物の処理及び清掃に関する法律」(the Waste Disposal and Public Cleansing Law). **廃棄物分別収集** sorted waste collection. **廃棄物・リサイクル対策部** ⇨環境(環境省)

バイコロジー 〔自転車利用による環境保護〕bikecology.

廃材 scrap wood; timber waste. ¶**廃材チップ** wood-waste chips.

廃酸 (a) waste acid; (a) spent acid. ★⇨廃塩酸

廃車 an end-of-life vehicle [car]; a disused vehicle [car]; a scrapped vehicle [car];〔登録を抹消すること〕deregistration;〔登録を抹消した車〕a deregistered vehicle. ★⇨廃棄(廃棄車両), 英和 ELV Directive ▶廃車寸前の車 a car that is ready for the scrap heap. / 廃車にする scrap a vehicle; take a vehicle out of service.

排出 〔CO_2 などの〕emission; discharge; exhaust;(排泄) excretion. ★⇨温室(温室効果ガス), 削減, CO_2(和英篇末尾), 二酸化炭素 ▶～する〔CO_2 などを〕emit; discharge; excrete; release; eject; egest. / 老廃物を体外へ～する eliminate waste matter from the system. ¶**排出移動登録制度** 〔化学物質・環境汚染物質の〕the Pollutant Release and Transfer Register. ★⇨英和 PRTR **排出ガス** exhaust (gas). ★⇨排気(排気ガス) **排出ガス規制** ⇨排ガス(排ガス規制). **排出管** 〔機械の〕a discharge pipe; an exhaust pipe;〔生物の〕an efferent duct. **排出器** 〔皮膚・腎臓・肺など〕an emunctory. **排出器官** 〔排泄器官〕an excretory organ. **排出基準** 〔排水の〕⇨排水(排水基準);〔排気の〕⇨排ガス(排ガス基準) / 二酸化炭素の～基準 emission standards for carbon dioxide. **排出権** 〔温室効果ガスの〕the [a] right to emit《carbon dioxide》;〔排出許容量〕allowance. / ～権取引 emission(s) trading; allowance trading. ★⇨英和 emission trading / ～権取引所 an emissions-trading [emissions] exchange. / ～権取引制度 an emission(s) trading system (略: ETS). / ～権ビジネス emission(s) business **排出源[元]** 〔廃棄物・排気ガス・汚染物質などの〕an emission source; an emitter; a source《of pollution》. **排出原単位** 〔二酸化炭素の〕a (basic) unit of CO_2 emission(s). **排出口** an outlet; an issue.　**排出削減**　〔温室効果ガス・ダイオキシンなどの〕(an) emission(s) reduction. / 5.2%の〜削減量 an emissions reduction of 5.2%.　**排出削減対策**　an emissions reduction strategy.　**排出削減目標**　an emissions reduction target.　**排出(事業)者責任**　〔廃棄物を出す人や企業の責任〕waste disposer('s) liability.　**排出物**　〔工場廃液・煤煙など〕effluent;〔廃棄物〕industrial waste; waste matter [fluid];〔汗・糞・尿など〕excrement; excreta.　**排出弁**　an exhaust valve.　**排出抑制効果**　a reduction in emissions. / 高温燃焼にはダイオキシンの〜抑制効果がある．Combustion at high temperature is effective in reducing [curbing] emissions of dioxin.　**排出量**　the amount discharged; emissions.　**排出量許可証**　an emissions permit.　**排出量取引**　〔京都メカニズムの1つ〕Emissions Trading (略: ET).　**排出量取引所**　=排出権取引所．　**排出枠**　〔CO_2 などの〕an emissions quota [cap, allowance]《for CO_2》.

廃食(用)油　waste cooking oil. ★⇨廃油 ▶化学薬品で廃食油を固めてから捨てる dispose of waste cooking oil after solidifying it with chemicals.　¶**廃食(用)油燃料**　(a) fuel made from waste cooking oil.　**廃食(用)油燃料化**　turning [converting] waste cooking oil into fuel.

煤塵（ばいじん）　particles of soot; smuts. ★⇨降下(こう)煤塵, ディーゼル(ディーゼル煤塵), 煤煙　¶**煤塵濃度**　soot and dust concentration.

排[廃]水　**1**〔不要な水・液体を排出すること，排水〕draining; drainage; pumping out; bailing. **2**〔廃棄された水・液体，廃水〕wastewater. ★⇨暗渠(あんきょ), 温排水, 家庭(家庭排[廃]水), 工場(工場排[廃]水), 処理(排水処理), 生活(生活廃[排]水), 下水, 垂れ流す, 廃液, 汚水 ▶排水する drain (off);〔溝で流す〕dike;〔ポンプで〕pump out;〔水あかを〕bail. / 排水がよい[悪い] drain well [badly]. / 低地の水は排水が遅い．Water drains off slowly from low ground. / 産業廃水 industrial waste(water). / 自然排水法 free drainage. / 載貨排水 load displacement.　¶**排[廃]水管**　a drainpipe; a watershoot; a waste pipe.　**排水器**　a drainer.　**排水基準**　an effluent standard.　**排水渠**（きょ）　a drain; a culvert.　**排[廃]水口**　a drain; an overflow.　**排水孔**　〔甲板の〕a scupper;〔海綿の〕an osculum

《複数形 oscula》. 排水坑 a wastewater tunnel. 排水溝 a drainage ditch [channel, canal, ditch, pipe]; a drain; a drainageway. 排水工事 drainage construction. 排水作業 draining (work); pumping out. 排[廃]水処理 wastewater treatment; wastewater disposal; liquid-waste treatment. 排[廃]水処理設備 wastewater-treatment equipment. 排[廃]水処理装置 a wastewater treatment facility [plant]. 廃水堰(ぜき) a wasteweir. 排水設備[施設] drainage facilities. 排水トン数 displacement (tonnage). 排水能力 〔下水道などの〕(a) drainage capacity. / 下水道の排水能力を上回る豪雨 heavy rains that exceed the drainage capacity of [heavy rains that overload] the sewers. 排[廃]水パイプ ＝排水管. 排水ポンプ a drain [drainage, draining] pump. 排水路 an overflow; a spillway.

排泄(はいせつ) (an) excretion; (a) discharge; removal (from the body). ★⇨排出 ▶〜する excrete; evacuate; purge; discharge; eject; egest. ¶排泄管 an emunctory; a nephridium 《複数形 nephridia》. 排泄器官 an excretory [excretive] organ. 排泄腔 a cloaca 《複数形 cloacae》; an atrium 《複数形 atria》. 排泄作用 〔人体の〕(the body's) excretory process; 〔器官の〕(an) excretory function. /〜作用がある〔食物繊維などが〕help the body remove waste matter; help remove [flush] waste matter from the body; 〔薬剤などが〕help the body eliminate 《excess sodium》. 排泄腺 an excretory gland. 排泄物 excrement; excreta; bodily waste(s); excretion(s); f(a)eces; 〔糞便以外の, 新陳代謝による〕output; 〔尿〕liquid (body) wastes; 〔糞便〕solid (body) wastes. 排泄路 an excretory passage.

廃船 a scrapped vessel; a decommissioned ship; a vessel retired from service; (倉庫などに利用されている) a hulk. ▶〜にする scrap [decommission] a vessel.

廃タイヤ ⇨古タイヤ

ハイテク 〔先端技術〕high technology; high tech. ¶ハイテク汚染 high-tech pollution. ハイテク産業 a high-tech [high-technology] industry. ハイテク製品 a high-tech product. ハイテクごみ 〔電子機器廃棄物, 電子ごみ〕high-tech [hi-tech] waste; electronic waste; e-waste; electronic scrap; e-scrap.

売電 〔自家発電電力を売ること〕 selling electricity (to the power company). ★⇨太陽(太陽光発電), 英和 feed-in tariff

買電 〔自家発電電力を電力会社が購入すること〕buying electricity (from a private producer).

廃電気・電子機器 waste electrical and electronic equipment. ★⇨廃家電, 英和 WEEE

廃土 〔土木工事などで排出される土〕waste soil [earth].

廃[排]熱 waste heat. ★⇨温排水 ▶エンジンから出される廃熱 waste heat produced by [from] engines. / 廃熱を回収して利用する recover and utilize waste heat. ★⇨英和 cogeneration, ごみ(ごみ発電) / 都市廃熱 urban waste heat; waste heat from city buildings. / 高[低]温廃熱 high[low]-temperature waste heat. / 高[低]温廃熱回収 high[low]-temperature waste heat recovery; the recovery of high[low]-temperature waste heat. ¶**廃熱回収装置** a waste heat recovery system; a system to recover [use] waste heat. ★⇨熱(熱回収(効率)) **廃熱口** a heat vent; a heat exhaust port. **廃熱効率** (waste) heat exhaust efficiency. **廃熱装置** a waste heat exhaust system; a system to exhaust [disperse] waste heat. **廃熱利用** waste heat utilization; utilization of waste heat. **廃熱利用システム** a system utilizing waste heat.

廃農地 disused [abandoned] farmland. ★⇨農地

廃品 waste [unwanted, discarded] articles [materials]; waste products; junk. ★⇨廃棄物, ごみ ▶〜の山 a pile of unwanted items [materials]. ¶**廃品回収** collection [reclamation] of waste [unwanted, used] articles [materials, items]; collection [recovery] of scrap. ★⇨実施 **廃品回収運動** a campaign to collect used goods. **廃品回収業者** a junk dealer; a rag-and-bone merchant; a ragman; a junkman. **廃品回収車** a used goods [junk] collection truck. **廃品再生** reclamation (of waste articles); reuse of unwanted articles [materials]. ★⇨再生

廃物 a useless article [thing]; waste [discarded] material [products]; waste. ★⇨廃棄物 ▶〜になる become useless; go [run] to waste; go to the scrap heap; outlive *one's* usefulness. ¶**廃物利用** the utilization of waste material [products].

バイフューエル車 〔天然ガスとガソリンなど，二種類の燃料で走る自動車〕a bi-fuel vehicle.

廃プラスチック waste plastic.

ハイブリッド ⇨雑種 ¶**ハイブリッド・エンジン** a hybrid engine. **ハイブリッド(自動)車[カー]** 〔ガソリン・電気併用の〕a hybrid car; a hybrid (electric) vehicle (略: HV). ★⇨燃料電池 **ハイブリッド個体** 〔交雑種〕a hybrid individual. **ハイブリッド胚** a hybrid embryo. **ハイブリッド米** 〔雑種強勢による高収量一代雑種〕hybrid rice.

廃油 waste [rejected, defective] oil. ★⇨オイル，流出 ▶**廃植物油** waste vegetable oil. ¶**廃油塊[ボール]** balls of waste oil; a waste-oil ball. **廃油凝固剤** a used cooking oil coagulant [coagulating agent, hardener]. **廃油回収船** waste oil recovery ship. **廃油処理** waste oil disposal. **食用[家庭]廃油** waste cooking oil.

廃炉 the decommissioning of a nuclear power plant. ▶〔原子炉を〕~にする decommission.

曝[暴]露(ばく) 〔放射線や有害物質などに直接触れること〕exposure. ★⇨被曝，粉塵(粉塵暴露) ▶〔作業員などの〕アスベストへの職業(的)[間接]~ occupational [indirect] exposure to asbestos. ★⇨アスベスト ¶**曝露・リスク評価大気拡散モデル** the Atmospheric Dispersion Model for Exposure and Risk Assessment. (略: ADMER).

箱舟計画 〔絶滅危惧種の動植物のための〕an ark project.

破砕 ▶~する crush; smash; shatter; break to pieces; break up; fragment. ¶**破砕屑** rubble; detritus; residue; shivers; crushed waste; shredder dust; 〔自動車の〕an automobile shredder residue (略: ASR). **破砕ごみ** shredded waste [refuse]. ★⇨ごみ，廃棄物

バスケット方式 〔温室効果ガスの排出量を削減目標に一括して合算する方式〕a [the] basket approach.

爬虫類(はちゅう) 〔個々の爬虫類動物〕a reptile;〔総称的に〕the reptiles. ¶**爬虫類学** herpetology. **爬虫類学者** a herpetologist.

白化(はっ)現象 〔植物の〕chlorosis;〔動物の〕albinism. ▶サンゴの~ ⇨英和 coral bleaching. / 生命のゆりかごであるサンゴ礁の~が進めば，生態系は大きな打撃を受けることになる．If coral reefs, a cradle of life, undergo further bleaching, the ecosystem will be enormously impacted.

発癌(はつがん) ¶**発癌遺伝子** an

oncogene. **発癌因子** a carcinogen; a carcinogenic factor; a factor in causing cancer. **発癌作用** a carcinogenic effect [action]; carcinogenic activity [behavior]. **発癌性** carcinogenicity. ▶〜性の(ある) carcinogenic《chemicals, substances, agents》; cancer-causing[-forming, -producing]《substance》; oncogenic. / ヒトに対して〜性がある cause cancer [be carcinogenic] in humans. **発癌性試験** experimental carcinogenesis. **発癌(性)物質** a carcinogen; a carcinogenic [cancerigenic] substance; a cancer-causing substance. **発癌促進物質** a cancer accelerator; a (cancer) promoter. **発癌リスク** a《high》risk of (developing) cancer; carcinogenic risk.

曝気(ばっき) aeration. ★⇨英和 air stripping ¶ **曝気槽**〔浄化槽〕an aeration tank.

バックエンド対策〔放射性廃棄物の処理処分対策および原子力施設の廃止措置〕management of the back end of the nuclear fuel cycle; back-end management.

バックエンド費用〔核燃料サイクルでの使用済み燃料の再処理・廃棄物処理などに要する費用〕back-end costs.

発現率〔副作用・奇形・症状などの〕incidence rate.

伐採(ばっさい) (timber-)felling; logging; lumbering. ★⇨択伐(たくばつ) ▶〜する cut down《a tree, a forest》; fell《a tree》; log; lumber. / 違法〜 illegal timber-felling; illegal cutting. / 過(剰)〜 overcutting; overharvesting; overlumbering. ¶ **伐採跡地** a cutover area. **伐採検査** a stamp inspection. **伐採小屋** a logging cabin. **伐採材積** cut volume. **伐採時期** a felling season; a cutting period. **伐採者** a feller; a logger; a woodcutter; a lumberer; a lumberjack. **伐採順序** a cutting order [sequence]. **伐採地** (a) cutover. **伐採適期** a suitable period for felling. **伐採適量** a normal cut. **伐採点** a cutting height [point]; the height for cutting. **伐採道路** a logging road. **伐採標準量** a standard cutting volume. **伐採歩合** a felling [cutting] rate [percentage]. **伐採方法** a felling [cutting] method. **伐採木指定** marking trees for felling. **伐採面積** the area of a cutover. **伐採量** the amount of timber [lumber] (to be) felled [for felling]; the number of trees (to be) felled [for felling]; a fall. **伐採齢** a felling age. **伐採列** a cutting series. **伐採割当** a felling [cutting] allocation.

パッシブ・スモーキング ⇨英和 passive smoking

パッシブ・ソーラー・システム[ハウス] 〔外部動力なしの太陽エネルギー利用形態〕a passive solar system [house].

発生 〔事件・病気などの〕occurrence; an outbreak; 〔生物の個体形成〕development; growth; 〔出現〕appearance; (熱・電気などの) generation; production. ★⇨自然(自然発生), 大量(大量発生), 青粉(ぁぉこ), 赤潮 ▶～する occur; happen; crop up; break out; come into existence [being]; be generated; be produced; originate《from...》; come《from...》; spring《from...》;〔突然〕erupt. / 事故が～する an accident occurs [takes place, arises, comes about]. / 台風が発生する a typhoon forms [develops]. / 有毒ガスが～するおそれがありますので酸性の薬品と混ぜないでください．Please do not use in combination with acid chemicals, which may produce poisonous fumes. | Noxious fumes may develop if this product mixes with acids. / コレラの～ an outbreak of cholera. / ダイオキシンの～〔焼却炉からの〕the appearance of dioxin. / 蚊の～を防ぐ prevent the breeding of mosquitoes. /〔生物の〕異常～ an abnormal proliferation《of...》; an epidemic; a plague. / 個体～は系統～を繰り返す．Ontogeny recapitulates phylogeny. ¶ **発生医学** medical embryology. **発生遺伝学** developmental genetics. **発生確率** an [the]《earthquake, accident》probability. **発生器** 〔オゾン・蒸気・信号などの〕a generator. **発生源** a source; the source《of an outbreak of cholera》. ★⇨汚染(汚染源) / 悪臭の～源 a [the] source of a (bad) smell; where a smell comes from. / 騒音[汚染]～源 a [the] source of noise [pollution]. / 大気汚染の～源 a [the] source of air pollution; an air pollution source. / 室内に発生するダニやカビがアレルギーの～源として問題になっている．Mites and mold developing in homes [Domestic mites and mold] are becoming a concern as a source of [giving rise to] allergies. **発生工学** development(al) engineering. **発生生物学** developmental biology. **発生生理学** developmental physiology. **発生責任** 〔公害・

薬害などの〕responsibility for an outbreak; the 《government's》responsibility for the outbreak of 《drug-induced hepatitis》. **発生装置** 〔ガスなどの〕a producer. **発生遅延** hypoplasia; arrested development. **発生能** potency; development potency. **発生反復説** (the) recapitulation theory. **発生頻度** the frequency of occurrence 《of earthquakes》. / 網膜色素変性症は遺伝病としては〜頻度が高く，国内には約3万人の患者がいる．Retinitis pigmentosa occurs frequently as a hereditary disease, with about 30,000 affected individuals in Japan. **発生部位** 〔腫瘍などの〕the site of origin [occurrence] 《of the tumor》. **発生密度** 〔事故・病気などの〕(the) occurrence density; the incidence [number of 《accidents》] per unit area. **発生予察** 〔病害虫の〕forecasting plant disease epidemics; a plant disease forecast. **発生率** (an) incidence 《of two per thousand》; an incidence rate; a rate of occurrence. **発生例** an outbreak; a case; an instance 《of a disease》; 〈集合的に〉(an) incidence. / 6〜8月のインフルエンザの集団〜例はきわめて少ない．The incidence of large-scale influenza epidemics is extremely low between June and August. **発生炉** 〔ガスの〕a gas generator [producer]. **発生炉ガス** air gas; producer gas.

発電 〔電気の発生〕the generation of electricity; the production of electric power. ★⇨温度(温度差発電)，火力発電，原子(原子力発電)，ごみ(ごみ発電)，コンバインド・サイクル発電，自家(自家発電)，地熱(地熱発電)，水路(水路発電)，太陽(太陽光発電)，潮力(潮力発電)，波力(波力発電)，風力(風力発電)，RDF(和英篇末尾) ▶〜する generate [produce] electricity; produce electric power. / 川を〜に利用する use a river to generate electricity. / 水力〜 hydropower [hydroelectric power] generation. / 小水力〜 small hydropower. / 水力〜所 a hydroelectric power plant [station]; a hydropower plant [station]. / 自流式(水力)〜所 a run-of-the-river power plant. / ダム[貯水池]式(水力)〜所 a reservoir power plant. / 揚水式(水力)〜所 a pumped storage power plant. ¶ **発電器官** 〔生物の〕an electric organ. **発電原価** power generation costs; the cost of generating electricity. **発電効率** power generation efficiency. **発電床** 〔通る人の震動で発電する〕a power-generating floor; flooring incorporating piezoelectric materials that convert the tread of pedestrians into electricity. **発電素子** a power generation device. **発電単価** a power generation cost《of 6 yen/kWh》. **クリーン石炭発電** 〔ガス化した石炭を燃料として行う発電〕clean-coal power generation. **クリーン石炭発電所** a clean-coal power plant [station].

発泡スチロール expanded [expandable] polystyrene;〔商標名〕Styrofoam.

発泡プラスチック[樹脂] plastic foam; foam(ed) plastics; expanded plastics.

バラスト水 ⇨英和 ballast water

パラチオン 〔農薬〕parathion (insecticide). ¶**パラチオン中毒** parathion poisoning. **パラチオンメチル** 〔農薬〕parathion-methyl.

バリ・ロードマップ ⇨英和 Bali roadmap

波力 wave power; wave force. ¶**波力発電** wave-powered electrical generation; wave power generation. **波力発電所** wave-power station.

バルディーズ原則 ⇨英和 Valdez Principles

パルプ pulp. ▶〜にする reduce to pulp; pulp. / 広葉樹[針葉樹]〜 (a) hardwood [softwood] pulp. / 人絹[木材, 砕木, 製紙]〜 rayon [wood, ground, paper] pulp. / 代用〜 a substitute for wood pulp. / 無塩素〜 ECF paper. ★ECF は elemental chlorine-free の略. ¶**パルプ工場** a pulp mill. **パルプ材** pulpwood. **パルプ・シート** (a) pulp sheet. **パルプ成形** pulp molding. **パルプ製造機** a macerator. **パルプ廃液** pulp waste liquor; black liquor.

半減期 〔放射性元素の〕a half-life; a half-life period. ▶このアイソトープの〜は 23 分だ. This isotope has a half-life of twenty-three minutes. | The half-life of this isotope is 23 minutes. / その粒子は半減期 10^{-8} 秒で崩壊する. The particle decays with a half-life of 10^{-8} second.

繁殖 breeding; reproduction; propagation; multiplication; (増殖) increase;〔培養〕culture. ▶〜する breed; multiply; propagate [reproduce] itself;〔殖える〕increase;〔培養する〕cultivate. / バクテリアの〜 the propagation of bacteria. / 異種〜

broad breeding; crossbreeding; outbreeding. / 人工〜 artificial breeding [reproduction]. / 同種〜 narrow [close, in-and-in] breeding; inbreeding. ¶繁殖期 a breeding season. 繁殖行動 reproductive [breeding] behavior. 繁殖貸与 〔繁殖のために動物園間で行われる〕a breeding loan. 繁殖地 a breeding place. 繁殖率 a breeding coefficient. 繁殖力 propagating [procreative] power; fertility;〔潜在的〕breeding [biotic] potential. /〜力のある fertile;〔多産な〕prolific. 繁殖(容)器, 繁殖箱 a hatching container [box].

バンドン会議 〔1955年のアジア・アフリカ会議〕the Bandung Conference. ¶バンドン宣言 the Bandung Declaration.

反応 (a) reaction; (a) response. ▶〜する react《to...》; respond《to...》. / 過敏に〜する overreact; react oversensitively. / 身体がアレルゲンに〜してくしゃみや微熱などさまざまな症状を起こす. The body reacts to allergens and produces various symptoms such as sneezing and slight fever. ★⇨アレルギー(アレルギー反応) / 尿検査で薬物使用の〜が出た. A urine test proved positive for drugs. /(細菌免疫検査などに対して)〜を示す人[動物] a sensitive subject [person, animal]; a reactor. / アルカリ性の〜を示す show an alkaline reaction. / この薬品は酸に〜して有毒ガスを発生します. This chemical reacts with acid to produce a poisonous gas. / 拒絶[拒否]〜 rejection. / 陽性[陰性]〜 a positive [negative] reaction. ¶反応度 〔原子炉の〕reactivity. 反応度事故 〔原子炉の〕a reactivity accident. 反応炉 〔原子炉〕a reactor.

氾濫(はんらん) 〔洪水〕a flood. ★⇨水害 ▶〜する overflow; flood; flow [run] over《the banks》; inundate《a district》. /(川が)〜している be flooding (its banks); have overflowed (its banks); be in flood. / 利根川が〜した. The river Tone has overflowed (its banks) [risen above its banks]. / その雨で川の水が村一面に〜した. The river flooded [inundated] the whole village. ¶氾濫原 a flood plain.

ひ

日当たり ⇨日照 ▶わが家は南向きで～がいい．Our house faces south and gets plenty of sun (shine). / 南側にビルが出来て～が悪くなった．A building has gone up to the south and blocked out our sunshine [stopped the sun coming in, deprived us of sunlight]. / ～のよい部屋 a sunny room; a room which gets the sun. / ～の悪い家 a house with no [little] sun; a house that doesn't get the sun [much sunlight]. / このキノコは～の悪い場所に生える．This mushroom grows in the shade [in places that don't get the sun]. / お宅の～はどうですか．How much sunlight does your house get? | How sunny is your house? / ～良好．〔不動産広告〕Sunny. | Plenty of sunlight.

ピーク ¶ピークオイル論〔このまま石油を使い続ければ石油産出量がそう遠くない将来にピークに達し，その後は徐々に減少するという説〕the peak oil theory. ★⇨英和 peak oil　ピーク・カット〔最大需要電力の抑制〕peak cut. ピーク逆電圧　peak inverse [reverse] voltage.　ピーク時 (a) peak time. ▶～時に(は) at the [*sth's*] peak; during peak periods [hours]. / ～時の電力消費量 electricity [energy] consumption [usage] during peak (time) periods; (on-)peak electricity [energy] usage; peak-time electricity [energy] consumption.　ピーク出力 (a) peak capacity.　ピーク電圧 (a) peak voltage.　ピーク負荷 (a) peak load.　ピーク・ロード〔最大需要〕a peak load.　ピーク・ロード・プライシング〔ピーク時割高の料金制〕peak-load pricing.

ビーチ ⇨海岸 ¶ビーチ・グラス，ビーチ・ガラス〔海岸に漂着したガラスのかけら〕sea glass; beach glass.　ビーチコーミング〔漂着物拾い〕beachcombing ⟪for fun⟫. 動 beachcomb.

ヒート ⇨熱 ¶ヒート・アイランド a heat island. ★⇨英和 heat island　ヒート・アイランド現象 a [an urban] heat island effect [phenomenon]. ▶臨海部の無秩序な高層化によって海風が

さえぎられ，都心の〜アイランド現象は悪化の一途をたどっている．The uncontrolled construction of high-rise buildings along the bay blocks sea breezes and contributes to the worsening of the heat island effect in the center of the city.

ヒート・ポンプ 〔熱を低温側から高温側に移す〕a heat pump. **ヒートポンプ給湯器** a heat pump water heater. **ヒート・ポンプ蓄熱** (combined) heat pump/thermal storage. **ヒートポンプ・蓄熱センター** 〔財団法人〕the Heat Pump & Thermal Storage Technology Center of Japan (略: HPTCJ).

非営利組織[団体] ⇨ NPO(和英篇末尾)

ビオトープ 〔生物生活圏〕a biotope. ★⇨英和 biotope ▶学校〜 a school biotope. / 〜作りは，町に小さな自然を呼び戻すだけでなく，環境教育の手段としても有効である．The creation of biotopes not only brings nature back into towns but also helps with environmental education. ¶**ビオトープ管理士** 〔日本生態系協会の認定資格〕a biotope planner/builder. **日本ビオトープ管理士会** the Association of Biotope Planners and Builders of Japan.

被害 damage; harm. ★⇨海面, アスベスト(アスベスト被害), 外来(外来種被害防止法), 健康(健康被害), 公害, 黄砂(ﾎﾞ), 鉱毒, 消費(消費者被害), 森林(森林被害), 騒音(騒音被害), ¶**被害算定型影響評価手法** the life-cycle impact assessment method based on endpoint modeling (略: LIME). ★⇨環境(環境被害) **農作物被害** 〔台風・鳥獣などによる〕damage to crops [agricultural products]. ★⇨作物, 農作物, 全滅

日影規制 off-site shadow control.

日傘(ﾋﾞｶﾞｻ)**効果** 〔火山噴煙による日照減〕an [the] umbrella effect.

東アジア海域環境管理パートナーシップ Partnerships in Environmental Management for the Seas of East Asia. (略: PEMSEA).

東アジア地域ガンカモ類重要生息地ネットワーク the Anatidae Site Network in the East Asian Flyway.

干潟(ﾋｶﾞﾀ) tidal flats [wetlands]; (tidal) mudflats;〔海岸部の〕a beach at ebb tide. ★⇨人工(人工干潟), 潮(ｼｵ) ▶〜に住む生き物 the life of the (tidal) mudflats; animals that live on the tidal wetlands. / 各地で〜を守る運動が起きている．There are campaigns in many regions to protect the tidal mud flats.

光触媒(ひかりしょくばい) a photocatalyst. ¶ **光触媒加工** photocatalyst processing. **光触媒工業会** the Photocatalysis Industry Association of Japan（略：PIAJ）. **光触媒物質** (a) photocatalytic material.

光分解 〔生化学的な〕photolysis; photodissociation; photodecomposition;〔原子核の〕(nuclear) photodisintegration. ¶ **光分解性プラスチック** (a) photodegradable plastic. ★⇨分解

ピグー税 ⇨英和 Pigo(u)vian tax.

飛散 ▶～する disperse［scatter, fly off］(in all directions). ／風に～する be scattered in the wind. ／花粉～情報 the pollen count information［forecast］. ／花粉の～量 an amount of pollen (suspended) in the air. ¶ **飛散開始日** 〔花粉の〕the start of the pollen season (when there is more than 1 grain of pollen per square centimeter). **飛散性石綿** ⇨石綿(せきめん)

ビジター・センター 〔観光地の〕a visitor center.

微小粒子状物質 〔直径2.5マイクロメーター以下の浮遊粒子状物質〕fine particulate matter（略：PM2.5）; a respirable particle. ★⇨粒子状物質

ビスファチン 〔内臓脂肪から出るホルモン〕visfatin.

ビスフェノール A 〔樹脂原料〕bisphenol A.

非政府組織 ⇨NGO（和英篇末尾）

微生物 a microorganism; a microscopic organism; a microbe; a germ; minute forms of life. ▶～がうようよしている水槽の水 cistern water swarming with microbes［microorganisms］. ／～の働きによって発酵する ferment［undergo fermentation］by microbial action. ¶ **微生物遺伝学** microbial genetics. **微生物学** microbiology. **微生物学者** a microbiologist. **微生物系統保存施設** (a facility housing) a collection of microorganisms. **微生物試験** 〔食品・医薬品・化粧品などの〕microbial［microbiological］testing;（1回の）a microbial［microbiological］test. **微生物生態学** microbial ecology. **微生物センサー** 〔微生物を利用した〕a microbial (bio)sensor; a microbiosensor. **微生物たんぱく質** a single-cell protein（略：SCP）. **微生物電池** a microbial fuel cell. **微生物農薬** a microbial biopesticide. **微生物発電** microbial power generation.

非接触給電 〔電気自動車などへの〕inductive [induction] charging; non-conductive power transmission. ¶非接触給電システム an inductive [induction] charging system.

砒素(ひそ) arsenic. ¶砒素化合物 an arsenic compound; an arsenide. 砒素中毒 arsenic poisoning;〔慢性の〕arsenism.

人里植物 a ruderal plant; a plant that grows near human settlements.

避妊手術 〔ペットなどの〕⇨去勢, 不妊

被曝 〔放射線を浴びること〕exposure to radioactivity; radiation exposure;〔原水爆の被害〕⇨被爆. ★⇨放射線 ▶〜する be exposed to radiation. / 職業〜〔放射線技師・医師・看護師などの被曝〕occupational exposure to radiation. / 胎内〜 f(o)etal exposure to radiation; f(o)etal radiation damage;〔原水爆による〕⇨被爆(胎内被爆). ¶被曝管理 radiation exposure control; radiation protection. 被曝事故 a radioactive leak (affecting people); a case of accidental exposure to radiation; a radioactivity accident. 被曝者 〔放射線を浴びた人〕a person (who was) exposed to radiation; a radiation [radioactivity] victim;〔特に原水爆の〕⇨被爆(被爆者). 被曝線量 an exposed [exposure] dose; a radioactive dose; a dose of radiation [radioactivity].

被爆 ▶彼は広島で〜している. He was at Hiroshima when the A-bomb dropped. | He went through the A-bomb attack on Hiroshima. / 胎内〜〔原水爆による〕f(o)etal radiation damage resulting from atomic bombing. ¶被爆者 〔原水爆の放射線を浴びた人〕a person who was exposed to radiation from an A-[H-]bomb;〔原水爆の被災者〕a survivor [victim] of an A-[H-]bomb [an atomic air raid];〔広島・長崎の〕a *hibakusha*. ★⇨認定(認定被爆者) 被爆者健康手帳 a *hibakusha* health book. 被爆者団体 an A-bomb survivors' group. 被爆二世 a second-generation *hibakusha* [survivor of the Hiroshima or Nagasaki A-bombs].

非附属書Ⅰ国 〔気候変動枠組条約の附属書に記載されていない国; 温室効果ガス削減の義務がない国〕a non-Annex I country [party]. ★⇨英和 Annex Ⅰ Countries

非メタン炭化水素 a non-methane hydrocarbon（略:

NMHC).

百選 ▶日本の渚(なぎ)～に選ばれる be designated one of Japan's 100 [One Hundred] Most Beautiful Beaches. ★⇨名水百選

日焼け a sunburn; a (sun)tan. ★普通, 前者は赤い炎症を起こす日焼け, 後者は浅黒くなる日焼けをいう. ▶～する get sunburned; get tanned (by the sun); get a tan. /～した肌 (a) sunburned [tanned] skin; a tan. ¶**日焼け止め** (a) sunscreen; (a) sunblock oil [cream, cosmetics]. ★⇨英和 sun protection factor

費用 expenditure(s); (a) cost. ★⇨負担, 企業, 撤去

氷河 a glacier. ▶地球温暖化の結果, アルプス地方の～がかなり後退した. As a consequence of global warming, glaciers in the Alps have receded considerably. ¶**氷河期** a glacial stage [age, epoch]. **氷河湖** a glacial lake. **氷河時代** an ice age; the Ice Age; the glacial period. /〔動植物の種の〕～時代からの生き残り a survivor from the Ice Age. **氷河性海面変動** glacier-related sea-level change. **氷河成層** (a) glacial sediment. **氷河堆積物** (a) glacial deposit; a boulder drift. **氷河地形** a glacial landform.

病害虫 a [an agricultural] pest; an insect or a disease that damages crops;〔虫〕⇨害虫.

氷山 an iceberg.

表示〔食品・製品などの〕a label; labeling; (an) indication; (an) expression; (a) manifestation. ★⇨禁煙(禁煙表示), 警告表示, 原料(原料表示), 地理的表示, 人工(人工着色料), 食品(食品表示), 環境(環境ラベル) ▶瓶入り蜂蜜のラベル～を読む read the label on a jar of honey. / 食品は(ラベルの)～をよく見てから買います. I always read [check] the label carefully before buying food. / 虚偽[不当]～ mendacious [fraudulent, misleading, false] labeling. ★⇨偽装(偽装表示) / 注意～〔利用法を指示する〕a (printed) warning; a warning label. ¶**表示価格** a list [posted] price. **表示記載事項**〔医薬品の効能などの〕labeling claims. **表示使用期限**〔医薬品のラベルに表示された〕an [a printed] expiration date. **表示内容**〔医薬品の用法・使用期限など〕a labeling statement.

標識 ¶**標識魚**〔放流調査の〕a tagged fish. **標識再捕法**〔個体数推定法〕the mark-and-recapture method. **標識鳥** a bird marked with a band [ring]; a banded bird; a ringed bird. ★ ⇨放鳥(標識放鳥)　**標識放流** release of tagged fish.

標準木〔桜の開花宣言をする際などの〕a standard tree (for determining when 《cherry》 trees are in bloom). ★⇨開花

氷床(ひょうしょう)　an ice sheet.　¶**氷床コア**〔掘削試料〕an ice core sample.

漂着　▶〜する be cast [thrown, washed] ashore; drift ashore.　¶**漂着物**　driftage; something thrown [washed] up on the shore.　★⇨ごみ(漂着ごみ), 海岸(海岸漂着物), ビーチコーミング

漂白　▶〜する bleach.　¶**漂白液** (a) bleaching solution.　**漂白粉** (a) bleaching powder; chloride of lime.　**漂白剤** (a) bleaching agent; (a) bleach; a bleacher; a decolorant; a decolorizer.　**無漂白パン**　brown bread.

標本〔生物個体やその一部〕a specimen; a sample;〔生体の一部〕a preparation.　▶陳列用の〜 a specimen for display; a museum specimen / 動物[植物]の〜 a zoological [botanical] specimen.

漂流　▶〜する drift; float; be adrift.　¶**漂流作用**　driftage; drifting.　**漂流瓶**　a drift bottle; a floater.　**漂流物**　flotsam; floatage; driftage.　**漂流木**　driftwood.

飛来　¶**飛来経路**〔渡り鳥などの〕a migration route; a route of migration;〔黄砂・花粉などの〕a dispersion route. ★⇨黄砂(こうさ), 渡り鳥, 変化　**飛来地**　a migration [migratory] destination; the destination of migrating birds.　▶オオハクチョウの〜地 the destination of the whooper swans.

肥料　manure; (a) fertilizer; (a) plant food;〔堆肥〕compost. ★⇨堆肥(たいひ), コンポスト　▶〜の3要素〔窒素・燐酸・カリ〕the three major nutrients (nitrogen, phosphorus, and potassium). / 〜を施す put fertilizer [manure]《on [in]...》; manure; fertilize.　¶**化学肥料** chemical fertilizer [manure].　**カリ肥料**　potassium [potash] fertilizer.　**完全肥料**　complete [general, normal] manure.　**人造[合成]肥料**　artificial [synthetic] fertilizer [manure].　**窒素肥料**　nitrogen [nitrogenous] fertilizer.　**燐酸肥料** phosphate [phosphatic] fertilizer.

微量　▶〜の in minute [tiny, minuscule] quantities [amounts, numbers]; a minute [tiny, minuscule] quantity [amount, number] of...; 〔計量できない〕imponderable. / その牛乳から〜の農薬が検出された. A minute [tiny] amount of pesticide was found in the milk. | Agricultural chemicals, in almost imperceptible quantities, were discovered in the milk. ¶**微量栄養素** 〔ビタミンなど〕a micronutrient.

ビル風 (a) (strong) pedestrian-[street-]level wind caused by (high-rise) buildings. ▶〜の風洞実験[数値シミュレーション] a wind tunnel [numerical] simulation of wind environment around (high-rise) buildings.

貧栄養 ▶〜の oligotrophic. ¶**貧栄養湖**[**湿原**] an oligotrophic lake [moor].

貧酸素水塊 〔溶存酸素量がきわめて少ない水塊；魚介類などの大量死を誘発する〕an oxygen-deficient water mass.

品質 quality. ▶高[低]〜の製品 good-[poor-]quality products; high-[low-]quality products; products of good [bad, poor] quality. / 〜を検査する check [inspect] the quality of ... / 〜を改良[保証]する improve [guarantee] the quality of ... ¶**品質管理** quality control (略：QC). **品質証明書** a certificate of quality. **品質表示** quality labeling. **品質保持期間** a shelf life《of one week》. **品質保持期限** ⇨賞味期限

品種 a kind; a sort; a description; 〔等級〕a grade; 〔変種〕a variety; 〔家畜の〕a breed. ★⇨種(しゅ), 地理的品種 ▶改良〜 an improved grade [caliber]. / その〜のリンゴは岩手県以南では栽培されていない. That variety of apples is not grown in Iwate or other prefectures to the south of Iwate. ¶**品種改良** breed [variety] improvement; 〔家畜〕improvement of (the) breed; selective breeding《of cattle》; 〔植物〕improvement of plants; plant breeding; selective breeding of plants. / 〜改良する improve the breed [variety] 《of...》. **品種登録** 〔農林水産省に対して行う〕(plant) variety registration.

ふ

ファームステイ 〔農場ホームステイ〕a farmstay. ►〜する stay on a farm. ★⇨農家（農家民宿）

ファウナ ⇨英和 fauna

ファクターX 〔環境効率指標〕factor X.

ファサード保存 〔歴史的建造物の正面外壁のみの保存〕facade preservation; facadism.

フィードインタリフ制度 ⇨英和 feed-in tariff

フィールド・マーク 〔野鳥観察で種類判別に役立つ識別特徴〕a field mark.

フィッシュ・ロンダリング 〔獲ったマグロなどの魚種，捕獲漁場，捕獲船名等の付け替え〕fish laundering.

風化 weathering; aeration; efflorescence. ►〜する weather; effloresce. / 岩は風雨にさらされて〜する．Rocks are weathered by exposure to wind and rain.

風害 wind damage; damage from the wind. ►その地方は〜が甚だしかった．The storm [wind] did a lot of damage in that district.

風景 ⇨景観 ►残しておきたい日本の原〜 an archetypal image of Japan one wishes to preserve. ¶**風景式庭園** a landscape garden. **風景指示板** 〔展望台などでの〕a sign pointing out features of the landscape. **風景写真** a landscape [scenic] photograph.

風致 ¶**風致地区** an area of scenic beauty; a scenic area. **風致保存地区** a scenic preservation area [district]. **風致保安林** a scenery conservation forest. **風致林** a forest given official recognition and protection as part of a scenic area.

フード 〔食べ物〕food. ►ファースト[スロー]〜 fast [slow] food. ★⇨英和 slow food ¶**フード・ガイド** 〔食生活上の指針〕a guide to healthy eating; a dietary guide. **フード・ガイド・ピラミッド** 〔米農務省が作成した食生活指導の三角図〕a [The] Food Guide Pyramid. **フード・システム** 〔食料生産から消費までの諸産業の連鎖関係〕a food system. **フード・マイレージ[マイル]** 〔食料の輸送距離に輸送量を加味した環境負荷の指標〕food mileage; a food mile. ★⇨英和

food miles

風力 the velocity [force] of the wind; wind force [velocity]. ▶～で動く, ～式の wind-powered. ¶**風力エネルギー** wind(-generated) power. **風力階級** a wind(-force) scale; a scale of wind velocity. **風力記号** 〔天気図の〕a wind flag; a wind-force symbol. **風力計** an anemometer; a wind gauge. **風力係数** a coefficient of wind force. **風力資源** wind-power resources. **風力図** a wind-force diagram. **風力選別機** 〔ごみ処理場などの〕an air classifier. **風力タービン**〔発電用風車〕a wind turbine; a windmill;〔発電機〕a wind (power) generator; a windmill generator. **風力発電** wind (power [powered]) generation. **風力発電所** a wind power plant [station]. **風力発電塔** a wind (turbine) tower.

フェアトレード ⇨英和 Fairtrade ¶**フェアトレード商品** a Fairtrade product.

フェニトロチオン〔殺虫剤〕fenitrothion.

フェロモン〔性誘因物質〕a pheromone. ▶多くの昆虫は～を出して交尾の相手を引きつける. Many insects release pheromones to attract mates.

富栄養 ～の eutrophic. ¶**富栄養化** eutrophication. **富栄養湖**[湿原] a eutrophic lake [moor].

孵化(ふか) hatching; incubation. ▶～する hatch; incubate; sit on《eggs》; set eggs《under a hen》. / ～したばかりのひな a chicken hatchling. / ひなを～させる hatch out chickens. / ひなが～した. The chicks have hatched. | The chicks are out. / 人工～ artificial incubation. ★ ⇨自然(自然孵化) ¶**孵化器** an (artificial) incubator. **孵化鶏卵**〔発生途上の卵〕an embryonated egg. **孵化細胞** a germ cell. **孵化施設** a hatchery facility. **孵化場** a hatchery. **孵化放流** raising and releasing《salmon (fry)》. **孵化用卵** a hatching egg.

不快指数 the discomfort index (略: DI);〔温湿指数〕the temperature-humidity index (略: THI).

不活性 inactivity; inertness. ▶～の inactive; inert; noble《metal》; neutral. / エイズウイルスを～化する inactivate HIV. / 触媒を～化する deactivate a catalyst. ¶**不活性ガス**[気体] an inert gas. **不活性充填**(じゅうてん)**剤** an inert filler. **不活性溶媒** an inert solvent; an inactive solvent.

複合 ¶**複合汚染** multiple pollution [contamination]; complex contamination [pollution]. **複合(サイクル)発電** ⇨コンバインド・サイクル発電 **複合(型)災害** a compound [multiple] disaster. **複合毒性** ⇨毒(毒性) **複合肥料** a compound fertilizer.

輻射(ふく) ⇨放射 ¶**輻射暖房機** an electric radiator. **輻射電力** radiation power. **輻射熱** radiant[radiation] heat. **輻射冷暖房** radiant [panel] heating and cooling.

複層ガラス double glazing. ▶高断熱〜ガラス low-E glass; low-emissivity glass.

複層林 a multistoried forest.

覆土(ふく) covering up seeds with soil.

伏流 an underflow. ¶**伏流水** riverbed [subsoil, infiltration] water.

不耕起栽培 〔田畑を耕さずに作物を栽培する農法〕no-till(age) [natural] farming.

不作 a bad [poor] harvest [crop]; a scanty harvest; a crop [an agricultural] failure; a short [lean] crop [yield]. ▶5年も〜が続いてその国は食糧難だ. After five whole years of crop failures, the country is in a food crisis. / ひどい冷夏のため〜になった. Due to the most unseasonably cool summer, the crops have failed. / 〜の年 a lean year. / 今年は米が〜だ. The rice crop has failed this year. | We have had a short crop of rice this year. | This has been a bad year for the rice crop. / ひでりに〜なし. Drought never brings dearth.

富士山憲章 〔富士山の環境保全のために静岡・山梨両県が制定したもの〕the Mount [Mt.] Fuji Charter.

腐植 humus. ▶その土壌は〜に富んでいる. The soil is rich in humus. ¶**腐植栄養** dystrophy; dystrophia. **腐植栄養化** dystrophication. **腐植栄養湖** a dystrophic lake. **腐植酸** humic acid. **腐植土** humus (soil); mold. ★⇨腐葉土

浮水植物 a floating plant.

腐生 〔死物寄生〕saprophagy. ¶**腐生細菌** saprotrophic [saprophytic] bacteria. **腐生者** a saprotroph [saprobe, saprovore, saprophage]; a saprophyte; a saprozoite. **腐生植物** a saprophyte. **腐生生物** a saprotrophic [saprophytic] organism; a saprotroph [saprophyte].

附属書Ⅰ国 〔気候変動枠組条約の

附属書に記載されている国；温室効果ガス削減を義務づけられている国〕an Annex I country [party]. ★⇨英和 Annex Ⅰ Countries

ブタクサ花粉 ragweed pollen. ★⇨花粉

フタル酸 phthalic acid. ¶ フタル酸塩[エステル] phthalate. **フタル酸ジエチルヘキシル** 〔内分泌撹乱作用の疑いのある可塑剤〕diethylhexyl phthalate (略: DEHP).

負担 〔分担するもの〕a burden; (賦課) an imposition; (責任) a responsibility; a liability; (義務) an obligation; (支出) defrayment; (租税の) incidence. ▶汚染者〜原則⇨英和 PPP / 費用を〜する bear [shoulder, stand] the expense; meet the cost. / その費用は利用者の(自己)〜になる. The expenses are charged to the user. | The cost falls on [is borne by] the user. / この法律は消費者にリサイクル費用の一部〜を求めるものである. This law requires the consumer to pay [bear, defray] some of the cost of recycling. / ペットボトルのリサイクルにかかる費用はそれを生産した企業が〜するべきだ. It should be the responsibility of the producers to bear the costs of plastic bottle recycling. | The producers should be liable for the costs of recycling plastic bottles.

普通地域 〔国立公園内などの保護区分の1つ〕an ordinary zone.

物質 matter; substance; material. ¶ **物質主義** materialism. **物質循環システム** a material(s) recycling system. **物質代謝[交代]** metabolism. **物質文化[文明]** material culture [civilization].

弗素(ふっそ) fluorine. ▶(虫歯予防のため)水道水に〜を少量入れる fluoridate a water supply. / 〜は虫歯を予防すると信じられている. Fluorine is believed to prevent [be a deterrent to] tooth decay. ¶ **弗素(中毒)症** fluorosis.

物理的封じ込め ⇨P4施設(和英篇末尾)

不妊 ▶一時[永久]〜〔放射線の影響などによる〕temporary [permanent] infertility [sterility]. / 犬[猫]の〜去勢手術 cat [dog] spay or neuter surgery. ★⇨去勢, 英和 TNR

不燃 ▶〜(性)の noninflammable; uninflammable; nonflammable; noncombustible; incombustible; 〔耐火〕fireproof. ¶ **不燃化** 〔木材・家屋などの〕fireproofing; rendering [making] 《urban areas》 fireproof. / 都市

の〜化を進める further fireproofing in urban areas; make progress in fireproofing a town. / 〜化促進事業 a fireproofing project.　**不燃(性)ガス** uninflammable gas;〔不活性ガス〕inert gas.　**不燃建材** noninflammable building material.　**不燃建築(物)** a non(in)flammable [fireproof] building.　**不燃構造** noncombustible construction.　**不燃ごみ** ⇨ごみ　**不燃材料** (a) noncombustible [(an) incombustible] material.　**不燃性フィルム** safety [uninflammable] film.　**不燃都市** a fireproof town [city]; an incombustible city.　**不燃物** incombustibles; nonflammables.

浮漂(ふひょう)植物[生物] pleuston;〔浮き草〕duckweed.　★⇨浮水植物，浮遊

不法　▶〜な unlawful; illegal; illegitimate; unjust; iniquitous; injurious; wrong(ful); unwarrantable; unwarranted; unjustifiable; unjustified. / (漁船で)〜操業する engage in illegal fishing operations.　¶**不法係留** illegal mooring [docking]《of motorboats》.　**不法係留船** an illegally moored [docked] vessel [ship, boat].　**不法行為** an unlawful [illegal] act; an illegality; a tort. / 組織的な〜行為 organized lawlessness. / 〜行為を働く commit [perpetrate] an illegal act. / 〜行為者 a wrongdoer; an offender; a misfeasor; a malfeasant.　**(廃棄物などの)不法投棄** illegal dumping [disposal]《of waste》. / 産業廃棄物を〜投棄する dump [dispose of] industrial waste illegally. / 〜投棄現場 an illegal disposal site; an illegal dump. / その会社は〜投棄したごみの回収を約束して告発を免れた．The company promised to remove the illegally dumped refuse and escaped being denounced to the authorities.

フミン〔腐植土物質の一種〕humin.　¶**フミン酸** ⇨腐植酸　**フミン質** humic substances.

浮遊　▶〜する float. / 〜性生物 floating organism. / (魚類の)〜性の卵 pelagic eggs.　¶**浮遊カビ菌** floating mold spores.　**浮遊ごみ** floating garbage [waste].　**浮遊塵** suspended dust.　**浮遊塵埃(じんあい)** floating dust; dust particles in suspension in air.　**浮遊微粒子**〔大気中の浮遊粒子状物質〕suspended particulate matter (略: SPM);〔水中の〕suspended solids (略: SS).　**浮遊物** floating [suspended] matter [particles];

flo(a)tage. **浮遊物質(量)** 〔水中に浮遊し溶解していない物質；水質指標の1つ〕suspended solids (略: SS). **浮遊粉塵** airborne particles. **浮遊粒子状物質** suspended particulate matter. ★⇨英和 SPM

冬日(ふゆび) 〔最低気温0°C未満の日〕a "winter day" during which the temperature falls below 0°C. ★⇨真冬日, 夏日(なつび)

冬水(ふゆみず)**田んぼ** ⇨湛水(たんすい)(冬季湛水水田)

腐葉土 leaf mold; humus. ★⇨腐植

フライ・アッシュ 〔燃焼ガス中の浮遊灰；セメントなどの混入材料に再利用される〕fly ash. ¶**フライアッシュ・セメント** fly ash cement.

ブラキストン線[ライン] 〔本州・北海道間の動物分布境界線〕Blakiston's line.

プラスチック (a) plastic. ★⇨ごみ(資源ごみ), 分解 ▶葉状〜 a laminate. / グリーン〜 ⇨グリーン(グリーン・プラ) / 植物系〜 ⇨バイオ(バイオマス・プラスチック) / 石油系〜 (a) petroleum-derived plastic. / 埋めても腐らず，燃やすとダイオキシンが出る〜ゴミは始末が悪い．Plastics are a real problem [very difficult to deal with]. They won't decompose if you bury them and they produce dioxin if you burn them. / 〜製の食器から有害化学物質が出て食物といっしょに人間の体内に入る危険がある．There is a danger that harmful chemicals may leach [migrate] out of plastic eating utensils and enter the human body together with food. ¶**プラスチック製品** a plastic; plastic goods. **プラスチック製容器包装** plastic containers and packaging. / 容器包装用〜 plastic for the packing [packaging, wrapping] of containers. ★⇨容器包装 **プラスチックフィルム包装** plastic film wrapping [packaging] **プラスチック容器** a plastic container. **プラスチック容器包装リサイクル推進協議会** the Plastic Packaging Recycling Council (略: PPRC).

ブラック・バス a black bass《複数形 black bass》. ▶〜を密放流する release black bass illegally.

プランクトン plankton. ¶**細菌プランクトン** bacterioplankton. **植物プランクトン** phytoplankton. ▶海水の異常高温で植物〜が大量に発生した．Unusually high ocean temperatures led to a massive proliferation of phytoplankton. ★⇨赤潮 **動**

物プランクトン zooplankton.

プリオン 〔感染性のたんぱく質粒子〕a prion. ★⇨英和 Bovine Spongiform Encephalopathy

ブルーギル a bluegill.

プルサーマル計画 〔余剰プルトニウムを熱中性子炉で利用する日本の原子力発電計画〕the "plutonium-thermal project." ★⇨プルトニウム

プルサーマル発電 "pluthermal" [plutonium-thermal] power generation; power generation which uses mixed plutonium-uranium oxide as fuel.

古雑誌 an old magazine. ★⇨古紙

古新聞 old newspapers. ★⇨古紙

古タイヤ ▶積まれた～ a mound [pile] of old [waste] tires. ★⇨撤去

プルトニウム plutonium. ★⇨プルサーマル計画, ウラン(ウラン・プルトニウム混合酸化物), 回収, 集中管理, 余剰(余剰プルトニウム) ▶原子炉[兵器]級～ reactor-[weapons-]grade plutonium. ¶ **プルトニウム海上輸送** plutonium marine transportation; plutonium transportation by sea. **プルトニウム・サイクル** a plutonium cycle. **プルトニウム抽出** plutonium extraction. **プルトニウム抽出実験** a plutonium extraction experiment; experimental plutonium extraction. **プルトニウム爆弾** a plutonium bomb. / ～型核爆弾 a plutonium-based [plutonium-fueled] nuclear bomb. **プルトニウム・ピット** 〔核爆発を引き起こすプルトニウム 239 を含む金属球; 核兵器の最重要部品〕a plutonium pit. **プルトニウム分離工場** a plutonium separation plant. **プルトニウム・リサイクル[再利用]** plutonium recycling [recycle, reuse]. **プルトニウム利用計画** a plan for plutonium usage.

ブルントラント委員会 the Brundtland Committee. ★⇨英和 World Commission on Environment and Development

触れあい contact; a touch(ing); mutual touching. ▶他者[自然]との～ contact with other people [with nature]. ★⇨自然 ¶ **触れ合い体験** a hands-on experience;〔動物園での〕a petting zoo experience.

フレックス(燃料)車 〔アルコールとガソリンの併用車〕a flex-fuel vehicle (略: FFV); a flex-fuel car.

プロシューマー 〔生産者でもあるセミプロの消費者〕a prosumer.

フロラ ⇨英和 flora.

フロン 〔クロロフルオロカーボンの慣用名〕a CFC; a chlorofluorocarbon. ★⇨自動車(自動車フロン券), 英和 chlorofluorocarbon ▶脱[ノン]～冷蔵庫 a CFC-free refrigerator. / 特定～〔オゾン層破壊効果が特に大きい五種類のフロン類〕specified CFCs; controlled CFCs; five CFCs especially harmful to the ozone layer. ★1996年から生産禁止. ¶**フロン・ガス** CFC gas. **フロン(ガス)削減策** CFC reduction measures. **フロン回収破壊法** the CFC Recovery and Destruction Law. **フロン規制** CFC regulations. **フロン代替物質** 〔ブタンなど〕a CFC substitute. ★⇨代替(代替フロン)

分煙 ¶separation of smoking and nonsmoking areas. ▶職場の～化 division of a workplace into smoking and nonsmoking areas. ★⇨喫煙, 禁煙 / 公共空間の～化を推進する promote the division [separation] of public spaces into smoking and nonsmoking areas. ¶**完全分煙** complete [total] separation of smoking and nonsmoking areas. **不完全分煙** semi-separation [incomplete separation] of smoking and nonsmoking areas. / この店は不完全～だ. The place doesn't separate smoking and nonsmoking areas completely [thoroughly, properly].

分化 specialization; 〔派生〕differentiation; divergence; divergency. ▶～する specialize; differentiate. / ～していない, 未～の(器官などが) unspecialized; undifferentiated; (細胞などが) undifferentiated. / ヒトとチンパンジーは500万年前に～した. Man and the chimpanzee diverged five million years ago. / 幹細胞の～ differentiation of a stem cell.

分解 〔化合物などの〕decomposition; resolution; dissolution; (複合化合物の) degradation; 〔生物学上の〕disassimilation; disintegration. ▶～する resolve (itself)《into…》; dissolve《into…》; decompose; be decomposed. ★⇨酵素, アルコール / 化合物を元素に～する reduce [dissolve, resolve] a compound into its elements. / 水を酸素と水素に～する resolve water into oxygen and hydrogen. / この物質は空気に当てるとすぐ～する. This substance will readily decompose itself in the air. / ～しにくい, 難～性の indecomposable; indissoluble; insoluble; simple; nondegradable; stable. / ～しやすい, ～性

の degradable; unstable. ★⇨生分解性，光(ひか)分解，難分解性，英和 biodegradable / 〜性〔物質の〕degradability. / 〜性ポリマー a degradable polymer. / プラスチックごみは地中で〜しないので処理に困っている．Plastic waste is difficult to dispose of because it doesn't decompose in the earth.

文化遺産 〔一般の〕(a) cultural heritage;〔世界文化遺産〕a Cultural Heritage Site; a cultural asset recognized by the UN. ★⇨世界(世界遺産) ▶このままでは貴重な歴史的〜が失われてしまう．If things go on like this, our precious cultural heritage will be lost.

文化財 a cultural asset [property]; (a) cultural treasure. ★⇨史跡，天然(天然記念物)，伝統的建造物群保存地区 ▶〜の流出を防ぐ prevent the exodus of cultural assets overseas. / 国指定[県指定]〜 a nationally-[prefecturally-]designated important cultural property [asset]. / 重要〜に指定される be designated [registered] as an important cultural asset [property]. ¶**文化財保護法** the Cultural Properties Protection Law.

文献調査 a document study;〔過去の地質調査のデータをもとに高レベル放射性廃棄物の最終処分地〔⇨地層処分〕としての適否を決めるための原子力発電環境整備機構による調査〕a "document study"; a suitability study by the Nuclear Waste Management Organization of Japan, used as the first stage in the selection process, to establish whether a site would be safe for a high-level radioactive waste dump. ★文献調査の後，概要調査 (a rough outline study), 精密調査 (a detailed study) と進む．

分散 ▶〜(化)する break up; scatter; disperse;〔産業などを〕decentralize. / 原油供給源の〜化を図る必要がある．We should aim to diversify oil sources. / 首都機能を〜させる distribute [decentralize] functions of the central government. ¶**分散型電源** ⇨電源

分子 a molecule ¶**分子生物学** molecular biology.

分収林 a profit-sharing forest. ¶**分収育林〔造林〕** profit-sharing forestry; a profit-sharing forestry system. **分収林特別措置法** the Profit-Sharing Afforestation Special Measures Law.

粉塵(ふんじん) (mine [mineral]) dust; fine particles. ★⇨浮遊

(浮遊粉塵) ▶ディーゼル[アスベスト]〜 ⇨ディーゼル, アスベスト ¶ **粉塵計** a dust counter. **粉塵公害** dust pollution. **粉塵爆発** a dust explosion. **(職業的)粉塵暴露** (occupational) exposure to 《mineral》 dust [particles].

分水 the diversion [shedding] of water;〔分水嶺などで分かれる水〕diverted water. ¶ **分水栓**〔水道の〕a corporation cock [stop]. **分水装置** a diversion device. **分水嶺[界, 山脈]** a watershed; a divide; a dividing ridge. ▶この山脈はこの国の〜嶺を成している. This mountain range forms the watershed of this country. **分水路** a flood-control [diversion] channel.

糞土(ふんど)〔糞の混じった土〕soil with manure; composted soil. ▶ミミズの〜 soil with castings (and worms).

糞尿(ふんにょう) human excrement; human waste; excreta; f(a)eces and urine; night soil;〔肥料〕manure. ★⇨屎尿(しにょう), ペット ¶ **糞尿汲み取り人** a night-soil collector; a sewage and sanitation worker. **糞尿被害**〔犬猫などの〕problems with cats and dogs doing their business [shitting and peeing] (in *one's* garden). ▶〜被害を受ける have [suffer from] problems with animal excrement.

分泌(ぶんぴつ) secretion. ★⇨内分泌 ▶〜する secrete. /〜性の, 〜を促す secretory; secretive; secernent. ¶ **分泌液** secreting fluid; juice. **分泌器官** a secretory organ; a secernent. **分泌腺[細胞]** a secretory [secreting] gland [cell]. **分泌物** a secretion; an exudate.

分布 distribution. ★⇨温帯, 生息, 住む, 生物(生物分布学), 垂直[水平]分布 ▶〜する be distributed; range 《from one place to another》. / アザラシはオットセイよりも広い範囲に〜している. True seals range more widely than the eared varieties. / 高度〜 a vertical distribution. / 動植物の(地理的)〜 the geographical distribution of plants and animals. ¶ **分布域**〔動植物の〕a distributional range; a range. **分布上限[下限]**〔動植物などの〕an upper [a lower] limit of distribution. **分布南限[北限](地[域])**〔動植物などの〕a southern [northern] limit of distribution. ★⇨南限, 北限 **分布表** a distribution chart [diagram].

分別 ▶ごみを〜する separate rubbish. ¶ **分別ごみ** rubbish [trash] to be kept separate [separated]. **分別収集** 〔ごみの〕collection of rubbish by type; sorted trash collection. **分別処理** 〔ごみの〕separated processing.

分流式下水管[下水道] 〔雨水と汚水を別々の管で処理場へ送る方式〕the separate sewer system.

分類 classification; grouping; assortment. ▶～する classify; divide into [reduce to] classes. / クジラは哺乳類に～される．Whales are classified as mammals. / リンネの植物～法 Linnean system [classification] of plants. / 人為[自然]～法 an artificial [a natural] classification system; artificial [natural] classification. / ～の困難な種 a critical species. ¶**分類階級** levels [ranks, categories] of [in] the hierarchy. **分類学** taxonomy; systematics; the science of classification. **分類学者** a taxonomist; a systematist. **分類群** a taxon《複数形 taxa》; a taxonomic group.

へ

平均 ⇨雨量(平均雨量), 海面(平均海面), 加重等価平均感覚騒音レベル, 気温(平均気温), 企業(企業平均燃費), 健康(健康寿命), 降雨(平均降雨量), 日照(日照時間), 燃費(平均燃費), 変化

閉鎖 ¶**閉鎖型生態系実験施設** a closed ecology experiment facility (略: CEEF); a biosphere. ★⇨英和 biosphere **閉鎖(性)海域** enclosed coastal waters; an enclosed coastal sea. **閉鎖生態系** a closed ecosystem [ecological system]. **閉鎖生態系循環式養殖システム** a closed ecological recirculating aquaculture system (略: CERAS). **閉鎖生態系生命維持システム** a controlled ecological life support system (略: CELSS).

平和利用 (the) peaceful [commercial, industrial] use (of …). ★⇨核, 原子(原子力) ▶宇宙の〜原則 the principle of peaceful use of space.

ベースライン・アンド・クレジット 〔温室効果ガスの排出量取引方式〕baseline and credit; a baseline-and-credit system.

ペーハー ⇨ pH (和英篇末尾)

ヘキサクロロフェン 〔殺菌消毒薬〕hexachlorophene.

壁面緑化 ⇨緑化

ベジタリアン 〔菜食主義者〕a vegetarian. ★⇨菜食 ¶**ベジタリアン食** vegetarian food; a vegetarian meal. ▶〜食をご希望のお客様はご予約の際にお申し出ください. Passengers requiring vegetarian meals, please indicate this at the time of reservation.

鼈甲(べっこう) tortoiseshell.

ペット a pet. ★⇨去勢, 不妊, 地域(地域猫) ▶家で飼う〜 a house pet. /〜可のアパート an apartment (house) where pets are permitted [that allows pets]. /〜可. 〔表示〕Pets Permitted. /〜不可. 〔表示〕No Pets. /〜歓迎. 〔表示〕Pets Welcome. /〜専用出入り口 a pet entrance [door]. /〜に餌をやる feed a pet. /ハムスターを〜にする keep a hamster as a pet; keep a pet hamster. /野鳥をただかわいいからといって〜にするのはよくない. It's not a good thing to keep a wild bird as a

ペットボトル

pet just because it's cute. / 〜の世話をする look after [take care of] a pet. / 〜のフンの始末 cleaning up a pet's mess [after a pet]; cleaning out a pet's cage. / 何か〜を飼っていますか. Do you keep [Have you got] any pets? / 〜を飼えないマンションが多い. There are many condominiums where pets may not be kept [are prohibited]. ¶ペット愛好家 a pet lover; a petlover.　ペット・アレルギー an allergy to pets [pet dander]. / 〜アレルギーである have (a) pet (dander) allergy.　ペット感染症[病] a pet(-communicated) infection; an infection caught from a pet. ★⇨感染　ペット・キャリア〔ペット運搬用のかばん・かご〕a pet carrier.　ペット業者 a pet dealer.　ペット共生マンション a pet-friendly apartment [condominium].　ペット産業[業界] the pet industry.　ペットシッター a pet sitter; a petsitter.　ペット条例 a pet ordinance.　ペット植物 a pet plant.　ペット・ショップ a pet shop.　ペット美容室 a pet beauty salon.　ペット・フード〔ペットのえさ〕pet food; petfood.　ペットフード工業会 the Pet Food Manufacturers Association, Japan.　ペット服〔ペット用の衣類〕《an item of》pet clothing [apparel].　ペット・ホテル pet accommodation; a pet hotel [motel].　ペット用品 items for pets.　ペット霊園 a pet cemetery.　ペット・ロス(症候群)(grief over) the loss of a pet; "pet(-)loss syndrome."

ペットボトル a PET bottle; a plastic bottle. ★PETは, ポリエチレン・テレフタレート(polyethylene terephthalate)の頭字語. ⇨回収, 再生, 資源, 負担 ¶ペット(ボトル)飲料 a drink in a plastic bottle.　ペットボトル再生樹脂 recycled PET (bottle) resin.　ペットボトル成型機 a molder [moulder] for PET bottles.　ペットボトルリサイクル推進協議会 the Council for PET Bottle Recycling.

へどろ sludge; ooze; slime; colloidal sediment. ▶〜状の slimy; oozing. / ダムには大量の〜がたまっている. A large amount of sludge has accumulated in the dam. / 河口のほぼ半分が〜に埋まっている. About half the estuary is filled with sludge. / 〜を除去する remove [get rid of, pump away] sludge.

ベランダ菜園 ⇨菜園

ヘルシンキ宣言〔1964年に世界医

師会で採択された，ヒトを対象とする医学研究の倫理的原則〕the Declaration of Helsinki.

ベルリン・マンデート the Berlin Mandate. ★⇨英和 Berlin Mandate

ペレット・ストーブ 〔粒子状に固めた木材燃料を使う灰・煙の少ないストーブ〕a pellet stove.

変異 (a) variation; ⇨突然変異．▶〜する vary; be variable. / 動植物体には〜が起こることもある．Variations can occur in plants and animals. / 一時的〜 modification. ¶ **変異株** a mutant strain. **変異原** a mutagen. **変異原性試験** a mutagenicity test; a mutation test. **変異原物質** a mutagenic substance. **変異細胞** a mutant cell. **変異種** a variant species; a mutant species. **変異性** variability. **変異生成** mutagenesis. **変異体** a variant; a mutant.

変化 (a) change; (a) variation; 〔変更〕(an) alteration; 〔変転〕(a) mutation; 〔移行〕a transition; 〔多様性〕variety; diversity. ▶〜する〔変わる〕change; (形態が変わる) transform (itself)《into...》; be transformed; be metamorphosed; (一定しない) vary. / 平均気温が2度上がるだけで，環境は劇的に〜する．With a rise in temperature of only two degrees, the environment will change dramatically. / 環境の〜 a change in the environment; environmental change. / 日本は四季の〜がはっきりしている．In Japan the changes from one season to the next are clear. / 湖周辺の〜に伴い，白鳥の渡来数が減ってきている．In the wake of changes in the lakeside, the numbers of migrating swans are falling. / 水質の〜の影響を受けやすい生物 a form of life that is susceptible to changes in (the) water quality.

変種 〔分類上の〕a variety《of dahlia》; 〔突然変異による〕a mutation; a sport; 〔奇形〕a monster; a freak of nature; (植物) a freak plant. ★⇨種(しゅ)，新種 ▶〜の varietal. / 亜〜 a subvariety. / 人工〜 an artificial variety; a strain. / 風土［地方］的〜 a climatic [geographical] variety.

ベンゼンヘキサクロリド benzene hexachloride. ★⇨英和 BHC

ベンゾピレン 〔発ガン性物質〕benzopyrene.

ベントス ⇨底生生物，英和 benthos

ほ

保安林 a protection [protective, conservation] forest. ★⇨保全(保全林) ¶魚(うお)つき保安林 a fish-breeding forest. ★⇨魚付き林 干害防備保安林 a drought prevention forest. 航行目標保安林 a navigation landmark forest. 水害防備保安林 a flood damage prevention forest. 水源涵養(かんよう)保安林 a headwater conservation forest. 潮害防備保安林 a tide and salty wind prevention forest. 土砂崩壊防備保安林 a landslide prevention forest. 土砂流出防備保安林 a soil run-off prevention forest. なだれ防止保安林 a snow avalanche prevention forest. 飛砂防備保安林 a shifting-sand prevention forest. 風致保安林 a scenery conservation forest. 保健保安林 a public health forest. 防火保安林 a fire prevention forest. 防雪保安林 a snow drift prevention forest. 防風保安林 a windbreak forest. 防霧保安林 a fog inflow prevention forest. 落石防止保安林 a forest planted for protection against falling rocks.

ポイ捨て 〔ごみ・たばこなどの〕careless tossing away 《of...》; littering. ★⇨捨てる ¶ポイ捨て禁止条例 an ordinance against littering. ★⇨空き缶 ▶たばこの～禁止条例を設けている自治体も少なくない. Quite a number of local authorities have passed ordinances prohibiting people from discarding [throwing away] cigarette butts.

防疫 control [prevention] of epidemics [infectious disease]; communicable disease control; disinfection;〔検疫〕quarantine. ★⇨検疫, 家畜(家畜防疫), 植物(植物防疫法) ▶～に努める strive hard to prevent an epidemic. / 伝染病の～措置を講ずる take preventive measures against an epidemic. / (検疫による)～体制を取る[強化する] take [step up, strengthen, heighten] quarantine measures; impose [strengthen] a quarantine. ¶防疫課 an infectious disease prevention section. 防疫官 an epidemic prevention commissioner; a health official. **防疫訓練** infection prevention training. **防疫線** a sanitary cordon. **防疫服** a biohazard suit.

防煙林 an antismoke forest; a smoke shelter forest.

防音 sound isolation [insulation]; soundproofing. ▶〜性のある sound absorbing. /〜が施してある be soundproofed. ¶**防音カーペット** a soundproof carpet. **防音ガラス** soundproof glass. **防音効果** a soundproofing effect. **防音工事** soundproofing. **防音校舎** a soundproof schoolhouse [school building]. **防音構造** soundproof construction. **防音材料** deadening [soundproofing, sound-insulating] material. **防音室** a soundproof chamber. **防音漆喰**(しっくい) acoustic plaster. **防音車輪** 〔鉄道の〕noise-reducing wheels. **防音装置** soundproofing; soundproofing equipment; a sound arrester;〔内燃機関の〕a silencer. **防音タイル** an acoustic tile. **防音壁** a noise-blocking wall. **防音林** a sound-absorbing forest.

放棄 ▶〜する abandon; relinquish. ★⇨核(核放棄), 投棄, 放置 / 産業廃棄物の大量〜 mass dumping of industrial waste. ¶**放棄田** an abandoned [a disused, an uncultivated] rice field.

防蟻(ぼうぎ) 〔白蟻被害の予防〕termite prevention. ¶**防蟻剤** (a) termiticide; an antitermite chemical. **防蟻シート** a termicide-treated sheet [film]. **防蟻処理** antitermite treatment. ★⇨防腐 **防蟻対策** antitermite measures.

蜂群(ほう)**崩壊症候群** 〔ミツバチの群れが巣箱から突然消え失せる現象〕colony collapse disorder. ★⇨英和 CCD

防護 protection; guard. ▶〜する protect; guard. ¶**防護基準** 〔有毒物質を扱う職場などの〕a protection standard. ★⇨放射線(放射線防護) **防護服** a protective suit; a hazmat [Hazmat] suit. ★ hazmat は hazardous material (危険物質) の略 (⇨化学(化学防護服)).

防砂堤 〔海岸の砂の移動を防ぐ〕a groin [groyne]；an erosion control bank. ★⇨砂防

防砂林 〔砂の飛来や海岸の砂の移動を防ぐ〕a stand of trees (planted) to prevent sand shifting.

放射 emission;〔輻射〕(光・熱の) radiation; (ラジウムなどの) emanation. ★⇨放射性, 放射

放射性

線, 放射能 ▶〜する radiate; emit; emanate. ¶**放射エネルギー** radiant energy. **放射化学**〔放射性同位体の化学〕radiochemistry; radiation chemistry. **放射霧**(む)〔放射冷却によって発生する〕radiation fog. **放射計** a radiometer. **放射熱** radiant heat. **放射冷却** radiative [radiational] cooling.

放射性 radioactivity. ★⇨放射線, 放射能 ▶〜の radioactive. ¶**放射性医薬品** radiopharmaceuticals; radioactive medicine. **放射性(壊変)系列** a radioactive series. **放射性核種** a radioactive nuclide; a radionuclide. **放射性ガス** (a) radioactive gas. **放射性元素** a radioactive element; a radioelement. **放射性降下物** radioactive [radiation] fallout. **放射性鉱物** a radioactive mineral. **放射性コバルト** radiocobalt. **放射性指示薬** a radioactive indicator. **放射性塵** radioactive dust; radiodust. **放射性炭素** radiocarbon. **放射性沈殿物** a radioactive deposit. **放射性同位元素[同位体]** a radioisotope; a radioactive isotope. **放射性トレーサー** a radioactive tracer. **放射性廃棄物** radioactive [atomic] waste; radioactive waste matter [products]. ★⇨核(核廃棄物), ガラス(ガラス固化), セメント固化, バックエンド対策, 文献調査 / 〜廃棄物の処理 the disposal of radioactive waste. / 低[高]レベル〜廃棄物 low-level [high-level] radioactive waste. / 〜廃棄物は処理を誤ると始末に負えない. Any mishandling of radioactive waste can lead to it going out of control [can have disastrous and unforeseeable consequences]. **放射性物質** a radioactive substance [material];《口》a hot material. / 強い〜物質を取り扱う実験室《口》a hot laboratory. / 医薬用〜物質 an atomic cocktail. **放射性崩壊** radioactive decay; radioactive disintegration; radiative decay. **放射性薬剤** a radioactive drug [pharmaceutical]; a radiopharmaceutical. **放射性ヨウ素[ヨード]** radioiodine; radioactive iodine (略: RAI); iodine-131. **放射性ヨード治療** radioiodine [radioactive iodine, RAI, iodine-131] therapy. **放射性粒子** radioactive particles;《口》Geigers.

放射線 radiation;〔光線〕radial [radiant] rays;〔放射能による〕radioactive rays. ★⇨線量, 被曝 ▶〜に敏感な sensitive to radiation; radiosensitive. / 〜の強度[レベル] radiation intensity; a radiation level; a level of radiation. / 残留～ residual radiation. ¶ **放射線医** a radiologist. **放射線医学** radiology; radiation medicine. **放射線育種場** the Institute of Radiation Breeding（略：IRB）. **放射線遺伝学** radiation genetics. **放射線影響研究所**〔広島・長崎にある日米共同研究機関〕the Radiation Effects Research Foundation（略：RERF）. **放射線化学**〔放射線の化学作用を扱う〕radiation chemistry; radiochemistry. **放射線カプセル** a radioactive seed [capsule, pellet]. **放射線強化兵器** an enhanced radiation weapon（略：ERW）. **放射線業務** radiation work. **放射線業務従事者** a radiation worker. **放射線計数管** a radiation [Geiger] counter. **放射線外科** radiosurgery; radiation surgery; stereotactic external beam irradiation. **放射線検知器** a radiation detector. **放射線殺菌**〔食物などの〕radiation sterilization. **放射線遮蔽**(しゃへい) radiation shielding. **放射線腫瘍学** therapeutic radiation oncology. **放射線障害**《suffer》radiation damage; radiation hazard [injury, sickness]; radiolesion. **放射線障害防止法** the Radiation Injury Prevention Law. ★正式名は「放射性同位元素等による放射線障害の防止に関する法律」(the Law Concerning Prevention of Radiation Injury due to Radioisotopes, Etc.). **放射線照射** irradiation; exposure to radiation; radiation exposure. / ～照射食品 irradiated food. / ～誤照射 mistaken exposure to irradiation; mis-irradiation. / ～過剰照射 excessive irradiation; over-irradiation. / ～照射量 the amount of irradiation; (the degree of) exposure to radiation [radiation exposure]. / 1200 レントゲンの～照射(量) (an) irradiation of 1200 roentgens. **放射線診断** radiodiagnosis. **放射線診断医** a diagnostic radiologist. **放射線生態学** radioecology; radiation ecology. **放射線生物学** radiation biology; radiobiology. **放射線測定** radiation measurement; radiometry. **放射線損傷** radiation damage. **放射線治療[療法]** radiation treatment; radiotherapeutics; radiation therapy. / ～化学療法 chemoradiotherapy. **放射線治療医** a radiotherapist; a radiotherapeutist. **放射線透過検査** a radiograph examination. **放射線熱傷** a radiation burn. **放射線被曝** radiation [radioactive]

放射能

exposure; exposure [being exposed] to radiation [radioactivity]. **放射線病** radiation sickness; a radiation disease. **放射線防護** radiation protection. / ～防護基準 radiation protection standards [norms]; the radiation protection guide. / 電離～の防護基準を定める set standards of protection against ionizing radiation. / ～防護学 radiation protection science. / ～防護服 a radiation protection suit.

放射能 radioactivity. ★英語の radioactivity は「放射線を出す能力・性質」を意味するので，内容的に放射線そのものを指している場合には radiation の方が正確 (⇨放射線, 放射性). ▶空気中の～ atmospheric radioactivity. / ～の強さ[レベル] a radioactive [radioactivity, radiation] level; the intensity of radioactivity; a level of radioactivity [radiation]. / 核実験による～の影響 radiation effects from nuclear tests. / ～のある[を含む]物質 radioactive element. / ～のない[を含まない] inactive. / 大量の～を浴びる be showered with a large amount of radioactivity. / 人工[自然, 残存]～ artificial [natural, residual] radioactivity. ¶ **放射能雨**[雪] radioactive rain [snow]. **放射能汚染** radioactive contamination. ★⇨汚染 **放射能汚染図** a radioactive contamination map. **放射能汚染物質** a radioactive contaminant. **放射能雲** a radioactive cloud. **放射能シェルター** a fallout shelter. **放射能事故** a radiation accident. **放射能遮断壁** a radiation shield. **放射能障害** radiation damage; 〔症候群〕a radiation syndrome. **放射能除去**[除染] radioactive decontamination. **放射能泉** 〔ラジウム鉱泉〕a radium [radioactive] (hot) spring [mineral spring]. **放射能測定** 〔環境の〕radiological monitoring. **放射能測定機** a radiation detector. **放射能探知機** a radiation detector. **放射能調査** 〔環境の〕a radiological survey. **放射能調査隊** a radiological survey team [party]; 〔軍の〕a monitoring team. **放射能塵** radioactive dust; radiodust. **放射能濃度** radioactivity concentration. **放射能灰** radioactive ashes [fallout]. **放射能兵器** a radiological [radioactive] weapon. **放射能防護服** a radiation (-proof) suit; protective gear. **放射能漏れ** a radiation leak; radiation [radioactive] leakage.

放獣 〔捕獲した動物などを自然に帰すこと〕release into the wild; wild release. ▶捕獲した熊を〜する release captured bears (back) into the wild. / 住民たちは捕獲した熊の〜に反対している．The residents oppose the release of captured bears into the wild. / 捕獲した熊の学習〜〔人間を恐れるように条件付けしてから放す〕release of captured bears back into the wild after conditioning them to fear humans and human settlements.

防臭 ⇨消臭 ¶**防臭弁** a stink [stench] trap; a gas trap;〔下水溝の〕a drain trap.

防獣 animal-proofing; protection against wild animals; keeping out [off] wild animals. ★⇨防鳥, 防鹿(ぼうろく)柵 ¶**防獣ネット[網]** an animal net;〈集合的に〉animal netting.

防除 ▶蚊の〜 prevention of the breeding of, and the extermination of mosquitoes. / 害虫の〜 control of insect pests. ★⇨駆除, 害虫, 虫 ¶**生物的防除** biological control; biocontrol. ★⇨英和 biological control

放水 ⇨放流 ¶**放水路** a flood control channel; a sluiceway; a drainage canal;〔水車の〕a tail-race. ▶荒川〜路 the Arakawa Canal. / 〜路建設計画 a water diversion project.

包装 wrapping; packaging. ▶〜する pack; package; wrap... up;〔贈答用に〕gift-wrap. ★⇨簡易包装, 容器包装 / (デパートなどの)過剰〜 excessive packaging (at department stores); over-packaging; using too much wrapping.

暴騒音 blaring loud noise. ★⇨騒音 ¶**暴騒音規制条例** noise control ordinances [regulations].

放置 ⇨放棄, 投棄, 危険(危険物) ¶**放置自転車** an illegally parked bicycle; an abandoned [a discarded] bicycle. ★⇨撤去, 迷惑 ▶駅前の〜自転車には困ったものだ．The bicycles left in front of the station have become a real problem. **放置自転車税** a tax on illegally parked [abandoned, discarded] bicycles. **放置車両** an illegally parked vehicle. **放置林** a neglected forest.

放鳥 〔鳥を放つこと〕releasing a bird; setting a bird free; letting a bird fly (around) freely;〔鳥を自然に帰すこと〕returning a bird to nature;〔式典などの〕a ceremonial bird release;〔放たれる鳥〕a caged bird to be set free. ▶～する release birds; set birds free (as an act of mercy)./養殖したキジを猟場に～する release artificially raised pheasants into a hunting ground; stock a hunting ground with artificially raised pheasants./立山で捕えたライチョウが7羽富士山に～された. Seven snow grouse caught on Mt. Tateyama were released on Mt. Fuji. ★⇨英和 rewild /トキの試験～ an experimental release of Japanese crested ibis./自然～ releasing (artificially-raised) birds into the wild./標識～ release of a tagged bird. ★⇨標識 ¶**放鳥式** a bird releasing ceremony; a ceremony for releasing birds to the wild.

防潮 ¶**防潮(水)門** a tide gate. **防潮堤** a tide embankment; a seawall. **防潮林** trees [a forest] planted to protect the shoreline.

防鳥 birdproofing; protection from birds; keeping out [off] birds. ★⇨防獣 ¶**防鳥ネット[網]** a bird net;〈集合的に〉bird netting.

防腐 ▶(…に)～処理を施す apply antiseptic treatment to … ★⇨防蟻(ぼう) ¶**防腐剤** an antiseptic (substance);〔液〕an antiseptic solution; a preservative. ★⇨食品(食品防腐剤), 保存(保存料)

防風 protection against wind; breaking the force of [resisting] the wind. ¶**防風柵** a windbreak fence; windbreak fencing. **防風性** windproofness; windproof qualities. **防風ネット** windbreak netting; a windbreak net. **防風林** a windbreak (forest); a shelterbelt.

放流 ▶～する〔ダムなどの水を〕discharge;〔稚魚を〕release. ★⇨標識放流, 密放流, 放水/多摩川にアユを～する release *ayu* into the Tama River./～アユ released *ayu*./(釣った)魚を再～する catch and release fish. ★⇨キャッチ・アンド・リリース ¶**放流警報**〔ダムの〕a dam release [discharge] warning; a warning to people downstream of a dam that water is going to be discharged. **放流施設**〔ダムの〕dam discharge facilities. **放流瓶** a drift bottle. **放流量**〔ダムなどからの〕the discharge volume; the volume of water discharged;〔稚魚などの〕the number of fish released. **自然放流**〔ダムの〕natural discharge [overflow]. **事前放流**〔大雨が予想される場合のダムか

らの放水〕(a) predicted discharge.

防鹿柵(ぼうろくさく)〔鹿の食害防止用の柵〕a deer fence; 〈集合的に〉deer fencing.

ホエール ⇨鯨(くじら) ¶**ホエール・ウオッチング** whale watching.

ポータブル・トイレ 〔携帯用トイレ〕a portable toilet;【商標】a Porta Potti. ★⇨トイレ

北限 the northern limit; the northernmost point. ★⇨南限 ▶ドイツはブドウが栽培できる〜だ. Germany is the northernmost country for growing grapes. | Grapes can be grown no further north than Germany.

撲滅 eradication; extirpation; extermination; destruction; suppression; annihilation. ▶〜する eradicate; extirpate; exterminate; destroy; suppress; annihilate; wipe out (of existence); make an end of...; put down; stamp out. / 害虫を〜する exterminate vermin. / 伝染病を〜する stamp [crush] out an epidemic.

捕鯨 whaling; whale fishing. ★⇨鯨(くじら), 改訂管理制度, 国際(国際捕鯨委員会) ▶商業〜 commercial whaling. / 調査〜 whaling for scientific (research) purposes; scientific whaling; research whaling. / 調査〜枠 a [the] quota of whaling for scientific research. / 南極〜 Antarctic whaling; whaling in the Antarctic Sea. ¶**捕鯨会社** a whaling company. **捕鯨期間** the whaling season. **捕鯨基地** a whaling station. **捕鯨業** the whaling industry. **捕鯨禁止** a ban [moratorium] on whaling. **捕鯨禁止運動** an antiwhaling campaign. **捕鯨国** a whaling country [nation]. **捕鯨支持国** a pro-whaling nation [country]. **捕鯨場** a whaling ground; a whale fishery. **捕鯨水域** a whaling area; whaling grounds. **捕鯨船** a whaling ship; a whale catcher [chaser]; a whaler; a killer boat [ship]. **捕鯨船員** a whaling man; a whaleman. **捕鯨船団** a fleet of whalers. **捕鯨船長** a whaling master [captain]. **捕鯨砲** a whaling gun; a harpoon gun. **捕鯨砲手** a harpooner. **捕鯨母船** a whaling mother ship; a whale factory ship. **捕鯨モラトリアム** a whaling moratorium.

保健 〔健康(の維持)〕(preservation of) health; 〔衛生〕hygiene; hygienics; sanitation. ▶公害病患者を対象とする〜手帳 a health handbook for a patient with a pollution-related disease. ¶**保健医** a (public) health doctor. **保健課** the Health Section. **保健機能食品** (a) functional health food. **保健師[士]** a public health nurse; a hygienist. ★現在は男女とも「保健師」を用いる. **保健指導** health guidance. **保健指導員** a health worker. **保健所** a (public) health center.

保護 protection; shelter;〔鳥などの〕preservation;〔林などの〕reservation;〔天然資源・自然などの〕conservation. ★⇨環境(環境保護), 自然(自然保護), 世界(世界遺産, 世界自然保護基金), 鳥獣, 越冬, 保全 ▶〜する protect; safeguard; shelter;〔鳥などを〕preserve;〔林などを〕reserve;〔天然資源などを〕conserve. / 国際〜鳥[動物]に指定される be designated as an internationally protected bird [animal] (species). / 野生動物〜政策 a policy for [of] wildlife protection. / コウノトリの〜増殖計画 a project to breed storks in protective captivity. /〔張り紙などで〕迷子の子犬を〜しています. I'm looking after a stray puppy. ¶**保護区** a (nature [wildlife]) preserve [reserve, refuge, sanctuary]. ★⇨海洋(海洋保護区), 生物(生物圏保護区) **保護犬**〔保健所に保護された犬〕a rescued dog. ★⇨動物(動物保護センター) **保護色** protective coloring [coloration]; a protective color; cryptic [concealing, apatetic] coloration. /(体が)〜色になっている be protectively [cryptically] colored. **保護水域[水面]** protected waters; waters designated for environmental protection. **保護地(域)**〔先住民族などの政府指定居留地〕a reservation;〔野生動物の〕a sanctuary; a protected area; a (nature) reserve. / 野生動物の〜地 a sanctuary [reservation] for wild animals; a wildlife reservation [sanctuary]. **保護林** a reserved forest; a forest reserve.

保障措置協定〔原子力を平和利用にとどめるなどと保障する国際原子力機関(⇨英和 IAEA)と当該国間の協定〕a nuclear safeguards agreement.

捕食 ▶鳥を〜する prey upon birds. ★⇨英和 predator-prey ¶**捕食行動** predatory behavior. **捕食者[動物]** a predator; a predatory animal. / 超〜動物〔肉食動物を捕食する〕a hyperpredator.

保水 water retention; moisture retention. ▶森林の〜(能)力 the

water [moisture] retaining ability of a forest. ★⇨緑(緑のダム) ¶**保水機能** (a) water [moisture] retention function. ★⇨雨水(あまみず) **保水剤** (a) water retention agent. **保水性** water [moisture] retentivity [retention]. **保水性舗装** water-retentive paving [pavement].

ポスト ¶**ポスト化石燃料経済[時代]** a post fossil [post-fossil] fuel economy [era]. **ポスト工業化** postindustrialization. **ポスト工業化社会** (a) postindustrial society. **ポスト産業資本主義** post-industrial capitalism. **ポストフォーディズム** 〔多品種少量生産主義〕post-Fordism. **ポスト・ハーベスト** 〔収穫後の農産物への処理〕postharvest treatment of crops.

保全 ⇨保護, 環境(環境保全), 湿地(湿地保全), 水質(水質保全), 眺望(ちょうぼう)(眺望保全), 里地, 国土 ▶棚田を～する maintain rice terraces. ¶**保全林** a forest preserve; a preserved forest. ★⇨保安林 / 生活環境～林 a forest preserved to protect the living environment; a community forest preserve. / 水土～林 a forest preserved for water and soil protection.

母川(ぼせん) ¶**母川回帰** homing to the mother river; the return (of salmon) to the river of their birth. **母川国** 〔溯河性魚種の〕a state of origin (of anadromous stock). **母川国主義** the state of origin principle; the principle that fish belong to the country in whose rivers they spawn.

保存 ⇨保全, 保護, 開発, ファサード保存 ▶文化財を～する conserve [preserve] cultural assets [properties]. ★⇨文化財 / 古い街並みがよく～されている. The old townscape has been well preserved. / 種(しゅ)の ～ conservation [preservation] of (a) species. ★⇨種 / ～がきく keep (well). ★⇨常温 / 冷凍～ (preservation by) freezing. ¶**保存運動** a conservation movement. **保存液** 〔動物標本などの〕a preservative medium. **保存温度** storage temperature. **保存会** a conservation society; a society for the preservation《of the coastline》. **保存可能期間** how long *sth* can be kept;〔店頭での〕a shelf life. **保存科学** 〔文化財の〕conservation science. **保存工事** 〔文化財などの〕conservation work. **保存料** a preservative. / 人工～料 an artificial preservative. ★⇨合成(合成保存

料），防腐（防腐剤）　**保存緑地帯** a protected [preserved] green space.

北極　the North Pole. ▶〜の arctic. ¶北極気候影響評価 Arctic Climate Impact Assessment（略：ACIA）．　**北極気団** the arctic air mass.　**北極区**〔動物地理学上の〕the Arctic Region.　**北極圏[地帯]**　the Arctic Circle [zone]．　**北極圏（国立）野生生物保護区**〔米国の〕the Arctic National Wildlife Refuge (⇨英和；略：ANWR).

ホット・エア〔温室効果ガスの達成余剰分〕"hot air" (reductions).

母乳　mother's milk; breast milk. ▶〜中のダイオキシン類濃度 dioxin concentration(s) in breast milk. ¶**母乳育児** breast-feeding; raising a child on breast milk.　**母乳汚染** contamination of mother's milk.

哺乳類（ほにゅうるい），**哺乳動物**　a mammal; a mammalian.

ボランティア　a volunteer. ¶**ボランティア・ガイド**　a volunteer guide.　**ボランティア活動**　volunteer activities; voluntary [volunteer] work; volunteerism;〔地域での〕community service. ▶〜活動を促進する promote volunteer activities.　**ボランティア休暇**　volunteer leave; time off for voluntary work.　**ボランティア体験**　volunteer experience; experience as a volunteer.

ポリ　¶ポリエチレン・テレフタレート polyethylene terephthalate（略：PET）．★⇨ペットボトル　**ポリ塩化ビニル**　polyvinyl chloride（略：PVC）．　**ポリ塩化ビフェニル**　⇨PCB(和英篇末尾)　**ポリ臭化ジフェニルエーテル[ビフェニル]**　⇨英和 polybrominated diphenylether [biphenyl]　**ポリ袋**　a plastic bag.

ホルムアルデヒド　formaldehyde. ▶高い〜放散量〔建築材料からの〕a high level of formaldehyde emission(s). ★⇨シック・ハウス，英和 sick house syndrome

ボン合意〔2001年7月の京都議定書に関する政治合意〕the Bonn Agreement.

ボン条約　⇨英和 Bonn Convention

本能　(an) instinct. ▶種を保存する〜が備わっている be endowed with an instinct to preserve the species. / 自己防衛〜を持つ have [possess] the instinct of self-defense. / 自己保存〜 the instinct of self-preservation. / 〜的な instinctive; instinctual.

ま

マイ 〔自分専用の〕my; *one's* own; personal; for *one's* own use. ¶**マイカー** *one's* own car; a private car. ▶マイカーで…へ行く go to... in [driving] *one's* own car. / マイカーを手放し，カーシェアリングに参加する人がじわじわと増えている．Gradually more and more people are giving up their private cars and taking part in car sharing. / マイカー通勤する commute (to work) by [in *one's* own] car; drive to work. / マイカー族〔1人〕a car owner; a person who drives everywhere [uses his or her car rather than public transport]; 《口》a car person; 〈集合的に〉car owners (who drive everywhere [use their cars rather than public transport]); 《口》car people. / マイカー規制 private car restrictions. / マイカー利用を制限する cut down on the use of private cars. **マイ箸**(はし) *one's* own [personal] chopsticks (for use when dining out). / マイ箸を持ち歩く bring along *one's* own chopsticks. **マイバッグ** 〔レジ袋の代わりに，自宅から持参する買い物袋〕*one's* own shopping [carrier] bag. ★⇨エコ(エコバッグ) / マイバッグ持参のお客様には5ポイントプレゼント！Five extra points to customers who bring their own bags!

マイクログリッド 〔複数の源からなる電力供給の比較的小規模なネットワーク〕a microgrid.

埋蔵 ▶(土地が)石油を〜している have [hold] underground deposits of oil. ★⇨天然(天然ガス) / 金[石炭]〜地帯 a goldfield [coalfield]. / (…における)石炭[石油]の〜量 (estimated) coal [oil] reserves; (the estimated amount of) coal [oil] deposits (in...) / 可採〜量 (the amount of) recoverable reserves. ¶**埋蔵文化財包蔵地** land that holds buried cultural properties. **埋蔵量置換率** 〔石油の〕a reserve replacement ratio (略：RRR).

マイナス・イオン a negative ion. ▶〜発生器を搭載したエアコン an air conditioner containing a mechanism to produce negative ions.

前(まえ)**処理** pretreatment;

マグロ

preprocessing. ¶前処理建屋〔使用済み核燃料再処理工場の〕a pretreatment building.

マグロ，鮪 〔クロマグロ・ホンマグロ〕a bluefin (tuna); a horse mackerel; a tuna; tuna (meat). ¶マグロはえなわ漁船 a tuna long-liner. 責任あるまぐろ漁業推進機構 the Organization for the Promotion of Responsible Tuna Fisheries (略：OPRT). ★⇨英和 ICCAT

町並[街並]条例 a townscape ordinance. ★⇨保存

マテリアル ¶マテリアル・フロー〔物質の廃棄までの流れ〕(a) material flow. マテリアル・フロー・コスト会計〔環境会計の手法の1つ〕⇨英和 material flow cost accounting. マテリアル・リサイクル〔素材原料としての再利用〕material recycling.

真夏日(まなつび) 〔最高気温30°C 以上の日〕a "tropical day" on which the temperature is 30°C or above. ★⇨夏日 ▶38日連続の〜 38 consecutive days with temperatures of 30°C or above; 38 consecutive "tropical days."

マニフェスト 〔産業廃棄物管理票〕a manifest. ¶マニフェスト制度〔廃棄物の〕a (hazardous) waste manifest system; an industrial waste manifest system. ▶電子〜〔産廃電子管理票〕を使って産廃の流れを追う use an electronic manifest to track waste shipments.

真冬日 〔最高気温0°C 未満の日〕a "midwinter day"; a day when the temperature stays below 0°C. ★⇨冬日

マルポール条約 〔国際海洋汚染防止条約〕the MARPOL Convention. ★ MARPOL は Marine Pollution の略(⇨海洋(海洋汚染防止条約)).

マングローブ a mangrove.

慢性 ▶水不足が〜化している． The water shortage is chronic. ¶慢性毒性 chronic toxicity. ★⇨毒 慢性病患者 a person [patient] with a chronic illness [disease].

み

未開拓[開発]　▶〜の undeveloped; unreclaimed; unexploited; untapped; wild. / 〜地 undeveloped [unreclaimed, waste] land; untapped territory; virgin [maiden] soil; wilds. ★⇨開発

ミクロシスチン　〔青粉(あお)の産生する毒素〕microcystin. ★⇨青粉

水　water. ★⇨水源, 水質, 水道, 節水, 断水, 治水, 貯水, 保水, 世界(世界水の日), 青の革命　▶川の水かさが増している. The river is high [swollen, in flood]. / 川の水かさが減った. The river has fallen [sunk, gone down]. / そこは水はけがよい[悪い]. The place is well [poorly] drained. / (水道の)水が出しっぱなしだ. Somebody left the water running. ★⇨つけっぱなし / 地球は水の惑星だ. Earth is the water planet.
¶**水争い**　an irrigation [a water-rights] dispute. / 水争いをする dispute about the water rights. ★⇨灌漑(かんがい)　**水草**　⇨水生植物　**水栽培**　⇨水耕栽培

水資源開発公団　the Water Resources Development Public Corporation (略: WARDEC).

水資源機構　the Japan Water Agency (略: JWA).　**水処理**　water treatment; water processing. / 水処理剤 chemicals for water treatment. / 水処理施設[膜] a water treatment facility [membrane].　**水鳥**　a waterfowl; a water [an aquatic] bird; a palmiped; a swimming [natatorial] bird; a wading bird.　**水飲み場**　a (public) drinking fountain; a bubbler; 〔動物の〕a watering place. ★⇨バードバス　**水不足**　a water shortage. / 昨年は水不足が深刻だった. There was a serious water shortage last year. ★⇨旱魃(かんばつ), 慢性　**水辺(地域)**　waterfront (area). ★⇨沿岸, 湿地

密閉バケツ　〔生ゴミの発酵処理用の〕a compost bucket [container]. ★⇨コンポスター

密放流　▶魚の〜(the [an]) illegal release of fish. ★⇨ブラックバス, 放流 / 湖にある種の生物を〜する release a species into a lake illegally.

密猟[漁]　poaching. ▶キジ[鮭]

の〜をする poach (for) pheasants [salmon]. ¶密猟[漁]者 a poacher. 密漁船 a poaching boat [vessel].

ミティゲーション ⇨英和 mitigation.

ミトコンドリア mitochondria. ¶ミトコンドリア **DNA** mitochondrial DNA.

緑 green; verdure. ★⇨グリーン, 緑地, 環境 ▶木々の緑 the shiny green of the trees. / この辺りは緑が多い[少ない]. There's lots of [very little] green around here. / その丘は豊かな緑におおわれている. The hill is covered with rich green. / 緑の週間 Arbor [Green] Week. / 窓辺に緑を置く put greenery by the window. / 砂漠に緑を！ Green the deserts! ★⇨砂漠 ¶緑資源機構 the Japan Green Resources Agency（略：J-Green）. 緑のオーナー制度 ⇨分収育林. 緑の革命〔高収量品種の穀類の導入により食料危機に対処しようとした農業革命〕a green revolution. ★⇨英和 green revolution 緑の国勢調査 ⇨自然(自然環境保全基礎調査) 『緑の循環』認証会議〔森林認証の審査機関〕the Sustainable Green Ecosystem Council（略：SGEC）. ★⇨森林(森林認証) 緑のダム〔森林の保水機能をさす言葉〕the "green dam" effect. ★⇨英和 green dam, 保水 緑の地球ネットワーク〔中国山西省の黄土高原で緑化協力活動を行う日本の NGO〕the Green Earth Network（略：GEN）. 緑の党〔ドイツ・英国などの地球環境保護政党〕the Green Party. ★⇨英和 Green Party みどりの日〔5月4日〕Greenery Day. 緑の募金〔国土緑化のための〕contribution to the Green Fund.

水俣(みなまた)病 Minamata disease; Minamata mercury poisoning. ★⇨水銀, 認定 ▶先天性[胎児性]〜 congenital [fetal] Minamata disease.

未臨界 ▶〜の subcritical. ★⇨臨界 ¶**未臨界核実験** a subcritical test. ★⇨核(核実験) **未臨界(原子)炉** a subcritical reactor.

ミレニアム開発目標〔2000年9月国連首脳会議で宣言された〕the Millennium Development Goals（★⇨英和 MDGs）.

む

無煙 ▶〜の smokeless. ¶無煙環境〔たばこの煙のない〕a smokefree [smokeless] environment. ★⇨たばこ, 煙 **無煙たばこ** a smokeless cigarette. **無煙炭** anthracite (coal); smokeless [hard, blind] coal; stone coal; glance coal. **無煙燃焼** smokeless combustion. **無煙燃料** smokeless fuel. **無煙灰皿** a smokeless ashtray. **無煙ロースター**〔焼肉店などの〕a smokeless grill.

無灰(むかい) ▶〜の ash-free; ashless. ¶**無灰炭** ashless coal. **無灰燃料** ash-free fuel.

無害 ▶〜な harmless; innocent; innoxious《food》; innocuous《drug》; inoffensive《person》;〔毒性のない〕nonpoisonous; nontoxic. ★⇨無毒 / この殺虫剤は人間には〜です．This pesticide is not harmful to humans. / 有害物質を〜化する技術を開発する develop a technique for rendering harmful materials harmless [neutralizing harmful materials].

無過失 ¶**無過失(損害)賠償責任** no-fault liability for compensation. **無過失責任** strict [absolute, no-fault] liability; liability without fault. **無過失責任主義** the principle [doctrine] of strict [absolute, no-fault] liability. **無過失補償制度**〔医療事故についての〕a no-fault compensation system.

無花粉杉 a pollen-free Japanese cedar. ★⇨杉

無機 ▶〜の inorganic; unorganized; mineral. ★⇨有機 / 土壌の〜成分 the mineral content of the soil. ¶**無機化合物** an inorganic compound. **無機的環境**〔生物群集とともに生態系をなす〕the physical [inorganic] environment. **無機肥料** an inorganic fertilizer. **無機物[体]** an inorganic substance [matter, body]; a mineral.

無菌 asepsis; sterility. ▶〜の without bacilli [germs]; germ-free; sterile;〔殺菌した〕sterilized; pasteurized; aseptic. ★⇨殺菌 ¶**無菌牛乳** sterilized [pasteurized] milk. **無菌飼育** germfree feeding. **無菌室** a clean [bioclean] room. **無菌状態** an aseptic [a germfree] condition; asepsis. **無菌動物** a germfree [an axenic] animal. **無菌培養** sterile culture.

無公害 ▶～の clean; non-polluting. ★⇨クリーン, 低公害 ¶**無公害エンジン** a pollution-free engine. **無公害産業** a nonpolluting industry. **無公害車** 〔有毒物質を排出しない〕a zero-emission vehicle (略: ZEV).

無臭 ▶～の odorless; inodorous; scentless. / 無色～の猛毒ガス a colorless, odorless, (and) highly poisonous gas. / この防虫剤は～だから服に匂いはつかない. This mothball is odorless, so clothes won't smell.

無脊椎(むせきつい)動物 an invertebrate (animal).

虫 ⇨昆虫, 害虫 ¶**虫よけ** 〔薬品〕an insect repellent; (粉状の) insect powder; (球状の) a mothball; 〔設備〕a protection against moths. ▶**虫よけスプレー** an insect repellent spray. ★⇨防蟻(ぼう)

無袋(たい) ▶リンゴの～栽培〔袋掛けをしない〕growing of apples without bagging. ¶**無袋リンゴ** an unbagged apple.

無毒 ▶～の harmless; non-toxic; nonpoisonous; avirulent;〔動物が〕nonvenomous. ★⇨無害, 毒 ¶**無毒化** detoxification. / ～化する detoxify.

無農薬 ▶～の農産物 pesticide-free farm produce. ★⇨低農薬, 農薬 / うちの野菜は～です. We don't use insecticide [pesticide] on vegetables. ¶**無農薬栽培** farming without insecticides [pesticides]. **無農薬農業** pesticide-free farming. **無農薬・無化学肥料栽培** 〔いわゆる有機栽培〕farming without chemical fertilizers and pesticides. **無農薬野菜** vegetables grown without pesticides; pesticide-free vegetables.

無リン[無燐]洗剤 a phosphate-free detergent. ★⇨洗剤

群れ 〔一般に〕a group; a troop;〔人の〕a crowd; a throng; (ひとかたまりの) a knot; (多人数の) a multitude;〔獣の〕(牛馬などの) a herd; (羊の) a flock; (オオカミ・猟犬などの) a pack; (ライオンの) a pride 《of lions》; (猿の) a troop; (アザラシ・オットセイ・ペンギンなどの) a rookery; (クジラの) a gam; (オットセイ・クジラ・小鳥などの少数の) a pod; (追われていく家畜の) a drove; (移動中の大群) a horde; (一列になって移動中の) a string;〔鳥の〕a flock; (飛んでいる) a flight; (ウズラなどの) a bevy; (シャコなどの) a covey; (ガチョウの) a gaggle; (シギの) a wisp; (サギ類の) a siege; (飛んでいる

野鳥などの) a skein; (水上で休んでいる水鳥の) a raft;〔魚の〕a school; a shoal; (移動中の) a run;〔虫の〕a swarm; (ミツバチなどの) a cluster; (イナゴなどの) a cloud. ▶イルカの〜が泳いでいる．The school [pod] of dolphins is swimming together. / 〜をなす form groups; be found in groups; flock; band together;〔魚が〕school. ★⇨群居, 群集 / (魚などが)いくつも〜をなして泳ぐ swim in schools. / 10 頭から 20 頭の〜をなして移動する move in herds of ten to twenty. / 群れを作る動物 a gregarious [herd] animal. / (鳥が)〜をなして飛んで行く〔まばらに〕travel in loose flocks [companies];〔かたまって〕travel in tight flocks. / ニホンザルは 20 匹から 100 匹ほどの〜を作って住んでいる．Japanese macaques live in a troop of 20 to 100.

め

名水 〔飲み水〕(a) famed mineral water;〔河川〕a famous river; a renowned beautiful stream. ¶**名水百選** 〔環境省による〕100 famed mineral waters. ★⇨百選 ▶日本の〜百選 water from one of Japan's 100 [One Hundred] Best Springs; a drinking water selected as one of the best one hundred in Japan.

迷鳥 a straggler; a stray (bird); a strayer; an accidental [occasional] visitor.

命名 ▶(…に)〜する give a name (to…); name; call; designate. ¶**命名規約** 〔学名の〕the rules of nomenclature. ★⇨学名 / 国際動物〜規約 the International Code of Zoological Nomenclature (略: ICZN). / 国際植物〜規約 the International Code of Botanical Nomenclature (略: ICBN).

迷惑 a nuisance; a bother; a pest; (a) trouble; annoyance;〔迷惑な行為〕an annoying act. ★⇨近所迷惑, 公害, 騒音 ▶〜な音を出す make an annoying [irritating] noise. / (…に)〜をかける trouble; give [cause]… trouble;〔邪魔する〕disturb;〔悩ます・煩わせる〕annoy; bother; worry; cause annoyance to …; be a nuisance to …. / 君のためにみんなが〜しているんだよ. You're making a nuisance of yourself to everybody. / 他のお客さまのご〜になりますので since it is a nuisance to other customers [passengers]. / 大層ご〜をおかけしてすみません. I'm sorry to be such a nuisance to you. | I am afraid I have put you to a great deal of trouble. / 少しは周囲の〜も考えろ. Consider a little the bother you're causing to people around you. / 人の〜も顧みない放置自転車が多い. There are many bicycles abandoned by people who don't care if they inconvenience others. ★⇨放置 ¶**迷惑施設** a nuisance [NIMBY] facility; an eyesore. ★⇨ニンビー **迷惑駐車** nuisance (car) parking. **迷惑駐輪** nuisance caused by parked bicycles. **迷惑防止条例** a public nuisance ordinance. / 〜行為防止条例 Regulations for the Prevention of Public and Private Nuisances.

メガソーラー 〔特に欧州の，大規模太陽光発電所〕a mega solar power plant. ★⇨太陽(太陽光発電)

恵み ▶大地の～ the kindly fruits of the earth. / 秋は自然の～がいっぱいだ．Autumn is full of the blessings [bounty] of nature. ★⇨自然/自然の～に頼って生活する live off the bounty of Nature. / ～の雨 a welcome [blessed] rain. / 川は私たちに豊かな～をもたらしてくれる．Rivers bring us bountiful blessings.

メジャー 〔国際石油会社〕a major; an oil major. ★⇨穀物(穀物メジャー), 資源(資源メジャー)

雌化(めす か) 〔オスの〕feminization [demasculinization] of males. ★⇨雌雄(しゆう) ▶～する feminize; demasculinize. ¶メス化現象 feminization; demasculinization.

メタノール methanol; methyl [wood] alcohol. ¶メタノール自動車 a methanol-fueled car [vehicle].

メタン methane. ¶メタンガス methane; marsh gas. メタン・ハイドレート methane hydrate.

メチル methyl. ¶メチル・アルコール methyl [wood] alcohol; spirit(s) of wood; pyroligneous alcohol [spirit]; methanol. メチル・ジメトン 〔有機リン殺虫剤〕methyl demeton. メチル水銀 methylmercury.

免疫 immunity (from a disease); immunization. ▶(…の)～ができる[になる] gain (an) immunity [become immune] to [from] … / 病気に～がない have no immunity [be nonimmune] to a disease. / 一度この病気にかかると後は～になる．Those who have had this disease once are immune from it ever afterward. / 自己～ autoimmunity. / 体内の～機構 the body's immune mechanism. ¶免疫異常 immunopathy. 免疫遺伝学 immunogenetics. 免疫化学 immunochemistry. 免疫学 immunology. / ～学者 an immunologist. 免疫期間 a period of immunity. 免疫グロブリン immunoglobulin (略: Ig). / ～グロブリン製剤 an immunoglobulin preparation. 免疫系[システム] an immune system. 免疫血清 an immune serum; an antiserum. 免疫原 an immunogen. 免疫抗体 an immune antibody. 免疫細胞 an immune cell; an immuno-

cyte. **免疫制御** immune regulation; immunomodulation.
免疫生物学 immunobiology.
免疫毒性 immunotoxicity. **免疫不全症** an immunodeficiency disease. / 後天性〜不全症候群〔エイズの正式名〕acquired immunodeficiency syndrome (略：AIDS). **免疫抑制** immunosuppression. **免疫抑制薬[剤]** an immunosuppressant (drug [agent]). **免疫療法** immunotherapy. **免疫力** immune strength. / 〜力の低下 a decrease in immune strength. / 近年，感染症に対するわれわれの〜力が低下してきていると考えられている．Our immune strength [power of immunity] to infection is thought to have decreased [weakened, been lowered] in recent years. **免疫(力)増強** increase in immunity; immunopotentiation. **免疫(力)増強剤** an immunopotentiator.

面的汚染 wide-area [areal] pollution. ★⇨汚染，越境汚染

も

藻 an alga《複数形 algae》;〔アオウキクサ属の〕duckweed;〔海藻〕seaweed. ★⇨藻類，褐虫藻(かっちゅう)，英和 algal bloom, killer seaweed ¶**藻場**(ば) a seaweed bed.

猛禽(きん) a bird of prey; a predatory bird; a raptor(e); a raptorial (bird).

猛暑 ▶〜に襲われる［見舞われる］be hit by a heat wave. ¶**猛暑日**〔最高気温 35°C 以上の日〕a day on which the maximum temperature is 35 degrees or more. ★⇨真夏日

燃え滓(か)［殻］ cinders; embers. ▶灰になった燃え殻 white embers.

燃える［燃えない］ごみ ⇨ごみ，焼却，英和 RDF

モーダル ▶鉄道輸送への〜シフト〔大量輸送方法の転換〕a modal shift to rail transportation. ¶**モーダル・ミックス**〔複合輸送〕a modal mix.

木材 wood;〔建築用〕(house) timber; lumber. ★⇨伐採(ばっさい)，森林(森林認証)，パルプ ¶**木材防腐剤** a wood preservative.

木質 ▶〜の woody; ligneous. ¶**木質感**〔プラスチック製品などの〕a wood-like feel. **木質(系)バイオマス** ⇨バイオ(バイオマス) **木質系廃棄物** wood [woody] waste. **木質ペレット**〔無煙燃料〕wood pellet. ⇨ペレット・ストーブ

持ち帰る ⇨ドギー・バッグ，ごみ

もったいない wasteful; wasting. ★⇨英和 Maathai ▶まあ〜．What a waste! / この靴はまだはける．もったいなくて捨てられない．These shoes are still wearable. It's a waste to throw them away. / 必要がないものに金を使うのは〜．Spending money on things that you don't need is just like throwing it away. / そんなに紙を何枚も使っては〜．It is a sheer waste [simply a waste, bad economy] to use so many sheets of paper. / 母はもったいながって包装紙でもひもでも何でも取っておく．My mother keeps all kinds of wrapping paper and piece of string as she thinks it is too wasteful to throw them away. / 食べ物を捨てることに現代人はもったいなさを感じなくなったの

モニター

だろうか. Have people of the modern day lost the sense of wastefulness regarding throwing food away?

モニター，モニタリング ¶消費者モニター a consumer monitor. ⇨環境(環境モニター[モニタリング]), 生物(生物モニタリング), 地球(地球環境モニタリング・システム), 国民生活モニター調査, ODA民間モニター(和英篇末尾) **モニタリング・ステーション**〔環境放射線量測定施設〕an environmental radiation monitoring station. **モニタリング調査[検査]** a monitoring study [survey]. ▶輸入果実のモニタリング検査〔厚生労働省の命令により輸入業者が行う検査〕a monitoring inspection of imported fruit. **モニタリング・ポスト**〔環境放射線量測定施設〕an environmental radiation monitoring post.

モノカルチャー〔単作農法〕monoculture. ¶モノカルチャー経済 monocultural economy.

藻場(ば) ⇨藻

森 ⇨森林

モルモット〔天竺鼠(てんじくねずみ)〕a (domestic) guinea pig; Cavia porcellus. ★「モルモット」の名は英語の"marmot"から出たものであるが，これはまったく別の動物. ▶医学実験の～にされるのはごめんだ. I refuse to be used as a guinea pig in a medical experiment. ★⇨動物(動物実験, 実験動物), 実験

漏れる leak; spill. ▶座礁した船から重油が漏れ出した. Crude oil spilled into the sea from the grounded ship. ★⇨重油, タンカー, 油(油漏れ), 燃料(燃料漏れ), 放射能(放射能漏れ) / 漏れを止める stop [plug, seal] a leak.

モントリオール議定書 ⇨英和 Montreal Protocol

や

野外 〔野原〕the fields;〔屋外〕the open air; the outdoors. ▶～で out in the fields; in the open (air); alfresco. / 外来生物の～放出 release of non-native species into the wild. ★⇨放鳥, 放流 / ～鳥類観察ハンドブック[図鑑] a field guide to birds. ¶ **野外活動** an outdoor activity. **野外活動指導者** an outdoor activity leader. **野外識別目印**〔野生動物の〕a field mark. **野外焼却** ⇨野焼き **野外調査** field research [work].

焼畑 land cleared by the slash-and-burn method. ¶ **焼畑農業** slash-and-burn farming; the slash-and-burn method (of agriculture).

薬用 ▶～の medicinal; medical; officinal; for medicine. / この草は～になる. This plant is used for medicinal purposes. | This herb has healing qualities. ¶ **薬用化粧品** medicated cosmetics. **薬用酵母** medicinal yeast. **薬用酒** medicinal drinks [spirits, liquor]; an alcoholic beverage that has medicinal qualities. **薬用植物**〔薬草〕a medicinal plant [herb];〔ハーブ〕a(n) herb. **薬用植物学** pharmaceutical botany. **薬用石けん[クリーム]** medicated soap [cream]. **薬用炭** medicinal carbon; medicinal charcoal.

野犬 ⇨野良犬(のらいぬ), 野生 ▶～化する return to a wild state; become feral. / ～化した捨て犬 an abandoned dog that has gone feral; ownerless dogs turned feral. / 無責任に捨てられた犬が～化して養鶏場を襲う事故が頻発している. There are frequent instances of irresponsibly abandoned dogs going feral and attacking poultry farms. ¶ **野犬狩り** a roundup of [hunting up] homeless [ownerless] dogs. / ～狩りをする round [hunt] up homeless [ownerless] dogs. **野犬収容所** a dog pound. **野犬捕獲員** a dogcatcher.

夜行性 ▶～性の動物[鳥] a nocturnal animal [bird].

野生 ▶～の wild; ferine; feral; savage; undomesticated; uncultivated. ★⇨因果関係, 自

然, 種(しゅ), トラフィック, 英和 CITES, Red List / ～の馬は扱いが難しい. Wild horses are hard to manage. / ～動物を檻に閉じ込める close up [confine] a wild animal in a cage; cage a wild animal; put a wild animal in a cage. / ～ザルの餌付けに成功した. We managed to make the wild monkeys take food from us [eat the food we gave them]. / ～する〔植物が〕grow wild;〔動物が〕live in the wild state. ★⇨自生 / これらの植物は日本に～する. These plants grow wild in Japan. / この種は～では数年に一度しか開花しない. In the wild state this variety blooms only once in several years. / ～の鹿は最近数えたところでは500頭いる. At [According to] the last count there are 500 wild deer. / 保護した動物を～に返す return a captive animal to the wild. ★⇨放鳥, 放流, 野外 / 動物の～復帰 return [restoration] of an animal to the wild. / ～化訓練 reintroduction training. / 保護したパンダの～化訓練 training for returning [reintroducing] captive pandas to the wild. / ある種の生物の～順化 adaptation [rehabilitation] of a species to the wild. / ～化する go wild; become [turn] feral; revert to the wild state. ★⇨野犬, 野良猫(のらねこ) / ～化した家畜 cattle gone wild; feral livestock. / ～化した栽培種 cultivated species gone wild. / かごから逃げ出して～化した鳥もいる. Some caged birds escaped to establish themselves in the wild. ★⇨野鳥 ¶ **野生型** a wild type. **野生型株** a wild-type strain. (…の)野生種 a wild species (of …). **野生状態** the state of nature; a [the] feral state. / ～状態での観察記録はきわめて珍しい. Records of observations in the wild are extremely rare. / 飼育環境をできるだけ～状態に近づける make the captive environment as close as possible like the natural environment. ★⇨行動(行動展示) **野生植物** a wild plant [fruit]; a wilding. **野生生物** 〈集合的に〉wildlife. ★⇨生息(生息地), 保護, 鳥獣 **野生絶滅** ⇨絶滅 **野生鳥獣[動物]管理学** wildlife [wild animal] management (studies). **野生動植物保護区域** a wildlife sanctuary [preserve]. **野生動物** a wild [feral] animal.

野草 wild grass [herbs].

野鳥 a wild bird;〈集合的に〉wild fowl. ★⇨ペット, 渡り鳥 ¶ **野鳥の会** a wild bird(s) society. **野鳥観察** bird-watching; birding. ▶〜観察をする bird-watch. /〜観察家 a bird-watcher; a birder. /〜観察用ハンドブック〔軽便な図鑑など〕a field guide to birds. **野鳥捕獲者** a fowler. **野鳥保護** wild bird protection.

谷津田(やつだ)〔谷間の湿地にある田〕a rice paddy in valley wetlands [a marshy valley].

ゆ

誘引剤[物質] an attractant.

有害 ▶〜な bad; pernicious; detrimental; deleterious; harmful; injurious; noxious; malignant; baleful; baneful; inimical;〔毒性の〕toxic. ★⇨毒 / 健康に〜な detrimental [injurious] to health; bad for the health. / たばこは体に〜だ. Smoking is harmful to your health. / 作物に〜な harmful [injurious] to the crops. / 〜な植物 a noxious plant ¶ **有害化学物質** a harmful [hazardous] chemical (substance). ★⇨遺伝子, 無毒 **有害ガス** noxious gases. **有害昆虫** a noxious insect. ★⇨害虫 **有害(産業)廃棄物** harmful [toxic] (industrial) waste; hazardous waste. ★⇨廃棄物 **有害紫外線** harmful [hazardous] ultraviolet rays. ★⇨紫外線, UV(和英篇末尾) **有害食品** a harmful food. **有害重金属** a harmful [hazardous] heavy metal. **有害鳥獣** harmful birds and animals; bird and animal pests. ★⇨食害 / 〜鳥獣駆除 harmful bird and animal control; pest control. **有害添加物** 〔食品の〕a harmful food additive. ★⇨添加物 **有害物質** a hazardous [harmful] material [substance]; a toxic substance. **有害物質規制法** 〔米国の〕the Toxic Substances Control Act (略: TSCA). **有害物質を含有する家庭用品の規制に関する法律** the Law For the Regulation of Household Products Containing Harmful Substances. **有害ミネラル** a harmful mineral.

有機 ▶〜の organic. ★⇨無機, 残留(残留性有機汚染物質) ¶ **有機栄養** heterotrophism; heterotrophy. **有機栄養生物** a heterotroph. **有機栄養素** an organic nutritive element. **有機塩素系殺虫剤[農薬]** an organochlorine pesticide. **有機化学** organic chemistry. **有機化合物** an organic compound. / 揮発性〜化合物 a volatile organic compound (略: VOC). **有機栽培** organic growing [cultivation]; organic farming. / 〜栽培の農作物⇨有機農作物 / 〜無農薬栽培 organic cultivation without insecticides [pesticides]. ★⇨無農薬 **有機質** organic matter [substance]. **有機JASマーク** an Organic [organic] JAS Mark [mark]. **有機食品** organic food(s). ★⇨オーガニック **有機処理** organic treatment; organic disposal. **有機水銀中毒** organic mercury poisoning. ★⇨濃度, 垂れ流す **有機スズ** an organotin (compound). **有機(性)汚泥** organic mud. **有機生物学** organismal biology. **有機太陽電池** an organic solar cell. **有機土** organic soil. **有機農業[農法]** organic farming [agriculture]. ★⇨有機栽培 **有機農産物[農作物]** organically grown (farm) produce. **有機農産物加工食品** processed food(s) from organic agricultural products. **有機廃棄物** organic waste. **有機肥料** an organic fertilizer. **有機物(質)** organic matter [substance]. **有機米** organic rice; organically grown rice. / 〜無農薬栽培米 organic rice produced without insecticides [pesticides]. **有機野菜** organically grown [organic] vegetables. **有機溶剤[溶媒]** an organic solvent. **有機溶剤中毒** organic solvents poisoning. **有機燐** an organophosphate. **有機燐化合物** an organophosphorus [organic phosphorus] compound; an organophosphate. **有機燐系殺虫剤** (an) organophosphate [organophosphorus] insecticide. **有機燐剤** 〔農薬〕 an organophosphorus pesticide. **有機燐中毒** organophosphate poisoning.

遊休 ▶〜地の有効利用を進める promote the efficient use of idle land. ¶**遊休化** getting out of use [unused]; falling into disuse. **遊休農地** idle [unused] farmland; farmland lying fallow.

湧水(ゆうすい) a spring; spring water ¶**湧水池** a spring-fed [groundwater] pond. **湧水量** 〔地下水・温泉などの〕a spring flow (rate) [discharge rate].

遊水 ¶**遊水機能** water storage function. **遊水池, 遊水地** a retarding basin; a flood retarding basin.

優性 〔遺伝形質の〕dominance; 〔遺伝力〕prepotency. ▶〜の dominant. ¶**優性遺伝** dominant inheritance; prepotence. **優性遺伝子[因子]** a dominant gene. **優性形質** a dominant trait [character].

優占 ¶**優占種[個体]** a dominant species [individual]; a dominant. **優占度** dominance.

誘致 ▶〜する lure; decoy; entice;

attract; invite. ★⇨原発／地元が工場～に積極的でないのは産廃問題があるからだ．Local people are reluctant to invite a manufacturing company to set up its plant in their district because of their industrial waste.

有毒 ▶～な[の]〔毒のある〕poisonous; toxic; venomous; mephitic;〔有害な〕noxious. ★⇨毒，有害／それは空気に触れると～の気体を発する．It gives off [emits] a poisonous [toxic] gas on contact with the air. ¶**有毒化学薬品** toxic chemicals. **有毒ガス** a poisonous [a toxic] gas;〔鉱山の〕damp. **有毒植物[動物]** a poisonous plant [animal]. **有毒廃棄物** toxic waste. ★⇨廃棄物 **有毒物質** a poisonous [toxic] substance.

遊歩道 a promenade; an esplanade; a parade; a public walk;〔樹蔭のある〕a mall;〔特に板張りの〕a boardwalk. ▶湖岸に沿って～が続いている．A promenade runs along the lake.／緑陰の～を散策する take a stroll along the promenade in the shade of trees. ¶**自然遊歩道** a nature trail.

遊離〔分離〕separation; isolation; extrication; liberation. ▶フロンから～した塩素が触媒反応によってオゾンを破壊する．Chlorine released from CFCs destroys ozone in catalytic reactions.／アレルゲンが IgE に結合すると免疫細胞がヒスタミンを～し，それが今度はアレルギー症状を引き起こす．Binding of allergens to IgE causes the immune cell to release histamine, which in turn causes allergy symptoms.

輸出 exportation; export. ★⇨農産物，農業(農業輸出国)，石油(石油輸出国機構)，英和 CITES ▶～する export; ship abroad.／公害の～ export of pollution. ★⇨英和 Basel Convention

油井(ゆせい) an oil(-producing) well; an oiler. ¶**油井火災** an oil well fire. **油井ガス** casinghead gas. **油井掘削機** an oil (drilling) rig; a rig. **油井噴出事故** an oil blowout; a blowout at an oil well. ★⇨重油，油濁，油(あぶら)

油濁(ゆだく) polluting (a body of) water with oil; oil pollution. ★⇨国際(国際油濁補償基金)，重油 ¶**油濁事故対策協力条約** ⇨英和 OPRC convention. **油濁防止装置** oil pollution control equipment. **油濁防除** oil pollution control. **船舶油濁損害賠償保障法，油濁法** the Law on Liability for Oil Pollution Damage.

輸入 import; importation. ★⇨

輸出 ▶〜する import. / わが国はエネルギー資源のほとんどを〜に頼っている．Our country depends on (foreign) imports for most of its energy resources. ★⇨自給(自給率) / 食料〜大国 (a nation that is) a major importer of foodstuffs. ¶ **輸入感染症** an imported infectious disease; an afferent infection disease. **輸入停止** (a) suspension of imports. / 米国からの鶏肉の〜停止に踏み切る take the step of halting the import of chicken meat from the United States. **輸入届出制度** 〔輸入動物による人の感染症を防ぐために厚生労働省が定めた制度〕the notification system for the importation of animals into Japan (requiring a health certificate from the exporting country authorities, etc).

ユネスコ 〔国連教育科学文化機関〕UNESCO; the United Nations Educational, Scientific and Cultural Organization. ¶ **ユネスコ・アジア文化センター** 〔財団法人〕the Asia/Pacific Cultural Centre for UNESCO (略: ACCU). **日本ユネスコ協会連盟** the National Federation of UNESCO Associations in Japan. **ユネスコ協力会議** a UNESCO cooperation conference. **ユネスコ・クーポン** a UNESCO coupon. **ユネスコ憲章** the Constitution of UNESCO. **ユネスコ政府間海洋学委員会** the Intergovernmental Oceanographic Commission (略: IOC). **ユネスコ世界遺産センター** (the) UNESCO World Heritage Centre (略: UNESCO/WHC, UNESCO-WHC).

よ

容器包装 containers and packaging. ▶紙製～ paper containers and packaging. / プラスチック製～ ⇨プラスチック ¶**容器包装ゴミ** container and packaging waste [garbage]. **容器(包装)リサイクル法** the Container and Packaging Recycling Law. ★正式名は、「容器包装に係る分別収集及び再商品化の促進等に関する法律」(the Law for Promotion of Sorted Collection and Recycling of Containers and Packaging).

溶剤 a solvent; a resolvent; a flux. ¶**溶剤吸入** solvent inhalation. **溶剤(系)塗料** solvent paint. **有機溶剤** an organic solvent.

養殖 aquaculture; cultivation; breeding; raising; rearing. ▶真珠の～ pearl culture. / ～する rear; raise; breed; cultivate. / 海水[淡水, 囲い, 陸上]～ seawater [freshwater, enclosed, inland] aquaculture. / 海洋～ ocean culture; mariculture. / マスの池水～ the cultivation of trout in ponds. ¶**養殖いかだ** a culture raft. **養殖池** a culture pond. **養殖学** 〔動植物の〕 thremmatology. **養殖魚** a farm-raised fish; a farmed fish; a hatchery fish. **養殖漁業** aquaculture [aquiculture]; fish-farming; mariculture; the fish-raising industry. **養殖場** a nursery; a farm. / ウナギの～場 an eel farm [nursery]. **養殖物** 〔魚〕 a farm-raised fish; a farmed fish; a hatchery fish; 〔海藻〕 cultivated [farmed] seaweed; 〔真珠〕 a cultured pearl. ★⇨天然(天然物) / このフグは～ものだ. This is farmed *fugu*. / ～もののカキ[ウナギ] a farmed oyster [eel]. **完全養殖** 〔魚介類の世代循環の全過程を人工管理下で行うこと〕 complete aquaculture. / 魚の完全～を行う produce fish by carrying out the entire life cycle artificially.

用水 ⇨灌漑(かんがい), 公共, 工業, 生活, 農業 ▶防火～ water for fighting fires. ¶**用水池** a reservoir. **(農業)用水期** the time of year when the consumption of irrigation water is at its peak. **用水権** water rights; 〔水車などの〕 water power. **用水槽[桶(おけ)]** a rainwater tank [barrel]. **用水路** 〔灌漑の〕an irrigation channel [canal];〔発電所などの〕a flume.

揚水(ようすい) ¶**揚水機[装置]** a pump; pumping equipment. **揚水式ダム** a dam for pumped storage; a pump-up dam. **揚水(式)発電** pumping-up power generation; pumping-up hydraulic power generation. **揚水(式)発電所** a pumping-up (electric) power plant [station]. **揚水所** a pumping plant. **揚水ポンプ** 〔発電所の〕a storage pump;〔一般に〕a scooping pump; a water pump.

溶存酸素 dissolved oxygen (略: DO). ★⇨貧酸素水塊 ▶5ppmの〜濃度 a dissolved oxygen concentration of 5 ppm. ¶**溶存酸素計** a dissolved oxygen analyzer. **溶存酸素飽和度** the proportion of dissolved oxygen. **溶存酸素量** an amount [amounts] of dissolved oxygen.

溶存水素 dissolved hydrogen (略: DH).

要注意外来生物 ⇨外来(外来生物)

洋島 an oceanic island.

用途地域 〔都市計画の〕a specific use district [zone]. ¶**用途地域規制** a zoning regulation [restriction, control]. **用途地域制** land-use zoning. **用途地域変更** land-use rezoning; change of the land use specification.

溶融 ¶**溶融塩(原子)炉** molten-salt reactor (略: MSR). **溶融還元製鉄法** 〔高炉を使わない〕the direct iron ore smelting reduction process (略: DIOS). **溶融炭酸塩型燃料電池** ⇨燃料電池

葉緑素 chlorophyll. 形 chlorophyllous. ▶〜入りの歯磨き toothpaste with chlorophyll in it. ¶**葉緑体** a chloroplast.

余剰(よじょう) ¶**余剰電力** surplus electricity [power]. **余剰農産物** surplus farm produce [agricultural products]; farm [agricultural] surpluses. **余剰物資[食糧, 資源]** surplus goods [food, resources]. **余剰プルトニウム** surplus plutonium. ▶〜プルトニウムの処理 surplus plutonium disposition. ★⇨プルトニウム **余剰米** surplus rice.

四日市喘息(よっかいちぜんそく) Yokkaichi asthma (from factory smoke).

余熱 ▶〜利用設備 a waste heat utilization facility. ¶**余熱除去** 〔原子炉などの〕residual heat removal (略: RHR). /〜除去系統 a residual heat removal system (略: RHRS). / ～除去系配管 residual heat removal piping. / ～除去ポンプ a residual heat removal pump.

ら

ライトアップ illumination. ▶橋を〜する light up a bridge. / ビルを〜する illuminate a building.

ライドシェア 〔自動車の共同使用・相乗り〕ride-sharing. ▶〜を実行する participate in [organize] a rideshare. ★⇨カー・シェアリング, 相乗り, マイ(マイカー)

ライトレール(・トランジット) 〔次世代路面電車〕light-rail transit (略: LRT).

ライフサイクル・アセスメント 〔製品の原料調達・製造から流通・消費・廃棄処理までの環境への影響の評価〕life cycle assessment (略: LCA).

ライフサイクル・コスト 〔建築費・メンテナンス費・冷暖房費・解体費など建造物の一生にかかる諸費用の総体〕(a) life cycle cost (略: LCC).

落葉 〜する defoliate; 〈木が主語〉shed [cast] 《its》 leaves; 〈葉が主語〉fall; be shed. ▶木々はすっかり〜している. The leaves are all (gone) off the trees. | The trees are (quite) bare of leaves. | The trees have dropped all their leaves. / 常緑樹は冬でも〜しない. Evergreen trees don't lose their leaves even in winter. ★⇨常緑 ¶**落葉期** defoliation. **落葉(広葉)樹[樹林]** a deciduous (broad-leaved) tree [forest].

ラ・ニーニャ ⇨英和 La Niña, エル・ニーニョ

ラムサール条約 〔国際湿地条約〕the Ramsar Convention. ★⇨英和 Ramsar Convention, 湿地 ¶**ラムサール条約登録地** a Ramsar site. **ラムサール賞** 〔国際的に重要な湿地の保全に貢献した個人・団体に贈られる賞〕the Ramsar Award.

乱開発 unbridled [unrestrained] land development; environmentally destructive development. ★⇨開発

乱獲, 濫獲 reckless [excessive, indiscriminate] fishing [hunting]; overfishing; overhunting; overshooting. ▶〜する fish [hunt] recklessly [excessively, indiscriminately]; overfish; overhunt. ★⇨絶滅 / 〜によるクジラの生息数の激減 a sharp decrease in the whale popula-

乱獲

tion as a result of excessive hunting. / 〜を禁じる[防ぐ] prohibit [prevent] indiscriminate [reckless] fishing [hunting].

り

リオ宣言 〔1992年の〕the Rio Declaration (on Environment and Development). ★⇨英和 Rio Declaration

陸域環境学 terrestrial environmental studies; study of the terrestrial environment.

陸域観測(技術)衛星 the Advanced Land Observing Satellite (略: ALOS).

陸水 inland water. ¶**陸水学** ⇨湖沼(しょう)学　**陸水生物** limnophilous life.

陸生[棲] ▶〜の terrestrial; terraneous; living on land; land-dwelling[-living]. ¶**陸生昆虫[植物]** a land insect [plant]. **陸生生物** land life.　**陸生動物** a land animal [dweller]. / 日本の〜動物〈集合的に〉the terrestrial fauna of Japan.　**陸生哺乳類** a land mammal.

陸封 a landlock. ▶〜された魚 landlocked fish. ¶**陸封型** 〔魚などの〕a landlocked form.

利権 〔鉱山・鉄道の〕a concession; concessions. ▶中東にある米国の石油〜 American oil interests in the Middle East. ¶**鉱山開発利権** mining concessions.　**森林利権** a forest license [concession].

リコール 〔欠陥製品についてメーカーが自主的に回収や無償修理を行う意向の公表〕a recall (of faulty products). ▶〜する recall. ★⇨食品(食品リコール), 自主(自主回収) / 自動車メーカーはその車種の〜を発表した. The auto manufacturer announced a recall of that model. ¶**リコール運動** a recall campaign.　**リコール隠し** 〔自動車メーカーなどによる〕a recall cover-up. / 欠陥車の〜隠しをする cover up customer complaints about defective vehicles. ★⇨情報(情報隠し)　**リコール制** the recall system.

リサイクル 〔廃物・資源などの再利用〕recycling. ★⇨回収, 再生, 企業, 負担, 家電リサイクル法, 環境(環境省), 建設(建設リサイクル法), 自動車(自動車リサイクル法), 省エネ(省エネ・リサイクル支援法), 粗大ごみ, デポジット制度, 内部リサイクル制度, 容器包装リサイクル法, 英和 recycle ▶服を〜する recycle clothes. / 使用済み核燃料を〜する recycle

551

spent nuclear fuel. / 〜可能なプラスチック recyclable plastic. / 〜可能なもの recyclables. / 〜用の識別表示マーク a recycling symbol. / 資源の〜を促進する promote the recycling of resources. ★⇨資源(資源ごみ,資源リサイクル) / 空き缶〜運動 the empty can recycling movement [campaign]. ★⇨空き缶 / 〜指向の会社[社会]recycling-oriented businesses [society]. ¶リサイクル市(いち)　a recycled-goods [second-hand] market [bazaar].　リサイクル型社会 ⇨循環(循環型社会)　リサイクル技術　recycling technology.　リサイクル効率　recycling efficiency.　リサイクル産業　the recycling industry.　リサイクル紙[家具]　recycled paper [furniture].　リサイクル・ショップ　a recycled-goods shop.　リサイクル製品　a recycled product.　リサイクル・デザイン　〔リサイクルのことをあらかじめ考慮した設計〕design for recycling.　リサイクル率　a recycling rate. / 家電製品の〜率 a recycling rate of home electrical appliances.　リサイクル料金　a recycling fee.

利水　water utilization; irrigation.　★⇨灌漑(かん),ダム

リスク　a risk.　★⇨危険　▶肺がんになる〜 the possibility [risk] of getting [developing] lung cancer. / 感染〜 a risk of infection [contagion].　¶リスク・アセスメント　risk assessment (略: RA).　リスク管理[マネージメント]　risk management. / 病院の〜管理者[マネージャー]a hospital risk manager.

リゾート　▶〜地 a resort (area);〔海辺の〕a seaside resort;〔山の〕a mountain resort. / 〜開発 development of recreational [leisure] facilities; resort development. ★⇨開発　¶リゾート法　the Resort Law. ★正式名は「総合保養地域整備法」(the Law for Development of Comprehensive Resort Areas)という.

リターナブル　〔容器が,再加工することなく再利用できる〕returnable.　▶〜びん[ボトル]a returnable bottle. ★⇨回収

リチウム　lithium.　¶リチウム・イオン電池　〔二次電池〕a lithium-ion [Li-ion] battery.

立地　▶〜条件がよい[悪い] be conveniently [inconveniently] located; be located favorably [unfavorably]; be in a favorable [an unfavorable] situation. / 計画中のビルの〜環境 the site environment [surroundings] of a proposed building. / 高齢者にふさわしい〜環境を整備する create surroundings suitable for elderly people. / 地元住民との原発〜交渉 negotiations with local residents over the siting of a nuclear plant. ¶立地調査 a location (suitability) investigation [survey]. / 発電所の〜調査 an investigation into the suitability of a site for a power station.

リデュース ⇨削減, 英和 three R's

離農 ▶〜する give up [abandon] farming; leave the land. ¶離農農家 a farmer who has given up farming (in favor of another occupation). ★⇨農家

略奪農法[農業] exploitive farming; a slash-and-burn method of agriculture.

流域 a basin; a valley; a watershed. ★⇨水系, 雨量(流域雨量指数), 英和 Catchment Board ▶利根川の〜[下流〜] the basin [lower reaches] of the Tone River. / 揚子江の〜 the Yangtze valley. ¶流域下水道 a river-basin sewerage (system). 流域面積 the size of a catchment area [basin].

硫化(りゅうか)水素 hydrogen sulfide; sulfuret(t)ed hydrogen.

硫酸 sulfuric [sulphuric] acid. ★⇨亜硫酸 ▶廃〜 spent [waste] sulfuric acid. ★⇨廃酸 / 〜ピッチ入りのドラム缶 oil drums containing sulfuric acid pitch. ¶硫酸雨 sulfuric acid rain. 硫酸スラッジ ⇨英和 acid sludge 硫酸ミスト〔大気汚染の〕sulfuric acid mist.

粒子状物質 particulate matter (略: PM). ★⇨微小粒子状物質, 浮遊(浮遊粒子状物質), PM(和英篇末尾), 英和 particulate matter ▶〜減少装置 a particulate emission control device.

流出 ▶〜する〔石油などが〕⇨漏れる, 重油, 石油(石油流出); flow [run] out; issue; discharge; escape;〔川・水路が〕drain out; debouch. / 石油の〜が海洋生物の大量死を引き起こした. The oil spill caused a massive death toll among sea creatures. ¶流出原油 spilled crude oil. 流出口〔河川の〕a debouchment. 流出物 effluence. 流出油 spilled oil; an oil spill. / 〜油の除去 an oil-spill cleanup. / 〜油災害 an oil-spill disaster. (…からの)流出量 the volume of (the) effluent [liquid flowing out (of…)].

リユース ⇨再使用, 英和 reuse ¶リユース・カップ[コップ]〔洗って繰り返し使用できるコップ〕a reusable cup. リユース食

器 a reusable food container.

流水 〔流れる水〕running [flowing] water; a stream;〔下水だめ・工場などから出る水〕⇨排水, 英和 effluent

流氷 floating [drift, pack] ice; an ice floe. ▶今年も〜が接岸した．Ice floes have come ashore again this year. / 〜に囲まれる be surrounded by floating [drifting] ice. ¶ 流氷初日〔陸から肉眼で流氷が確認できた最初の日〕the first date on which ice is visible to the naked eye from the shore. 流氷接岸初日〔流氷がその冬初めて接岸した日〕the first date of drift ice on shore. 流氷終日〔陸から肉眼で流氷が確認できた最後の日〕the last date on which ice is visible to the naked eye from the shore. 流氷帯 a drift ice region. 流氷除け〔埠頭の〕an icebreaker.

流木 driftwood; drifting wood [logs]. ¶ 流木止め〔ダム・河川等に設置する〕a (log) boom.

流量 ⇨ダム, 河川(河川流量)

利用 utilization; use; exploitation. ★⇨平和利用, 廃物(廃物利用) ▶〜する use; utilize; make (good) use of …; exploit. / 天然資源を〜する exploit [tap, harness] natural resources. / 自然を〜したクリーン・エネルギー clean energy produced [generated] by harnessing nature. / 地形をうまく〜した建築 a building where geographic features are used to good advantage. / 限られた資源を有効に〜する make effective use of limited resources. / ビールびんの再〜 reuse [recycling] of a beer bottle. / プラスチックの再生〜 recycling of plastics. ★⇨再生(再生紙, 再生利用), 古紙(古紙利用率), リサイクル, 交通(交通機関)

猟 〔鳥獣を捕ること〕(銃猟) shooting;(狩猟) hunting;(出猟) a shooting [hunting] expedition. ★⇨狩猟, 銃猟 ▶猟をする shoot; hunt. / 猟に行く go shooting [hunting]; go on a shooting [hunting] expedition. / イノシシ猟 wild boar shooting [hunting]. / かすみ網猟 catching (of) birds with a fine-mesh net. / わな猟 trapping [ensnaring] (of) wild game. / 猟の季節, 猟期 the shooting [hunting, open] season. / 4月から10月までは猟は禁じられている．The closed [close] season is from April to October. ★⇨休猟, 禁猟 / 鹿の猟期が終わった．The hunting season for deer has ended. | The open season on deer is now over. / 猟が解禁になった, 猟期に入った．The shooting [hunting] season has opened [begun]. ★⇨解禁／猟期はいつからいつまでですか. When does the shooting season begin and end? ¶**猟区** a hunting area [zone, ground]. **猟師** a hunter. **猟獣** a game animal;〈集合的に〉game. **猟場** a (game) preserve; a hunting ground. **猟鳥** a game bird;〈集合的に〉game.

漁(りょう) ▶海［川］へ漁に出る go fishing in the sea [in a river]. ★⇨漁業(ぎょう), 解禁, 禁漁, 休漁／漁をする (catch) fish. ／昆布漁 kelp gathering. ¶**漁師** a fisherman.

領海 territorial waters [seas];〔国際法上の〕the closed sea; the marginal sea; a marine belt. ★⇨公海 ▶日本の〜内［外］で within [outside of] Japanese (territorial) waters. ／〜3海里内で within the three-mile territorial limit. ¶**領海侵犯** (a) violation of [(an) intrusion into] territorial waters.

涼感(りょうかん)**素材** clothing materials that make the wearer feel cool.

量産, 量販 ⇨大量(大量生産, 大量販売)

両生類 an amphibian. ¶**両生類学** amphibiology.

緑化(りょくか) ⇨緑化(りょっか)

緑地 a green tract of land;〔村の共有緑地〕a village green;〔草原〕a (wide) green space;〔砂漠のオアシス〕an oasis《複数形 oases》. ¶**緑地化** afforestation. **緑地計画** green space planning. **緑地帯** a greenbelt (★⇨英和 greenbelt); a green zone;〔街路と歩道の間の〕a green [grass] strip; a tree lawn. **緑地帯都市** a greenbelt town. **緑地保全区域** a green conservation zone; an urban green space conservation area; a green preserve. **緑地率** ⇨緑被(りょひ)率

緑道 a greenway.

緑被(りょくひ)**率**〔都市計画上の〕a green coverage ratio; a ratio of green space (to total area).

緑化(りょっか) tree planting; afforestation; greening. ★⇨屋上緑化, 乾燥, 国土(国土緑化), 建造物(建造物緑化), 都市緑化 ▶ある地域を〜する plant trees in an area; plant an area with trees. ／壁面〜や屋上〜は, 夏の日差しをさえぎり, 植物の蒸散作用による気化熱によって建物の表面温度を下げてくれるし, 都市部のCO_2吸収源としても期待できる. Raising plants on walls and rooftops not only blocks the summer sun and, through the removal of the heat of evaporation through the plants' transpiration, lowers the surface temperatures of the buildings; it is also a potential absorber of carbon dioxide in urban areas. / 環境問題を考える当社は〜事業を行っています. With the environment in mind, we are carrying out a green program. ¶緑化運動 a tree-planting campaign [drive]. 屋内緑化 indoor greenery [greening]; decorating indoors with greenery [plants]; displaying greenery [plants] indoors. 壁面緑化 wall greening.

燐, リン phosphorus; phosphor; 形 phosphorous. ★⇨無リン洗剤, 有機(有機燐) ▶燐と化合させる, 燐を加える phosphorate. / 燐を除去する dephosphorize. ¶燐化物 a phosphide. 燐酸 phosphoric acid. ★⇨肥料 燐酸塩 phosphate. 燐酸型燃料電池 a phosphoric acid fuel cell (略: PAFC). 燐酸肥料 a phosphatic fertilizer; a phosphorus [phosphorous] fertilizer. 燐中毒 phosphorism; phosphorus [phosphorous] poisoning.

臨界 (nuclear) criticality. ★⇨未臨界 ▶〜に達する[なる] go critical. / 即発[遅発]〜〔原子炉の〕prompt [delayed] criticality. ¶臨界事故 〔原子炉の〕a criticality accident; an inadvertent criticality. 臨界実験 〔原子炉の〕(a) critical experiment. 臨界前核実験 a subcritical (nuclear) experiment [test]. ★⇨核(核実験)

林学 〔林業学〕forestry; 〔森林科学〕dendrology. ¶林学者 a dendrologist; a forestry expert.

林冠 a tree [crown] canopy; a canopy of trees. ★⇨英和 crown canopy

林業 forestry. ★⇨林野庁, 森林(森林・林業基本法) ¶林業機械化センター 〔林野庁の〕Forestry Mechanization Center. 林業試験場 〔都道府県の〕a forestry experiment station.

林床(りんしょう) 〔林の地表面〕the forest floor. ¶林床植生 forest floor vegetation [cover]. 林床植物 a forest floor plant.

臨床環境医学 〔環境汚染などと関連する臨床症状の研究治療を行う〕clinical ecology. ¶臨床環境医 a clinical ecologist.

林地 forest land. ¶林地残材 forest-thinning waste.

林野 forests and fields; a forest land. ★⇨山林 ¶林野庁 〔農水省の一部局〕the Forestry Agency.

る・れ

類型 a type; a pattern. ★⇨分類 ¶ **類型学** typology.

ルイサイト 〔びらん性毒ガス〕lewisite.

類人猿 an anthropoid (ape); a troglodyte. ¶ **大型[小型]類人猿** a great [lesser] ape.

ルール ⇨基準, 規制 ▶遺伝子研究を進める上での～(作り)が必要だ. Rules must be established for conducting genetic research. ★⇨遺伝子(遺伝子組み換え)

レア・アース 〔希土類元素〕a rare earth element [metal].

レア・メタル 〔希有金属〕rare metal.

冷夏 a [an unusually] cool summer. ★⇨不作, 暖冬

冷害 cold-weather damage; damage from [caused by] cold weather. ▶作物を～から守る protect crops from being damaged by cold weather; take measures [steps] to protect farm crops from cold-weather damage.

冷却 cooling; refrigeration. ▶～する〔冷やす〕cool; chill; refrigerate;〔低温になる〕cool (down); chill; get cool. / 空気[液体]～式の air-[liquid-]cooled. ¶ **冷却器** a refrigerator; a freezer; a cooler. **冷却材**〔原子炉の〕a coolant; a refrigerant. / ～材喪失事故 a loss of coolant accident (略: LOCA). **冷却水** cooling water; a coolant. / 1次[2次]～水〔原子炉などの〕primary [secondary] cooling water. **冷却装置** a cooling device [apparatus]. / 1次[2次]～装置〔原子炉などの〕a primary [secondary] cooling system.

冷蔵 cold storage; refrigeration. ★⇨冷凍 ▶～する refrigerate; keep … cold [in cold storage]. / 魚を～する keep fish on ice. / 要～.〔食品の表示〕Keep refrigerated. ¶ **冷蔵庫** a refrigerator;《口》a fridge;〔冷凍用〕a freezer; an ice chest; an icebox. ★⇨フロン / 冷凍～庫 a refrigerator/freezer. / 食べ物を～庫に入れる put food in a refrigerator. / ～庫に魚肉を入れておく keep fish in the refrigerator. / ～庫の開け閉めはなるべく短時間ですますと電気代の節約になる.

冷暖房

You will reduce your electricity bill if you leave the refrigerator door open only as briefly as possible. ★⇨消費(消費電力), 庫内温度　**冷蔵車**　a refrigerator car [van]; a cold-storage car; a chill car.　**冷蔵船**　a cold-storage ship; a refrigerator boat.　**冷蔵倉庫**　a cold store.　**冷蔵肉**　chilled meat.

冷暖房　〔室内などの空気調節〕air conditioning (and heating);〔冷暖房装置〕an air conditioner (and heater). ★⇨エアコン, 暖房, 冷房 ▶～完備〔掲示〕Fully air-conditioned (and heated). ¶**冷暖房効率**　air-conditioning efficiency.　**冷暖房費**　heating and cooling [air conditioning] costs.

霊長類　〔一頭の〕a primate. ★総称としては複数形 primates を使う.

冷凍　freezing; refrigeration. ★⇨冷蔵 ▶～する freeze; refrigerate. /急速～する quick-freeze; sharp-freeze. /～乾燥する freeze-dry; lyophilize. /～保存する keep ... in a freezer. /魚[肉]を～輸送する transport fish [meat] in a refrigerated [refrigerator] car. ¶**冷凍(運搬)船**　a refrigerator ship [boat].　**冷凍器[装置]**　a refrigerator; a refrigerating machine; a freezer.　**冷凍庫**　a freezer.　**冷凍コンテナ**　a reefer container.　**冷凍剤**　a refrigerant.　**冷凍室**　a freezer;〔冷蔵庫の〕a freezer [freezing] compartment.　**冷凍車**　a refrigerator car [van]; a chill car.　**冷凍受精卵**　a frozen fertilized egg.　**冷凍食品**[肉, 魚, 野菜, 食材] frozen [deep-frozen] food [meat, fish, vegetables, foodstuff, ingredients].

冷熱発電　〔液化天然ガスを利用した〕(LNG) cold energy power generation.

冷房　air-conditioning; air cooling. ★⇨冷暖房, エアコン ▶部屋に～を入れる air-condition [air-cool] a room. /すべての工場は夏には～, 冬には暖房がしてある. All the factories are air-conditioned [air-cooled] in summer and heated in winter. /ただ今～中です. ドアを開け放しにしないでください. The air-conditioning is on. Please do not leave the doors open. /エアコンを暖房から～に切り替える switch the air conditioner from warm [the heating mode] to cool [the cooling mode]. /～のきいている部屋 an air-conditioned [air-cooled] room. /～完備.〔掲示〕(Fully) Air-conditioned. ¶**冷房効率**　cooling efficiency.　**冷房車**　an air-conditioned [air-cooled] car [train]. / 弱～車 a railway car with the air conditioning turned down [set to a higher temperature]. **冷房病** air-conditioning sickness; illness induced [brought on] by air-conditioning. **雪冷房** snow cooling.

レーザー ¶**レーザー核融合** laser (nuclear) fusion. **レーザー核融合炉** a laser fusion reactor. **レーザー濃縮装置** 〔原発用ウランの〕laser enrichment equipment. **レーザー法** 〔ウラン濃縮法の1つ〕laser uranium enrichment.

レジームシフト 〔急激な気候変動と(それに伴う)生態系の変化〕(a) regime shift. ¶**気候レジームシフト** (a) climate [climatic] regime shift. ★⇨気候(気候ジャンプ)

レジ袋 a (supermarket) checkout bag. ▶～有料化の動きは，店側が客離れを恐れるためか，なかなか進展しない．The movement to charge for checkout bags isn't making much progress, perhaps because stores are afraid of losing customers. ★⇨マイ(マイバッグ) /〔店員が客に対し〕～はご入用ですか？ Do you need a bag?

レスポンシブル・ケア 〔化学物質の責任管理〕responsible care (略：RC).

劣化ウラン 〔化学毒性・放射性物質〕depleted uranium. ¶**劣化ウラン装甲** depleted-uranium armor. **劣化ウラン弾** a depleted-uranium shell; a DU shell.

劣性 ¶**劣性遺伝** recessive heredity. ★⇨優性 **劣性遺伝子** a recessive gene. **劣性形質** a recessive (character).

レッド ⇨英和 REDD ¶**レッド・リスト** 〔絶滅の恐れのある動物種名〕the Red List of endangered species. ★⇨英和 red data book, Red List

連鎖 ▶生命の～ a chain of life. / ～反応を起こす cause [touch off, set off, trigger] a chain reaction. ¶**連鎖球菌** a streptococcus《複数形 streptococci》. / ～球菌感染症 streptococcal infection; streptococcosis. **連鎖群** 〔遺伝子の〕a linkage group. **連鎖地図** a linkage map.

レンジャー 〔森林や国立公園などの管理人〕a forest ranger. ▶国立公園の～ a ranger of a national park. ★⇨自然(自然保護官), 英和 forest ranger

ろ

老朽 ¶**老朽化水田** a degraded paddy. **老朽建築物[校舎]** a(n) antiquated [dilapidated] building [school building]. **老朽施設** decrepit [antiquated] equipment [facilities]. **老朽車** a superannuated [decrepit] car. **老朽船** an old [a superannuated, a timeworn] vessel; an overage ship.

浪費 a waste; a wasteful use; wastefulness; dissipation; extravagance; prodigality;〔浪費した物・量〕waste. ★⇨消費, もったいない ▶～する waste; use [spend] wastefully; use to no purpose; dissipate; throw [chuck, fritter] away; squander; be prodigal of.... / 資源を～する waste resources. ★⇨資源, 電力 / 時間の～ waste [dissipation] of time. / 燃料の～を防ぐ avoid [prevent] wasting fuel. ★⇨節約 / ～型社会 a throw-away [wasteful] society. / ～的な生活 a wasteful [extravagant] lifestyle.

ロード・プライシング 〔特定地域の公道を走行する自動車への課金〕road pricing; a road pricing system.

ローマ・クラブ ⇨英和 Club of Rome.

ローマ宣言 〔1996年, 世界食料サミットの〕the Rome Declaration on World Food Security and World Food Summit Plan of Action.

濾過(ろか) filtration; filtering. ★⇨急速濾過, 限外濾過 ▶～する (pass through a) filter; filtrate. / 飲料水をフィルターで～する filter drinking water. / 不純物を～して取り除く filter out impurities. / ～して沈殿物を分離する filter the precipitate. ¶**濾過集塵装置** a filter dust separator; a dust filter. **濾過除菌** sterilization by filtration. **濾過装置** a filter; a filtering device. **濾過池** a filter bed [basin].

六弗化(ろくふっか)**ウラン** uranium hexafluoride.

炉心 〔原子炉の〕a (reactor) core. ¶**炉心隔壁** a reactor core shroud. **炉心プラズマ** core plasma. **炉心崩壊** core destruction (略: CD). **炉心溶融(事故)** a (core) meltdown.

六価クロム hexavalent chromium. ▶工場跡地の土壌から環境基準を超える〜化合物が検出された．Hexavalent chromium compounds were found on the site of the former factory in quantities exceeding environmental standards.

ロッテルダム条約 the Rotterdam Convention. ★ 2004 年発効．正式名は「国際貿易の対象となる特定の有害な化学物質及び駆除剤についての事前のかつ情報に基づく同意の手続に関するロッテルダム条約」(the Rotterdam Convention on the Prior Informed Consent Procedure for Certain Hazardous Chemicals and Pesticides in International Trade) で，Prior Informed Consent の略から PIC 条約とも呼ばれる．

露天掘り open-air mining; openwork (mining); opencut (mining); strip mining; opencast mining. ¶**露天掘り採掘場** a strip mine; an open pit mine. **露天掘り炭鉱** an open-pit coal mine; an open mine.

ロハス ⇨英和 LOHAS

路面電車 a streetcar. ★⇨ライトレール

ロンドン条約 the London Convention. ★正式名称は「廃棄物その他の物の投棄による海洋汚染の防止に関する条約」(the Convention on the Prevention of Marine Pollution by Dumping of Wastes and Other Matter).

わ

ワークショップ 〔研究集会〕a workshop.

ワールドウォッチ研究所 〔環境問題を扱う米国の研究機関〕the Worldwatch Institute（略：WWI）．

枠組み ▶相互協力の〜を作る draw up a framework [an outline] for mutual cooperation. / 〜合意〔国家間などの大筋の合意〕a framework agreement. ★⇨英和 United Nations Framework Convention on Climate Change

ワクチン a vaccine. ★⇨感染 ▶子供にジフテリア予防の〜を注射[接種]する inoculate [vaccinate] a child against diphtheria. / 経口[食べる]〜 an oral vaccine. / ポリオの生〜 a live polio vaccine. / 不活化[未承認, 輸入]〜 (an) inactivated [unapproved, imported] vaccine.

ワシントン条約 the Washington Convention. ★⇨英和 CITES ▶〜で指定されている希少鳥 rare birds listed in the Washington Convention.

渡瀬線(わたせせん) 〔屋久島・種子島と奄美諸島との間の動物分布境界線〕the Watase [Watase's] line.

渡り 〔鳥などの〕passage; transit;〔鳥・魚などの〕migration; a migratory movement. ▶(鳥や蝶が)南へ[越冬のため]〜をする migrate south [for the winter]. ¶**渡り鳥** a migratory [passage, migrating] bird; a bird of passage;〈集合的に〉migrants. / 〜鳥の経路 a flyway. / 〜鳥条約 a treaty on the protection of migratory birds; a migratory bird protection treaty. / 飛来する〜鳥が最近急に増えてきて野鳥観察者にも数え切れなくなった．There has been such an increase in arrivals of migratory birds that observers have lost count of them. | Arrivals of migratory birds have seen such an increase that even birdwatchers have become unable to keep count (of them). ★⇨飛来，越冬，足環

罠(わな) a trap; a snare;〔虎挟(とらばさ)み〕a leghold trap. ▶罠にかかる[落ちる] be caught in a trap [web]; fall into a snare

[trap]; be ensnared; be (en)trapped. / (…をねらって)罠をしかける set [lay] a snare [trap] (for ...) / イノシシの通り道に罠をしかけた. They set [laid] a trap on the wild boar trail. ¶**罠猟** trapping; hunting with traps. ★⇨猟

割り当て a quota; an allotment. ★⇨英和 AAUs ▶電力〜量 an allocated amount of power.

割り箸(ばし) half-split disposable [throwaway] chopsticks. ★⇨マイ(マイ箸)

ワンス・スルー 〔使用済み核燃料を再処理・再利用せずに最終処分する方式〕a once-through system.

A〜Z

ADI ⇨英和

AIM モデル 〔アジア太平洋地域の温暖化対策統合評価モデル〕Asian-Pacific Integrated Model.

BHC ⇨英和 benzene hexachloride

BOD 〔生物化学的酸素要求量〕⇨英和 biochemical oxygen demand

Bt 〔殺虫性毒素たんぱく質生産細菌〕⇨英和 Bacillus thuringiensis ¶**Bt コーン** 〔Bt の遺伝子を組み込んだトウモロコシ〕Bt corn. **Bt 農薬** a Bt-based insecticide.

CAS 〔化学情報検索サービス機関〕CAS. ★Chemical Abstracts Service の略. ¶**CAS 登録番号** a CAS registry number.

CDM ⇨英和 CDM ¶**CDM 理事会** 〔国連の〕the CDM Executive Board; the CDM EB.

CERES ⇨セリーズ原則

CITES ⇨英和

CO_2 ⇨二酸化炭素, 温室効果ガス ¶**CO_2 回収** CO_2 recovery. **CO_2 排出量** 《reduce》CO_2 emissions; emissions of CO_2. **CO_2 封じ込め** 〔地中などへの〕《underground》CO_2 sequestration; CO_2 containment. ★⇨英和 carbon sequestration **CO_2 吸収源** ⇨緑化, 吸収

CoC 認証 ⇨英和 chain of custody

COD ⇨英和 chemical oxygen demand

COP ⇨英和

CSR ⇨英和 corporate social responsibility

DAC 〔OECD の開発援助委員会〕the DAC; the Development Assistance Committee.

DDGS 〔穀物からバイオエタノールを製造した残りかす〕DDGS; distillers dried grains with solubles.

DDT ⇨英和

DNA 〔デオキシリボ核酸〕DNA; deoxyribonucleic acid. ▶…から〜を抽出する extract DNA from…. / 鯨肉の〜登録 DNA registration of whale meat. ¶**組み換え DNA 技術[研究]** recombinant DNA technology [research]. ★⇨遺伝子 **組み換え DNA 実験指針** guidelines for recombinant DNA experiments.

E10 (イー)**燃料** ⇨英和 E
EM菌 ⇨英和 EM
EPA 〔米国環境保護局〕Environmental Protection Agency.
EPN 〔有機リン殺虫剤〕EPN; ethyl para-nitro-phenyl.
ESCO事業 ⇨英和 ESCO
ETC 〔高速道路のノンストップ自動料金収受システム〕ETC; the Electronic Toll Collection System. ¶**ETCカード** an ETC card. **ETC車載器** an on-board electronic toll collection [ETC] unit; ETC on-board equipment (略: ETC OBE).
EuP指令 〔EUによる省エネ設計に関する指令〕the EuP Directive; the Directive on Eco-Design of [for] Energy-using Products. ★ EuP は Energy-using Products の略.
EV ⇨電気 (電気自動車)
FAO 〔国連の食糧農業機関〕the FAO; the Food and Agriculture Organization.
FSC ⇨英和
GAP ⇨英和
GLP ⇨英和
GM 〔遺伝子組み換え〕genetic modification. ▶~作物[食品, 大豆] ⇨遺伝子 (遺伝子組み換え)
GMO 〔遺伝子組み換え生物〕a GMO; a genetically modified organism. ★⇨遺伝子 (遺伝子組み換え)
GRAS 〔米国の食品医薬品局 (FDA) が定める食品添加物の安全基準〕GRAS. ★ generally recognized as safe (一般に安全と認められる) から. ¶**GRAS認証** GRAS certification; a GRAS exemption. **GRAS物質** 〔GRAS が使用を認めている物質〕a GRAS substance.
GRI 〔国際的な持続可能性報告書策定をめざす団体; 本部オランダ〕the GRI; the Global Reporting Initiative. ¶**GRIガイドライン** 〔GRI が策定する, 持続可能な社会の実現に向けた指標〕the GRI (Sustainability Reporting) Guidelines.
IPCC ⇨英和
ISO 〔国際標準化機構〕ISO; the International Organization for Standardization. ¶**ISO 14001規格** 〔環境に配慮していることを示す国際規格〕the ISO 14001 standard. ★14001は "fourteen thousand one" と読む.
ITTA ⇨英和 International Tropical Timber Agreement
ITTO ⇨英和
K値 〔大気汚染物質排出指標の一つ; また魚肉・食肉などの鮮度指標〕the K-value.
LD50 〔50% 致死量〕LD50; median lethal dose; lethal dose 50%.

LDC 〔後発開発途上国〕an LDC; a least [less] developed country.

LNG 〔液化天然ガス〕LNG; liquefied natural gas. ¶**LNG 船** an LNG carrier [tanker].

MARPOL 条約 ⇨マルポール条約

MOX 〔原子炉の混合酸化物燃料〕MOX; mixed oxide (fuel).

MSC 〔海洋管理協議会〕⇨英和 Marine Stewardship Council. ¶**MSC 漁業認証** 〔持続可能で自然環境に配慮した漁業を行う漁業者団体に MSC が与える認証〕MSC Fisheries Certification. **MSC ラベル** 〔海の環境に配慮した水産物であることを示す〕an [the] MSC label. ★通称「海のエコラベル」⇨エコ (エコラベル)

MSDS ⇨英和 material safety data sheet

MSY 〔(資源の再生力範囲内の)(年間)最大産出[生産]量〕maximum sustainable yield .

NGO 〔非政府組織〕an NGO; a nongovernmental organization.

NOx 〔窒素酸化物〕NOx; (a) nitrogen oxide. ★⇨自動車 (自動車 NOx・PM 法)

NPO 〔民間の非営利組織〕a nonprofit organization. ★⇨環境 (環境政策) ¶**NPO 法人** an NPO corporation; an incorporated NPO. ▶認定〜法人 an approved specified nonprofit [NPO] corporation. **NPO 学校** an NPO school. **NPO 銀行** [バンク] a nonprofit bank (for NPO's). **NPO 支援税制** the NPO preferential tax [tax support] system. **NPO 法** the NPO Law. ★正式名称は「特定非営利活動促進法」(the Law concerning the Promotion of Specific Non-Profit Organization Activities).

ODA 〔政府開発援助〕Official Development Assistance. ¶**ODA 白書** 〔外務省発行〕the Japan's Official Development Assistance [ODA] White Paper 《2005》. **ODA 民間モニター** a civilian ODA monitor; an aid monitor (for ODA).

OPRC 条約 ⇨英和 OPRC convention

P4 施設 〔きわめて危険な病原体や遺伝子組み換え生物を扱う、封じ込め度合いが最も厳しい実験施設〕a P4 facility. ★ P は physical containment (物理的封じ込め) の略.

PC ¶**PC グリーンラベル** 〔パソコンの環境ラベル〕a PC green label. ★⇨環境 (環境ラベル) **PC グリーンラベル制度** the PC green label system. **PC リサイクルマーク** a PC recycling mark.

PCB 〔ポリ塩化ビフェニール〕PCB; polychlorobiphenyls; polychlorinated biphenyls. ★⇨英和 polychlorinated biphenyl, coplanar PCB ▶PCBで汚染された土壌 soil contaminated with PCBs.

PDCA 〔計画・実施・点検・行動；品質管理の実践手順〕PDCA; plan, do, check, act(ion). ¶**PDCAサイクル** the PDCA [plan-do-check-act] cycle.

pH 〔水素イオン指数〕pH. ★ potential of hydrogen の略. ▶pHが4.4の溶液 a solution with a pH of 4.4. / pHの大きな変化 a large change in pH; a large pH change. / ほぼ一定のpHを維持する maintain an approximately constant pH. / その溶液のpHはいくらですか. What is the pH of the solution? / 色の変化が見られるpH範囲 the pH range of an indicator during which a color change can be seen. ¶**pH計[メーター]** a pH meter. **pH試験紙** pH test paper. **pH指示薬** a pH indicator. **pH調整剤** 〔食品添加物の1つ〕a pH adjuster.

PIC条約 ⇨ロッテルダム条約

PL ⇨製造 (製造物責任)

PM 〔粒子状物質〕PM; particulate matter. ★⇨粒子状物質 ¶**PM濃度** PM [particulate matter] concentration; the concentration of PM [particulate matter]. **PM (排出)量** (the amount of) PM [particulate (matter)] emissions. ▶排ガス中のPM量 the amount of PM [particulate matter] in gas emissions. **超低PM (排出ディーゼル)車** an ultra-low-PM emission (diesel) vehicle. **超低PM排出ディーゼル車認定制度** 〔国土交通省の〕an ultra-low-PM emission diesel vehicle certification system.

POPs ⇨残留 (残留性有機汚染物質)

PPM 〔百万分率[比]〕ppm. ★ parts per million の略. ▶大気中の窒素酸化物の濃度が0.031 PPMを上回る. Nitrogen oxide concentration levels exceed 0.031 ppm. / 廃棄物の水銀含有量が10 PPMである. The mercury content of the waste is 10 ppm [10 parts per million].

PPP ⇨英和

PRTR 〔化学物質・環境汚染物質の排出移動量届出制度〕a PRTR; a pollutant release and transfer register. ★⇨英和 PRTR ¶**PRTR法** the PRTR Law; the Pollutant Release and Transfer Register Law. ★正式名称は「特定化学物質の環境への排出量の把握等及び管理の改善の促進に関す

PSE

る法律」(the Law Concerning Reporting, etc. of Releases to the Environment of Specific Chemical Substances and Promoting Improvements in Their Management).

PSE 〔電気用品安全法 (⇨電気)の規格に適合するものであることを証するマーク〕PSE. ★PSE は Product Safety of Electrical Appliances and Materials の略. ¶**PSE マーク** a PSE mark.

RDF 〔ごみ固形燃料〕⇨英和 RDF; refuse-derived fuel. ¶**RDF 発電** RDF (power) generation. **RDF 発電所** an RDF power plant.

REACH (リーチ)**規制** 〔EU の新化学物質規制; 2007 年施行〕the REACH Regulation. ★⇨英和 REACH

RGGI ⇨地域 (地域温室効果ガス・イニシアチブ)

RMU ⇨英和

RoHS 指令 ⇨英和 RoHS directive

RPS 法 the RPS Law. ★「新エネルギー利用特別措置法」の通称. RPS は Renewables Portfolio Standard (代替エネルギー使用割合の基準) の略. ★⇨英和 renewable portfolio standard

SG マーク 〔日本の製品安全マーク〕an SG mark. ★SG は和製英語 safety goods の略.

SGEC 〔森林認証を行う『緑の循環』認証会議〕the SGEC; the Sustainable Green Ecosystem Council. ★⇨森林 (森林認証)

SRI ⇨英和

SUA 条約 〔海洋航行の安全に対する不法な行為の防止に関する条約〕the SUA Convention; the Convention for the Suppression of Unlawful Acts Against the Safety of Maritime Navigation.

TBT ⇨英和, トリブチルスズ

TDI 〔耐容一日摂取量〕a TDI; the TDI《of a product》; a [the] tolerated daily intake.

TOC 〔総[全]有機炭素〕TOC; total organic carbon.

UNEP ⇨英和

UPOV 〔植物新品種保護国際同盟〕UPOV; the International Union for the Protection of New Varieties of Plants. ★⇨英和 UPOV ¶**UPOV 条約** the UPOV Treaty.

UV 〔紫外線〕UV; ultraviolet; ultraviolet rays [light]. ★⇨紫外線 ▶UV カットの化粧品 UV-blocking cosmetics. / この窓ガラスは UV カットで紫外線を防いでいる. These windowpanes block ultraviolet rays by means of (a) UV filter. / このガラスには UV カット加工が施され

ている．This glass is UV-filtered [UV-blocked]. / UV カット・ガラス[フィルム]〔自動車などの〕UV-blocking[-filtering] glass [film]. / UV カット・サングラス UV-blocking[-filtering] sunglasses.

VOC VOC. ★ volatile organic compound (揮発性有機化合物) の略． ¶**VOC 対策** 〔住宅建材などの〕VOC measures.　**VOC 対策建材** a low-VOC building material.

WBCSD ⇨持続(持続可能な開発のための世界経済人会議)

WECPNL ⇨英和

WEEE ⇨英和

瀬川至朗（せがわ　しろう）

早稲田大学大学院 政治学研究科教授，ジャーナリズムコース プログラムマネージャー（地球環境問題，ジャーナリズム論，メディア産業論）．1954年岡山生まれ．東大教養学部卒（科学史・科学哲学専攻）．元毎日新聞編集局次長，科学環境部長．記者時代に，熱帯雨林や地球温暖化の問題に取り組み，1992年にリオで開催された地球サミットを取材した．

著書『心臓移植の現場』(1988年・新潮社)，『カードの科学』(1993年・講談社ブルーバックス)，『健康食品ノート』(2002年・岩波新書)

共編著『アジア30億人の爆発～迫り来る食糧危機，資源戦争』(1996年・毎日新聞社)，『理系白書』(2006年・講談社文庫)，『ジャーナリズムは科学技術とどう向き合うか』(2009年・東京電機大学出版局) など．

論文「地球環境破壊とのゴールなき競争が始まった」(『エコノミスト』1992年6月30日号)，「目指すは行動派シンクタンク～幅広い視野と長期的視点が不可欠な環境問題」(『新聞研究』2003年5月号) など．

KENKYUSHA
〈検印省略〉

英和・和英 エコロジー用語辞典
Kenkyusha English and Japanese Dictionary of Ecology

| 2010年8月20日　印　刷 | 2010年9月1日　初版発行 |

監修者　瀬　川　至　朗
編　者　研究社辞書編集部
発行者　関　戸　雅　男
印刷所　研究社印刷株式会社

発行所　株式会社　研究社

〒102-8152
東京都千代田区富士見2-11-3
電話 （編集）03(3288)7711(代)
　　 （営業）03(3288)7777(代)
振替 00150-9-26710

Printed in Japan / ISBN 978-4-7674-3468-1　C0582
http://www.kenkyusha.co.jp/
装幀：Malpu Design（清水良洋）/ 装画：きたざわけんじ